Plant Responses to Abiotic and Biotic Stresses

Plant Responses to Abiotic and Biotic Stresses

Guest Editors

Longxing Tao
Tingting Chen

MDPI

Basel • Beijing • Wuhan • Barcelona • Belgrade • Novi Sad • Cluj • Manchester

Guest Editors

Longxing Tao
Research and Development Center of Rice Cropping Technology
China National Rice Research Institute
Hangzhou
China

Tingting Chen
National Key Laboratory of Rice Biology
China National Rice Research Institute
Hangzhou
China

Editorial Office
MDPI AG
Grosspeteranlage 5
4052 Basel, Switzerland

This is a reprint of the Special Issue, published open access by the journal *International Journal of Molecular Sciences* (ISSN 1422-0067), freely accessible at: https://www.mdpi.com/journal/ijms/special_issues/8C138K44L9.

For citation purposes, cite each article independently as indicated on the article page online and as indicated below:

Lastname, A.A.; Lastname, B.B. Article Title. *Journal Name* **Year**, *Volume Number*, Page Range.

ISBN 978-3-7258-5619-0 (Hbk)
ISBN 978-3-7258-5620-6 (PDF)
https://doi.org/10.3390/books978-3-7258-5620-6

Contents

International Journal of
Molecular Sciences

MDPI

Editorial

Special Issue "Plant Responses to Abiotic and Biotic Stresses"

Longxing Tao and Tingting Chen *

State Key Laboratory of Rice Biology and Breeding, China National Rice Research Institute, 359 Tiyuchang Road, Hangzhou 310006, China; taolongxing@caas.cn
* Correspondence: chentingting@caas.cn; Tel.: +86-057-163-370-370

During their growth and development, plants often encounter a variety of biotic and abiotic stresses, such as drought, high salinity, extreme temperatures, heavy metal pollution, and pests and diseases. These stresses disrupt the normal physiological metabolism of plants, leading to inhibited growth, reduced yield, and even death [1,2]. In agricultural production, these stress factors directly affect food security. For instance, globally, crop yield reduction caused by drought reaches 10–30% every year, while losses resulting from pests and diseases amount to 20–40%. In recent years, agricultural stress events have occurred with increasing frequency, posing severe multidimensional hazards to agricultural production [3]. Global warming has led to an increase in extreme weather events. Data from the World Meteorological Organization shows that the period from 2015 to 2024 was the warmest decade on record, and 2024 was the hottest year ever recorded globally, with the average global surface temperature being approximately 1.5 °C higher than that before the industrial revolution [4]. Under such environments, meteorological fluctuations have intensified significantly; extreme weather events have occurred more frequently, heatwaves have become more common, and phenomena such as alternating freezing and drought periods, as well as sudden shifts between droughts and floods, have happened repeatedly [5]. Moreover, the range of pests and diseases has expanded. Climate warming has reduced the mortality rate of overwintering pests and diseases, significantly increasing the base number of surviving pests and diseases after winter. This has led to an increase in the number of generations of pests and diseases, an extension of their damage period, and the expansion of their infestation range to the north. In the past, the cotton bollworm rarely appeared in the cotton-producing areas of North China, but after climate warming, it broke out on a large scale in this region. There have also been extreme shifts between droughts and floods. Global warming has intensified the transition between droughts and floods. Taking the Poyang Lake Basin as an example, the water area reached a historical maximum in 2020. However, within just two years, by 2022, the basin experienced an extreme drought event that resulted in a severe contraction of the lake area, significantly impacting agricultural activities in the surrounding regions.

The hazards of stress factors to agricultural production are manifested in the following aspects. Firstly, there is a significant reduction in crop yield [6]. From 2023 to 2024, Ghana experienced droughts and high temperatures, which created harsh growth conditions for cocoa trees and exacerbated pests and diseases. As a result, cocoa production plummeted by 40% from the expected 820,000 tons to less than 500,000 tons. In May 2024, due to heavy rain and flooding in Rio Grande do Sul, Brazil, the soybean yield of this season dropped from the predicted 22.24 million tons to 19.53 million tons, with a loss of approximately 10% to 15%. In 2023, Henan Province was hit by "harvest-season rain" (prolonged rain during the wheat harvest period), which caused wheat lodging, germination, and mildew, leading to a nearly 6.9% year-on-year decline in summer grain yield. Secondly, the quality of agricultural products is impaired [7]. In September 2024, the grape-growing area at the eastern foot of Helan Mountain in Ningxia experienced rare, continuous rainy weather for half a month. The sugar content of grapes dropped from the normal 230–240 g per liter to below 220 g per liter. This not only reduced the alcohol content of wine and decreased flavor substances but also caused grape cracking and the spread of pests and diseases due to the rain. In Thailand, long-term high temperatures of around 40 degrees Celsius have restricted the growth of durians, resulting in reduced durian yield, increased planting costs, and a shortened harvesting period. Thirdly, the farmland ecosystem is damaged [8]. Extreme weather conditions can cause soil degradation and soil erosion, exerting a negative impact on the farmland ecological environment. For example, long-term drought reduces the activity of soil microorganisms, destroys the soil aggregate structure, and leads to a decline in soil fertility. During floods, the scouring effect of floodwater on the topsoil is significant, thinning the soil layer and causing a large loss of nutrients. Furthermore, the agricultural product supply chain is disrupted. Extreme weather can also damage agricultural infrastructure, disrupt transportation and distribution, break the agricultural product supply chain, trigger market supply shortages, drive up prices, and exacerbate inflation. During the 2025 floods in South China, logistics in major banana-producing areas were interrupted, leading to a 45% weekly increase in the market price of bananas. To address the challenge of such stresses, it is crucial to understand how plants effectively cope with high-intensity extreme weather conditions, and to elucidate the underlying molecular and physiological mechanisms governing their responses to abiotic stresses [9,10]. Thus, we organized a Special Issue titled "Plant Responses to Abiotic and Biotic Stresses". This issue comprises fifteen contributions that supply novel insights into the mechanisms underlying plants' responses to abiotic and biotic stresses. It is precisely this knowledge that can facilitate the development of technologies to assist plants in withstanding abiotic and biotic stresses, thereby alleviating the impacts of external stresses and ensuring crop security [11,12].

Niedziela et al. (Contribution 1) studied Aluminum (Al) stress and its relation to epigenetic mechanisms in triticale, whose stress tolerance is linked to *ALMT* (7R) and QTLs (3R, 5R, 6R) with 36% heritability. The results showed that different methods for quantifying Al-induced DNA methylation changes led to detailed differences, and DN-CG, DM-CG, and DN-CHG markers were key to explaining Al tolerance. Additionally, they found that epigenetic markers had different densities but comparable distribution patterns in root tip and leaf tissues (supporting Al stress transmission via somatic memory), and that methylation changes were genome-wide, unrelated to Al stress-specific genes. Thus, the study indicated that Al stress may spread through somatic memory, with its induced methylation changes occurring across the genome rather than targeting specific Al-related genes.

Research by Ming et al. (Contribution 2) provides the first genomic identification of SPL transcription factors in *Ginkgo biloba*. Thirteen *GbSPL* genes were identified and phylogenetically grouped. Dispersed duplication contributed to family expansion. Promoter analysis indicated roles in flavonoid biosynthesis and stress adaptation via light, hormone, and stress-responsive elements. Expression analyses showed tissue-specific patterns, differential expression in flavonoid-rich leaves, and regulation under water stress. Nuclear localization of GbSPL2/11 was confirmed. This study establishes a functional and evolutionary foundation for GbSPLs in secondary metabolism and drought response.

Gaonosi et al. (Contribution 3) found that *Aphis craccivora* reduces cowpea yield. This study evaluated host-plant resistance. Tswana and B261-B were resistant, with low aphid growth and damage. Tswana likely suppressed aphid reproduction. Several genotypes were susceptible, showing high aphid counts, sooty mold, and plant damage. Resistance markers SNP1_0912 and CP 171F/172R validated resistance in Tswana and ER7, but failed in some controls, possibly due to QTL–environment interactions or aphid biotypes. Findings support breeding resistant cowpea lines; Tswana will be sequenced to develop new markers.

Ginori et al. (Contribution 4) evaluated heat and light stress tolerance in four wax begonia genotypes. Tolerant types (FB08-059, OPGC 5104) exhibited thicker cuticles, acute leaf folding, and elevated anthocyanin, aiding photoprotection. FB08-059 showed a 25.83% cuticle increase and 3× anthocyanin rise, limiting ROS and sustaining photosynthesis. Susceptible genotypes (Cocktail Vodka, Sprint White) had higher ion leakage, chlorophyll loss, and impaired gas exchange. Findings of the research highlight cuticle thickness, anthocyanin accumulation, and leaf folding as key indicators of heat and light stress resilience, and could support breeding programs for stress-resilient cultivars.

Yuan et al. (Contribution 5) identified 72 *CaBTB* genes in pepper (*Capsicum annuum* L.), revealing phylogenetic collinearity with related species. Expression and RT-qPCR analyses showed that multiple *CaBTB* genes responded significantly to various treatments. *CaBTB25* was highly expressed in leaves, contained light-responsive promoters, and was localized in the nucleus without transcriptional activation. These results provide a foundation for understanding *CaBTB* gene functions in pepper development and environmental interactions.

In *Eucalyptus grandis*, twelve phosphate starvation responses (EgPHRs) were identified by Xu et al. (Contribution 6) and phylogenetically classified into three groups. EgPHRs form a species-specific interaction network, with EgPHR6 binding multi-species SPX proteins, while woody plant PHRs uniquely interact with TRARAC-kinesin ATPase. Promoter analysis revealed 59 transcription factors regulating EgPHR expression linked to development, stress, and hormones. Transcriptomic and RT-qPCR analyses showed that all EgPHRs respond to phosphate deficiency, with EgPHR2/6 strongly induced; they also react to salt, cold, jasmonic acid, and boron deficiency. Notably, nitrogen starvation suppressed most EgPHRs, indicating nutrient signaling crosstalk. These results underscore EgPHRs' multifaceted role in abiotic stress adaptation in woody plants.

Chlorine (Cl), an essential plant nutrient, is underutilized in agriculture due to toxicity concerns at high levels. The study of Su et al. (Contribution 7) identifies 3 mM as the optimal Cl^- concentration in nutrient solutions, significantly enhancing tomato plant biomass, yield, and chlorophyll content. Applying 3 mM Cl^- (as $CaCl_2$ and KCl) boosted photosystem I and II activities, increasing the net photosynthetic rate by over 25%. In fruits, Cl^- elevated soluble sugars by 12.3 to 16.5% by promoting glucose, fructose, and sucrose accumulation. This resulted from enhanced invertase activity, sucrose synthase, and sucrose phosphate synthase during ripening. The findings reveal the critical role of Cl^- in improving fruit quality and photosynthesis, providing valuable insights for optimized fertilization in protected horticulture.

Cyanobacteria from extreme environments are biotechnologically valuable due to their adaptability and natural products. Bilova et al. (Contribution 8) compared the primary metabolomes of twelve such strains using GC-MS/LC-MS. Results showed distinct metabolite profiles across all strains, with greater differences among those from highly contrasting habitats. Terrestrial and freshwater extremely tolerant strains showed fewer uniquely accumulated metabolites than extremophiles from saline–alkaline environments, which displayed high metabolic diversity. This suggests divergent survival mechanisms in high-salinity/alkalinity habitats. Strain-specific metabolites are linked to both persistence during storage and original environmental adaptations.

Drought stress severely impacts the growth, quality, and yield of *Codonopsis pilosula* seedlings. Wang et al. (Contribution 9) evaluated nine cultivars under well-watered (BF) and drought-stressed (SF) conditions using a growth–physiology–yield composite and drought tolerance indices. Water treatment significantly affected emergence rate, while both water treatment and cultivar influenced root length, proline, ascorbate peroxidase activity, and chlorophyll content. Cultivar G1 outperformed others across multiple metrics. Drought stress reduced total root length, single root fresh weight, and yield by 18.33%, 28.4%, and 33.9%, respectively. Cultivars exhibited distinct physiological responses to drought, with varying tolerance levels identified. The findings support breeding efforts for drought-resistant cultivars of *C. pilosula*.

Teng et al. (Contribution 10) examined temperature effects on polyamine metabolism in the diatom *Skeletonema dohrnii*. Under three temperature regimes, physiological parameters, polyamine composition, and polyamine oxidase (PAO) gene expression were analyzed. Low temperatures reduced growth rate, increased biogenic silica, raised putrescine and spermine content, decreased spermidine, and down-regulated PAO expression. High temperatures increased growth rate and spermidine, and altered putrescine and spermine concentrations. The findings indicate that temperature shifts influence algal growth and silica content, induce specific polyamine profile changes, and implicate PAO gene involvement in temperature response regulation.

Spychala et al. (Contribution 11) investigated the molecular basis of adult plant resistance (APR), including miRNA expression and their target genes, in wheat hybrids derived from resistant cultivar Glenlea and Polish varieties. Under controlled inoculation with *Puccinia triticina* (Pt), RT-qPCR analysis revealed time-specific differential expression of APR genes and miRNAs, with peak expression at 6, 24, and 48 h post-inoculation. Notably, *Lr46-Glu2* was upregulated, indicating its role in the immune response. Gene ontology analysis identified hundreds of potential targets for tae-miR5384-3p and tae-miR9653b, advancing understanding of APR-based defense mechanisms. The results reveal that wheat APR genes (e.g., Lr46) and miRNAs collaboratively regulate resistance to leaf rust through temporal expression.

Xiong et al. (Contribution 12) found that nano-silicon foliar application alleviates salt stress in rice seedlings (variety 9311) by improving root architecture, enhancing photosynthesis, boosting antioxidant enzyme activities, increasing hormone contents of indole-3-acetic acid, cytokinin, gibberellin, jasmonic acid, salicylic acid, reducing abscisic acid and ROS, promoting Si^{4+}, K^+, and Ca^{2+} uptake, and decreasing Na^+ and Cl^- accumulation, ultimately supporting better growth under NaCl stress.

Li et al. (Contribution 13) identified 122 subtilase (*SBT*) genes in cultivated peanut (*Arachis hypogaea* L.), classifying them into six phylogenetic subgroups. Structural and collinearity analyses revealed evolutionary conservation, with numerous homologs in *Glycine max*. Cis-element, GO, and KEGG analyses indicated roles in hormone response and abiotic stress. Tissue-specific expression profiling highlighted genes like *AhSBT2* and *AhSBT102* in seed development, while *AhSBT39* and AhSBT115 were upregulated under drought, cold, hormone treatments, and *Ralstonia solanacearum* infection in resistant varieties. The results provide foundational insights into *AhSBT* functions for improving stress tolerance and agronomic traits in peanut.

Tobiasz-Salach et al. (Contribution 14) investigated the role of foliar silicon (Si) application in alleviating salt stress (200 mM NaCl) in wheat. Si at 0.1% and 0.2% concentrations significantly improved physiological parameters, including chlorophyll content (CCI), gas exchange (P_N, C_i, E, gs), and chlorophyll fluorescence (PI, RC/ABS, F_V/F_m, Fv/F_0), compared to the lower dose (0.05%). The mitigating effect was sustained over time. Additionally, MSAP analysis revealed that Si application induced dose-dependent changes in DNA methylation patterns, suggesting epigenetic regulation as a key mechanism in enhancing salt adaptation. These findings demonstrate that foliar Si can effectively mitigate salinity impact through physiological and epigenetic modifications.

Wang et al. (Contribution 15) explored mechanisms behind the balance between high yield and quality by using two varieties of ZZY1 and ZZY8. ZZY1 showed higher yield due to greater seed-setting and grain weight, while ZZY8 exhibited superior quality with lower chalkiness and higher head rice rate. Although ZZY1 had lower total dry matter, it allocated more to panicles and nonstructural carbohydrates, with higher grain-filling rates, ATP/ATPase levels, and key enzyme activities early in filling. Transcriptome analysis highlighted carbohydrate and energy metabolism as central to trade-offs, indicating that limited energy in ZZY1 hindered simultaneous high yield and quality development.

Plant responses to abiotic and biotic stresses involve intricate physiological adjustments, molecular regulation, and epigenetic modifications, as revealed by fifteen studies focusing on crops like rice, wheat, peanut, etc. The abiotic stresses involved in our thematic study include water/drought stress, salt stress, aluminum and chloride ion stress, heat and light stress. In addition, three studies were conducted from the aspect of genome-wide identification and analysis, and one study was carried out from the aspect of primary metabolome analysis. Wang et al. (Contribution 15) linked rice yield-quality discordance to energy metabolism genes. Through these abiotic stress research, epigenetic regulation is deepened. The study on triticale under aluminum stress compares three DNA methylation quantification models, identifying DN-CG, DM-CG, and DN-CHG as core epigenetic markers. These markers show similar distribution in roots and leaves, supporting somatic memory-mediated stress signal transmission, beyond prior focus on genetic variation. Transcription factor functional expansion is notable and morpho-physiological mechanisms are refined. In *Ginkgo biloba*, 13 *GbSPL* genes were found to expand via dispersed duplication, with only 2 miR156-targeted, differing from angiosperms. GbSPL2/11 localize to the nucleus, regulating flavonoid biosynthesis and water stress responses. Stress-tolerant *Begonia semperflorens* (FB08-059) uses 25.83% thicker cuticles, leaf folding, and tripled anthocyanins to maintain photosynthesis, providing new resistance evaluation indices. Research on biotic stress, as well as the study on the miRNA–gene regulatory network of wheat resistance to leaf rust, cover the precise excavation and marker validation of pest resistance resources. In summary, these studies break through the limitation of the "single stress–single gene–single mechanism" in traditional stress research. From multiple dimensions, such as transcriptional regulation and morphological–physiological integration, they construct stress response networks with both species-specificity and universality, providing new genetic resources and theoretical basis for breeding stress-resistant crops.

Conflicts of Interest: The authors declare no conflicts of interest.

List of Contributions:

1. Niedziela, A.; Orłowska, R.; Bednarek, P.T. DNA Methylation Changes Reflect Aluminum Stress in Triticale and Epigenetic Control of the Trait. *Int. J. Mol. Sci.* **2025**, *26*, 4995.
2. Ming, M.; Yi, M.; Qin, C.; Yan, L.; Sun, Y.; Zhang, J.; Cao, F.; Fu, F. Genome-Wide Identification of *Ginkgo biloba* SPL Gene Family and Expression Analysis in Flavonoid Biosynthesis and Water Stress. *Int. J. Mol. Sci.* **2025**, *26*, 4932.
3. Gaonosi, G.G.; Lekgari, L.; Mosupiemang, M.; Sehularo, M.N.; Tshegofatso, A.B.; Tiroesele, B.; Motlhaodi, T.; Ngwako, S.; Pholo-Tait, M. Phenotypic Screening and Marker-Assisted Validation of Sources of *Aphis craccivora* Koch Resistance in Cowpea (*Vigna unguiculata* L.). *Int. J. Mol. Sci.* **2025**, *26*, 4406.

4. Ginori, J.; Nguyen, C.D.; Wilson, S.; Deng, Z.; Huo, H. Investigating Morphological and Physiological Responses to Stress in *Begonia semperflorens*. *Int. J. Mol. Sci.* **2025**, *26*, 3514.
5. Yuan, Q.; Wang, J.; Liu, F.; Dai, X.; Zhu, F.; Zou, X.; Xiong, C. Genome-Wide Identification of the BTB Domain-Containing Protein Gene Family in Pepper (*Capsicum annuum* L.). *Int. J. Mol. Sci.* **2025**, *26*, 3429.
6. Xu, H.; Xing, Y.; Li, G.; Wang, X.; Zhou, X.; Lu, Z.; Ma, L.; Yang, D. Decoding PHR-Orchestrated Stress Adaptation: A Genome-Wide Integrative Analysis of Transcriptional Regulation Under Abiotic Stress in *Eucalyptus grandis*. *Int. J. Mol. Sci.* **2025**, *26*, 2958.
7. Su, L.; Lu, T.; Li, Q.; Li, Y.; Wan, X.; Jiang, W.; Yu, H. Chlorine Modulates Photosynthetic Efficiency, Chlorophyll Fluorescence in Tomato Leaves, and Carbohydrate Allocation in Developing Fruits. *Int. J. Mol. Sci.* **2025**, *26*, 2922.
8. Bilova, T.; Golushko, N.; Frolova, N.; Soboleva, A.; Silinskaia, S.; Khakulova, A.; Orlova, A.; Sinetova, M.; Los, D.; Frolov, A. Strain-Specific Features of Primary Metabolome Characteristic for Extremotolerant/Extremophilic Cyanobacteria Under Long-Term Storage. *Int. J. Mol. Sci.* **2025**, *26*, 2201.
9. Wang, H.; Chen, Y.; Guo, F.; Wu, D.; Liang, W.; Dong, P.; Cheng, J. Evaluating Drought Tolerance in *Codonopsis pilosula* Seedlings: Combining Growth, Physiology, Yield, and Tolerance Indices. *Int. J. Mol. Sci.* **2025**, *26*, 1600.
10. Teng, W.; Sun, J. Effect of Temperature on Polyamine Oxidase Genes in *Skeletonema dohrnii*. *Int. J. Mol. Sci.* **2025**, *26*, 1048.
11. Spychała, J.; Noweiska, A.; Tomkowiak, A.; Bobrowska, R.; Szewczyk, K.; Kwiatek, M.T. Unraveling Effects of miRNAs Associated with APR Leaf Rust Resistance Genes in Hybrid Forms of Common Wheat (*Triticum aestivum* L.). *Int. J. Mol. Sci.* **2025**, *26*, 665.
12. Xiong, J.; Yang, X.; Sun, M.; Zhang, J.; Ding, L.; Sun, Z.; Feng, N.; Zheng, D.; Zhao, L.; Shen, X. Mitigation Effect of Exogenous Nano-Silicon on Salt Stress Damage of Rice Seedlings. *Int. J. Mol. Sci.* **2025**, *26*, 85.
13. Li, S.; Fu, H.; Sharif, Y.; Abdullaziz, S.; Wang, L.; Zhang, Y.; Zhuang, Y. Genome-Wide Identification, Functional Characterization, and Stress-Responsive Expression Profiling of Subtilase (*SBT*) Gene Family in Peanut (*Arachis hypogaea* L.). *Int. J. Mol. Sci.* **2024**, *25*, 13361.
14. Tobiasz-Salach, R.; Stadnik, B.; Mazurek, M.; Buczek, J.; Leszczyńska, D. Foliar Application of Silicon Influences the Physiological and Epigenetic Responses of Wheat Grown Under Salt Stress. *Int. J. Mol. Sci.* **2024**, *25*, 13297.
15. Wang, Y.; Li, G.; Ma, J.; Su, H.; Hu, W.; Lin, J.; Fu, W.; Zeng, Y.; Tao, L.; Fu, G.; et al. Energy Deficiency and Misdistribution Leads to Disrupted Formation in Grain Yield and Rice Quality. *Int. J. Mol. Sci.* **2024**, *25*, 12751.

References

1. Leisner, C.P.; Potnis, N.; Sanz-Saez, A. Crosstalk and trade-offs: Plant responses to climate change-associated abiotic and biotic stresses. *Plant Cell Environ.* **2023**, *46*, 2946–2963. [CrossRef] [PubMed]
2. Mishra, G.; Mohapatra, S.K.; Rout, G.R. Plant membrane transporters function under abiotic stresses: A review. *Planta* **2024**, *260*, 125, Erratum in *Planta* **2024**, *260*, 137. [CrossRef] [PubMed]
3. Yang, Y.; Tilman, D.; Jin, Z.; Smith, P.; Barrett, C.B.; Zhu, Y.; Burney, J.; D'Odorico, P.; Fantke, P.; Fargione, J.; et al. Climate change exacerbates the environmental impacts of agriculture. *Science* **2024**, *385*, 1058. [CrossRef] [PubMed]
4. McCulloch, M.T.; Winter, A.; Sherman, C.E.; Trotter, J.A. 300 years of sclerosponge thermometry shows global warming has exceeded 1.5 °C. *Nat. Clim. Change* **2024**, *14*, 171–177. [CrossRef]
5. Lee, C.; Zeng, M.; Luo, K. How does climate change affect food security? Evidence from China. *Environ. Impact Assess. Rev.* **2024**, *104*, 107324. [CrossRef]
6. Zhang, H.; Zhu, J.; Gong, Z.; Zhu, J. Abiotic stress responses in plants. *Nat. Rev. Genet.* **2022**, *23*, 104–119. [CrossRef] [PubMed]
7. Sato, H.; Mizoi, J.; Shinozaki, K.; Yamaguchi-Shinozaki, K. Complex plant responses to drought and heat stress under climate change. *Plant J.* **2024**, *117*, 1873–1892. [CrossRef] [PubMed]
8. Rillig, M.C.; van der Heijden, M.G.A.; Berdugo, M.; Liu, Y.; Riedo, J.; Sanz-Lazaro, C.; Moreno-Jiménez, E.; Romero, F.; Tedersoo, L.; Delgado-Baquerizo, M. Increasing the number of stressors reduces soil ecosystem services worldwide. *Nat. Clim. Change* **2023**, *13*, 478–483. [CrossRef] [PubMed]

9. Mittler, R.; Zandalinas, S.I.; Fichman, Y.; Van, F. Reactive oxygen species signalling in plant stress responses. *Nat. Rev. Mol. Cell Biol.* **2022**, *23*, 663–679. [CrossRef] [PubMed]
10. Musazade, E.; Mrisho, I.I.; Feng, X. Auxin metabolism and signaling: Integrating independent mechanisms and crosstalk in plant abiotic stress responses. *Plant Stress* **2025**, *18*, 101034. [CrossRef]
11. Ali, S.; Ulhassan, Z.; Shahbaz, H.; Kaleem, Z.; Yousaf, M.A.; Ali, S.; Sheteiwy, M.S.; Waseem, M.; Ali, S.; Zhou, W. Application of magnesium oxide nanoparticles as a novel sustainable approach to enhance crop tolerance to abiotic and biotic stresses. *Environ. Sci. Nano* **2024**, *11*, 3250–3267. [CrossRef]
12. Avendaño, V.A.; Sampedro-Guerrero, J.; Gómez-Cadenas, A.; Clausell-Terol, C. Modern approaches to enhancing abiotic stress tolerance using phytoprotectants: A focus on encapsulated proline. *J. Plant Physiol.* **2025**, 154602. [CrossRef]

International Journal of
Molecular Sciences

MDPI

Article

DNA Methylation Changes Reflect Aluminum Stress in Triticale and Epigenetic Control of the Trait

Agnieszka Niedziela, Renata Orłowska and Piotr Tomasz Bednarek *

Plant Breeding and Acclimatization Institute—National Research Institute, Radzików, 05-870 Błonie, Poland; a.niedziela@ihar.edu.pl (A.N.); r.orlowska@ihar.edu.pl (R.O.)
* Correspondence: p.bednarek@ihar.edu.pl

Abstract: Aluminum (Al) stress is typical for acidic soils and may affect cereals' yield. Al tolerance in triticale is mostly affected by the aluminum-activated malate transporter (ALMT) gene (7R) and some other QTLs (3R, 5R, and 6R). The trait is heritable in about 36% of cases, indicating that epigenetic factors may impact the phenomenon. This study demonstrates that utilizing different methods to quantify DNA methylation changes induced by Al stress results in detail differences, and the results evaluated should be compared critically. The Common and the basic General approaches are sufficient if general information is needed. The General (extended variant) approach may deliver data on methylation changes affecting symmetric sequence contexts. The markers assigned to DN-CG, DM-CG, and DN-CHG were suggested as the most important in explaining Al tolerance in triticale. Analysis of the maps constructed based on root tips and leaf tissues showed different densities of the epigenetic markers but reflected the comparable patterns of their distribution, supporting the hypothesis that Al stress could be transmitted to other plant tissues due to somatic memory. Methylation changes occur throughout the genome and are not associated with specific genes related to aluminum stress.

Keywords: triticale; aluminum stress; epigenetic alterations; DNA methylation

1. Introduction

Aluminum (Al) toxicity is a significant abiotic stress affecting plant growth and productivity in over 50% of the world's arable land where soil acidity is a problem [1]. In acidic soils (pH < 5.5), aluminum is usually present in its soluble Al^{3+} ions, which are readily available for uptake by plant roots [2]. Al^{3+} ions absorbed in roots inhibit cell division and elongation, disrupt nutrient uptake, and lead to growth cessation [3].

Triticale (x *Triticosecale* Wittmack, AABBRR, 2n = 6x = 42), a man-made hybrid between wheat (*Triticum* ssp.) and rye (*Secale* ssp.), was developed to inherit the desirable qualities of both parents, making it an adaptable and resilient crop. Its energy-rich grain with high protein content and higher amount of lysine content compared to other cereals makes it a valuable component of animal diets [4]. While its primary application is in feeding animals, triticale is also used in human food production for bread, pasta, and cereals [4], and in industrial applications like biodegradable films [5,6] and bioethanol production due to its high starch content [7,8]. Triticale has emerged as a model species for studying aluminum tolerance, particularly due to its diverse genotypic responses to Al stress [9–11]. However, understanding the genetic and epigenetic mechanisms underlying Al tolerance in triticale is far from complete [12–14].

As a complex trait, Al tolerance is affected by many genes, and different mechanisms are triggered to mitigate the toxic effects of the metal. The most recognized is the exudation of organic acids (OA) (i.e., as malate or citrate) from the root apex [2,3]. These acids are transported from the root cells into the rhizosphere, where they bind to Al^{3+} ions, forming non-toxic complexes that are less likely to enter the root. Organic acids distribution is facilitated by specific transporters, mostly belonging to the aluminum-activated malate transporters (ALMT) [15–17] and multidrug and toxin efflux (MATE) families [15,18,19]. The genes coding ALMT and MATE proteins have been mapped to chromosomes 3R, 4R, 6R, and 7R in rye [20–23]. Similarly, Al-tolerant genes were assigned to 2H, 3H, and 4H in barley [24,25] and to 4D, 4B, and 3B in wheat [15,18,19]. Regarding triticale, the ALMT gene encodes a malate transporter, which plays a critical role in Al^{3+} detoxification and is located in the QTL region on the 7R chromosome and explains 36% of phenotypic variance [26].

In parallel to genetic aspects of Al tolerance, epigenetic input, defined as heritable changes in gene expression without DNA sequence changes [27], regarding abiotic stresses, is discussed [28–30]. Such changes, among others, could be reflected at the histone modification level [31,32] or as DNA methylation alterations [33]. The latter affects symmetric (CG and CHG) or asymmetric (CHH) sequence contexts, influencing chromatin structure and transcription [33,34].

It was also evidenced that transposable elements (TEs) methylation in or near genes influenced their expression [35,36]. Moreover, regulating TEs by small interfering RNAs (siRNAs) is believed to be an epigenetic mechanism for balancing the expression and repression of stress-responsive genes of plant survival [28,37] Still, this aspect of Al tolerance in cereals is poorly understood, even though it may significantly affect the trait. Studies in wheat, maize, and rice suggest that the expression of aluminum-related genes is regulated by DNA methylation of transposable elements (TEs) [29,38,39]. DNA methylation changes due to Al stress may, at least, modulate the expression profiles of the genes coding organic acids transporters, as was demonstrated in barley for the HvAACT1 gene responsible for citrate secretion [35]. A multiretrotransposon-like (MRL) sequence insertion detected in the upstream genomic region of the HvAACT1 gene significantly enhanced its expression in the root tips of Al-tolerant accessions. Furthermore, in accessions with low levels of HvAACT1 expression, this MRL insertion was present but highly methylated [35]. Similarly, Al stress promotes demethylation in the coding region of wild tobacco plants' glycerophosphodiesterase-like protein gene (NtGPDL), enhancing expression [40]. Another study showed that overexpression of the S-adenosylmethionine (SAM) in AvSAMS1-transformed Arabidopsis plants impacts changes in DNA and histone H3 methylation after exposure to Al [41]. In rice exposed to Al stress, there were 26 genes that, at the transcriptomic level, differentiated tolerant and non-tolerant plants. The genes appeared to be 5 to 10 times more methylated in a tolerant cultivar than in a susceptible one [29].

Interestingly, transgenerational inheritance of epigenetic modifications has been observed in plants, suggesting that epigenome-targeted breeding strategies might enhance crop resistance to abiotic stresses [29,42,43]. Furthermore, epigenetic changes may not necessarily affect genes encoding the trait. They may impact multiple genes, influencing various cellular processes, such as ion transport, reactive oxygen species (ROS) detoxification, or hormone signaling [29,44]. Thus, further studies are necessary to gain a better understanding of how plants respond to abiotic stresses and how epigenetic regulation can be harnessed to improve crop tolerance [30,42].

Several molecular methods are available for studying epigenetic aspects of DNA methylation changes due to abiotic stresses. Two of them, namely Methylation Sensitive

Amplification Polymorphism (MSAP) [45] and methylation Amplified Fragment Length Polymorphism (metAFLP) [46], are based on the AFLP technique [47] and utilize restriction enzymes with different specificity to DNA methylation [45]. The methods may utilize different approaches to quantify changes. MSAP, in its initial form, uses the so-called transitions for quantification, whereas semi-quantitative solutions may quantify methylation changes based on quantifying individual methylation changes affecting restriction sites. In its original form, the methods are dedicated to DNA methylation changes. The exception is the metAFLP approach, which may quantify sequence and DNA methylation alterations in all symmetric and asymmetric contexts. Unfortunately, the methods miss information on marker sequence. The disadvantage could be overcome by utilizing the DArTseqMet approach [48], which uses the same enzymes as MSAP and additionally delivers sequence data that could be processed in further analyses when combined with the semi-quantitative MSAP approach. Quantitative information regarding symmetric sequence contexts could be incorporated into data analysis.

This study aims to verify whether distinct ways of quantifying markers into variation types using the MSAP approach combined with DArTseq markers deliver comparable information on DNA methylation changes in plant materials affected by Al stress to illustrate putative distinctions or similarities of the systems. Furthermore, we wish to identify which types of DNA methylation changes related to symmetric DNA methylation contexts mainly affect Al tolerance and whether such markers are located close to the Al-tolerant genes or not.

2. Results

2.1. *Plant's Response to Aluminum Stress*

Based on the Al physiological test, eight and seven lines were classified as tolerant (T) or non-tolerant (NT). The average root regrowth ranged from 0.9 to 2.9 cm for T lines, and trace or no regrowth was observed in the case of NT lines (Table 1). According to the physiological Al test results, there were seven NT and eight T lines. Furthermore, seven lines were classified as spring (S) and eight as winter (W). The root tips (1) and leaves (2) tissues were used in further analyses.

Table 1. The aluminum test results illustrating the average root regrowth (RG) in cm measured after 24 h of Al treatment followed by 48 h of growing under unstressed conditions.

Line	L195	L198	L201	L203	L1	L17	L27	L34	L145	L190	L422	L451	L291	L438	L444
Cl *	S/T	S/T	S/T	S/T	S/NT	S/NT	S/NT	S/NT	W/T	W/T	W/T	W/T	W/NT	W/NT	W/NT
RG	1.8 ± 0.4	2.9 ± 0.4	2.8 ± 0.5	0.9 ± 0.2	0.1 ± 0.1	0.0 ± 0.0	0.0 ± 0.0	0.0 ± 0.0	2.2 ± 0.3	1.7 ± 0.2	2.1 ± 0.3	1.9 ± 0.4	0.1 ± 0.1	0.0 ± 0.0	0.0 ± 0.0

* Cl—classification; RG—root regrowth; S and W state for spring and winter lines; T and NT reflect tolerant and non-tolerant lines.

2.2. *MSAP Analysis*

The raw number of DArTseqMet markers equaled 17,177. Based on the frequency of the marker presence among plants representing the line identified in SbfI/HpaII and SbfI/MspI digests, the presence or absence of a marker is indicated. As an outcome, four-digit codes were generated and used for DNA methylation changes affecting HpaII-MspI site quantification, following the reasoning illustrated in Supplementary Table S1. Depending on the quantification model, fluctuations in quantitative characteristics were evidenced (Table 2).

Table 2. The MSAP-based quantitative characteristics utilizing DArTseqMet markers reflecting analyzed plant materials and tissues.

Plant Materials			Common Model						General Model									
									Basic						Extended			
Tolerance	S/W	Plant Part	DM (%)	DNM (%)	MP (%)	NMP (%)	M (%)	NM (%)	DM (%)	DNM (%)	MP (%)	NMP (%)	M (%)	NM (%)	DN-CHG (%)	DN-CG (%)	DM-CHG (%)	DM-CG (%)
T	S	R	4.25	3.26	33.38	59.92	36.34	63.66	9.79	9.11	36.27	44.82	45.38	54.62	0.86	8.22	1.32	8.48
T	S	R	4.64	2.74	33.69	60.09	36.01	63.99	9.56	8.79	37.24	44.40	46.03	53.97	0.55	8.19	0.87	8.69
T	S	R	2.14	3.89	33.47	62.46	36.64	63.36	8.53	9.43	36.88	45.16	46.31	53.69	0.88	8.45	0.60	7.93
T	S	R	1.82	5.37	35.00	60.24	39.41	60.59	8.53	10.56	37.82	43.08	48.38	51.62	1.41	9.01	0.44	8.10
NT	S	R	2.55	7.65	34.29	59.21	40.45	59.55	8.88	11.70	36.84	42.59	48.53	51.47	2.03	9.55	0.65	8.23
NT	S	R	1.78	9.49	35.74	57.15	43.40	56.60	8.52	12.57	37.25	41.66	49.82	50.18	2.13	10.31	0.45	8.07
NT	S	R	6.65	4.63	38.86	51.64	42.73	57.27	12.42	10.37	40.45	36.76	50.82	49.18	1.28	9.01	2.68	9.73
NT	S	R	7.04	3.92	38.94	51.06	42.45	57.55	12.14	9.95	40.33	37.58	50.28	49.72	1.04	8.78	2.40	9.74
T	W	R	1.74	4.23	35.32	60.81	38.73	61.27	8.50	9.89	37.80	43.81	47.69	52.31	1.11	8.70	0.32	8.18
T	W	R	2.45	5.73	35.09	59.59	39.69	60.31	8.73	10.63	37.58	43.06	48.21	51.79	1.32	9.21	0.58	8.15
T	W	R	2.19	6.66	35.02	57.17	41.25	58.75	8.78	11.71	38.25	41.26	49.96	50.04	2.14	9.40	0.51	8.27
T	W	R	1.52	4.82	36.84	60.01	40.38	59.62	8.52	10.03	39.23	42.21	49.27	50.73	0.60	9.40	0.31	8.21
NT	W	R	12.92	9.34	35.26	44.04	43.92	56.08	14.93	12.40	38.46	34.22	50.85	49.15	2.18	10.09	4.26	10.67
NT	W	R	10.06	4.97	34.23	49.52	39.69	60.31	12.59	11.02	40.53	35.86	51.55	48.45	1.62	9.31	3.08	9.51
NT	W	R	6.75	4.41	39.27	51.21	42.97	57.03	12.40	10.20	40.65	36.75	50.85	49.15	1.09	9.02	2.65	9.75
NT	S	L	3.19	3.48	35.75	58.06	39.04	60.96	9.23	9.64	38.94	42.18	48.58	51.42	0.84	8.72	0.69	8.55
T	S	L	3.11	3.85	33.01	61.34	36.39	63.61	8.87	9.35	36.82	44.96	46.16	53.84	0.77	8.51	0.76	8.11
T	S	L	8.16	2.98	33.73	56.81	36.10	63.90	12.18	8.55	37.04	42.24	45.58	54.42	0.29	8.22	2.94	9.24
NT	W	L	3.91	5.85	35.88	54.76	41.61	58.39	9.79	11.30	38.95	39.96	50.25	49.75	1.84	9.37	0.88	8.91

Lines are encoded as NT (Al non-tolerant) and T (Al tolerant). Spring lines are indicated as S, whereas winter as W. Plant parts reflect root, where R is the root tip regrowth, and leaf tissue, indicated by L. DM—DNA demethylation, DNM—DNA de novo methylation, MP—methylation preservation (sites that were methylated in control and treated lines); NMP—non-methylated sites (as MP but regarding non-methylated sites); M—MP plus DNM; NM—NMP and DM; CG and CHG indicate two types of symmetric sequence contexts. The General model assumes that the whole genetic background explaining the MSAP four-digit code was used to quantify DN, DNM, M, MP, NM, and NMP characteristics employed for calculations. The sequence contexts model is similar to the General model. However, DM and DNM characteristics were evaluated in symmetric sequence contexts. Finally, the Common model is similar to the General one except that the most abundant explanation of the four-digit MSAP profile was used for calculation.

In general, the widely used model (Common) [45] resulted in lower values of DNA demethylation (DM), de novo DNA methylation (DNM), DNA methylation status preservation (MP), and total methylation (M) of the HpaII-MseI restriction sites compared to the data gained in the General model of its basic variant. However, non-methylation preservation (NMP) and non-methylation (NM) contained higher values in the Common than the General basic model, independent of whether root tips or leaf tissues were analyzed. As an extended variant of the General model is part of a basic one but delivers information on DNA de novo methylation and DNA demethylation affecting CG and CHG sequence contexts of the restriction sites, the respective values were the lowest compared to other quantitative data. The highest standard deviation values affected DNA methylation quantitative characteristics of the Common model, whereas the lowest was typical for the extended variant of the General one (Table 3).

Table 3. The arrangement of basic statistics concerning MSAP quantitative characteristics.

Model		Variation Type (%)	Minimum	Maximum	Mean	SD
Common		DM	1.52	12.92	4.57	3.21
		DNM	2.74	9.49	5.12	1.96
		MP	33.01	39.27	35.41	1.89
		NMP	44.04	62.46	56.58	4.93
		M	36.01	43.92	39.85	2.64
		NM	56.08	63.99	60.15	2.64
General	Basic variant	DM	8.50	14.93	10.15	1.96
		DNM	8.55	12.57	10.38	1.16
		MP	36.27	40.65	38.28	1.42
		NMP	34.22	45.16	41.19	3.36
		M	45.38	51.55	48.66	1.99
		NM	48.45	54.62	51.34	1.99
	Extended variant	DN-CHG	0.29	2.18	1.26	0.56
		DN-CG	8.19	10.31	9.06	0.59
		DM-CHG	0.31	4.26	1.40	1.20
		DM-CG	7.93	10.67	8.76	0.78

Significant Pearson's correlation coefficients (Supplementary Table S2) were evaluated for DNA demethylation (DM) characteristics evaluated in all models. They were negatively correlated with NMP and positively correlated with MP. The correlation between NMP and MP was lower (from −0.503 to −0.803) than between DM and NMPs (from −0.783 to −0.937). Furthermore, a high correlation level was evaluated for all DNM (Common and basic variants of the General model) and DN affecting CG and CHG sequence contexts (from 0.892 to 0.962). DN and DNM were negatively correlated with NM, whereas NMs were positively correlated in each model.

2.3. *ANOVA Analysis*

ANOVA indicated some differences between the models regarding the extent to which tolerant/non-tolerant and winter/spring lines, as well as root tips/leaf tissues, explain quantitative variables describing different aspects of DNA methylation due to Al treatment. In all models tested, the ANOVA was significant ($\alpha \leq 0.05$). In the Common model, four main effects due to the winter–spring trait of the lines were detected, but only those associated with M and NM were significant. Similarly, in the General model (basic variant), four main effects related to winter–spring lines in the cases of DNM, MP, M, and NM were also revealed. Finally, in the extended variant of the General model, the main effects were also due to winter–spring lines in the cases of de novo DNA methylation of both symmetric sequence contexts.

Furthermore, the quantitative characteristics were explained by either two-way or three-way interactions. Two-way interactions were preferentially represented by tolerance and tissue and rarely by tolerance and winter–spring lines. The highest value of explained variance was 0.753 (M and NM variables of the basic general model), whereas the lowest was 0.281 (DNM-Common model) (Table 4).

Table 4. The arrangement of ANOVA results with quantitative variables of each model treated as dependent variables and qualitative ones as explanatory variables.

Model	Dependent Variable	ANOVA Statistics Description	Main Effects/Interactions	Statistics					
				MSE	*MS*	*F(2,16)*	*p*	R^2	R^2_{adj}
Common	DM%	Model		4.95	53.08	10.73	0.001	0.573	0.519
		interactions	tolerance * tissue		27.66	5.59	0.008		
			tolerance * W-S * tissue		50.08	10.12	0.006		
	DNM%	Model		2.76	12.52	4.53	0.028	0.361	0.281
		interactions	W-S		7.81	2.83	0.110		
			tolerance * W-S * tissue		19.53	7.07	0.017		
	MP%	Model		2.3	13.72	5.96	0.011	0.427	0.355
		Main effects	W-S		1.94	0.84	0.370		
		interactions	tolerance * W-S		24.58	10.68	0.005		
	NMP%	Model		8.05	153.97	19.12	0.0001	0.700	0.668
		interactions	tolerance * tissue		94.65	11.75	0.0002		
			tolerance * W-S * tissue		72.61	9.02	0.008		
	M%	Model		2.45	43.00	17.55	0.0001	0.686	0.647
		Main effects	W-S		22.66	9.25	0.008		
		interactions	tolerance * W-S * tissue		66.84	27.28	0.00008		
	NM%	Model		2.45	43.00	17.54	0.00009	0.686	0.647
		Main effects	W-S		22.66	9.25	0.008		
		interactions	tolerance * W-S * tissue		66.85	27.28	0.00008		
General (basic variant)	DM%	Model		1.81	20.03	11.08	0.001	0.581	0.528
		interactions	tolerance * tissue		11.71	6.48	0.004		
			tolerance * W-S * tissue		13.6	7.52	0.014		
	DNM%	Model		0.77	5.98	7.75	0.004	0.492	0.428
		interactions	W-S		4.96	6.42	0.022		
			tolerance * W-S * tissue		8.25	10.69	0.005		
	MP%	Model		1.13	9.10	8.07	0.004	0.5	0.44
		Main effect	W-S		5.1	4.53	0.049		
	NMP%	Model		3.26	75.7	23.22	0.00001	0.743	0.712
		interactions	tolerance * tissue		62.28	19.1	0.00005		
			W-S * tissue		32.88	10.08	0.006		
	M%	Model		0.98	27.85	28.49	0.0001	0.781	0.753
		Main effects	W-S		16.61	16.99	0.001		
	NM%	Model		0.98	27.85	28.49	0.0001	0.781	0.753
		Main effects	W-S		16.61	16.99	0.001		

Table 4. *Cont.*

Model	Dependent Variable	ANOVA Statistics Description	Main Effects/Interactions	*MSE*	*MS*	*F(2,16)*	*p*	R^2	R^2_{adj}
General (extended)	DN-CHG%	Model		0.23	1.21	5.15	0.019	0.391	0.315
		Main effects	W-S		0.70	2.99	0.100		
		interactions	tolerance * W-S * tissue		1.71	7.3	0.016		
	DN-CG%	Model		0.19	1.69	8.91	0.003	0.526	0.468
		Main effects	W-S		1.19	6.28	0.023		
		interactions	tolerance * W-S * tissue		2.24	11.79	0.003		
	DM-CHG%	Model		0.69	7.43	10.79	0.001	0.574	0.521
		interactions	tolerance * tissue		9.4	13.65	0.002		
			tolerance * W-S * tissue		5.46	7.93	0.012		
	DM-CG%	Model		0.31	3.06	9.96	0.001	0.555	0.499
		interactions	tolerance * tissue		4.28	13.94	0.002		
			tolerance * W-S * tissue		1.83	5.97	0.026		

* interaction.

2.4. Elastic Net Regression Analyses

The employment of elastic net regression showed that in the Common model, DNM and DM were insignificant in explaining Al tolerance in contrast to non-tolerance, spring in contrast to winter lines, and in describing the difference between root and leaf tissue (Table 5). However, NMP and M described tolerant and non-tolerant materials, with the prevailing positive effect of M in non-tolerant materials and the negative effect of NMP. When the basic variant of the general model data was employed in the analysis, DNM proved significant in describing tolerant and non-tolerant lines. Furthermore, the negative effect of NMP and M's positive effect in non-tolerant lines were evaluated. The extended variant of the general model showed that DM-CG, DN-CHG, and DN-CHG positively affect non-tolerant lines. Thus, demethylation and de novo methylation negatively impact Al tolerance. The observed effects are mostly related to the CG sequence contexts.

Table 5. The arrangement of the elastic net regression analyses, indicating variables and respective regression coefficients that were significant in explaining the classification of non-tolerant/tolerant lines and their winter–spring belonging.

Classification	DNA Methylation Quantitative Models								
	Common			General					
				Basic Variant			Extended Variant		
	T-NT	S-W	Root-Leaves	T-NT	S-W	Root-Leaves	T-NT	S-W	Root-Leaves
Optimal Lambda	0.0511	0.1931	0.1285	0.0578	0.0833	0.1244	0.0419	0.2083	0.1209
Intercept.	0.2028	−0.3184	−1.3218	5.1232	−19.9417	−1.3218	−38.316	−0.3185	−1.3217
DM	0	0	0	0	0	0			
DM-CHG							0	0	0
DM-CG							2.3539	0	0
DNM	0	0	0	0.2477	0	0			
DN-CHG							0.5351	0	0
DN-CG							1.8941	0	0
MP	0	0	0	0	0	0			
NMP	−0.2853	0	0	−0.4391	0	0			
M	0.4006	0	0	0.2128	0.4021	0			
NM	0	0	0	-1.6×10^{-16}	-3.9×10^{-16}	0			

2.5. Distribution of Markers Assigned to DN-CG, DN-CHG, and DM-CG Regarding Tolerance

The DArTseqMet markers assigned to the DNA de novo events and DNA demethylation affecting CG sequence contexts and mapped to the triticale chromosomes (based on reference maps) were not evenly distributed along chromosomes. Furthermore, they were differently represented in root tips (Figures 1 and 2) and leaves (Figures 3 and 4). Comparable distribution was observed in the case of DN-CHG markers (Figures 5 and 6). In all cases, the number of markers identified in root tips was much higher than the number of markers reflecting leaves, indicating that the Al stress is transmitted to leaves to a limited extent. The markers related to root tips are preferentially mapped to telomeric regions, sometimes forming tightly linked blocks, establishing putative hotspots of epigenetic changes. The same was observed for markers identified in leaves.

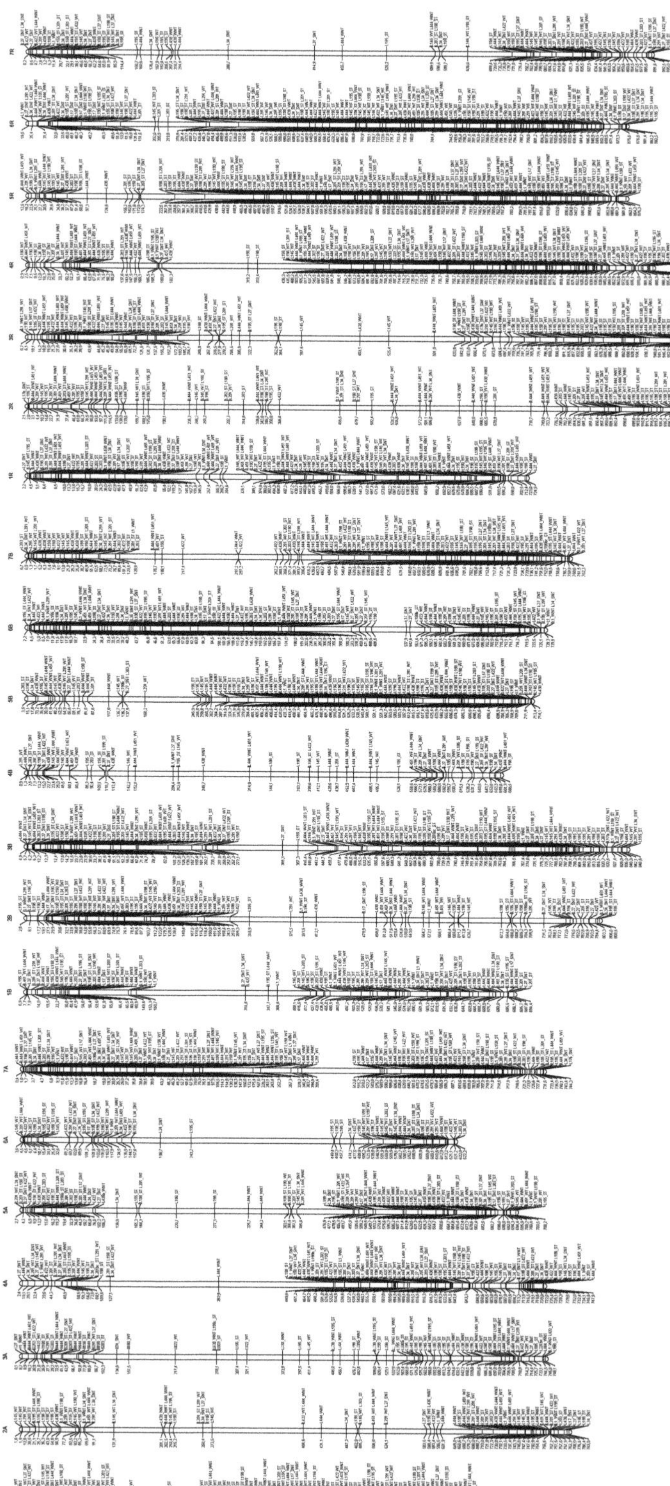

Figure 1. Representation of the genetic map of triticale illustrating the location of the DN-CG-assigned DArTseqMet markers univocally classified to the DN events within CG sequence contexts (1101) in root tips (<0.5 cm). Scale on the map: 1.0 = 1 mln bp. W and S indicate winter and spring forms, and NT and T reflect non-tolerant and tolerant lines.

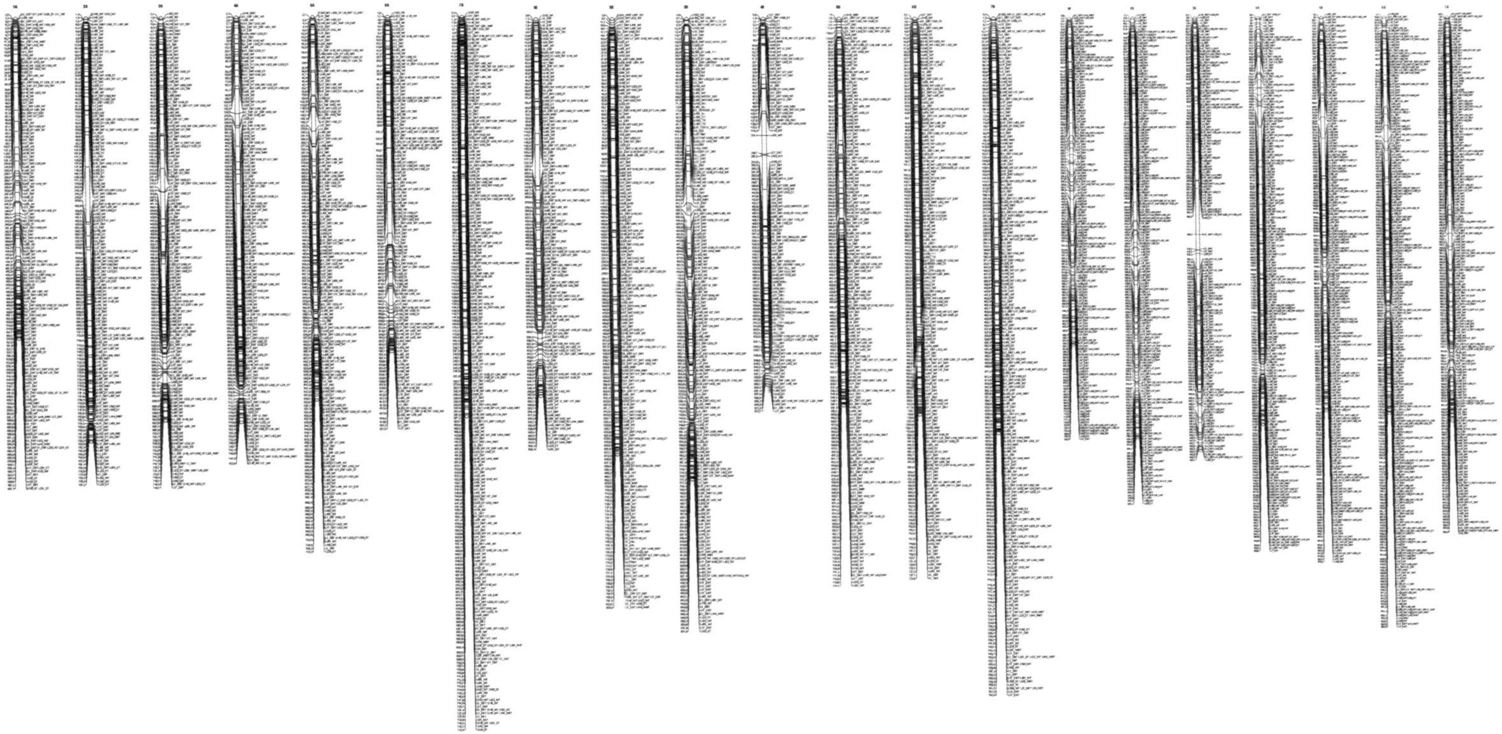

Figure 2. Representation of the genetic map of triticale illustrating the location of the DM-CG - assigned DArTseqMet markers univocally classified to the DM events within CG sequence contexts (0111) in root tips (<0.5 cm). Scale on the map: 1.0 = 1 mln bp. W and S indicate winter and spring forms, and NT and T reflect non-tolerant and tolerant lines.

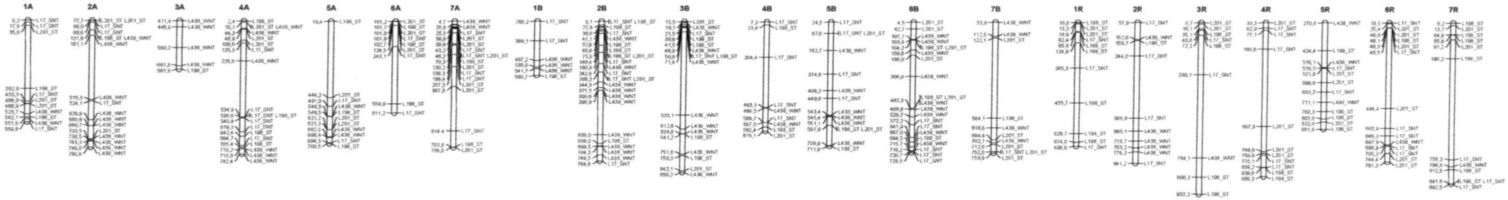

Figure 3. Representation of the genetic map of triticale illustrating the location of the DN-CG-assigned DArTseqMet markers univocally classified to the DN events within CG sequence contexts (1101) in leaves. Scale on the map: 1.0 = 1 mln bp. W and S indicate winter and spring forms, and NT and T reflect non-tolerant and tolerant lines.

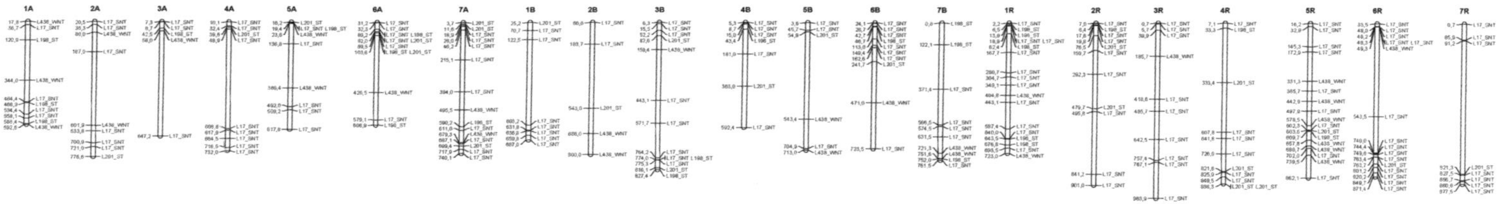

Figure 4. Representation of the genetic map of triticale illustrating the location of the DM-CG-assigned DArTseqMet markers univocally classified to the DM events within CG sequence contexts (0111) in leaves. Scale on the map: 1.0 = 1 mln bp. W and S indicate winter and spring forms, and NT and T reflect non-tolerant and tolerant lines.

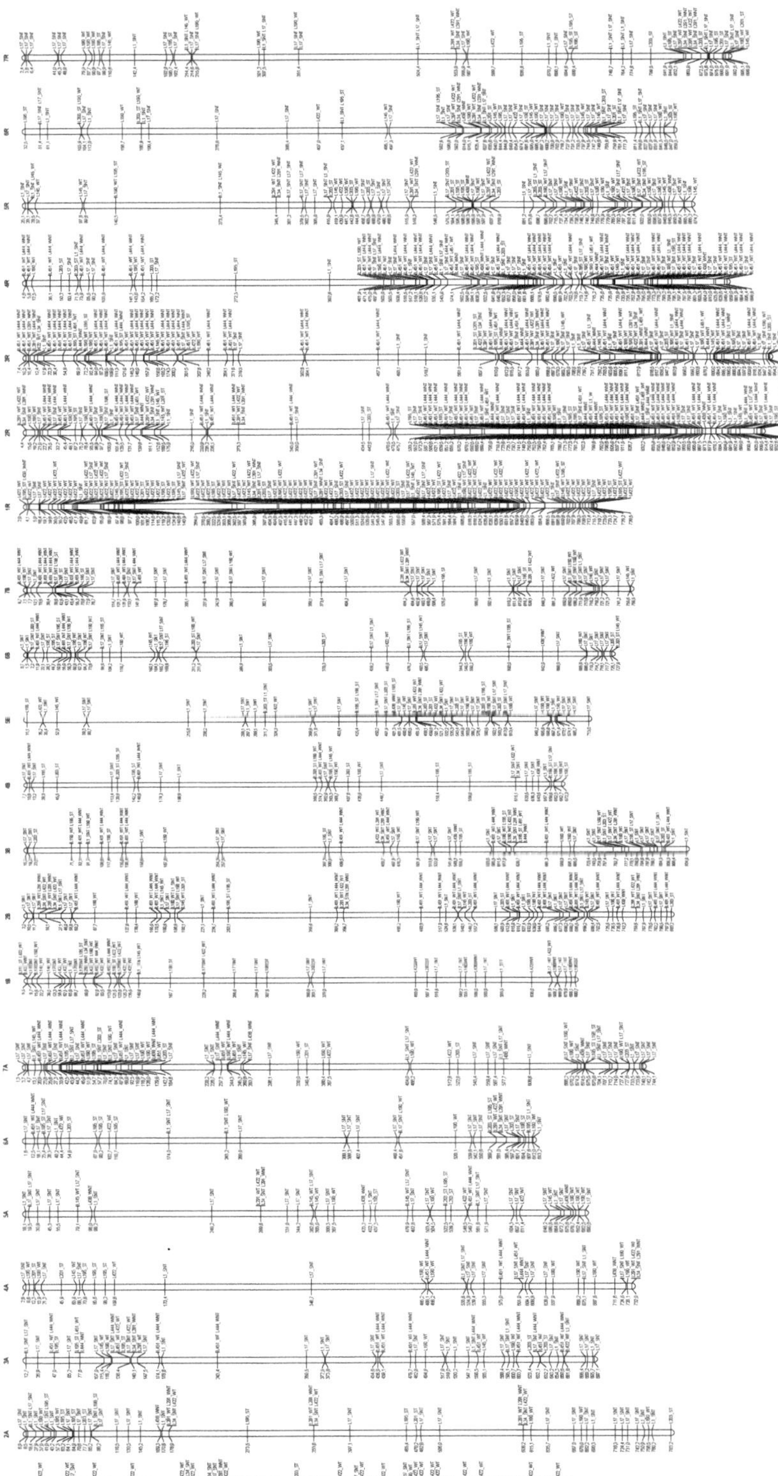

Figure 5. Representation of the genetic map of triticale illustrating the location of the DN-CHG-assigned DArTseqMet markers univocally classified to the DM events within CG sequence contexts (1110) in root tips (<0.5 cm). Scale on the map: 1.0 = 1 mln bp. W and S indicate winter and spring forms, and NT and T reflect non-tolerant and tolerant lines.

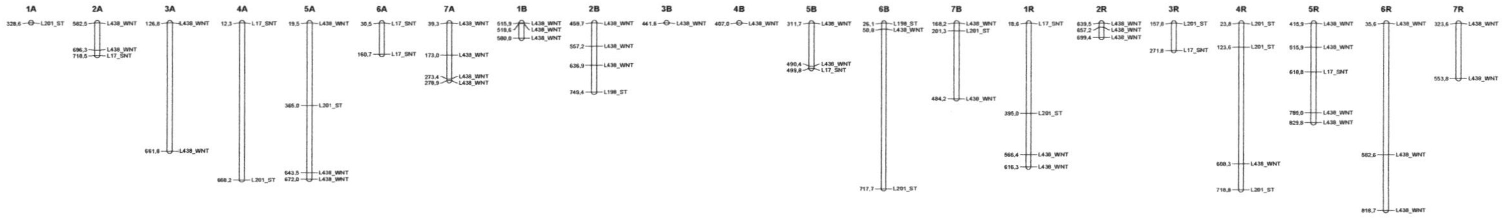

Figure 6. Representation of genetic map of triticale illustrating the location of the DN-CHG-assigned DArTseqMet markers univocally classified to the DM events within CG sequence contexts (1110) in leaves. Scale on the map: 1.0 = 1 mln bp. W and S indicate winter and spring forms, and NT and T reflect non-tolerant and tolerant lines.

3. Discussion

3.1. Uniformity of Plant Materials

The current study is based on at least 40 individual plants representing each line analyzed to ensure that a sufficient amount of root tissue was collected for the analysis. There were three biological repetitions of such experiments. We have chosen root tips as the tissue directly in contact with the Al-containing medium, whereas leaves were collected because stress could be transmitted to other tissues [49,50]. However, the extent of such transmission is not known. The lines were selected for at least seven generations via selfing each plant as a starting point for the next generation. Thus, all further analyses were expected to be based on genetically uniform materials. Furthermore, as all Al-treated lines were subjected to the same stress level (the experiment that allowed tissue collection), it was anticipated that the seventh (and subsequent) generations would have identical genotypes. Phenotypic analysis of the lines based on the Al test confirmed that tolerant and non-tolerant materials had stable phenotypes, as indicated by the root regrowth data.

3.2. Characteristic of DArTseqMet Approach for DNA Methylation Study

Currently, several approaches that allow quantification of DNA methylation changes based on the MSAP approach [51–53] are available. Depending on how the molecular profiles are interpreted, the approaches can be classified as Common and General (including basic and extended variants) (Supplementary Table S1). The Common approach assumes the most probable explanation, omitting other alternatives. The transition resulting in such an interpretation is classified into a specific type and counted. Contrarily, the general approach tries to interpret differences between profiles based on putative changes affecting the site's cytosines. It can also deliver information on changes affecting symmetric sequence contexts regarding DNA methylation changes, giving information on epigenetic aspects of Al stress. Unfortunately, the limitation of the approaches is the lack of information regarding DNA marker sequences if the AFLP markers are used. Therefore, an adequate marker system allowing marker classification to varying DNA methylation characteristics and their quantification is needed. Moreover, preferences should be given to the marker system that allows mapping utilizing its information on chromosomal location.

The DArTseqMet approach is based on the HpaII/MspI isoschizomer and delivers marker sequence information. If combined with MSAP or semi-quantitative MSAP, the changes could be quantified and normalized. Furthermore, the analysis allows the evaluation of the (epi) genetic background reflecting the four-digit codes regarding CG and CHG sequence contexts. Thus, detailed information on subtle DNA methylation changes related to the methylation alterations transferred during DNA replication and those reflecting some epigenetic aspects could be extracted.

The general approach, which utilizes many DArTseqMet markers, delivers reliable and extensive information on epigenetic changes affecting DNA methylation patterns. It was demonstrated that differences were present using varying MASP characteristics evaluated in the Common or General models. For example, higher DM, DNM, MP, and M values were evaluated compared to the Common model, whereas NMP and NM values were higher in the Common model. The differences are due to how the two models quantify respective differences, with the Common approach underestimating some changes. The values of DM and DN related to the CG and CHG contexts of the extended variant were lower than those of DN and DNM evaluated in the other cases, which is not unexpected, as only a tiny fraction of changes were considered (due to the low representation of changes related to CHH sequence context). It is also unsurprising that the data evaluated based on all methods exhibit varying correlations (Supplementary Table S2), with only a few cases above 0.8. Still, correlated data indicate that models reflect the same phenomenon with varying approximations.

3.3. Epigenetic Alterations in Triticale Under Al Stress

As could be seen, all quantitative characteristics analyzed in the General model were significant, whereas in the Common one, the model encompassing DNM% was insignificant. Furthermore, in the Common and General models, there were some differences in the presence of the ANOVA's main effects (Table 4); however, in nearly all cases, the main effect was due to the winter/spring lines. Regarding the extended variant of the general model, all characteristics resulted in significant ANOVA, with only DN-CG% and DN-CHG% having the main effect in winter/spring materials. In the case of all statistical models, at least two-way interactions, primarily due to tolerance and tissue, were found. Thus, although analyses based on the Common and General models reflect the same phenomenon, they differ in detail, suggesting that it is crucial to apply the best one that fits experimental requirements and scientific demands. Apparently, however, ANOVA supports the notion regarding epigenetic alterations in triticale DNA under Al stress.

As indicated earlier, phenotypic stability reflecting Al tolerance was evaluated for the materials tested. Such a result might reflect genotypic stability due to the Al-tolerant gene we mapped earlier to the 7R chromosome, which explains approximately 36% of phenotypic variance [26]. Besides, in triticale, the presence of the QTLs responsible for the trait on 3R, 4R, 5R, and 6R indicates that the other genes may influence the trait [54]. Moreover, the heritability coefficient in the broad and narrow sense was determined for twenty triticale genotypes presented at the International Maize and Wheat Improvement Centre (CIMMYT), and they varied from 63.77 to 91.80 and from 74.46 to 94.56, respectively [55]. Such results indicate that simple inheritance is governed by a dominant gene and the presence of additional factors that influence trait expression. Thus, an alternative explanation is that during subsequent inbreeding steps, epigenetic aspects of Al tolerance were also stabilized. The notion is supported by studies indicating the role of epigenetics expressed at the DNA methylation level regarding rice, wheat, barley, and maize and showing fluctuations in methylation patterns within differentially expressed genes [29] or transposons [35,38,39] due to Al treatment.

The question is whether or not DNA methylation changes regarding metal stress are related to the respective tolerant gene coding sequences. Based on available data, DNA hypomethylation was observed at the promoters of the Heavy Metal ATPase 2 (TaHMA2) and ATP-Binding Cassette (TaABCC2/3/4) metal detoxification transporters in the resistant wheat genotype Pirsabak 2004 compared with the control in response to Pb, Cd, and Zn [56]. Besides, regulation by DNA methylation under heavy metal stress is not restricted to the promoter regions of tolerant genes but is also observed in their coding regions and TEs. TEs have been implicated in Al stress responses in barley [35]. In the Al-tolerant genotype of barley, the multiretrotransposon-like (MRL) insertion and the expression of the HvAACT1 gene responsible for coding citrate transporters are due to demethylation processes. Additionally, transposon insertions close to genes have been proposed as a source of epialleles and a mechanism affecting specific genes' transcriptional regulation [35].

Interestingly, epigenetic changes are also present in the case of other stresses and appear in the gene body or its vicinity, affecting the gene's transcription. A set of 36 randomly selected genes that underwent DNA methylation changes under drought in barley showed a high modulation at the transcriptome level [44]. For example, the expression of MLOC_61723 decreased more than 200 times under water deficiency and returned to the basal level after the rewatering phase. Another example comes from in vitro tissue cultures. It was shown that even poorly androgenic DH rye lines may become androgenic if they pass several anther culture cycles of plant regeneration [57]. However, there is no information on whether methylation changes affect specific genes or are dispersed along species chromosomes.

3.4. Distribution of the Epigenetic Markers Linked to Al Tolerance on the Triticale Map

It is not apparent whether DNA methylation changes related to Al stress impact the whole genome or affect regions where respective Al-tolerant genes were mapped, suggesting that either the whole genome is involved in stress response or that the genome is tuned at the regions of trait coding genes linking epigenetic markers and genetic traits [14,35]. In the former case, epigenetic changes might be less specific and reflect a minute response to the stress. Thus, epigenetic differences between control lines and Al-treated counterparts could be evaluated. We have also suspected that epigenetic mechanisms preferentially act on a density level [52], meaning that DNA methylation may not always affect the same cytosines, but depending on the density of mapped markers, the two alternatives could be differentiated. Alternatively, genetic and epigenetic factors work cooperatively to support Al tolerance in triticale. If the latter hypothesis is valid, markers differentiating control and Al-treated lines should be mapped near Al-tolerant QTLs. Those DArTseqMet markers univocally assigned to the given type of event and found differentiating control and treated lines based on ANOVA were mapped based on their known chromosomal location [58] to distinguish exclusively between the alternatives discussed. Thus, only DNM-CG-, DNM-CHG-, and DM-CG-assigned markers were used for map construction. As can be seen, markers are distributed along all chromosomes, with some preferences towards telomeric regions. The distribution is comparable to markers detected for root tips and leaves. Furthermore, no evident increased marker density is found near known Al-tolerant QTLs related to the Al stress response in triticale [14,54]. Thus, DNA methylation changes reflect the global tuning of the triticale genome to Al stress rather than being associated with the functioning of Al-tolerant genes. Still, the presented data do not necessarily exclude directed effects, as the observed dispersion of markers may reflect the tuning of specific biochemical pathways or gene expression patterns supporting tolerance.

However, an alternative explanation regarding Al tolerance is also possible. It is well-known that aluminum stress may alleviate Reactive Oxygen Species (ROS) production. Studies on rice (*Oryza sativa*) revealed that Al^{3+} ions generate H_2O_2 or O_2 particles, which leads to DNA damage, lipid peroxidation, or even cell death in root cells [59,60], disturbing the proper functioning of organisms. However, plants' antioxidant system is capable of eliminating ROS. Several enzymes like peroxidase (POD), catalase (CAT), ascorbate peroxidase (APX), superoxide dismutase (SOD), glutathione reductase (GR), and glutathione peroxidase (GPX) take part in the recovery of ROS-mediated damages. The same function is also assigned to glutathione (GSH), ascorbate, and proline [61]. Additionally, Al ions, which are not bound in the cell wall and cell membrane, can enter the cell and directly cause DNA damage by binding to the phosphoric acid residues in DNA. Recent studies demonstrate that aluminum leads to the induction of the DNA Damage Response (DDR) pathway, which is highly specialized in detecting DNA damage [62,63].

Further analysis of marker sequences and identification of biochemical, transcriptional, and other cellular processes are needed to confirm any hypothesis.

3.5. Tissue-Specific Methylation Level in Triticale Under Al Stress

An interesting aspect of this study is that it reflects somatic memory. An analysis of maps constructed on DArTseqMet markers shows that the density of root tip-based maps is much higher than leaf-based maps. It is what we expect if a stress signal is transmitted to other tissues. Whether the markers identified for leaves reflect random effects or may represent the most significant genomic regions linked to Al tolerance needs further investigation. Contrary to our results, Liu et al. [50] showed that more DNA methylations occurred in leaves than in roots of sunflowers under salt and alkali stress. However, roots carried more CG or CHG variations under each stress, indicating that roots play the central part of a sunflower plant's response to various stresses [50]. Drought stress mainly induces demethylation events in leaves, whereas novel methylations are more abundant in roots [44]. But, after the rewatering phase, more new methylations were induced in leaves, whereas there were more demethylations in roots. Such organ-specific methylome changes might regulate the drought resistance in barley. Other researchers suggested that distinct genes regulated tissue-specific biological functions and eventually involved differential DNA methylations [49,64]. Further analysis of sequence data may be helpful in a deeper understanding of the phenomenon in triticale.

3.6. Elastic Net Regression Analyses

An essential aspect of this study is the employment of elastic net regression analysis, a valuable method used to detect the influence of different variables in explaining a trait. As expected, somewhat different variables' effects were evaluated depending on whether the Common or general (basic variant) model characteristics were used to explain differences between non-tolerant and tolerant or spring and winter lines. When the General (basic variant) approach data were analyzed to explain T/NT, DNM, and M, negative and positive effects on Al-non-tolerant lines were evaluated. Except for DNM, the Common and basic general models were congruent regarding effects and their signs. However, the extended model revealed that DNM was related to the CG context, and DN-CG and DN-CHG affected T-NT. Interestingly, the effect of DM-CG was close to that of DN-CG, whereas DN-CHG had the lowest effect. This indicates that most DNA methylation changes revealed in the analysis were due to the context of CG.

4. Materials and Methods

4.1. Plant Materials

Eight spring and seven winter triticale lines represented by at least 40 plants were subjected to the Al test. There were three biological repeats of each of the tests. The lines were highly homozygotic (F7–F10 generations) and screened for Al tolerance annually. Breeders kindly delivered plant materials from Strzelce Plant Breeders Ltd., Experimental Station Małyszyn, 66-400 Gorzów Wielkopolski, Poland.

4.2. Physiological Test

Triticale seeds were soaked successively in alcohol (1 min), 20% bleach solution (Domestos, Unilever, Polska), and distilled water (5, 10, and 15 min). The moist seeds were germinated on a paper filter in a petri dish for 24 h at 20 °C in the dark. Germinated seeds were transferred to nets floating on a basic medium (2.0 mM $CaCl_2$, 3.25 mM KNO_3, 1.25 mM $MgCl_2$, 0.5 mM $(NH_4)2SO_4$, and 0.2 mM NH_4NO_3, pH 4.5) in a plastic tray [65]. The seedlings were maintained under a controlled environment (temperature 25 °C, photoperiod 12/12 h day/night, light intensity 40 W/m^2) in a growth cabinet (POL-EKO-APARATURA, ST500 B40 FOT10). After three days, the seedlings were transferred onto the same medium containing aluminum ions (16 ppm; $AlCl_3 \times 6H_2O$) for 24 h. The plants were washed in water and placed in the basic medium for 48 h. Al tolerance is based on the ability of seedlings to continue root growth and was therefore evaluated by measuring root regrowth. The root tips 0.3–0.5 cm in length (1), and leaves (2) from 7-day-old seedlings, both exposed and non-exposed to Al^{3+} ions, were excised and used for DNA isolation.

4.3. DNA Isolation

Total genomic DNA was isolated from bulked samples representing all Al-tested plants, reflecting given root tips and leaves as separate samples of a given line, using the Plant DNeasy MiniKit 250 (Qiagen, Hilden, Germany) following the manufacturer's instructions. DNA quantity was measured spectrophotometrically (NanoDrop ND-1000, Wilmington, DE, USA), and its integrity and purity were tested on a 1.2% agarose gel in TBE buffer. The isolations from each biological repeat were conducted separately and subjected to further analysis independently.

4.4. Methylation Sensitive DArT Sequencing—DArTseqMet

DNA markers were generated at Diversity Arrays Technology Pty Ltd., Canberra, Australia. DArTseq is an efficient genotyping-by-sequencing technology based on next-generation sequencing (NGS), allowing the discovery of whole-genome sequencing markers [58]. For methylation analysis, the Methylation Sensitive Amplification Polymorphism (MSAP) method, which uses isoschizomers with different specificities for DNA methylation at the restriction site, was adapted to the DArTseq technology [48]. Two parallel libraries were prepared per sample using a DNA double-digestion restriction-based protocol, combining the enzymes SbfI and MspI into the first and SbfI and HpaII into the second platform. DNA samples were processed in digestion/ligation reactions as Kilian et al. [58] described, but with adaptors compatible with restriction enzyme (RE) in brackets. Then, SbfI/MspI and SbfI/HpaII fragments were effectively amplified through PCR. Amplification products were single-end sequenced in an Illumina HiSeq 2500. The raw DArTseqMet marker did not contain missing data.

4.5. DArTseqMet Markers Interpretation and Quantification Models

It is assumed that the presence of a given marker sequence means that the given marker has the respective restriction site that the given enzyme could digest and that its putative methylation does not affect digestion. Furthermore, the presence of a given digested DArTseqMet marker in any digests (control/Al-treated) means that the sequence is available in any case, and that it is missing should be interpreted in terms of cytosine methylation that blocks digestion. Following that reasoning, the presence of the marker is coded as one (1), whereas its absence is coded as zero (0). As control is compared to its Al-treated counterpart, and there are two types of digests, each marker is reflected by a four-digit code reflecting distinct restriction site methylation status that could be distinctly evaluated employing varying quantitative models of analysis or DArTseqMet markers based on the MSAP or semiquantitative MSAP approaches. Under such conditions, mutations are not considered a source of variation.

Thus, two quantitative models could be distinguished. One is based on the original MSAP procedure (the Common model), and the other is based on the semi-quantitative MSAP variant. The latter may be employed using essential quantitative characteristics, and if needed, more details related to sequence contexts could be extracted (extended variant). The main differences between the models are given below (Supplementary Table S1):

1. The Common model is based on the prevailing methylation/non-methylation state/status background explanations for the given transition types, which are used for methylation quantification purposes.
2. The General model:
    - The Basic variant assumes that all possible explanations regarding the methylation background that stands behind the MSAP profiles are used for calculations following the approach presented earlier [52];
    - The Extended variant is based on the same assumptions as the Basic one, but a detailed analysis of restriction site methylation status is conducted to address them in sequence context.

4.6. *Statistics*

Descriptive statistics, Pearson's correlation coefficients, 3-way ANOVA, and Elastic net regression analyses were performed using XlStat ver. 2024.3.0 software [66]. ANOVA was employed to examine how categorical variables (Al tolerance, winter–spring lines, and plant tissues) affected the quantitative variables evaluated based on the Common and General (basic and extended variants) models of quantifying DNA methylation. The Elastic net regression analysis with an alpha parameter set to one included a single categorical variable as a dependent variable, including all quantitative variables of the Common and general models as explanatory variables.

4.7. *Map Construction*

The MSAP markers were mapped to the triticale chromosomes based on their known positions delivered by Diversity Arrays Technology Pty Ltd., Canberra, Australia. Only markers showing the same methylation pattern (for example, '0111' for CG demethylation and '1101' for CG de novo methylation) were mapped, at least for two of the three experiment repetitions for the given triticale line. MapChart ver. 2.3 software was used to draw the genetic linkage maps [67].

5. Conclusions

The presented study demonstrates that utilizing different methods to quantify DNA methylation changes induced by Al stress, although congruent, differ in details, and the results evaluated based on them should be compared critically. The Common basic general approach is sufficient if general information is needed. However, when a deeper insight into the phenomenon is required, it is recommended to use an extended general approach. Using the General (extended variant) approach in combination with ANOVA, the differences between Al-treated and control lines regarding DN-CG, DM-CG, and DN-CHG were detected. Analysis of the maps constructed based on those markers showed that they were located in telomeric regions without grouping in the vicinity of the Al-tolerant QTLs, suggesting that Al-treatment induces minute changes in the genome. Still, other explanations are not excluded. Moreover, genetic maps constructed based on root tips and leaf tissues differed in density but reflected the comparable patterns of their distribution, supporting the hypothesis that Al stress could be transmitted to other plant tissues due to somatic memory. Further analyses of marker sequences are needed to ensure or disclose whether epigenetic changes due to Al stress are random or whether they activate/deactivate processes supporting genetic factors of tolerance.

Supplementary Materials: The following supporting information can be downloaded at: https://www.mdpi.com/article/10.3390/ijms26114995/s1.

Author Contributions: Formal analysis, data curation, and writing—original draft, A.N.; conceptualization, formal analysis, and writing—review and editing, P.T.B.; investigation and data curation, R.O. All authors have read and agreed to the published version of the manuscript.

Funding: This research did not receive any specific grant from funding agencies in the public, commercial, or not-for-profit sectors. It was supported by Plant Breeding & Acclimatization Institute, National Research Institute funds.

Institutional Review Board Statement: Not applicable.

Informed Consent Statement: Not applicable.

Data Availability Statement: Data will be made available upon request.

Acknowledgments: We would like to thank Andrzej Kilian (Diversity Arrays Technology Pty. Ltd., Canberra, Australia) for the discussion on interpreting the data and cooperation.

Conflicts of Interest: The authors declare that they have no known competing financial interests or personal relationships that could have appeared to influence the work reported in this paper.

References

1. Bartholomé, J.; Ospina, J.O.; Sandoval, M.; Espinosa, N.; Arcos, J.; Ospina, Y.; Frouin, J.; Beartschi, C.; Ghneim, T.; Grenier, C. Genomic selection for tolerance to aluminum toxicity in a synthetic population of upland rice. *PLoS ONE* **2024**, *19*, e0307009. [CrossRef] [PubMed]
2. Kochian, L.V. Cellular mechanisms of aluminium toxicity and resistance in plants. *Annu. Rev. Plant. Physiol. Plant Mol. Biol.* **1995**, *46*, 237–260. [CrossRef]
3. Ma, J.F.; Ryan, P.R.; Delhaize, E. Aluminium tolerance in plants and the complexing role of organic acids. *Trends Plant Sci.* **2001**, *6*, 273–278. [CrossRef]
4. Zhu, F. Triticale: Nutritional composition and food uses. *Food Chem.* **2018**, *241*, 468–479. [CrossRef]
5. Aguirre, A.; Borneo, R.; León, A.E. Properties of triticale protein films and their relation to plasticizing-antiplasticizing effects of glycerol and sorbitol. *Ind. Crops Prod.* **2013**, *50*, 297–303. [CrossRef]
6. Salvucci, E.; Rossi, M.; Colombo, A.; Pérez, G.; Borneo, R.; Aguirre, A. Triticale flour films added with bacteriocin-like substance (BLIS) for active food packaging applications. *Food Packag. Shelf Life* **2019**, *19*, 193–199. [CrossRef]
7. Liard, G.; Lesage, P.; Samson, R.; Stuart, P. Systematic assessment of triticale-based biorefinery strategies: Environmental evaluation using life cycle assessment. *Biofuels Bioprod. Biorefin.* **2018**, *12*, S60–S72. [CrossRef]
8. Sanaei, S.; Chambost, V.; Stuart, P. Systematic assessment of triticale-based biorefinery strategies: Sustainability assessment using multi-criteria decision-making (MCDM). *Biofuels Bioprod. Biorefin.* **2018**, *12*, S73–S86. [CrossRef]
9. Delhaize, E.; Ma, J.F.; Ryan, P.R. Transcriptional regulation of aluminium tolerance genes. *Trends Plant Sci.* **2012**, *17*, 341–348. [CrossRef]
10. Szewińska, J.; Różańska, E.; Papierowska, E.; Labudda, M. Proteolytic and structural changes in rye and triticale roots under aluminium stress. *Cells* **2021**, *10*, 3046. [CrossRef]
11. Niedziela, A.; Domżalska, L.; Dynkowska, W.M.; Pernisová, M.; Rybka, K. Aluminum stress induces irreversible proteomic changes in the roots of the sensitive but not the tolerant genotype of triticale seedlings. *Plants* **2022**, *11*, 165. [CrossRef] [PubMed]
12. Kim, B.; Baier, A.; Somers, D.; Gustafson, J. Aluminum tolerance in triticale, wheat and rye. *Euphytica* **2001**, *120*, 329–337. [CrossRef]
13. Zhang, X.; Jessop, R.; Alter, D. Organic acid exudation associated with aluminium stress tolerance in triticale and wheat. *Aust. J. Agric. Res.* **2003**, *54*, 979–985. [CrossRef]
14. Niedziela, A. The influence of Al^{3+} on DNA methylation and sequence changes in the triticale (×Triticosecale Wittmack) genome. *J. Appl. Genet.* **2018**, *59*, 405–417. [CrossRef]
15. Sasaki, T.; Yamamoto, Y.; Ezaki, B.; Katsuhara, M.; Ahn, S.J.; Ryan, P.R.; Delhaize, E.; Matsumoto, H. A wheat gene encoding an aluminium-activated malate transporter. *Plant J.* **2004**, *37*, 645–653. [CrossRef]
16. Fontecha, G.; Silva-Navas, J.; Benito, C.; Mestres, M.A.; Espino, F.J.; Hernández-Riquer, M.V.; Gallego, F.J. Candidate gene identification of an aluminum-activated organic acid transporter gene at the Alt4 locus for aluminum tolerance in rye (*Secale cereale* L.). *Theor. Appl. Genet.* **2007**, *114*, 249–260. [CrossRef]
17. Ligaba, A.; Maron, L.; Shaff, J.; Kochian, L.; Piñeros, M. Maize ZmALMT2 is a root anion transporter that mediates constitutive root malate efflux. *Plant Cell Environ.* **2012**, *35*, 1185–1200. [CrossRef]

18. Ryan, P.R.; Raman, H.; Gupta, S.; Horst, W.J.; Delhaize, E. A second mechanism for aluminum resistance in wheat relies on the constitutive efflux of citrate from roots. *Plant Physiol.* **2009**, *149*, 340–351. [CrossRef] [PubMed]
19. Garcia-Oliveira, A.L.; Benito, C.; Prieto, P.; de Andrade Menezes, R.; Rodrigues-Pousada, C.; Guedes-Pinto, H.; Martins-Lopes, P. Molecular characterization of TaSTOP1 homoeologues and their response to aluminium and proton (H^+) toxicity in bread wheat (*Triticum aestivum* L.). *BMC Plant Biol.* **2013**, *13*, 134. [CrossRef]
20. Ma, J.F.; Taketa, S.; Yang, Z.M. Aluminium tolerance genes on the short arm of chromosome 3R are linked to organic acid release in triticale. *Plant Physiol.* **2000**, *122*, 687–694. [CrossRef]
21. Matos, M.; Camacho, M.V.; Pérez-Flores, V.; Pernaute, B.; Pinto-Carnide, O.; Benito, C. A new aluminium tolerance gene located on rye chromosome arm 7RS. *Theor. Appl. Genet.* **2005**, *111*, 360–369. [CrossRef] [PubMed]
22. Benito, C.; Silva-Navas, J.; Fontecha, G.; Hernández-Riquer, M.V.; Eguren, M.; Salvador, N.; Gallego, F.J. From the rye Alt3 and Alt4 aluminum tolerance loci to orthologous genes in other cereals. *Plant Soil* **2010**, *327*, 107–120. [CrossRef]
23. Silva-Navas, J.; Benito, C.; Téllez-Robledo, B.; El-Moneim, A.; Gallego, F.J. The ScAACT1 gene at the Qalt5 locus as a candidate for increased aluminium tolerance in rye (*Secale cereale* L.). *Mol. Breed.* **2011**, *30*, 845–856. [CrossRef]
24. Raman, H.; Moroni, S.; Sato, K.; Read, J.; Scott, J. Identification of AFLP and microsatellite markers linked with an aluminium tolerance gene in barley (*Hordeum vulgare* L.). *Theor. Appl. Genet.* **2002**, *105*, 458–464. [CrossRef] [PubMed]
25. Navakode, S.; Weidner, A.; Varshney, R.K.; Lohwasser, U.; Scholz, U.; Börner, A. A QTL Analysis of aluminium tolerance in barley; using gene-based markers. *Cereal Res. Commun.* **2009**, *37*, 531–540. [CrossRef]
26. Niedziela, A.; Bednarek, P.T.; Labudda, M.; Mańkowski, D.R.; Anioł, A. Genetic mapping of a 7R Al tolerance QTL in triticale (x Triticosecale Wittmack). *J. Appl. Genet.* **2014**, *55*, 1–14. [CrossRef] [PubMed]
27. Bird, A. Perceptions of epigenetics. *Nature* **2007**, *447*, 396–398. [CrossRef]
28. Ma, L.; Xing, L.; Li, Z.; Jiang, D. Epigenetic control of plant abiotic stress responses. *J. Genet. Genom.* **2025**, *52*, 129–144. [CrossRef]
29. Gallo-Franco, J.J.; Sosa, C.C.; Ghneim-Herrera, T.; Quimbaya, M. Epigenetic Control of Plant Response to Heavy Metal Stress: A New View on Aluminum Tolerance. *Front. Plant Sci.* **2020**, *11*, 602625. [CrossRef]
30. Chinnusamy, V.; Zhu, J.K. Epigenetic regulation of stress responses in plants. *Curr. Opin. Plant Biol.* **2009**, *12*, 133–139. [CrossRef]
31. Kumar, S.V.; Wigge, P.A. H2A.Z-containing nucleosomes mediate the thermosensory response in Arabidopsis. *Cell* **2010**, *140*, 136–147. [CrossRef] [PubMed]
32. Nunez-Vazquez, R.; Desvoyes, B.; Gutierrez, C. Histone variants and modifications during abiotic stress response. *Front. Plant Sci.* **2022**, *13*, 984702. [CrossRef] [PubMed]
33. Zhang, H.; Lang, Z.; Zhu, J.K. Dynamics and function of DNA methylation in plants. *Nat. Rev. Mol. Cell Biol.* **2018**, *19*, 489–506. [CrossRef] [PubMed]
34. Law, J.A.; Jacobsen, S.E. Establishing, maintaining and modifying DNA methylation patterns in plants and animals. *Nat. Rev. Genet.* **2010**, *11*, 204–220. [CrossRef]
35. Kashino-Fujii, M.; Yokosho, K.; Yamaji, N.; Yamane, M.; Saisho, D.; Sato, K.; Ma, J.F. Retrotransposon Insertion and DNA Methylation Regulate Aluminum Tolerance in European Barley Accessions. *Plant Physiol.* **2018**, *178*, 716–727. [CrossRef]
36. Ramakrishnan, M.; Satish, L.; Kalendar, R.; Narayanan, M.; Kandasamy, S.; Sharma, A.; Emamverdian, A.; Wei, Q.; Zhou, M. The dynamism of transposon methylation for plant development and stress adaptation. *Int. J. Mol. Sci.* **2021**, *22*, 11387. [CrossRef]
37. Chen, L.; Wang, T.; Zhao, M.; Tian, Q.; Zhang, W.H. Identification of aluminum-responsive microRNAs in Medicago truncatula by genome-wide high-throughput sequencing. *Planta* **2012**, *235*, 375–386. [CrossRef]
38. Taspinar, M.S.; Aydin, M.; Sigmaz, B.; Yagci, S.; Arslan, E.; Agar, G. Aluminum-Induced Changes on DNA Damage; DNA Methylation and LTR Retrotransposon Polymorphism in Maize. *Arab. J. Sci. Eng.* **2018**, *43*, 123–131. [CrossRef]
39. Pour, A.H.; Özkan, G.; Nalci, Ö.; Haliloğlu, K. Estimation of genomic instability and DNA methylation due to aluminum (Al) stress in wheat (*Triticum aestivum* L.) using iPBS and CRED-iPBS analyses. *Turk. J. Bot.* **2019**, *43*, 27–37.
40. Choi, C.S.; Sano, H. Abiotic-stress induces demethylation and transcriptional activation of a gene encoding a glycerophosphodiesterase-like protein in tobacco plants. *Mol. Genet. Genom.* **2007**, *277*, 589–600. [CrossRef]
41. Ezaki, B.; Higashi, A.; Nanba, N.; Nishiuchi, T. An S-adenosyl methionine synthetase (SAMS) gene from *Andropogon virginicus* L. confers aluminum stress tolerance and facilitates epigenetic gene regulation in Arabidopsis thaliana. *Front. Plant Sci.* **2016**, *7*, 1627. [CrossRef] [PubMed]
42. Springer, N.M.; Schmitz, R.J. Exploiting induced and natural epigenetic variation for crop improvement. *Nat. Rev. Genet.* **2017**, *18*, 563–575. [CrossRef] [PubMed]
43. Haque, M.N.; Singh, R.K.; Biswas, A. Role of epigenetics in plant stress responses: Implications in crop improvement. *Front. Plant Sci.* **2022**, *13*, 873456.
44. Chwiałkowska, K.; Nowakowska, U.; Mroziewicz, A.; Szarejko, I.; Kwasniewski, M. Water-deficiency conditions differently modulate the methylome of roots and leaves in barley (*Hordeum vulgare* L.). *J. Exp. Bot.* **2016**, *67*, 1109–1121. [CrossRef]

45. Reyna-López, G.E.; Simpson, J.; Ruiz-Herrera, J. Differences in DNA methylation patterns are detectable during the dimorphic transition of fungi by amplification of restriction polymorphisms. *Mol. Gen. Genet.* **1997**, *253*, 703–710. [CrossRef]
46. Bednarek, P.T.; Orłowska, R.; Koebner, R.M.D.; Zimny, J. Quantification of the tissue-culture induced variation in barley (*Hordeum vulgare* L.). *BMC Plant Biol.* **2007**, *7*, 10.
47. Vos, P.; Hogers, R.; Bleeker, M.; Reijans, M.; van de Lee, T.; Hornes, M.; Friters, A.; Pot, J.; Paleman, J.; Kuiper, M.; et al. AFLP: A new technique for DNA fingerprinting. *Nucleic Acids Res.* **1995**, *23*, 4407–4414. [CrossRef]
48. Pereira, W.; Pappas, M.; Grattapaglia, D.; Pappas, G., Jr. A cost-effective approach to DNA methylation detection by Methyl Sensitive DArT sequencing. *PLoS ONE* **2020**, *15*, e0233800. [CrossRef]
49. Aceituno, F.F.; Moseyko, N.; Rhee, S.Y.; Gutiérrez, R.A. The rules of gene expression in plants: Organ identity and gene body methylation are key factors for regulation of gene expression in Arabidopsis thaliana. *BMC Genom.* **2008**, *9*, 438. [CrossRef]
50. Liu, J.; Dong, X.-X.; Zhao, Z.-K.; Sun, H.-R.; Yu, L.-X. Analysis of sunflower DNA methylation in response to salt and alkali stresses based on methylation-sensitive amplified polymorphisms. *Curr. Plant Biol.* **2023**, *34*, 100282. [CrossRef]
51. Fulneček, J.; Kovařík, A. How to interpret Methylation Sensitive Amplified Polymorphism (MSAP) profiles? *BMC Genet.* **2014**, *15*, 2. [CrossRef] [PubMed]
52. Bednarek, P.T.; Orłowska, R.; Niedziela, A. A Relative quantitative Methylation-Sensitive Amplified Polymorphism (MSAP) method for the analysis of abiotic stress. *BMC Plant Biol.* **2017**, *17*, 79. [CrossRef] [PubMed]
53. Hermawaty, D.; Meitha, K.; Esyanti, R.R. The Potential of Methylation Sensitive Amplification Polymorphism (MSAP) Technique to Study Global DNA Methylation Changes in Plants: Basic Principles and Perspective. *Rev. Agric. Sci.* **2024**, *12*, 93–110. [CrossRef]
54. Niedziela, A.; Bednarek, P.; Cichy, H.; Budzianowski, G.; Kilian, A.; Aniol, A. Aluminum tolerance association mapping in triticale. *BMC Genom.* **2012**, *13*, 67. [CrossRef]
55. Silva, A.; Paterniani, M.; Camargo, C. Genetic variability and inheritance to aluminum tolerance in nutrient solution in triticale. *Bragantia* **2014**, *73*, 8–13. [CrossRef]
56. Shafiq, S.; Zeb, Q.; Ali, A.; Sajjad, Y.; Nazir, R.; Widemann, E.; Liu, L. Lead, Cadmium and Zinc Phytotoxicity Alter DNA Methylation Levels to Confer Heavy Metal Tolerance in Wheat. *Int. J. Mol. Sci.* **2019**, *20*, 4676. [CrossRef]
57. Zimny, J.; Oleszczuk, S.; Makowska, K.; Zimny, A.; Michalski, K.; Bednarek, P.; Sowa, S. Identification of an elite rye genotype with high androgenesis efficiency. In Proceedings of the Abstracts, Eucarpia General Congress, "Global Chellenges for Crop Improvement", Leipzig, Germany, 18–23 August 2024; p. 82.
58. Kilian, A.; Wenzl, P.; Huttner, E.; Carling, J.; Xia, L.; Blois, H.; Caig, V.; Heller-Uszynska, K.; Jaccoud, D.; Hopper, C.; et al. Diversity Arrays Technology: A Generic Genome Profiling Technology on Open Platforms. In *Data Production and Analysis in Population Genomics, Methods in Molecular Biology*; Pompanon, F., Bonin, A., Eds.; Humana Press: Totowa, NJ, USA, 2012; Volume 888, pp. 67–89.
59. Bera, S.; De, A.K.; Adak, M.K. Modulation of Glycine Betaine Accumulation with Oxidative Stress Induced by Aluminium Toxicity in Rice. *Proc. Natl. Acad. Sci. India Sect. B—Biol. Sci.* **2019**, *89*, 291–301. [CrossRef]
60. Rahman, R.; Upadhyaya, H. Aluminium toxicity and its tolerance in plant: A review. *J. Plant Biol.* **2021**, *64*, 101–121. [CrossRef]
61. You, J.; Chanz, Z. ROS regulation during abiotic stress responses in crop plants. *Front. Plant Sci.* **2015**, *6*, 1092. [CrossRef]
62. Szurman-Zubrzycka, M.; Chwiałkowska, K.; Niemira, M.; Kwaśniewski, M.; Nawrot, M.; Gajecka, M.; Larsen, P.B.; Szarejko, I. Aluminum or low pH—Which is the bigger enemy of barley? Transcriptome analysis of barley root meristem under Al and low pH stress. *Front. Genet.* **2021**, *12*, 675260. [CrossRef]
63. Kocjan, A.N.; Kwaśniewska, J.; Szurman-Zubrzycka, M.E. Understanding plant tolerance to aluminum: Exploring mechanisms and perspectives. *Plant Soil* **2025**, *507*, 195–219. [CrossRef]
64. Docimo, T.; De Stefano, R.; De Palma, M.; Cappetta, E.; Villano, C.; Aversano, R.; Tucci, M. Transcriptional; metabolic and DNA methylation changes underpinning the response of Arundo donax ecotypes to NaCl excess. *Planta* **2019**, *251*, 34. [CrossRef] [PubMed]
65. Anioł, A. Induction of aluminium tolerance in wheat seedlings by low doses of aluminium in the nutrient solution. *Plant Physiol.* **1984**, *75*, 551–555. [CrossRef] [PubMed]
66. XLSTAT Statistical Software for Excel. 2007. Available online: https://www.xlstat.com (accessed on 1 May 2025).
67. Voorrips, R.E. MapChart: Software for the graphical presentation of linkage maps and QTLs. *J. Hered.* **2002**, *93*, 77–78. [CrossRef]

International Journal of
Molecular Sciences

MDPI

Article

Genome-Wide Identification of *Ginkgo biloba* SPL Gene Family and Expression Analysis in Flavonoid Biosynthesis and Water Stress

Meiling Ming †, Mulin Yi †, Chunyue Qin, Luyao Yan, Yuhan Sun, Juan Zhang, Fuliang Cao * and Fangfang Fu *

State Key Laboratory of Tree Genetics and Breeding, Co-Innovation Center for Sustainable Forestry in Southern China, Nanjing Forestry University, Nanjing 210037, China; mingmeiling@njfu.edu.cn (M.M.); yimumu@njfu.edu.cn (M.Y.); qinchunyue@njfu.edu.cn (C.Q.); yanluyao@njfu.edu.cn (L.Y.); jianjiao@njfu.edu.cn (Y.S.); juanzhang@njfu.edu.cn (J.Z.)

* Correspondence: flcaoginkgo@njfu.edu.cn (F.C.); fffu@njfu.edu.cn (F.F.); Tel./Fax: +86-025-85427099 (F.C.); +86-025-85427921 (F.F.)

† These authors contributed equally to this work.

Abstract: SQUAMOSA promoter-binding protein-like (SPL) transcription factors specific to plants are vital for regulating growth, development, secondary metabolite biosynthesis, and responses to both biotic and abiotic stresses. Despite their importance, no systematic investigations or identifications of the SPL gene family in *Ginkgo biloba* have been conducted. In this study, we identified 13 SPL genes within the *Ginkgo biloba* reference genome, spanning seven chromosomes, and categorized these genes into six groups based on their phylogenetic relationships with *Arabidopsis thaliana* SPL gene families. Our analysis of gene structure, conserved domains, motifs, and miR156 target predictions indicates that GbSPLs are highly conserved across evolutionary timelines. Furthermore, synteny analysis highlighted that dispersed duplication events have expanded the SPL gene family in *Ginkgo biloba*. Examination of the cis-regulatory elements revealed that many *GbSPL* genes possess motifs associated with light, hormones, and stress, implying their involvement in flavonoid biosynthesis and adaptation to environmental conditions. RNA-Seq and qRT-PCR expression profiles of *GbSPL* genes across various tissues and low- and high-flavonoid leaves and during both short-term and long-term water stress illustrated their roles in flavonoid biosynthesis and responses to water stress. Subcellular localization experiments showed that GbSPL2 and GbSPL11 proteins are situated within the nucleus. Our research offers the first systematic characterization of the SPL gene family in *Ginkgo biloba*, establishing a valuable foundation for understanding their evolutionary background and functional roles in flavonoid biosynthesis and water stress response.

Keywords: *Ginkgo biloba*; SPL; gene family; flavonoid biosynthesis; water stress

1. Introduction

Transcription factors (TFs) are a group of proteins that contain structural regions capable of binding to specific DNA sequences located in gene promoter regions. They act as molecular switches that either enhance or inhibit the transcription of various genes [1]. Previous studies have characterized multiple transcription factor families, including R2R3-myeloblastoma (MYB), basic helix-loop-helix (bHLH3), WD40, APETALA2/ethylene-responsive factor (AP2/ERF), basic leucine zipper (bZIP), MADS-box, WRKY, NAC, homeodomain-leucine zipper (HD-Zip), and SQUAMOSA promoter-binding protein-like

(SPL) transcription factors [2]. SPL proteins are plant-exclusive transcription factors characterized by their 76-residue SBP domain. This evolutionarily conserved domain incorporates dual zinc finger motifs (C3H and C2HC types) and a bipartite nuclear localization signal (NLS). Notably, the C-terminal C2HC zinc finger spatially overlaps with the NLS, forming a structural module that enables nuclear import and subsequent gene regulation [3].

The initial identification of SPL genes (*AmSBP1* and *AmSBP2*) took place in the floral meristems of snapdragon (*Antirrhinum majus*) in 1996 [4]. Genome-wide studies have subsequently annotated SPL homologs across plant species: 17 members in *Arabidopsis thaliana* [5], 19 members in rice (*Oryza sativa*) [5], 15 members in tomato (*Solanum lycopersicum*) [6], 28 members in poplar (*Populus trichocarpa*), 27 members in the apple (*Malus domestica Borkh.*) [7], 24 members in Tartary Buckwheat (*Fagopyrum tataricum*) [8], 20 members in *Betula platyphylla Suk.* [9], 19 members in pear (*Pyrus pyrifolia*) [10], 23 members in *Medicago truncatula* [11], 17 members in sugarcane (*Saccharum spontaneum*) [12], 23 members in quinoa (*Chenopodium quinoa*) [13], 49 members in *Nicotiana tabacum*, 43 members in *Nicotiana benthamiana* [14], 25 members in sunflower (*Helianthus annuus*) [15], 12 members in sweet cherry (*Prunus avium* L.) [16], 17 members in litchi (*Litchi chinensis Sonn.*) [17], and so on. The classification of SPL gene subfamilies is largely consistent across different species and can generally be divided into six to nine subfamilies.

The SPL gene family typically operates through independent mechanisms or pathways that miR156 targets. The evolutionarily conserved miR156 family post-transcriptionally regulates multiple SPL genes by specifically cleaving their mRNA transcripts, which is a hallmark regulatory mechanism in plant development, secondary metabolism, and stress response [18]. The 20 nucleotide miR156 was originally characterized in *Arabidopsis thaliana*, and subsequent functional studies elucidated its regulatory partnership with SPL transcription factors (miR156-SPL module) in modulating pigmentation patterns across plant tissues [19,20]. For instance, SPL transcripts act as direct targets of miR156 in European aspen (*Populus tremula* L.), with this regulatory interaction influencing the outputs of the phenylpropanoid pathway, including anthocyanins, flavonoids, and flavonols [21]. In addition to their role in plant secondary metabolism, the miR156/SPL module has also been shown to play a crucial part in plant growth and development. Emerging evidence identifies the miR156-SPL axis as a key regulator of root system architecture. Molecular phenotyping indicates that the depletion of miR156 inhibits both lateral root initiation and adventitious root formation through SPL-mediated transcriptional reprogramming [22]. Cutting-edge plant synthetic biology strategies now use miR156 network rewiring to create stress-resilient crops, combining target gene editing with advanced transformation platforms for abiotic adaptation. The overexpression of miR156 enhanced stress tolerance in *Arabidopsis thaliana* and rice, uncovering an evolutionarily conserved miR156-SPL-DFR regulatory axis that aligns developmental plasticity with abiotic stress tolerance, forming a molecular basis for environmental adaptation in plants [23]. In addition to the miR156-SPL module, six AtSPLs in Arabidopsis—*AtSPL1*, *AtSPL7*, *AtSPL8*, *AtSPL12*, *AtSPL14*, and *AtSPL16*—are unaffected by miR156. Importantly, *AtSPL7* binds directly to the Cu-response element (CuRE) featuring the core sequence GTAC, which is essential for regulating copper homeostasis in the plant [24]. Additionally, both targeted and non-targeted SBP-box transcription factors work together to promote male fertility in Arabidopsis [25]. Overexpressing LcSPL2 in *Arabidopsis thaliana* advanced flowering and reduced rosette leaf number. Additionally, elevated LcSPL2 expression in *A. thaliana* increased the transcript levels of four flower-development-related genes, suggesting the role of LcSPL2 in promoting flower development [26]. OsSPL10, a transcription factor implicated in enhancing drought resistance in rice, plays a crucial role in this process by regulating the production of reactive oxygen species (ROS) and the movement of stomata. Analysis of haplotype and allele frequencies revealed that the OsSPL10Hap1 allele is prevalent in upland and improved lowland rice varieties, while the OsSPL10Hap2 allele is primarily found in lowland and landrace rice cultivars [27]. Furthermore, MeSPL9 impacts drought resistance by regulating levels of protective metabolites and JA signaling [28], while wheat plants overexpressing TaSPL6-A showed increased susceptibility to drought stress [29].

While SPL transcription factors are recognized regulators of various developmental pathways in angiosperms, their functional characterization in *Ginkgo biloba* remains significantly understudied. As a paleoendemic species often referred to as a "living fossil", *Ginkgo biloba* is widely cultivated worldwide for its dual benefits: medicinal properties and distinctive fan-shaped leaves. The medicinal properties of *Ginkgo biloba* primarily rely on secondary metabolites such as flavonoids and terpenoids. Ginkgo leaf extracts are utilized in the healthcare and food sectors because *G. biloba* flavonoids and their glycosides exhibit notable effects against bacteria, viruses, inflammation, and cancer while also offering neuroprotection [30]. Ginkgo trees amass antimicrobial and antioxidant secondary metabolites in the xylem, such as flavonoid glycosides and terpenes, which contribute to their exceptional longevity [31]. Flavonoid compounds can enhance *G. biloba*'s resistance to soil salt stress. Exposure to UV-B radiation also boosts the overall content of flavonoid glycosides [32]. *Ginkgo biloba* also demonstrates remarkable resilience to biotic challenges and abiotic stressors, with exceptional drought tolerance rooted in its evolutionary background. Thirteen putative GbSPL genes were identified in this study. A comprehensive analysis was conducted on gene properties, phylogenetics, structure, collinearity, cis-acting elements, and gene expression patterns. Genes related to flavonoid biosynthesis and water stress responses were prioritized for detailed examination. The findings of this study establish a framework for unraveling the transcriptional regulation of flavonoid metabolism and water stress in *Ginkgo biloba*. Comparative analysis of stress adaptation genes across ginkgo accessions may uncover conserved genetic modules suitable for pyramiding breeding, thereby synergistically enhancing the yields of medicinal compounds and drought-resistant traits.

2. Results

2.1. Whole-Genome Identification of the GbSPL Genes in Ginkgo biloba

The SPL gene family in *Ginkgo biloba* was systematically characterized using two independent approaches: the BLAST method and the HMMsearch method. Both methods identified 13 candidate genes each. The 13 overlapping members between the BLAST and HMMsearch methods were considered GbSPL genes in *Ginkgo biloba*. Subsequently, we removed the member with incomplete SBP domains based on the NCBI CDD website. Finally, a total of 13 GbSPL genes with the conserved SBP domain were identified in *Ginkgo biloba* and renamed according to their chromosomal positions (Table 1). Substantial variation was observed in protein properties: amino acid lengths ranged from 170 aa (GbSPL4) to 1314 aa (GbSPL10), with corresponding molecular weights varying from 18,774.35 Da to 145,251.18 Da (Table 1). Protein isoelectric points (pIs) showed values between 5.57 (GbSPL13) and 9.27 (GbSPL9). Additionally, their grand average of hydropathicity indices ranged from −0.776 to −0.414, and the instability index and aliphatic index were also analyzed. The instability index ranged from 37.22 (GbSPL8) to 65.54 (GbSPL6); only GbSPL8 was a stable protein. Predictions for subcellular localization indicated nuclear-specific localization for all SPL proteins except for the chloroplast localization of GbSPL10. miR156 target predictions revealed that only two GbSPLs (GbSPL1 and GbSPL9) possess GbmiR156 binding sites (Table 1).

Table 1. Basic information of GbSPL gene family members in *Ginkgo biloba*.

Gene Name	Gene ID	Peptide (aa)	Molecular Weight MW (Da)	Theoretical pI	Grand Average of Hydropathicity (GRAVY)	Instability Index	Aliphatic Index	Subcellular Localization Predictions	miR156 Target
GbSPL1	evm.model.chr1.642	518	57,582.8	7.36	−0.776	58.94	58.90	nuclear	YES
GbSPL2	evm.model.chr1.643	596	64,192.75	7.10	−0.657	55.27	65.30	nuclear	NO
GbSPL3	evm.model.chr1.1404	230	25,387.53	8.98	−0.750	55.07	50.91	nuclear	NO
GbSPL4	evm.model.chr5.834	170	18,774.35	6.52	−0.514	51.55	79.24	nuclear	NO
GbSPL5	evm.model.chr5.835	633	70,406.88	7.61	−0.765	65.35	60.39	nuclear	NO
GbSPL6	evm.model.chr5.836	447	49,707.51	8.55	−0.654	65.54	64.41	nuclear	NO
GbSPL7	evm.model.chr6.1273	1112	123,372.1	8.56	−0.499	55.69	75.50	nuclear	NO
GbSPL8	evm.model.chr8.1250	367	40,125.93	7.52	−0.622	37.22	58.96	nuclear	NO
GbSPL9	evm.model.chr9.1444	569	63,284.01	9.27	−0.707	53.76	66.40	nuclear	YES
GbSPL10	evm.model.chr9.1866	1314	145,251.18	8.69	−0.414	58.05	78.76	chloroplast	NO
GbSPL11	evm.model.chr10.1109	670	72,574.69	8.92	−0.741	59.74	60.70	nuclear	NO
GbSPL12	evm.model.chr11.1430	711	77,686.16	8.58	−0.693	56.53	58.30	nuclear	NO
GbSPL13	evm.model.chr11.1518	866	97,933.52	5.57	−0.523	50.55	76.74	nuclear	NO

The chromosomal mapping of 13 *GbSPL* genes in *Ginkgo biloba* revealed their uneven genomic arrangement, visualized through TBtools-II software for spatial distribution analysis. Overall, these 13 *GbSPL* genes were mapped across seven chromosomes: chr1, chr5, chr6, chr8, chr9, chr10, and chr11 (Figure 1). Both chr1 and chr5 contain three *GbSPL* genes each, while chr6, chr8, and chr10 each contain one *GbSPL* gene.

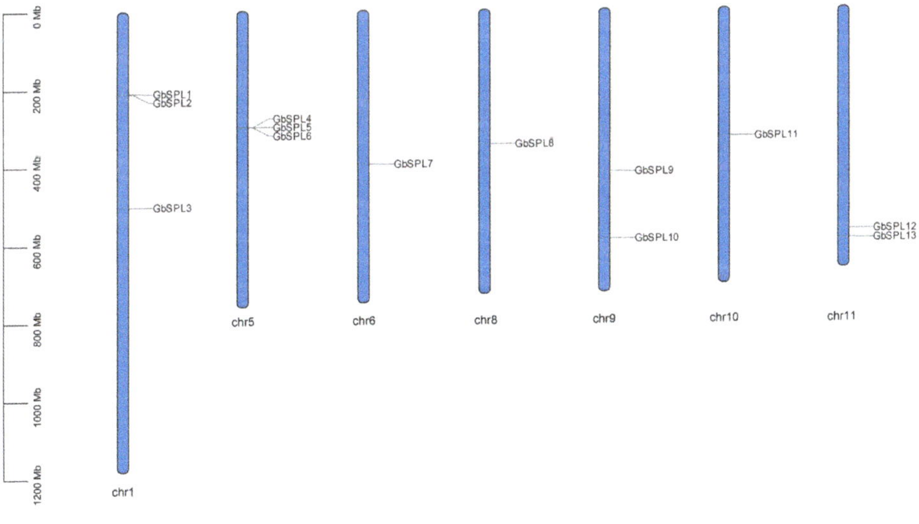

Figure 1. The chromosome distribution of *GbSPL* genes in the genome of *Ginkgo biloba*. The length of each chromosome was estimated in megabases (Mb).

2.2. Phylogenetic Analysis of GbSPL Genes

To investigate the evolutionary relationship of the identified *GbSPL* genes in *Ginkgo biloba*, a Maximum Likelihood (ML) phylogenetic tree was constructed using the protein sequences of 13 GbSPL genes from *Ginkgo biloba*, 16 AtSPL genes from *Arabidopsis thaliana*, and 19 OsSPL genes from *Oryza sativa* (Figure 2). The results revealed that the SPL gene families were classified into eight groups (Group I to Group VIII), consistent with a previous classification of SPL family distribution [8,17]. The GbSPL gene family was distributed across six groups (Group I, II, III, IV, V, VIII): one member in Group I, two members in Group II, five members in Group III, one member in Group IV, two members in Group V, and two members in Group VIII (Figure 2). This result suggested that the functions of SPL genes might have diverged during the evolution of *Arabidopsis thaliana* (dicotyledonous), *Oryza sativa* (monocotyledonous), and *Ginkgo biloba* (gymnosperm).

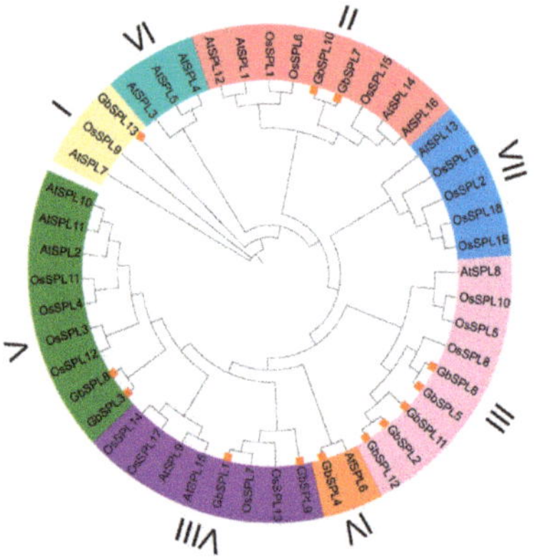

Figure 2. Phylogenetic analysis of GbSPL proteins with AtSPL proteins from Arabidopsis and OsSPL proteins from rice. The phylogenetic tree was constructed using IQ-Tree software, employing the Maximum Likelihood (ML) method, and bootstrap replications were set at 1000 times.

2.3. *Motif, Conserved Domain, and Gene Structure Analysis of GbSPLs*

To further provide insight into the GbSPLs, we investigate the motif, conserved domain, and gene structure analysis of GbSPL gene family members. The clustering was constructed based on the results of the phylogenetic analysis of SPL in *Ginkgo biloba* (Figure 3A). We utilized the MEME website tool to analyze the differences among the 10 conserved motifs (Motif 1–Motif 10) of the GbSPL proteins (Figures 3B and S1). Only motif 1 was present in all 13 *GbSPL* genes, whereas motif 2 was identified in 10 *GbSPL* gene members. Motif 1 and motif 2 may constitute parts of the conserved domain of SBP (Figure S1). The conserved motifs indicated that the same subgroup of GbSPL exhibited similar types, while different subgroups of GbSPL displayed varying motif types. For example, GbSPL3 and GbSPL8 in group V both contain motifs 1, 4, and 7. However, their motif composition shows notable differences compared to the members of group II, GbSPL7 and GbSPL10. Notably, despite being clustered together, the five members within group III (GbSPL2, 5, 6, 11, 12) exhibit discernible variations in their motif compositions. These variations within the same subgroup suggest that genes in the same subgroup might exhibit functional diversity.

We also identified the conserved domains of the GbSPL gene family members. The SBP domain was present in all 13 GbSPL family members. Structural conservation was noted in Group II members (GbSPL7 and GbSPL10), each containing both SBP and Ank domains (Figure 3C). Given the established role of the Ank domain in orchestrating protein interactomes through molecular recognition interfaces [25], GbSPL7 and GbSPL10 likely function as signaling nodes in cellular communication networks.

Gene structure is an important feature that can influence gene expression and function. In this study, we measured the exon–intron compositions of the 13 *GbSPL* genes (Figure 3D). The number of exons (CDSs) varied from 2 (GbSPL3 and GbSPL4) to 14 (GbSPL10), while the number of introns ranged from 1 to 13. This difference in gene structure indicates the functional diversity of *GbSPL* genes. Although the core coding sequences are well-defined, a comprehensive annotation of regulatory elements, including the 5′-UTR and 3′-UTR, remains a critical gap in the current *Ginkgo biloba* genomic resources [33].

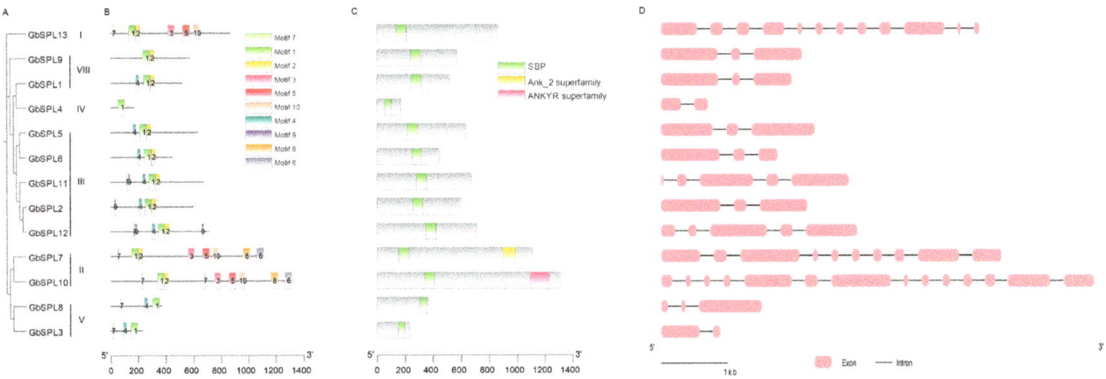

Figure 3. Phylogenetic trees, motifs, conserved domains, and gene structure analysis of GbSPL genes. (**A**) Phylogenetic trees of GbSPL proteins. (**B**) Motifs in the 13 GbSPL proteins, represented by ten colored boxes indicating different motifs. (**C**) Conserved domains present in the 13 GbSPL proteins. (**D**) Exon–intron structures of GbSPL genes.

2.4. Multiple Sequence Alignment Analysis and MicroRNA Target Prediction

Comparative SBP domain mapping identified three signature elements across the 13 GbSPLs: two zinc-binding clusters (Zn-1[C3H] and Zn-2[C2HC]) and an NLS (Figure 4A). Sequence conservation analysis showed that the SBP domains were highly conserved at the CQQC, SCR, and RRR sequences, with approximately 76 amino acid residues (Figure 4A). Considering the importance of the miR156/SPL module [20], the miR156 target sites of the 13 GbSPL genes were also predicted using the psRNATarget online tool. Surprisingly, only 2 out of 13 SPL genes contained miR156 binding sites. Each of these two genes harbored dual target loci: one recognized by miR156abc and the other by miR156f (Figure 4B).

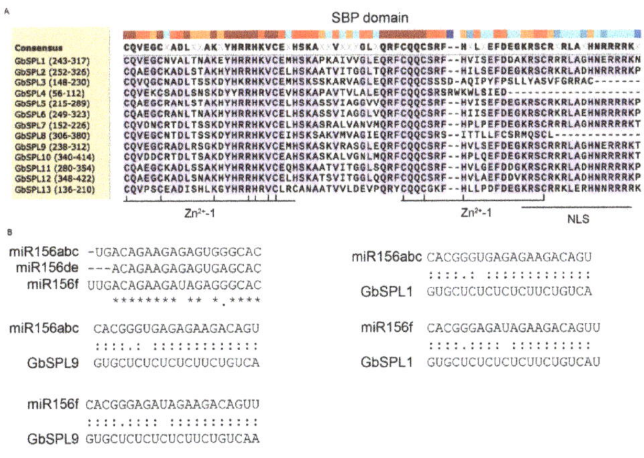

Figure 4. The sequence alignment of the SBP domain of GbSPL proteins and miR156 target gene binding sites. (**A**) The multiple sequence alignment of SBP domains. The two conserved zinc finger structures (C3H and C2HC) and NLS are indicated. The purple represents conserved bases. (**B**) The multiple sequence alignment of miR156 and its GbSPL target gene binding sites. The * represents conserved bases.

2.5. The Expansion and Collinearity Analysis of the GbSPL Gene Family

MCScanX software (v1.0.0) was utilized to identify the expansion and evolution history of the GbSPL gene family. A total of 7285 (26.2%) genes in the whole genome were generated through singleton duplication; 12,615 (45.3%) genes were generated by dispersed duplication; 1688 (6.1%) genes were generated through proximal duplication; 4905 (17.6%) genes were produced by tandem duplication; and 1343 (4.8%) genes were produced by whole-genome duplication (WGD) or segmental duplication (Figure 5A). Among the 13 GbSPL gene members, seven GbSPL genes resulted from dispersed duplication, four GbSPL genes originated from tandem duplication, and two GbSPL genes stemmed from WGD or segmental duplication. At the same time, none were produced by singleton or proximal duplication (Figure 5A,B). Two syntenic gene pairs (GbSPL3-GbSPL8, GbSPL2-GbSPL5) were identified within the GbSPL gene family (Figure 5C). These results indicate that dispersed duplication events were the primary force behind expanding the GbSPL gene family.

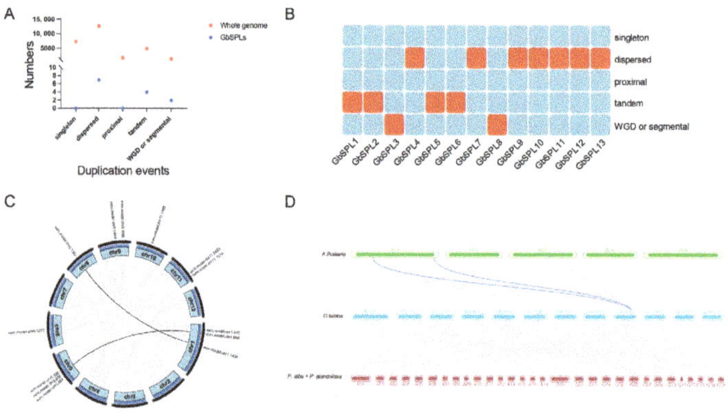

Figure 5. Evolutionary analysis of the GbSPL gene family. (**A**) The number of genes resulted from five duplication events, including singleton, dispersed, proximal, tandem, and WGD or segmental duplication. (**B**) The heatmap illustrates the duplication events of the GbSPL gene family. (**C**) The distribution and collinearity of the GbSPL gene family in the genome of *Ginkgo biloba*. The black line indicates the collinear gene pairs of GbSPL genes, while the gray line represents the collinear gene pairs in the *Ginkgo biloba* genome. (**D**) The distribution and collinearity of the GbSPL gene family in Arabidopsis, *Ginkgo biloba*, and Poplar. The blue line depicts the collinear gene pairs of GbSPL genes, whereas the gray line highlights the collinear gene pairs in the *Ginkgo biloba* genome with Arabidopsis and Poplar.

To elucidate the evolutionary dynamics of SPL gene duplication, cross-species synteny analyses were conducted comparing *Ginkgo biloba* with both a herbaceous model (*Arabidopsis thaliana*) and a woody relative, Poplar (*Populus alba* × *Populus tremula* var. *glandulosa* clone '84K'), utilizing MCScanX for whole-genome collinearity detection. Overall, ginkgo shows less genome-wide covariance with *Arabidopsis thaliana* and possesses two covariant SPL genes among them; it has no covariant SPL gene with the woody plant Poplar, although there are more homologous gene pairs with Poplar than with Arabidopsis in the whole genome (Figure 5D). Despite their more recent divergence as woody species, *Ginkgo biloba* and Poplar exhibit ancient SPL gene architectures, suggesting prolonged conservation of this regulatory family. Cross-species comparisons of GbSPL homologs could clarify functional divergence patterns while informing ancestral gene repertoire reconstructions in angiosperms and gymnosperms.

2.6. Cis-Acting Element Analysis in the Promoter of GbSPL Genes

Cis-regulatory elements within promoter regions mediate transcriptional modulation, thereby governing phenotypic expression through precise spatiotemporal control of gene activity. An in silico promoter analysis was conducted on GbSPL genes using the PlantCARE database, systematically screening for cis-acting elements within 2000 bp upstream of transcription start sites. A diverse distribution of cis-elements was observed in each GbSPL, including MYB binding sites, light, auxin, gibberellin, MeJA, abscisic acid, defense and stress responses, low-temperature elements, and others (Figure 6A), suggesting that GbSPLs may be involved in various biological processes. The cis-elements were categorized into four groups: plant growth and development-responsive elements (CAT-box, O2-site, MSA-like, and RY element), hormone-responsive elements (AuxRR-core, ABRE, TCA-elements, TATC-box, GARE-motif, CGTCA motif, TGACG motif, P-box, and TGA-element), light-responsive elements (TCT-motif, G-box, 3-AF1 site, GT1 motif, AAAC motif, ATCT motif, G-box, Sp1, GA motif, CAG motif, I-box, GATA motif, Gap box, ACE, LAMP element, MRE, ATC motif, TCCC motif, AE-box, chs CMTA, and AT1 motif), and stress-responsive elements (ARE, LTR, MBS, GC motif, and TC-rich repeats) (Figure 6B). The largest category of cis-elements was light-responsive, while hormone-responsive and stress-responsive elements were also prevalent in the promoters of GbSPL genes. Plant hormones, such as ABA and gibberellin, have been recognized as crucial components in various stress signal responses, and these cis-acting elements on the GbSPL promoters contribute to their functional diversity under stress conditions. The cis-acting analysis indicated that GbSPL genes play significant roles in both light and stress responsiveness.

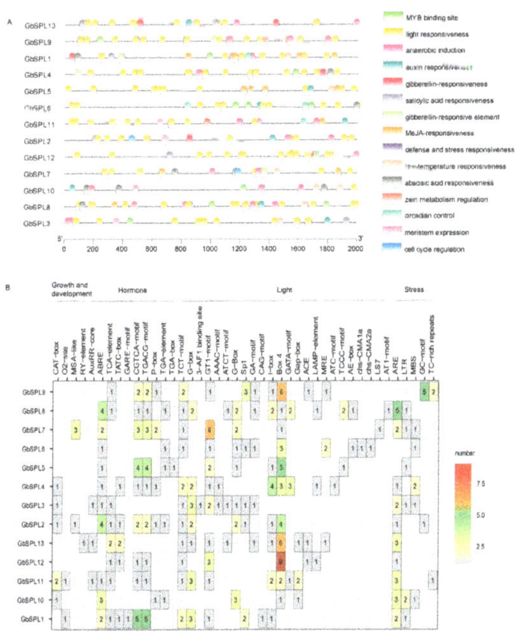

Figure 6. The cis-acting elements in the promoters of GbSPL genes. (**A**) The distribution of cis-acting elements in the promoters (upstream 2000 bp) of GbSPL genes. (**B**) The number of cis-acting elements among the four different types of response elements in the promoters of the GbSPL genes.

2.7. Expression Profiles of GbSPL Genes in Various Tissues, Low- and High-Flavonoid Leaves, and Water Stress in Ginkgo biloba

First, we investigated the tissue-specific expression patterns using RNA-Seq data from the reported study [34]. RNA-Seq-based expression profiles showed that *GbSPL* genes were differentially expressed in eight tissues, including mature fruit (MF), immature fruit (IF), root (R), stem (S), immature leaf (IL), mature leaf (ML), microstrobilus (M), and ovulate strobilus (OS) (Figure 7A). All 13 *GbSPL* genes were highly expressed in the stem, and 8 *GbSPL* genes (*GbSPL1*, *5*, *6*,*7*, *9*, *10*, *11*, *13*) exhibited high expression in the root (Figure 7A). The expression patterns of *GbSPL* genes in mature and immature fruits, as well as in mature and immature leaves, were highly consistent, indicating that the expression of the SPL gene is tissue-specific but not maturation-stage-specific. Additionally, most GbSPL genes were highly expressed in the ovulate strobilus but not in the microstrobilus, suggesting their role in the reproductive phase change.

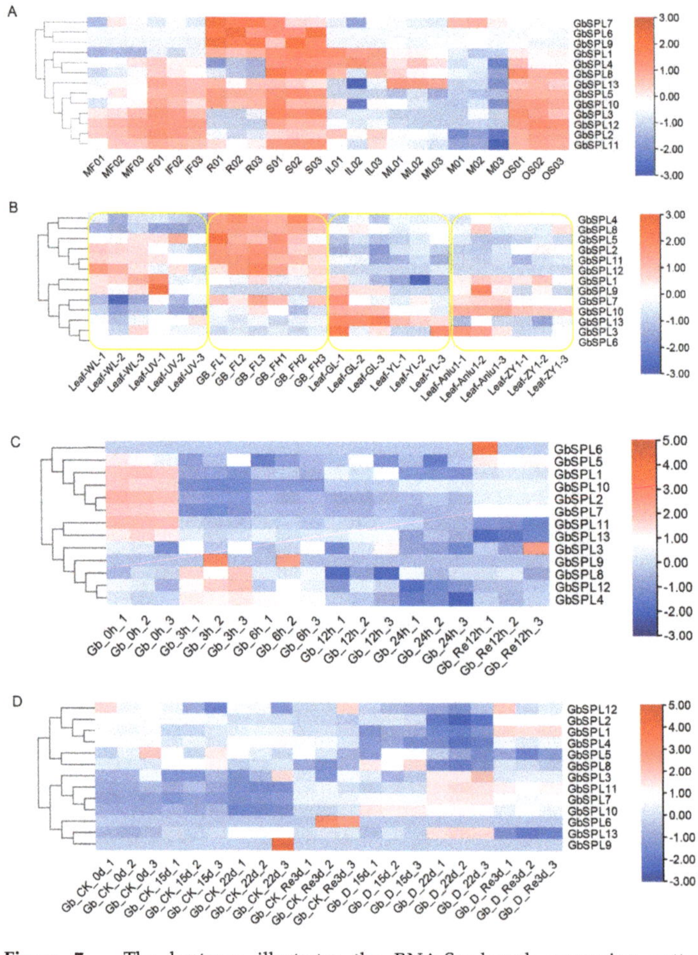

Figure 7. The heatmap illustrates the RNA-Seq-based expression patterns of 13 *GbSPL* genes. (**A**) Expression patterns of 13 *GbSPL* genes across eight tissues, including mature fruit (MF), immature fruit (IF), root (R), stem (S), immature leaf (IL), mature leaf (ML),

microstrobilus (M), and ovulate strobilus (OS). (**B**) Expression patterns of the 13 *GbSPL* genes grouped by contrasting low and high flavonoid content. The yellow boxes represent different groups. (**C**,**D**) Expression patterns of 13 *GbSPL* genes under short-term (**C**) and long-term (**D**) water stress. Red indicates high-expression genes, while blue represents low-expression genes.

Next, we explored the expression profiles of GbSPLs in four groups contrasting low and high flavonoid content [32,35–37]. Overall, the differentially expressed *GbSPL* genes varied across each group. For instance, *GbSPL1*, *GbSPL2*, *GbSPL11*, and *GbSPL12* were downregulated in the high-flavonoid leaves (UV-B treatment) of the first low- and high-flavonoid group (control and UV-B treatment) but not in the other groups (Figure 7B). Additionally, *GbSPL10* was specifically downregulated in the high-flavonoid leaves (yellow mutant) of the third low- and high-flavonoid group (green and yellow-mutant leaves), while *GbSPL3* was uniquely downregulated in the high-flavonoid leaves ('ZY1') of the fourth low- and high-flavonoid group ('Anlu1' and 'ZY1' leaves). These findings suggested that the biosynthetic and regulatory networks of flavonoids may differ based on the types of low and high flavonoids, possibly requiring the involvement of multiple SPL gene types.

Finally, we detected the expression patterns of *GbSPL* genes under both short-term and long-term water stress. During short-term water stress and rehydration (0, 3, 6, 12, 24, Re12 h) [38], the *GbSPL* genes were classified into two clusters: one induced by water stress (*GbSPL4*, *8*, *12*) and the other repressed by water stress (*GbSPL1*, *2*, *5*, *7*, *10*, *11*, *13*) (Figure 7C). In the long-term drought treatment (0, 15, 22 d, and re-watering 3 d) [39], the *GbSPL* genes were similarly divided into two clusters: one induced by drought stress (*GbSPL3*, *7*, *10*, *11*, *13*) and the other slightly repressed by drought stress (*GbSPL1*, *2*, *4*, *5*, *8*) (Figure 7D). Significant differences exist between short-term and long-term water stress in the upregulated and downregulated genes. This is likely due to distinct regulatory networks that govern responses to short-term and long-term water stress in *Ginkgo biloba*. For instance, *GbSPL11* was downregulated during short-term water shock but upregulated under long-term drought stress, while *GbSPL2* was downregulated in both short-term water shock and long-term drought stress (Figure 7C,D).

2.8. Quantitative Real-Time PCR (qRT-PCR) Analysis and Subcellular Localization Assay of the GbSPL Genes

To validate the results of RNA-Seq, four genes (*GbSPL2*, *10*, *11*, and *13*) that showed differential expression in short-term or long-term water stress responses were selected for qRT-PCR analysis. The qRT-PCR results were similar to the RNA-Seq findings, suggesting that *GbSPL2*, *10*, *11*, and *13* may function as negative regulators of short-term water shock in *Ginkgo biloba* (Figure 8A). However, the expression levels of *GbSPL2*, *10*, *11*, and *13* were upregulated during long-term drought stress compared to the control (Figure 8B), which is also consistent with the RNA-Seq results.

To investigate the location of GbSPL proteins, we examined the subcellular localization of GbSPL2 and GbSPL11 through transient expression in tobacco (*Nicotiana benthamiana*) leaves. Microscopic visualization revealed that the signal for the empty vector control, 35S-GFP, was distributed in both the nucleus and the membrane. We observed that the green fluorescent protein (GFP) signals of GbSPL2 and GbSPL11 overlapped with the nuclear marker mCherry, indicating their localization in the nucleus (Figure 8C), which is consistent with the prediction results from the WoLF PSORT online tools (Table 1). These findings indicate that GbSPL2 and GbSPL11 function as nuclear proteins, possibly serving as transcription factors.

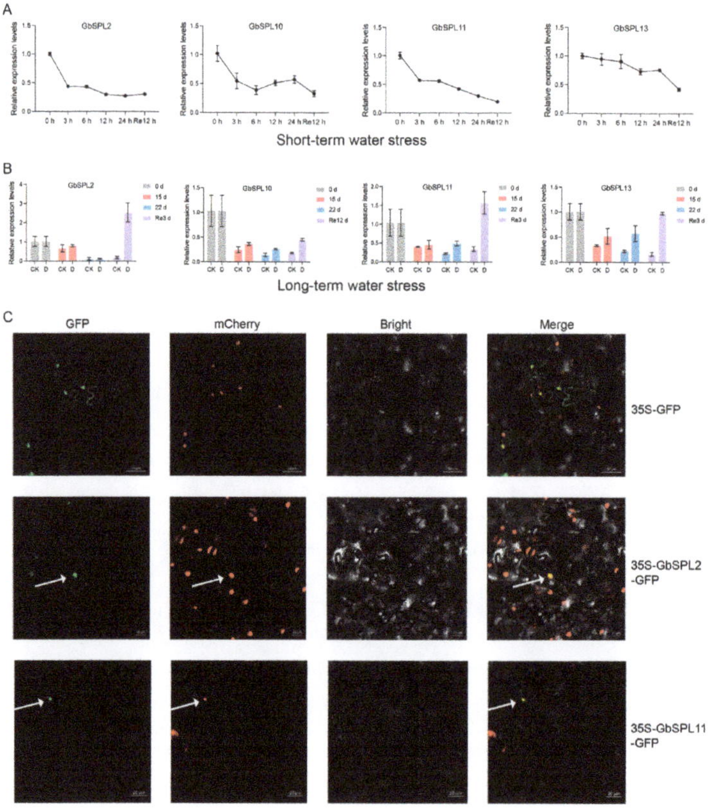

Figure 8. qRT-PCR analysis and subcellular localization of GbSPL genes. (**A**) qRT-PCR analysis of GbSPL genes under short-term water stress. All data are presented as means ± SD (n = 3). (**B**) qRT-PCR analysis of GbSPL genes under long-term water stress. (**C**) Subcellular localization of GbSPL2 and GbSPL11. mCherry, a nuclear marker. Arrows highlight fluorescent signals in the micrograph.

3. Discussion

3.1. Characteristics and Evolution of SPL Genes in Ginkgo biloba

As key plant-specific transcriptional regulators, SPL genes orchestrate multifaceted developmental programs encompassing secondary metabolite biosynthesis, morphogenesis (leaf and flower architecture), phase transition (juvenility–adulthood), reproductive timing control, and fruit morphogenesis through evolutionarily conserved molecular mechanisms [18,40]. The evolutionary trajectory of SPL genes dates back to their initial characterization in *Antirrhinum majus*, with subsequent phylogenetic analyses revealing conserved orthologs across Viridiplantae lineages—from basal streptophytes (Chlorophyta algae, Bryopsida mosses) to derived angiosperms (*Arabidopsis thaliana*, crop species, and woody plants) [17,41]. The functional conservation and divergence of SPL homologs in gymnosperms—evolutionarily significant non-flowering seeds plants—remain underexplored compared to angiosperm systems. *Ginkgo biloba* is an important gymnosperm whose SPL gene has not been reported. In this experiment, 13 *GbSPL* genes from *Ginkgo biloba* were screened using both a genome-level database constructed from the Pfam database (PF03110) and the BLAST protein database (Table 1). The number of GbSPL gene family members is comparable to that of *Arabidopsis thaliana* (16 members), tomato (15 members) [6], and Sweet

Cherry (12 members) [16]. Moreover, the chromosomal mapping of the 13 *GbSPL* genes in *Ginkgo biloba* revealed their uneven distribution across seven chromosomes (Figure 1), which is also similar to litchi [17]. The phylogenetic analysis showed that the GbSPL gene family was only distributed across six groups, whereas the AtSPL gene family was classified into eight groups (Figure 2), suggesting that the function of SPL genes may have diverged during the evolution of *Arabidopsis thaliana* (dicotyledonous) and *Ginkgo biloba* (gymnosperm).

The phylogenomic integration of sequence homology, motif analysis, gene structure, and molecular evolution patterns demonstrates conserved taxonomic clustering among *Ginkgo biloba* species. GbSPL genes within phylogenetic clades maintain conserved structural blueprints (Figure 3), implying evolutionary selection pressures linking gene structure plasticity to functional diversification [42]. GbSPL5 and GbSPL6 exhibit the same three motifs (Figure 3B), possibly due to internal tandem duplications on the same chromosome. Each GbSPL gene contains an SBP domain, whereas only GbSPL7 and GbSPL10 include an Ank domain (Figure 3C). In addition, target prediction identified 2 out of 13 GbSPL genes (GbSPL1 and GbSPL9) with miR156 target sites that may function within the miR156-SPL module (Figure 4B). However, neither GbSPL1 nor GbSPL9 possesses the Ank domain (Figure 3C). These results suggest functional divergence within the GbSPL family: while GbSPL1 and GbSPL9 likely coordinate with miR156, forming miR156-SPL modules to orchestrate growth-related processes, the majority of GbSPLs appear to operate through miR156-independent regulatory circuits. These findings differ from those of litchi [17] and *Medicago truncatula* [11], as the majority of SPL genes in litchi and *Medicago truncatula* contain the target sites for miR156.

We also investigated the duplication events of the GbSPL gene in the whole genome of *Ginkgo biloba* and found that the dispersed duplication is the lead duplication event, followed by tandem duplication and WGD or segmental duplication; however, no singleton or proximal duplication events were found (Figure 5A,B). This suggests that transposable element-mediated dispersed replication plays a major role in the expansion of the GbSPL gene family. Tandem and WGD or segmental duplications lead the gene family amplification in plants [43], which are also popular in expanding the GbSPL gene family. In addition, collinearity analysis with *Arabidopsis thaliana* and Poplar (*Populus alba* × *Populus tremula* var. *glandulosa* clone '84K') showed that the GbSPL gene had more homologous genes with *Arabidopsis thaliana* than with Poplar (Figure 5D), suggesting prolonged conservation of the SPL regulatory family.

3.2. Expression Patterns and Potential Functions of the GbSPL Genes

Gene expression profiles generally exhibit a strong correlation with biological functions. Cis-elements in the promoter regions of SPL genes are closely linked to expression profiles and function, as they are typically involved in regulating gene transcription [44]. Research shows that many cis-elements in the promoter regions of SPL genes are associated with plant growth and development, light, hormone, and stress responses (Figure 6), similar to those in other plants, such as *Betula luminifera* [45] and litchi [17]. Developmental time-course analyses in Arabidopsis revealed the stage-specific upregulation of *AtSPL15*, peaking during floral initiation phases with marked accumulation in inflorescence meristems [46]. Functional studies further demonstrated that *AtSPL15* accelerates the vegetative phase change while suppressing rosette leaf expansion. To predict the function of the GbSPL genes in plant growth and development, RNA-Seq-based expression patterns among various tissues and organs were first examined (Figure 7A). Most *GbSPL* genes exhibited high expression in the ovulate strobilus but not in the microstrobilus, suggesting their role in the reproductive phase change. Eight *GbSPL* genes (*GbSPL1, 5, 6, 7, 9, 10, 11, 13*) showed elevated expression in the root, indicating that they may play a novel role in root development.

Flavonoids are the key active compounds in *Ginkgo biloba*, and their biosynthesis is influenced by light and hormones [47]. Considering that many cis-elements in the promoter regions of SPL genes are associated with light and hormones, we speculated that the GbSPL genes are related to flavonoid biosynthesis. Analysis of targeted metabolite and flavonoid biosynthetic gene expression has demonstrated that *OsSPL17* regulates the expression of flavonoid biosynthesis genes, specifically CHI [48]. To predict the function of the *GbSPL* genes in flavonoid biosynthesis, expression patterns among the low- and high-flavonoid content were investigated in four groups (Figure 7B). Transcriptional profiling revealed many *GbSPL* genes that were differentially expressed in the four low- and high-flavonoid content groups (e.g., *GbSPL1*, *GbSPL2*, *GbSPL10*, *GbSPL11*, and *GbSPL12*). However, non-overlapping sets of differentially expressed *GbSPL* genes across these groups indicated group-specific regulatory dynamics.

Many important and diverse regulatory functions of SPL genes in plant growth and development have been well-understood [18]. However, the role of SPL genes in response to abiotic stress remains insufficiently explored. Given our interest in water stress, we also examined the gene expression of all GbSPL members under both short-term water shock conditions and long-term drought stress conditions (Figure 7C,D). The results revealed the reduced expression of some *GbSPL* genes (e.g., *GbSPL2*) in both short-term water shock and long-term drought stress, while the expression patterns of other *GbSPL* genes (*GbSPL4, 8, 11, 12*) were completely opposite in short-term and long-term drought conditions. This indicates that the *GbSPL4, 8*, and *12* genes may respond quickly and help ginkgo resist damage from adverse conditions in the short term, but they appear to lack function during prolonged drought conditions. The genes that respond to short-term and long-term water stress are not exactly the same, which is similar to beet (*Beta vulgaris* L.) [49] and foxtail millet (*Setaria italica*) [50].

4. Materials and Methods

4.1. Whole-Genome Identification of SPL Genes in *Ginkgo biloba*

The high-quality genomic data of *Ginkgo biloba*, as referenced in source [33], were procured from the Genome Sequence Archive database (https://ngdc.cncb.ac.cn/gwh/Assembly/18742/show, accessed on 1 April 2025). Initially, the amino acid sequences corresponding to the SPL gene family members in *Arabidopsis thaliana* were acquired from the Arabidopsis Information Resource (TAIR) database (https://v2.arabidopsis.org/, accessed on 1 April 2025). The complete protein sequences of *Ginkgo biloba* were utilized to establish a BLAST protein database using the makeblastdb software (V 2.10.1). By employing the Arabidopsis SPL genes as queries in a BLASTp (V 2.10.1) search against the genome-wide protein database of *Ginkgo biloba*, candidate SPL members were identified under stringent criteria (E-value $< 1 \times 10^{-5}$, Identity > 30%). Subsequently, the Hidden Markov models (HMMs) related to the SBP domain (PF03110) were retrieved from the Pfam database (http://pfam.xfam.org/, accessed on 1 April 2025). The HMMsearch software (V 3.3.2) was used to examine the complete protein sequences of *Ginkgo biloba* based on the Pfam-A models file containing the SBP domain, which also yielded a list of candidate SPL members of *Ginkgo biloba*. Thirdly, the overlapping candidate SPL genes were delineated by intersecting the results derived from both BLAST and HMM searches. Lastly, the NCBI CDD website tool (https://www.ncbi.nlm.nih.gov/Structure/bwrpsb/bwrpsb.cgi, accessed on 1 April 2025) was utilized to verify the completeness of the SBP domains within the SPL genes. The GbSPL genes with the conserved SBP domain were identified in *Ginkgo biloba* and renamed according to their chromosomal positions.

4.2. Physicochemical Properties, MicroRNA Target Prediction, and Chromosomal Location Analysis of SPL Genes

The protein-coding sequence length, molecular weight (MW), and isoelectric point (pI) were calculated using ExPASy ProtParam online tools (https://web.expasy.org/protparam/, accessed on 1 April 2025), and subcellular localization predictions of GbSPL genes were conducted with WoLF PSORT (https://wolfpsort.hgc.jp/, accessed on 1 April 2025). Using default settings, the miR156 target sequences were predicted with the online tool psRNATarget (https://www.zhaolab.org/psRNATarget/home, accessed on 1 April 2025). The chromosomal localization of the GbSPLs was obtained from *Ginkgo biloba* genome annotation information and visualized by TBtools-II (V2.210) [51].

4.3. Phylogenetic Analysis

Muscle software (V 3.8.1551) was used to perform multiple sequence alignments of SPL genes in *Arabidopsis thaliana* and *Ginkgo biloba*. We constructed the phylogenetic tree using IQ-Tree with the Maximum Likelihood (ML) method and 1000 bootstrap iterations. The final tree topology was visually rendered and annotated with the iTOL web server (https://itol.embl.de/, accessed on 1 April 2025).

4.4. Motif, Conserved Domain, Gene Structure, and Cis-Elements Analysis

MEME (https://meme-suite.org/meme/tools/meme, accessed on 1 April 2025) was utilized to search for the 10 motifs (6–50 width; any number of repetitions (anr)) in the *Ginkgo biloba* SPL family proteins. The NCBI Batch CD-Search Tool (https://www.ncbi.nlm.nih.gov/Structure/bwrpsb/bwrpsb.cgi, accessed on 1 April 2025) was employed to identify the conserved domains of SPL family proteins. The gene structure of the GbSPL family was extracted from the *Ginkgo biloba* genome annotation file using TBtools-II (V2.210). The PlantCARE website tool (https://bioinformatics.psb.ugent.be/webtools/plantcare/html/, accessed on 1 April 2025) was used to predict cis-acting elements within 2000 bp upstream of the promoter sequences of the GbSPL genes. TBtools-II (V2.210) software was utilized for visualization.

4.5. Gene Duplication and Syntenic Analysis of the GbSPL Gene Family

Genome assemblies and annotation files for *Arabidopsis thaliana* and Poplar (*Populus alba* × *Populus tremula* var. *glandulosa* clone '84K') genome data [52] were retrieved from Ensembl Plants (http://plants.ensembl.org/, accessed on 1 April 2025) and the National Center for Biotechnology Information (NCBI, https://www.ncbi.nlm.nih.gov/, accessed on 1 April 2025). Segmental duplication events between *Ginkgo biloba* and *Arabidopsis thaliana* or Poplar were identified and visualized through synteny analysis using TBtools-II (V2.210). The GbSPL gene replication events were analyzed using multiple collinear scanning toolkits (MCScanX, v1.0.0). The syntenic relationship between the GbSPL genes and SPL genes from *Arabidopsis thaliana* or Poplar was determined using Dual Synteny Plotter software (V2.210) in TBtools-II (V2.210).

4.6. RNA-Seq and Quantitative Real-Time PCR (qRT-PCR)-Based Expression Analysis of GbSPL Genes

The transcriptome data were downloaded from the National Center for Biotechnology Information database (NCBI, https://www.ncbi.nlm.nih.gov/, accessed on 1 April 2025) and then analyzed using the same method according to our previous study [39]. The accession number of different transcriptome data used in this study is listed in Supplementary Table S1. Heatmaps with the expression analysis of GbSPLs were generated by the TPM values with TBtools-II (V2.210). According to our previous study [38,39], a quantitative Real-Time PCR (qRT-PCR) assay was performed to confirm transcriptome sequencing results. Primer sequences used in this experiment are provided in Supplementary Table S2.

4.7. Subcellular Localization Analysis

Full-length coding sequences of *GbSPL4* and *GbSPL7* (excluding stop codons) were PCR-amplified from the cDNA of *Ginkgo biloba* leaves. The amplified products were directionally cloned into the pSAK277-GFP binary vector as C-terminal GFP fusions under the control of the CaMV 35S promoter. Constructs were heat-shocked into competent cells of *Agrobacterium tumefaciens* GV3101. The subcellular localization analysis followed our previously reported methods [38]. Primer sequences used in this experiment are provided in Supplementary Table S2.

5. Conclusions

The identification of 13 GbSPL genes categorized into six phylogenetic clusters indicates lineage-specific diversification, possibly reflecting adaptations to distinct environmental pressures faced by this ancient gymnosperm species. The prevalence of dispersed duplication events driving the expansion of the SPL family in *Ginkgo biloba* contrasts with the tandem duplication patterns typically observed in many angiosperms, suggesting different evolutionary mechanisms for gene family amplification in long-lived woody species. This divergence may contribute to Ginkgo's well-known resilience to various biotic and abiotic stresses over extended periods. The discovery of miR156 targeting two GbSPL genes is consistent with conserved post-transcriptional regulatory mechanisms but raises questions regarding the functional redundancy or neofunctionalization of non-targeted GbSPLs. The nuclear localization of GbSPL2 and GbSPL11 reinforces their roles as transcription factors, likely involved in regulating downstream targets related to flavonoid biosynthesis and water stress response pathways. The contrasting expression patterns of GbSPLs under flavonoid induction suggest a complex regulatory network where specific SPL members function as either activators or repressors depending on developmental or environmental cues. This research not only enhances our understanding of the functional genomics of *Ginkgo biloba* but also lays the groundwork for utilizing SPL genes to boost stress tolerance and secondary metabolite production in economically significant gymnosperms. Overall, our genome-wide analysis of the SPL gene family lays the groundwork for further investigation into the mechanisms behind flavonoid biosynthesis and water stress in woody plants and gymnosperms.

Supplementary Materials: The following supporting information can be downloaded at https://www.mdpi.com/article/10.3390/ijms26104932/s1.

Author Contributions: F.F., F.C. and M.M. conceived and designed the study. M.Y., C.Q., L.Y., Y.S. and J.Z. conducted the experiments. M.M. and M.Y. performed bioinformatics analyses and data visualization. M.M. and M.Y. wrote the manuscript. All authors have read and agreed to the published version of the manuscript.

Funding: This work was supported by the Jiangsu Provincial Key Research and Development Program (modern agriculture) (Grant No. BE2021367), the National Natural Science Foundation of China (32301613), and the China Postdoctoral Science Foundation (2023M741721).

Institutional Review Board Statement: Not applicable.

Informed Consent Statement: Not applicable.

Data Availability Statement: The public transcriptome data used in this research can be accessed in the inserted article.

Conflicts of Interest: The authors declare that they have no conflicts of interest.

References

1. Riechmann, J.L.; Heard, J.; Martin, G.; Reuber, L.; Jiang, C.; Keddie, J.; Adam, L.; Pineda, O.; Ratcliffe, O.J.; Samaha, R.R.; et al. Arabidopsis transcription factors: Genome-wide comparative analysis among eukaryotes. *Science* **2000**, *290*, 2105–2110. [CrossRef] [PubMed]
2. Manna, M.; Thakur, T.; Chirom, O.; Mandlik, R.; Deshmukh, R.; Salvi, P. Transcription factors as key molecular target to strengthen the drought stress tolerance in plants. *Physiol. Plant* **2021**, *172*, 847–868. [CrossRef] [PubMed]
3. Birkenbihl, R.P.; Jach, G.; Saedler, H.; Huijser, P. Functional dissection of the plant-specific SBP-domain: Overlap of the DNA-binding and nuclear localization domains. *J. Mol. Biol.* **2005**, *352*, 585–596. [CrossRef]
4. Klein, J.; Saedler, H.; Huijser, P. A new family of DNA binding proteins includes putative transcriptional regulators of the Antirrhinum majus floral meristem identity gene SQUAMOSA. *Mol. Gen. Genet. MGG* **1996**, *250*, 7–16. [CrossRef]
5. Yang, Z.; Wang, X.; Gu, S.; Hu, Z.; Xu, H.; Xu, C. Comparative study of SBP-box gene family in Arabidopsis and rice. *Gene* **2008**, *407*, 1–11. [CrossRef]
6. Salinas, M.; Xing, S.; Hoehmann, S.; Berndtgen, R.; Huijser, P. Genomic organization, phylogenetic comparison and differential expression of the SBP-box family of transcription factors in tomato. *Planta* **2012**, *235*, 1171–1184. [CrossRef] [PubMed]
7. Li, J.; Hou, H.; Li, X.; Xiang, J.; Yin, X.; Gao, H.; Zheng, Y.; Bassett, C.L.; Wang, X. Genome-wide identification and analysis of the SBP-box family genes in apple (Malus domestica Borkh.). *Plant Physiol. Biochem.* **2013**, *70*, 100–114. [CrossRef]
8. Liu, M.; Sun, W.; Ma, Z.; Huang, L.; Wu, Q.; Tang, Z.; Bu, T.; Li, C.; Chen, H. Genome-wide identification of the SPL gene family in Tartary Buckwheat (Fagopyrum tataricum) and expression analysis during fruit development stages. *BMC Plant Biol.* **2019**, *19*, 299. [CrossRef]
9. Hu, X.; Liu, C.; Tian, J.; Zhang, Y.; Xin, Q.; Chen, A.; Li, D.; Liu, X. Identification, molecular characterization, and expression analysis of the SQUAMOSA PROMOTER BINDING PROTEIN-LIKE (SPL) gene family in Betula platyphylla Suk. *Trees* **2019**, *34*, 229–241. [CrossRef]
10. Qian, M.; Ni, J.; Niu, Q.; Bai, S.; Bao, L.; Li, J.; Sun, Y.; Zhang, D.; Teng, Y. Response of miR156 SPLModule during the Red Peel Coloration of Bagging-Treated Chinese Sand Pear (Pyrus pyrifolia Nakai). *Front. Physiol.* **2017**, *8*, 550. [CrossRef]
11. Wang, H.; Lu, Z.; Xu, Y.; Kong, L.; Shi, J.; Liu, Y.; Fu, C.; Wang, X.; Wang, Z.Y.; Zhou, C.; et al. Genome-wide characterization of SPL family in Medicago truncatula reveals the novel roles of miR156/SPL module in spiky pod development. *BMC Genom.* **2019**, *20*, 552. [CrossRef] [PubMed]
12. Feng, X.; Wang, Y.; Zhang, N.; Zhang, X.; Wu, J.; Huang, Y.; Ruan, M.; Zhang, J.; Qi, Y. Systematic Identification, Evolution and Expression Analysis of the SPL Gene Family in Sugarcane (Saccharum spontaneum). *Trop. Plant Biol.* **2021**, *14*, 313–328. [CrossRef]
13. Ren, Y.; Ma, R.; Fan, Y.; Zhao, B.; Cheng, P.; Fan, Y.; Wang, B. Genome-wide identification and expression analysis of the SPL transcription factor family and its response to abiotic stress in Quinoa (Chenopodium quinoa). *BMC Genom.* **2022**, *23*, 773. [CrossRef]
14. De Paola, C.; Garcia-Carpintero, V.; Vazquez-Vilar, M.; Kaminski, K.; Fernandez-Del-Carmen, A.; Sierro, N.; Ivanov, N.V.; Giuliano, G.; Waterhouse, P.; Orzaez, D. Comparative analysis of the Squamosa Promoter Binding-Like (SPL) gene family in Nicotiana benthamiana and Nicotiana tabacum. *Plant Sci.* **2023**, *335*, 111797. [CrossRef] [PubMed]
15. Jadhao, K.R.; Kale, S.S.; Chavan, N.S.; Janjal, P.H. Genome-wide analysis of the SPL transcription factor family and its response to water stress in sunflower (Helianthus annuus). *Cell Stress. Chaperones* **2023**, *28*, 943–958. [CrossRef] [PubMed]
16. Sun, Y.; Wang, Y.; Xiao, Y.; Zhang, X.; Du, B.; Turupu, M.; Wang, C.; Yao, Q.; Gai, S.; Huang, J.; et al. Genome-Wide Identification of the SQUAMOSA Promoter-Binding Protein-like (SPL) Transcription Factor Family in Sweet Cherry Fruit. *Int. J. Mol. Sci.* **2023**, *24*, 2880. [CrossRef] [PubMed]
17. Xu, Z.; Wu, J.; Jing, X.; Khan, F.S.; Chen, Y.; Chen, Z.; Zhang, H.; Wei, Y. Genome-Wide Identification of Litchi SPL Gene Family and Expression Analysis in Pericarp Anthocyanin Biosynthesis. *Horticulturae* **2024**, *10*, 762. [CrossRef]
18. Jerome Jeyakumar, J.M.; Ali, A.; Wang, W.M.; Thiruvengadam, M. Characterizing the Role of the miR156-SPL Network in Plant Development and Stress Response. *Plants* **2020**, *9*, 1206. [CrossRef]
19. Gou, J.Y.; Felippes, F.F.; Liu, C.J.; Weigel, D.; Wang, J.W. Negative regulation of anthocyanin biosynthesis in Arabidopsis by a miR156-targeted SPL transcription factor. *Plant Cell* **2011**, *23*, 1512–1522. [CrossRef]

20. Wang, H.; Wang, H. The miR156/SPL Module, a Regulatory Hub and Versatile Toolbox, Gears up Crops for Enhanced Agronomic Traits. *Mol. Plant* **2015**, *8*, 677–688. [CrossRef]
21. Wang, Y.; Liu, W.; Wang, X.; Yang, R.; Wu, Z.; Wang, H.; Wang, L.; Hu, Z.; Guo, S.; Zhang, H.; et al. MiR156 regulates anthocyanin biosynthesis through SPL targets and other microRNAs in poplar. *Hortic. Res.* **2020**, *7*, 118. [CrossRef] [PubMed]
22. Xu, M.; Hu, T.; Zhao, J.; Park, M.-Y.; Earley, K.W.; Wu, G.; Yang, L.; Poethig, R.S. Developmental Functions of miR156-Regulated SQUAMOSA PROMOTER BINDING PROTEIN-LIKE (SPL) Genes in Arabidopsis thaliana. *PLoS Genet.* **2016**, *12*, e1006263. [CrossRef] [PubMed]
23. Cui, L.-G.; Shan, J.-X.; Shi, M.; Gao, J.-P.; Lin, H.-X. The miR156-SPL9-DFR pathway coordinates the relationship between development and abiotic stress tolerance in plants. *Plant J.* **2014**, *80*, 1108–1117. [CrossRef]
24. Yamasaki, H.; Hayashi, M.; Fukazawa, M.; Kobayashi, Y.; Shikanai, T. SQUAMOSA Promoter Binding Protein-Like7 Is a Central Regulator for Copper Homeostasis in Arabidopsis. *Plant Cell* **2009**, *21*, 347–361. [CrossRef] [PubMed]
25. Xing, S.; Salinas, M.; Hohmann, S.; Berndtgen, R.; Huijser, P. miR156-targeted and nontargeted SBP-box transcription factors act in concert to secure male fertility in Arabidopsis. *Plant Cell* **2010**, *22*, 3935–3950. [CrossRef] [PubMed]
26. Zhang, Y.; Hu, Q.; Zhai, X.; Tu, Z.; Wang, J.; Wang, M.; Li, H. Genome-wide investigation of SQUAMOSA promoter binding protein-like genes in Liriodendron and functional characterization of LcSPL2. *AoB Plants* **2024**, *16*, plae008. [CrossRef]
27. Li, Y.; Han, S.; Sun, X.; Khan, N.U.; Zhong, Q.; Zhang, Z.; Zhang, H.; Ming, F.; Li, Z.; Li, J. Variations in OsSPL10 confer drought tolerance by directly regulating OsNAC2 expression and ROS production in rice. *J. Integr. Plant Biol.* **2023**, *65*, 918–933. [CrossRef]
28. Li, S.; Cheng, Z.; Li, Z.; Dong, S.; Yu, X.; Zhao, P.; Liao, W.; Yu, X.; Peng, M. MeSPL9 attenuates drought resistance by regulating JA signaling and protectant metabolite contents in cassava. *Theor. Appl. Genet.* **2022**, *135*, 817–832. [CrossRef]
29. Zhao, Y.; He, J.; Liu, M.; Miao, J.; Ma, C.; Feng, Y.; Qian, J.; Li, H.; Bi, H.; Liu, W. The SPL transcription factor TaSPL6 negatively regulates drought stress response in wheat. *Plant Physiol. Biochem.* **2024**, *206*, 108264. [CrossRef]
30. Shen, N.; Wang, T.; Gan, Q.; Liu, S.; Wang, L.; Jin, B. Plant flavonoids: Classification, distribution, biosynthesis, and antioxidant activity. *Food Chem.* **2022**, *383*, 132531. [CrossRef]
31. Wang, L.; Cui, J.; Jin, B.; Zhao, J.; Xu, H.; Lu, Z.; Li, W.; Li, X.; Li, L.; Liang, E.; et al. Multifeature analyses of vascular cambial cells reveal longevity mechanisms in old Ginkgo biloba trees. *Proc. Natl. Acad. Sci. USA* **2020**, *117*, 2201–2210. [CrossRef] [PubMed]
32. Zhao, B.; Wang, L.; Pang, S.; Jia, Z.; Wang, L.; Li, W.; Jin, B. UV-B promotes flavonoid synthesis in Ginkgo biloba leaves. *Ind. Crops Prod.* **2020**, *151*, 112483. [CrossRef]
33. Liu, H.; Wang, X.; Wang, G.; Cui, P.; Wu, S.; Ai, C.; Hu, N.; Li, A.; He, B.; Shao, X.; et al. The nearly complete genome of Ginkgo biloba illuminates gymnosperm evolution. *Nat. Plants* **2021**, *7*, 748–756. [CrossRef]
34. Ye, J.; Cheng, S.; Zhou, X.; Chen, Z.; Kim, S.U.; Tan, J.; Zheng, J.; Xu, F.; Zhang, W.; Liao, Y.; et al. A global survey of full-length transcriptome of Ginkgo biloba reveals transcript variants involved in flavonoid biosynthesis. *Ind. Crops Prod.* **2019**, *139*, 111547. [CrossRef]
35. Wu, Y.; Guo, J.; Zhou, Q.; Xin, Y.; Wang, G.; Xu, L.-A. De novo transcriptome analysis revealed genes involved in flavonoid biosynthesis, transport and regulation in Ginkgo biloba. *Ind. Crops Prod.* **2018**, *124*, 226–235. [CrossRef]
36. Wu, Y.; Guo, J.; Wang, T.; Cao, F.; Wang, G. Metabolomic and transcriptomic analyses of mutant yellow leaves provide insights into pigment synthesis and metabolism in Ginkgo biloba. *BMC Genom.* **2020**, *21*, 858. [CrossRef] [PubMed]
37. Meng, J.; Wang, B.; He, G.; Wang, Y.; Tang, X.; Wang, S.; Ma, Y.; Fu, C.; Chai, G.; Zhou, G. Metabolomics Integrated with Transcriptomics Reveals Redirection of the Phenylpropanoids Metabolic Flux in Ginkgo biloba. *J. Agric. Food Chem.* **2019**, *67*, 3284–3291. [CrossRef]
38. Ming, M.; Zhang, J.; Tang, J.; Zhang, J.; Fu, F.; Cao, F. Transcriptome Profiling Revealed ABA Signaling Pathway-Related Genes and Major Transcription Factors Involved in the Response to Water Shock and Rehydration in Ginkgo biloba. *Forests* **2024**, *15*, 1690. [CrossRef]
39. Ming, M.; Zhang, J.; Zhang, J.; Tang, J.; Fu, F.; Cao, F. Transcriptome Profiling Identifies Plant Hormone Signaling Pathway-Related Genes and Transcription Factors in the Drought and Re-Watering Response of Ginkgo biloba. *Plants* **2024**, *13*, 2685. [CrossRef]
40. Chen, X.; Zhang, Z.; Liu, D.; Zhang, K.; Li, A.; Mao, L. SQUAMOSA Promoter-Binding Protein-Like Transcription Factors: Star Players for Plant Growth and Development. *J. Integr. Plant Biol.* **2010**, *52*, 946–951. [CrossRef]
41. Qin, S.W.; Bao, L.H.; He, Z.G.; Li, C.L.; La, H.G.; Zhao, L.F. Identification and regulatory network analysis of SPL family transcription factors in Populus euphratica Oliv. heteromorphic leaves. *Sci. Rep.* **2022**, *12*, 2856. [CrossRef]
42. Zhang, S.-D.; Ling, L.-Z.; Yi, T.-S. Evolution and divergence of SBP-box genes in land plants. *Bmc Genom.* **2015**, *16*, 787. [CrossRef] [PubMed]
43. Panchy, N.; Lehti-Shiu, M.; Shiu, S.-H. Evolution of Gene Duplication in Plants. *Plant Physiol.* **2016**, *171*, 2294–2316. [CrossRef] [PubMed]

44. Hernandez-Garcia, C.M.; Finer, J.J. Identification and validation of promoters and cis-acting regulatory elements. *Plant Sci.* **2014**, *217*, 109–119. [CrossRef] [PubMed]
45. Li, X.-Y.; Lin, E.-P.; Huang, H.-H.; Niu, M.-Y.; Tong, Z.-K.; Zhang, J.-H. Molecular Characterization of SQUAMOSA PROMOTER BINDING PROTEIN-LIKE (SPL) Gene Family in Betula luminifera. *Front. Plant Sci.* **2018**, *9*, 608. [CrossRef]
46. Schwarz, S.; Grande, A.V.; Bujdoso, N.; Saedler, H.; Huijser, P. The microRNA regulated SBP-box genes SPL9 and SPL15 control shoot maturation in Arabidopsis. *Plant Mol. Biol.* **2008**, *67*, 183–195. [CrossRef]
47. Liu, S.; Gu, X.; Jiang, Y.; Wang, L.; Xiao, N.; Chen, Y.; Jin, B.; Wang, L.; Li, W. UV-B promotes flavonoid biosynthesis in Ginkgo biloba by inducing the GbHY5-GbMYB1-GbFLS module. *Hortic. Res.* **2023**, *10*, uhad118. [CrossRef]
48. Sun, Y.J.; Fu, M.; Wang, L.; Bai, Y.X.; Fang, X.L.; Wang, Q.; He, Y.; Zeng, H.L. OsSPLs Regulate Male Fertility in Response to Different Temperatures by Flavonoid Biosynthesis and Tapetum PCD in PTGMS Rice. *Int. J. Mol. Sci.* **2022**, *23*, 3744. [CrossRef]
49. Xue, G.; Wu, W.; Fan, Y.; Ma, C.; Xiong, R.; Bai, Q.; Yao, X.; Weng, W.; Cheng, J.; Ruan, J. Genome-wide identification, evolution, and role of SPL gene family in beet (*Beta vulgaris* L.) under cold stress. *BMC Genom.* **2024**, *25*, 101. [CrossRef]
50. Lai, D.; Fan, Y.; Xue, G.; He, A.; Yang, H.; He, C.; Li, Y.; Ruan, J.; Yan, J.; Cheng, J. Genome-wide identification and characterization of the SPL gene family and its expression in the various developmental stages and stress conditions in foxtail millet (*Setaria italica*). *Bmc Genom.* **2022**, *23*, 389. [CrossRef]
51. Chen, C.; Wu, Y.; Li, J.; Wang, X.; Zeng, Z.; Xu, J.; Liu, Y.; Feng, J.; Chen, H.; He, Y.; et al. TBtools-II: A "One for All, All for One" Bioinformatics Platform for Biological Big-data Mining. *Mol. Plant* **2023**, *16*, 1733–1742. [CrossRef] [PubMed]
52. Qiu, D.; Bai, S.; Ma, J.; Zhang, L.; Shao, F.; Zhang, K.; Yang, Y.; Sun, T.; Huang, J.; Zhou, Y.; et al. The genome of Populus alba x *Populus tremula* var. *glandulosa clone 84K. DNA Res.* **2019**, *26*, 423–431. [CrossRef] [PubMed]

International Journal of
Molecular Sciences

MDPI

Article

Phenotypic Screening and Marker-Assisted Validation of Sources of *Aphis craccivora* Koch Resistance in Cowpea (*Vigna unguiculata* L.)

Galalea Gillian Gaonosi [1], Lekgari Lekgari [2], Marang Mosupiemang [1], Metseyabeng Nametso Sehularo [1], Aobakwe Boisy Tshegofatso [1], Bamphithi Tiroesele [1], Tiny Motlhaodi [2], Samodimo Ngwako [1] and Motlalepula Pholo-Tait [3,*]

1 Department of Crop and Soil Sciences, Botswana University of Agriculture and Natural Resources, Gaborone 00267, Botswana; gmmolutsi@buan.ac.bw (G.G.G.); marangmosupiemang@yahoo.com (M.M.); 201200327@buan.ac.bw (M.N.S.); atshegof@buan.ac.bw (A.B.T.); btiroese@buan.ac.bw (B.T.); sngwako@buan.ac.bw (S.N.)

2 Department of Field Crops and Horticulture, National Agricultural Research Development Institute, Gaborone 00267, Botswana; lekgari@nardi.org.bw (L.L.); tiny@nardi.org.bw (T.M.)

3 Centre for Bioeconomy, Faculty of Research and Graduate Studies, Botswana University of Agriculture and Natural Resources, Gaborone 00267, Botswana

* Correspondence: mtait@buan.ac.bw

Abstract: *Aphis craccivora* significantly affects cowpea (*Vigna unguiculata* L.) production, leading to yield reductions. Management strategies encompass physical barriers and biological and chemical methods, which can be costly and detrimental to the environment. Host-plant resistance offers a more sustainable alternative. This study evaluated cowpea genotypes in a screenhouse experiment. Tswana and B261-B were resistant, while B301, B259, and ER7 showed a tolerance phenotype. Tswana exhibited a low aphid population and minimal plant damage, probably due to suppression of reproduction and fecundity. Conversely, IT97K-556-6, SARI-21KTA-6, SARC 1-57-2, B013-F, B339, and Blackeye were susceptible to aphids, as shown by high aphid populations and dense sooty molds. Severe damage to plant vigor may be linked to direct aphid feeding and reduced photosynthesis efficiency. SNP1_0912 and CP 171F/172R markers confirmed aphid resistance in Tswana and ER7 as well as in the IT97k-556-6 and SARI-21KTA-6 controls. The inverse susceptible phenotype in the control group suggests that the markers may not function properly due to negative interactions between quantitative trait loci (QTL) and environmental factors. This could also indicate the presence of different aphid biotypes that severely damage Western African breeding lines. This study offers essential insights for breeding aphid-resistant cowpea varieties. Future efforts will involve sequencing Tswana to identify more resistance sources and create novel markers.

Keywords: *Aphis craccivora*; *Vigna unguiculata*; SNP1_0912; CP 171F/172R; quantitative trait loci; screening; resistance; susceptible; tolerance

1. Introduction

Cowpea (*Vigna unguiculata* L. Walp.) (2n = 2x = 22) is a member of the *Phaseoleae* tribe of the Leguminosae family. It belongs to the Fabaceae family and Faboideae sub-family [1]. It is a source of protein, minerals, and vitamins in thousands of low-income households' diets [1]. Statistics Botswana indicated that cowpea is the third most important crop after maize (*Zea mays* L.) and grain sorghum (*Sorghum bicolor* (L.) Moench) in Botswana [2]. It is a major source of dietary protein that nutritionally complements staple low-protein cereal [3]. The cowpea grain is a low-cost alternative to more expensive fish and cattle products on the African market [4]. The nutrients from cowpea are obtained from various parts such as dry seeds, green pods, green seeds, and tender green leaves. It is a vital cash crop that contributes to farmers' food security and income generation [4]. In addition to its nutritional value, cowpea is an essential component of the cropping systems in semi-arid regions of Sub-Saharan Africa [5,6]. This has been attributed to its ability to fix atmospheric nitrogen, which improves soil fertility for the success of cereal crops grown in rotation, particularly in smallholder farming systems. In this respect, cowpea is a high-potential strategic crop for addressing the complex challenges of hunger, malnutrition, environmental sustainability, climate change, and rising food prices, which the global community will confront in the coming decades [7].

The Southern African region, including Botswana, Namibia, Zambia, Zimbabwe, Mozambique, and the Republic of South Africa, is considered the centre of diversity of *V. unguiculata* [8]. In Botswana, diverse cowpea germplasm collections are conserved in the National Plant Genetic Resource Centre (NPGRC)-Ministry of Agriculture, and their duplicates are conserved in the Southern African Development Community (SADC) gene bank in Lusaka/Zambia [9,10]. The Botswana National Plant Genetic Resource Centre currently has a collection of over 1487 accessions of Vigna unguiculata [9]. Despite the large diversity of conserved genotypes, the genetic diversity within cowpeas remains relatively narrow, which further contributes to the large prevalence of low-yielding genotypes [11].

In addition, the yield potential and the stability of the grain quality of cowpeas are further restricted by the prevalence of abiotic and biotic stresses. Amongst biotic stresses, insects are considered primarily responsible for 90–100% of the yield reduction [12]. *Aphis craccivora* Koch (Homoptera: Aphididae), commonly known as cowpea aphid, is a cowpea pest of economic importance, causing significant 50–60% yield losses in Botswana [13]. Global reports indicated yield losses of 35% in Africa, and up to 20% to 40% yield losses were reported in Asia [14]. Cowpea aphids primarily infest the succulent parts of the plant, such as young shoots, leaves, and pods [15]. The aphid feeding damage includes sucking and removing plant sap, which reduces the nutrients and water available to the crop [16]. Aphid feeding eventually leads to chlorosis and stunting, which delays the commencement of flowering, and this can cause plant mortality when infestations are substantial, especially during the seedling stage [17]. Aphids excrete a sugary plant substance known as honeydew, and black sooty-like fungus (sooty mold) can develop on honeydew, reducing photosynthesis. This occurs by physically blocking sunlight from reaching the leaf surface, thereby hindering the plant's ability to produce food through the photosynthetic process [18]. Besides direct plant damage, aphids may transmit viruses such as Bean common mosaic virus (BcMV) and Pea seed-borne mosaic virus (PsbMV) [19]. The aphid's ability to reproduce rapidly and its potential to develop resistance to chemical pesticides makes it difficult to control [20]. When there is less food and the environment becomes less favourable due to overcrowding, winged forms occur, further expanding their territory and causing more damage [21].

Several management approaches, including physical barriers and biological or chemical methods, are employed to control aphids in cowpeas [22]. It was previously reported [16] that these approaches are costly, labour-intensive, and may result in environmental biases. Host-plant resistance (HPR), which describes the plant's innate resistance usually controlled by one or more genes, offers a sustainable and environmentally friendly approach to aphid management [23]. This pest management approach enables plants to withstand and survive significant pest damage due to their genetically inherited traits [24]. Developing host-plant pest-resistance varieties has become a crucial component in an integrated crop management system [25]. The concomitant aphid infestation screening and marker-assisted selection (MAS) have become the most promising tools for breeding, using molecular markers to identify and select genes or genomic regions of interest without environmental bias [26,27]. Meanwhile, phenotypic aphid infestation culture screening studies reported that the Blackeye commercial variety was highly susceptible to aphids due to lower fecundity compared to the Tswana cowpea variety [28]. In another study, landraces B383, B261-B, and Tswana exhibited the longest pre-reproductive periods, reduced aphid population growth, and reproductive success, suggesting an aphid resistance phenotype [29]. This was further confirmed in another study that demonstrated reduced aphid reproductive growth rates in B383 and B261-B landraces, while the inverse was observed for landraces B013-F, B339, and the Blackeye variety [30]. The findings were only based on the extended pre-reproductive reduction in population growth rate, survival, and fecundity [30]. However, the genetic mode of inheritance linked to aphid resistance, whether through simple inherited trait loci (SITLs) or quantitative trait loci (QTLs), remains unclear.

Considerable progress in West Africa has been made in the past years in developing cowpea varieties resistant to aphids using a marker-assisted selection strategy [31], confirming resistance to cowpea aphids in SARC 1-57-2 accession in Ghana. Resistance to aphids happens through antibiosis, which is controlled by a single dominant gene [32,33]. The aphid genes from SARC 1-57-2 were confirmed using a codominant SSR marker, CP 171F/172R, which is several centiMorgans in genetic distance from the resistance gene locus [34]. Genetic mapping of the aphid resistance in IT97K-556-6, an accession from Ghana, revealed two QTLs on cowpea linkage groups LG1 (minor) and LG7 (major), both with favourable alleles contributed from IT97K-556-6. The SNP1_0912 marker flanking the major aphid resistance QTL (QAc-vu7.1) was used to confirm a QTL conferring resistance to cowpea aphid in IT97K-556-6. This marker is located at 21.7 cM, and the QTL is at 22 cM on chromosome 2 of the cowpea genome [33]. In KVX295-2-124-99, two duplicate genes located on chromosomes 3 and 7 were reported to be involved in the aphid resistance ability, as demonstrated by segregating populations using marker MA61, which is located on chromosome 3, and marker MA70, on chromosome 7 [35]. Given the molecular variations in cowpea biotypes across different geographical locations in West Africa, this study aimed to evaluate cowpea's response to aphid infestation and validate sources of resistance to aphids that could serve as parental lines in a breeding program.

2. Results

2.1. Cowpea Adult Aphid Population Density Growth

The number of adult aphids on various genotypes was insignificant for all days except for days 11, 13, and 15 (Table 1). Landrace B261-B showed the significantly highest number of adult aphids (70.2) on day 11 after infestation. It was followed by B301 and IT97K-556-6, with an average number of adult aphids of 64.8 and 49.3, respectively. IT97K-556-6 demonstrated the significantly highest number of adult aphids (192.2) compared to other genotypes on day 13 after infestation. In addition to IT97K-556-6, a significantly higher number of adult aphids was observed in SARC 1-57-2 (189), B261-B (147.2), B339 (121.8), and Blackeye (115.7). The results demonstrated that Tswana had the lowest number of aphids (57.7) on day 15, which was significantly lower than those in B261-B (251.7), SARI-2KTA-6 (244.0), IT97K-556-6 (255.3), and KVX295-2-124-99 (377.8). The latter showed the highest number of adult aphids compared to other genotypes. Generally, most genotypes, including Tswana, ER7, B013-F, B261-B, B301, B359, SARI-2KTA-6, and IT97K-556-6, reached their peak number of adult aphids at day 17. However, SARC 1-57-2, KVX295-2-124-99, and B339 reached peak adult aphid counts on day 15 after infestation. Only the Blackeye variety attained the peak number of adult aphids 19 days after infestation.

Table 1. The number of adult aphics from 7 to 27 days after infestation.

Genotypes	Days After Infestation										
	7	9	11	13	15	17	19	21	23	25	27
Tswana	5.3 ± 0.6 ^{a}	4 ± 0.0 ^{bc}	47.5 ± 11.5 ^{abc}	93.3 ± 19.[illegible] ^{bc}	57.[illegible] ± 36.3 ^{b}	119 ± 100.8 ^{a}	118 ± 58.1 ^{a}	101.7 ± 51.0 ^{a}	94.8 ± 69.8 ^{a}	40.8 ± 26.[illegible] ^{a}	37.5 ± 16.7 ^{a}
Blackeye	4.7 ± 0.6 ^{a}	4.5 ± 0.5 ^{bc}	31.5 ± 23.3 ^{cd}	115.7 ± 58.[illegible] ^{abc}	179.3 ± 89.0 ^{b}	269.5 ± 168.1 ^{a}	288.7 ± 245.4 ^{a}	102.8 ± 4.6 ^{a}	34.8 ± 24.4 ^{a}	23.3 ± 15.[illegible] ^{a}	27.5 ± 19.3 ^{a}
ER7	5.8 ± 3.8 ^{a}	4.3 ± 1.2 ^{bc}	13.5 ± 11.2 ^{d}	44.8 ± 32.3 ^{c}	100.3 ± 57.5 ^{b}	140.5 ± 101.2 ^{a}	137.2 ± 41.9 ^{a}	127.3 ± 78.2 ^{a}	54.7 ± 35.5 ^{a}	33.7 ± 34.8 ^{a}	28.7 ± 29.7 ^{a}
B013-F	4.7 ± 0.8 ^{a}	5.2 ± 1.3 ^{b}	38.2 ± 14.2 ^{bcd}	101.7 ± 47.[illegible] ^{abc}	177.[illegible] ± 132.4 ^{ab}	316.2 ± 148.2 ^{a}	129.2 ± 108.0 ^{a}	55.5 ± 9.9 ^{a}	52.0 ± 31.1 ^{a}	20.2 ± 5.5 ^{a}	20.3 ± 16.4 ^{a}
B261-B	4.8 ± 0.3 ^{a}	3.5 ± 0.0 ^{bc}	70.2 ± 20.7 ^{a}	147.2 ± 36.3 ^{ab}	251.[illegible] ± 29.1 ^{ab}	364.0 ± 205.8 ^{a}	210.2 ± 108.0 ^{a}	125.0 ± 34.6 ^{a}	32.8 ± 5.5 ^{a}	37.7 ± 11.3 ^{a}	79.3 ± 26.5 ^{a}
B301	5.0 ± 0.0 ^{a}	5 ± 0.0 ^{bc}	64.8 ± 16.6 ^{ab}	93.3 ± 38.3 ^{bc}	173.5 ± 77.0 ^{b}	206.5 ± 102.4 ^{a}	136.3 ± 50.5 ^{a}	73.8 ± 23.4 ^{a}	28.3 ± 9.1 ^{a}	37.8 ± 5.5 ^{a}	76.4 ± 1.9 ^{a}
B339	4.2 ± 0.6 ^{a}	4 ± 0.87 ^{bc}	41.7 ± 13.9 ^{abcd}	121.8 ± 58.3 ^{abc}	156.3 ± 109.6 ^{b}	139.5 ± 87.3 ^{a}	128.2 ± 54.0 ^{a}	39.2 ± 29.2 ^{a}	37.7 ± 5.5 ^{a}	30.8 ± 5.[illegible] ^{a}	41.8 ± 29.6 ^{a}
B359	4.8 ± 0.3 ^{a}	3.2 ± 1.0 ^{c}	31.7 ± 6.0 ^{cd}	80 ± 19.1 ^{bc}	177.2 ± 87.2 ^{ab}	248.2 ± 221.0 ^{a}	187 ± 197.7 ^{a}	101.3 ± 60.9 ^{a}	66.2 ± 51.1 ^{a}	49.8 ± 39.3 ^{a}	51 ± 42.4 ^{a}
SARC 1-57-2	4.8 ± 1.2 ^{a}	3.8 ± 1.5 ^{bc}	40.7 ± 21.5 ^{abcd}	189 ± 95.0 ^{a}	198.[illegible] ± 100.6 ^{ab}	195.2 ± 53.8 ^{a}	124.2 ± 31.6 ^{a}	53.8 ± 12.4 ^{a}	83.8 ± 12.4 ^{a}	38.5 ± 0.0 ^{a}	25.5 ± 0.0 ^{a}
SARI-2KTA-6	4.5 ± 1.3 ^{a}	7 ± 2.6 ^{a}	26.3 ± 8.1 ^{cd}	109.3 ± 44.5 ^{abc}	24[illegible] ± 56.6 ^{ab}	305 ± 134.3 ^{a}	184.2 ± 72.4 ^{a}	107 ± 30.0 ^{a}	68.5 ± 54.9 ^{a}	34.5 ± 21.9 ^{a}	46.5 ± 19.1 ^{a}
IT97K-556-6	4.3 ± 1.2 ^{a}	4.8 ± 0.3 ^{bc}	49.3 ± 13.6 ^{abc}	192.2 ± 3[illegible].4 ^{a}	255.[illegible] ± 165.4 ^{ab}	366.5 ± 213.1 ^{a}	324.8 ± 170.8 ^{a}	55.0 ± 37.7 ^{a}	112.7 ± 47.6 ^{a}	33.5 ± 2.[illegible] ^{a}	21.0 ± 21.2 ^{a}
KVX295-2-124-99	4.8 ± 0.3 ^{a}	4.8 ± 0.3 ^{bc}	32.3 ± 17.9 ^{cd}	100.3 ± 32.5 ^{abc}	377.8 ± 143.2 ^{a}	343.8 ± 277.9 ^{a}	241.7 ± 188.0 ^{a}	145.7 ± 115.8 ^{a}	72.5 ± 45.9 ^{a}	48.3 ± 45.9 ^{a}	44.7 ± 47.1 ^{a}
p-value	0.95 ns	0.05 *	0.01 **	0.05 *	0.09 ns	0.56 ns	0.60 ns	0.42 ns	0.32 ns	0.96 ns	0.25 ns
F-value	0.38	2.96	2.96	2.3	1.89	0.89	0.85	1.09	1.25	0.34	1.34
CV%	25.7	22.21	39.2	42.1	52.47	[illegible]5.96	71.1	60.02	66.57	66.57	64.15

*, **, ns, Significant at $p \leq 0.05$, 0.01, or non-significant, respectively, within columns. Shown values are the means. Means followed by a dissimilar letter within a column are significant at 5% Fisher's LSD.

2.2. Number of Nymphs Produced

The reproduction rate of the number of nymphs significantly differed among genotypes on days 7, 9, 17, and 23 (Table 2). B301 showed the significantly highest number of nymphs, which was 64.2 and 149.3 on days 7 and 9, respectively. On day 17, Tswana had the lowest number of nymphs (248.5), and this was significantly lower than those observed for B261-B (1672.3), B359 (1672.3), SARI-2KTA-6 (1184.3), and IT97K-556-6 (1636.3). However, the significantly highest number of nymphs (1672.3) was shown on both B261-B and B359 landraces. Intriguingly, genotypes B013-F, B261-B, B301, B359, SARI-2KTA-6, SARC 1-57-2, and IT97K-556-6 reached the peak number of nymphs on day 17, which ranged between 981.5 and 1672.3. Genotype KVX295-2-124-99 took longer (23 days) to reach its peak number of nymphs compared to other genotypes. On day 23, IT97K-556-6 had a significantly higher number of nymphs (1573.5) than all other genotypes except for Blackeye and KVX295-2-124-99. Conversely, the significantly lowest number of nymphs was observed on day 23 in Tswana (267.7), which was significantly lower than in the Blackeye, IT97K-556-6, and KVX295-2-124-99 genotypes. There was a general decline in the number of nymphs produced from day 23 in all genotypes, although the decline was insignificant amongst genotypes on day 23 and day 27.

2.3. The Number of Alates (Winged Aphids)

Generally, the peak number of alates for most genotypes was reached on day 23 after infestation. Despite a non-significant variation in the number of aphids with wings on all other days, a significant variation was observed on day 23 (Table 3). On this day, Blackeye had the highest number of alates (219), which was significantly higher than those observed in Tswana (48.7), B261-B (53.5), and B359 (83.7). Despite a decreased trend in the number of alates after day 23, SARI-2KTA-6 and KVX295-2-124-99 showed an increased number of alates on day 27. The number of alates in those genotypes was significantly higher than in B013-F, B261-B, and IT97K-556-6.

Table 2. The number of nymphs from 7 to 27 days after infestation.

Genotypes	Days After Infestation										
	7	9	11	13	15	17	19	21	23	25	27
Tswana	39.5 ± 27.8 ^{abc}	94.5 ± 38.6 ^{abc}	137.8 ± 11.5 ^{ab}	240.2 ± 75.7 ^{ab}	210.2 ± 151.6 ^{c}	248.5 ± 136.0 ^{b}	312.3 ± 40.0 ^{b}	329.0 ± 320.1 ^{b}	267.7 ± 189.7 ^{b}	325.2 ± 150.5 ^{a}	386.8 ± 67.7 ^{abc}
Blackeye	32.2 ± 17.9 ^{bcd}	104.5 ± 56.9 ^{abc}	89.3 ± 35.2 ^{b}	347.7 ± 207.6 ^{ab}	767.2 ± 486.3 ^{abc}	914.7 ± 564.5 ^{ab}	1150.8 ± 883.4 ^{ab}	1056.5 ± 241.1 ^{a}	1016.5 ± 407.3 ^{ab}	538.8 ± 54.8 ^{a}	650.3 ± 464.9 ^{a}
ER7	8.0 ± 7.0 ^{d}	57.8 ± 39.1 ^{c}	121.5 ± 60.0 ^{ab}	135.0 ± 42.7 ^{b}	333.2 ± 198.5 ^{bc}	543.8 ± 351.2 ^{ab}	649.5 ± 494.0 ^{b}	655.7 ± 356.2 ^{ab}	309.7 ± 223.0 ^{b}	374.0 ± 288.2 ^{a}	137.7 ± 139.3 ^{bc}
B013-F	31.3 ± 9.8 ^{bcd}	102.3 ± 35.1 ^{abc}	116.3 ± 22.9 ^{ab}	486.8 ± 452.6 ^{ab}	609.7 ± 620.6 ^{abc}	981.5 ± 807.8 ^{ab}	587.2 ± 645.8 ^{b}	431.5 ± 75.7 ^{ab}	233.3 ± 138.9 ^{b}	160.3 ± 24.5 ^{a}	94.2 ± 62.6 ^{c}
B261-B	58.3 ± 20.0 ^{ab}	85 ± 28.5 ^{abc}	217.2 ± 127.9 ^{a}	696.3 ± 284.7 ^{a}	1059 ± 545.1 ^{ab}	1672.3 ± 750.3 ^{a}	1071 ± 409.0 ^{ab}	721.7 ± 168.9 ^{ab}	281.2 ± 26.8 ^{b}	321.3 ± 121.8 ^{a}	184.2 ± 87.4 ^{abc}
B301	64.2 ± 16.0 ^{a}	149.3 ± 10.5 ^{a}	170.8 ± 29.2 ^{ab}	333 ± 64.0 ^{ab}	727.7 ± 237.9 ^{abc}	755.2 ± 348.4 ^{ab}	621.3 ± 486.8 ^{b}	481.2 ± 277.1 ^{ab}	223.2 ± 66.8 ^{b}	273.7 ± 59.1 ^{a}	285 ± 24.8 ^{abc}
B339	23.7 ± 9.1 ^{cd}	100.8 ± 26.4 ^{abc}	183.8 ± 75.5 ^{ab}	350.7 ± 180 ^{ab}	487.5 ± 283.3 ^{abc}	665 ± 343.2 ^{ab}	854.8 ± 481.0 ^{ab}	751.5 ± 354.3 ^{ab}	771.5 ± 348.6 ^{ab}	393.8 ± 347.5 ^{a}	173.2 ± 120.9 ^{bc}
B359	32.0 ± 9.5 ^{bcd}	78.7 ± 28.0 ^{bc}	95.3 ± 25.0 ^{b}	416.2 ± 226.9 ^{ab}	571.5 ± 347.4 ^{abc}	1672.3 ± 750.3 ^{a}	498.5 ± 447.5 ^{b}	435.3 ± 301.3 ^{ab}	250.8 ± 128.0 ^{b}	390.7 ± 261.6 ^{a}	501 ± 343.7 ^{abc}
SARC 1-57-2	28.7 ± 18.6 ^{bcd}	104.8 ± 36.8 ^{abc}	159.2 ± 21.3 ^{ab}	407.0 ± 229.9 ^{ab}	484.3 ± 211.9 ^{abc}	1055.2 ± 231.0 ^{ab}	867.0 ± 254.6 ^{ab}	361.3 ± 111.4 ^{b}	375.8 ± 25.1 ^{b}	481.0 ± 0.0 ^{a}	322.0 ± 0.0 ^{abc}
SARI-2KTA-6	28.8 ± 7.3 ^{bcd}	132.8 ± 17.5 ^{ab}	104.3 ± 17.2 ^{b}	482.7 ± 235.3 ^{ab}	924.0 ± 143.6 ^{abc}	1184.3 ± 87.7 ^{ab}	1097.2 ± 33[illegible].4 ^{ab}	576.3 ± 85.8 ^{ab}	648.5 ± 384.7 ^{ab}	411.7 ± 209.4 ^{a}	608.5 ± 130.8 ^{ab}
IT97K-556-6	38.0 ± 7.7 ^{abcd}	139.0 ± 18.5 ^{ab}	84.8 ± 67.9 ^{b}	551.0 ± 224.1 ^{ab}	1163.7 ± 322.7 ^{a}	1636.3 ± 881.5 ^{a}	1810.3 ± 7[illegible]8.3 ^{a}	720.0 ± 550.3 ^{ab}	1573.5 ± 977.3 ^{a}	629.5 ± 75.7 ^{a}	330.5 ± 410.8 ^{abc}
KVX295-2-124-99	29.5 ± 20.8 ^{bcd}	82.2 ± 32.8 ^{bc}	116.0 ± 42.3 ^{ab}	539.8 ± 431.4 ^{ab}	886.3 ± 323.4 ^{abc}	947.5 ± 700.8 ^{ab}	1021.3 ± 743.4 ^{ab}	683.5 ± 421.5 ^{ab}	1127.2 ± 814.0 ^{ab}	688.5 ± 878.7 ^{a}	557.8 ± 504.5 ^{abc}
p-value	0.03 *	0.1 ns	0.09 ns	0.50 ns	0.26 ns	0.16 ns	0.22 ns	0.38 ns	0.03 *	0.83 ns	0.09 ns
F-value	2.59	1.86	1.88	0.97	1.55	1.8	1.45	1.15	2.61	0.57	2.04
CV%	46.26	32.9	39.4	63.42	58.85	57.57	64.73	52.43	78	79.91	62.1

*, ns, Significant at $p \leq 0.05$, 0.01, 0.001, or non-significant, respectively, within columns. Shown values are the means. Means followed by a dissimilar letter within a column are significant at 5% Fisher's LSD.

Table 3. The average number of Alates between 19 and 27 days after aphid infestation.

Genotypes	Days After Infestation				
	19	21	23	25	27
Tswana	57.8 ± 54.5 [a]	41.3 ± 12.2 [b]	48.7 ± 23.2 [b]	45.5 ± 30.5 [a]	54.0 ± 30.2 [bc]
Blackeye	117.0 ± 31.8 [a]	260.5 ± 171.8 [a]	219.3 ± 20.9 [a]	116.3 ± 39.2 [a]	174 ± 116.0 [abc]
ER7	22.2 ± 12.2 [a]	67.3 ± 38.6 [b]	131 ± 126.3 [ab]	68.7 ± 70.9 [a]	57.7 ± 64.3 [bc]
B013-F	44.5 ± 36.3 [a]	66.5 ± 12.7 [b]	131.3 ± 132.6 [ab]	44.5 ± 23.0 [a]	26.0 ± 20.8 [c]
B261-B	94.2 ± 53.5 [a]	62 ± 13.5 [b]	53.5 ± 4.8 [b]	92.2 ± 16.8 [a]	75.8 ± 10.3 [bc]
B301	125.8 ± 67.2 [a]	90.2 ± 35.3 [b]	148 ± 92.2 [ab]	101.5 ± 79.0 [a]	96 ± 22.6 [bc]
B339	97.5 ± 67.5 [a]	101 ± 29.1 [b]	99.3 ± 24.4 [ab]	55.75 ± 9.5 [a]	53 ± 20.1 [bc]
B359	66.7 ± 55.4 [a]	91.5 ± 44.0 [b]	83.7 ± 78.9 [ab]	83 ± 68.2 [a]	33 ± 31.4 [c]
SARC 1-57-2	111.5 ± 103.0 [a]	157.5 ± 60.1 [ab]	173.8 ± 87.3 [ab]	134 ± 0.0 [a]	63 ± 0.0 [bc]
SARI-2KTA-6	92 ± 45.0 [a]	143.7 ± 72.4 [ab]	190 ± 29.5 [ab]	107.33 ± 39.9 [a]	272.8 ± 6.0 [a]
IT97K-556-6	113.2 ± 40.6 [a]	171.7 ± 108.7 [ab]	169.7 ± 48.0 [ab]	112 ± 10.0 [a]	91.5 ± 113.8 [bc]
KVX295-2-124-99	76 ± 86.7 [a]	117.5 ± 80.0 [b]	131 ± 96.8 [ab]	81.2 ± 68.0 [a]	189.2 ± 177.3 [ab]
p-value	0.57 ns	0.08 ns	0.27 ns	0.48 ns	0.03 *
F-value	0.89	2.06	1.35	0.67	2.66
CV%	69.17	59.79	59.37	62.15	73.92

*, ns: significant at $p < 0.05$, 0.01 or non-significant (ns), respectively, within columns. Shown values are the means. Means followed by a dissimilar letter within a column are significant at 5% Fisher's LSD.

2.4. Aphid Damage and Plant Vigour Score

Figure 1 shows the plant damage and vigour following aphid infestation. Scoring was performed according to the method described in previous studies [14,36]. The scoring was based on a scale from 1 to 5, where ratings of 1 to 2 indicate some resistance, a rating of 3 indicates tolerance, and ratings of 4 to 5 represent susceptible cultivars. The local commercial varieties Tswana and landrace B261-B demonstrated resistant phenotypes (Figure 2A,B), with score ratings of 2.0 and 2.67, respectively (Figure 1). Furthermore, tolerant phenotypes were observed in the B301 and B359 landraces (Figure 2C,D), both of which scored 3 (Figure 1), as well as in the variety ER7, which scored 3.2 (Figures 1 and 2E). The breeding line KVX295-2-124-99 received a plant vigour score of 3.5, indicating an aphid-tolerant phenotype (Figures 1 and 2F).

Conversely, Blackeye and SARC 1-57-2 received the highest damage score ratings, 4.73 and 4.67, respectively (Figure 1), indicating that they are susceptible phenotypes (Figure 3A). Additionally, the susceptible genotypes included both B013-F and B339 landraces, which received a rating of 4.17. The breeding lines IT97K-556-6 and SARI-2KTA-6 also exhibited susceptible phenotypes, with scores of 4.0 and 4.4, respectively.

2.5. Genotypic Marker-Assisted Screening for Aphid Resistance

SNP1_0912 marker flanking resistance to aphid locus, with an expected aphid resistance band of ~165 bp and a susceptible band of ~150 bp, was used to segregate resistant and susceptible genotypes. The homozygous ~165 bp resistant band of interest was amplified in Tswana (Figure 4A) and ER7 (Figure 4C). However, the heterozygous ~165 bp resistant band and the ~150 bp susceptible band were amplified in Blackeye (Figure 4B). The SNP1_0912 marker also amplified a ~165 bp resistant band in ER7 (Figure 4C). Furthermore, the CP 171f/172r marker, which confers aphid resistance in sari-21kta-6, further confirmed homozygous aphid resistance in Tswana (Figure 4D) and ER7 (Figure 4E).

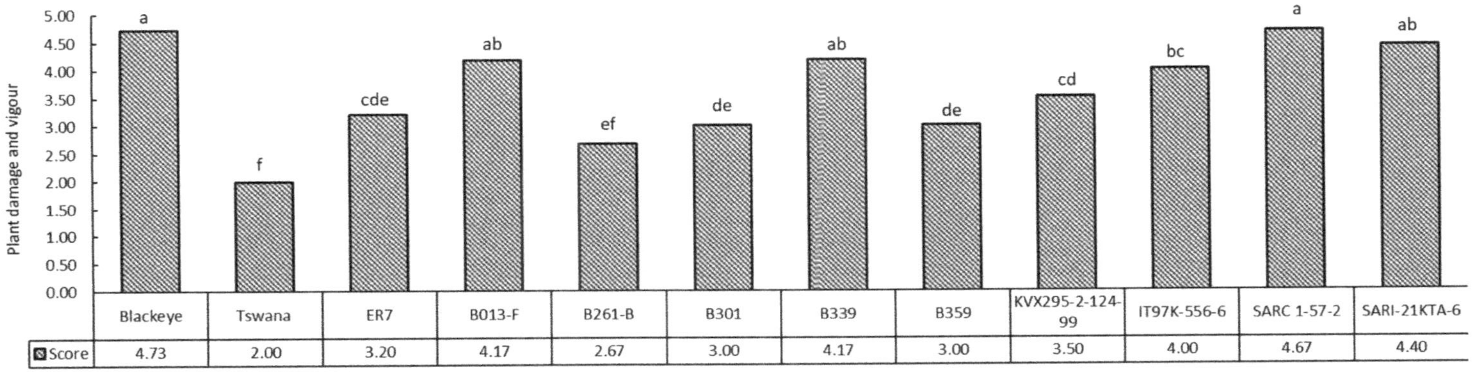

Figure 1. Plant damage and vigour scores at 27 days after infestation. The score rating represented by the different lowercase letters within the graph columns denotes a significant difference between genotypes at $p < 0.05$ (Fishers' LSD); Score rating of 1: highly resistant; score rating of 2: resistance genotypes; score rating of 3: tolerant phenotype; score rating of 4: moderately susceptible; score rating of 5: highly susceptible.

Figure 2. Resistant and tolerant cowpea plant vigour phenotype. The plant vigour for all genotypes was compared to Blackeye. Resistant phenotype: Tswana (**A**) and B261 (**B**); tolerant phenotypes: B301 (**C**), B359 (**D**), ER7 (**E**), and KVX295-2-124-9 (**F**).

Figure 3. Susceptible cowpea plant vigour phenotype following aphid infestation. Susceptible genotypes include SARC 1-57-2 (**A**), IT97K-556-6 (**B**), SARI-21KTA-6 (**C**), and B339 (**D**).

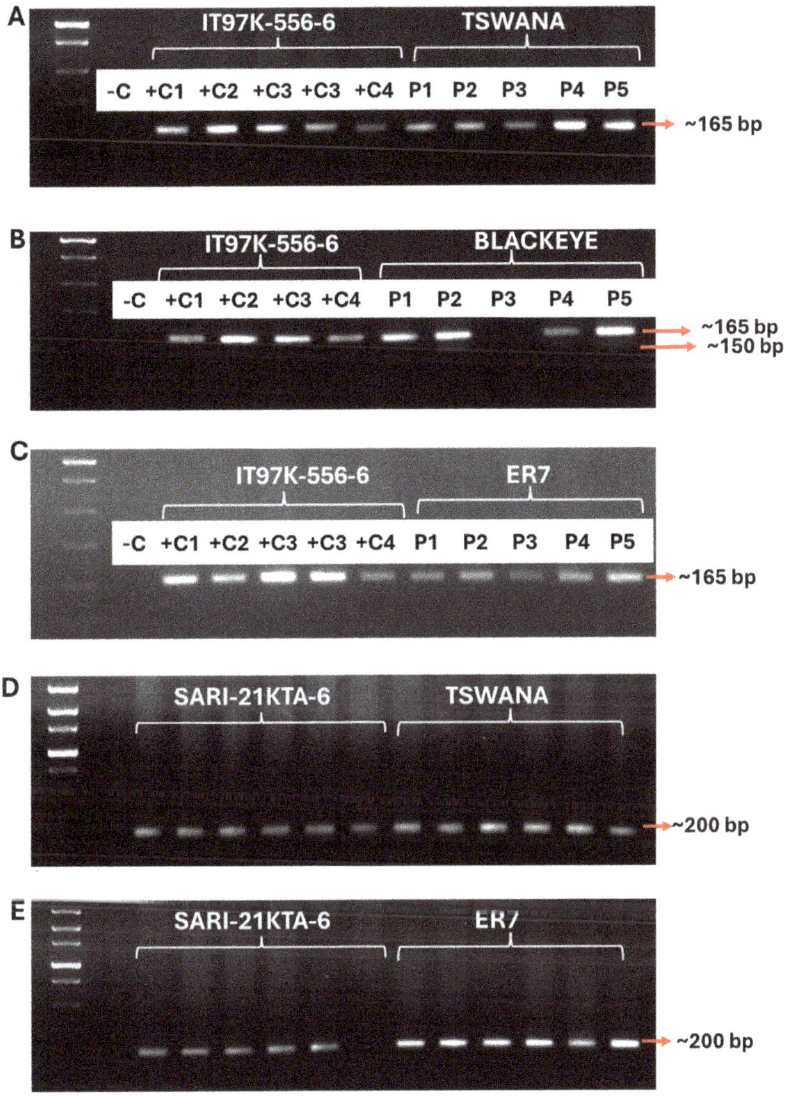

Figure 4. Genotyping of cowpea genotypes for aphid resistance. SNP1_0912 (**A–C**) molecular marker, which flanks the major QTL for aphid resistance in IT97K-556-6, was used to validate aphid resistance or susceptibility in Tswana (**A**), Blackeye (**B**), and ER7 (**C**). Resistance in Tswana (**D**) and ER7 (**E**) was further validated using the CP 171F/172R marker, which confers aphid resistance in SARI-21KTA-6. A 2.5% Agarose gel and a 1 kb molecular weight were used. −C: negative control without DNA template; +C1 to +C6: positive control plants conferring aphid resistance; P1 to P6: the targeted plants.

3. Discussion

Reproduction in *Aphis craccivora* happens through parthenogenesis, allowing populations to increase rapidly and making infestations difficult to control. High populations result from the aphids' prolific reproduction within a short period, and nymphs mature into reproductive adults, leading to a rapid population growth of aphids [37]. This was evident in this study, as the population of adults and nymphs rapidly increased 11 days after infestation, peaking between days 17 and 19. The general rapid increase in the number of nymphs and adult aphids could be attributed to the rise in temperature during the experimental period. High-temperature conditions have been reported to be associated with a rapid increase in colony expansion, especially during their larval stages and the time to reproductive maturity [30]. Meanwhile, previous studies reported a rapid increase in aphid populations between 7 and 10 days after infestation [38], while a high density of aphids was observed 14 days after infestation across most accessions [14]. In the current study, the earlier observed increase in aphid population may have resulted from high temperatures during the growth period and the use of different cowpea accessions, which could have had varying genetic makeup. Due to the high aphid population and probably food shortages, winged aphid offspring increased after day 17, suggesting the migration of aphids to better food sources [37]. While increased aphid density promoted wing formation, the diminishing food sources could have contributed to environmentally induced dimorphism (polyphenism) among parthenogenetic females [39].

The phenotypic screening for aphid resistance showed that Tswana had fewer adult aphid and nymph infestations than all genotypes. The plants exhibited slight signs of damage, as indicated by a slight yellowing of the lower leaves without any capping. Similar results were observed in a previous study when Tswana was compared to Blackeye [29]. According to the aphid resistance scores, the less damaged plant vigour and the low aphid infestation in Tswana demonstrated a moderately resistant phenotype [40]. Resistance to aphids in Tswana could be attributed to the antibiosis mechanism, which involves the production of poisonous metabolites, possibly through suppressed reproductive rate and fecundity [41]. In addition to Tswana, landrace B261-B exhibited a resistant phenotype despite the high aphid density. This was in agreement with previous studies, which reported that seedlings from cowpea-resistant varieties grew vigorously and survived despite aphid infestation [42]. Therefore, resistance to a high aphid population might be attributed to antixenosis, possibly through the suppression of aphid feeding intensity [40] and the probable redistribution of nutrients from damaged tissues, thereby reducing the impact of aphid feeding on overall plant growth [43].

Consistent with the previous study [30], landraces B301 and B359 were tolerant to aphid damage, exhibiting yellowing of the lower leaves and slight capping. Tswana and B359 showed comparable nymphs and adult aphids from 23 to 27 days after aphid infestation, as previously reported [29]. However, despite a comparable aphid population, the severity of plant damage resulted in Tswana being moderately resistant to aphid plant damage, while B359 showed tolerance. The difference could be attributed to damage caused by an increased population of nymphs on day 17 after aphid infestation in B359, as compared to a consistent resistance level achieved through suppression of the extended pre-reproductive period and a reduction in population growth rate in Tswana [30].

Consistent with a previous study [30], B301 and B359 were tolerant to aphid damage, exhibiting yellowing of the lower leaves and slight capping. Additionally, ER7 and KVX295-2-124-99 were also resistant to aphid damage. Tswana and B359 showed comparable nymphs and adult aphids from 23 to 27 days after aphid infestation, as previously reported [29]. However, despite a comparable aphid population, the severity of plant damage resulted in Tswana being moderately resistant to aphid plant damage, while B359 showed tolerance. The difference could have been attributed to damage caused by an increased population of nymphs on day 17 after aphid infestation in B359, as compared to a consistent resistance level achieved through suppression of the extended pre-reproductive period and a reduction in population growth rate in Tswana [30]. Considering their differences in the level of resistance against aphids, these genotypes could be less favourable/susceptible to aphid infestation, probably through antibiosis and antixenosis compounds that can disturb the functioning, growth, and development of aphids [44,45]. The possible antibiosis-related response to aphid damage could be attributed to the suppression of reproduction, as reflected by the lower number of adults, which subsequently showed a low reproduction rate, resulting in low fecundity (fewer nymphs). On the other hand, the lower damage to plant vigour and low plant mortality rate might have been due to antixenosis, probably through prevention or the reduction of the duration of aphid's stylet penetration beyond the epidermis and mesophyll [46,47]. This could have contributed to the decrease in aphid-salivary-related virus transmission. Beyond the aphid feeding effect, the decreased reproductive rate and fecundity might have been attributed to induced intricate aphid–host interaction mechanisms, including aphid feeding-induced phytohormones-mediated plant defence against aphids, as reported in various species-specific aphids [48–50]. Another possible mechanism could involve the detection of aphid salivary effectors by R proteins, initiating a stronger effector-triggered immunity through the induction of transcriptional factors. The upregulation of transcriptional factors would have subsequently induced the phytohormones and secondary metabolites [51].

On the contrary, Blackeye, B013-F, B339, IT97K-556-6, SARC1-57-2, and SARI-2KTA-6 were susceptible to aphids in the current study. The aphid susceptibility of Blackeye was positively associated with high aphid reproduction and high fecundity, which depleted food sources as reflected by an increased number of alates [21]. In IT97K-556-6, the high population of aphids resulted in damage to the plants, contradicting previously reported tolerance to aphids, which had been linked to a 40% survival rate of seedlings up to 21 days after infestation, even in the presence of a large number of aphids [38]. This was inconsistent with the reported highest level of antibiosis, which was attributed to the slowing of aphid development and multiplication [52–54]. The prevalence of the sooty mold in B013-F, B339, IT97K-556-6, SARC1-57-2, and SARI-2KTA-6 might have substantially damaged the plant's vigour beyond aphid damage. The sooty mold might have interfered with photosynthesis, thereby suppressing the plant's ability to effectively regulate light reactions in chloroplasts and disrupting the retrograde signalling necessary for the conversion of proplastids to chloroplasts. As a result, this might have led to inadequate energy supply for downstream metabolism and overall plant growth [55]. Furthermore, the severe wilting of the plants suggested the transmission and rapid spread of the plant virus within the plant beyond the direct feeding effect. This could have contributed to symptoms such as yellowing, wilting, and stunted growth, which ultimately caused more harm to the plants than the direct feeding of the aphids alone [56]. Meanwhile, variations in resistance levels to aphids have been previously reported among different lines in various regions of West Africa [54]. For instance, the stability of aphid resistance in IT97K-499-35 was reported in Nigeria and Burkina Faso [54]; however, it was susceptible to aphids in Ghana [12]. The high aphid infestation and sooty mold may be linked to a more virulent biotype of aphids in Botswana. This biotype could be damaging crops in West African varieties that contain host-plant resistance genes.

After aphid screening, DNA markers associated with genes that confer resistance in IT97K-556-6 and SARI-2KTA-6 were used to validate the sources of resistance in Tswana, Blackeye, and ER7. This validation was performed to genetically eliminate any false resistance that could have arisen from environmental biases [26,27]. The SNP1_0912 marker, which flanks the significant aphid resistance QTL (QAc-vu7.1) in IT97K-556-6 and SARI-2KTA-6 [33,57], confirmed the monogenic homozygous significant aphid resistance QTL (QAc-vu7.1). This marker confirmed aphid resistance in Tswana and ER7. Despite the confirmed resistance in IT97K-556-6, a contrary aphid susceptibility phenotype was observed in this variety. This suggests that, although the presence of the QAc-vu7.1 marker region, as indicated by the SNP1_0912 marker, is confirmed, phenotypic variation may indicate the presence of functional gene loci or QTL loci linked to resistance in Tswana and ER7. Furthermore, this suggested the significant negative effect of the QTL-by-environment interaction effect on the expression of QAc-vu7.1. Meanwhile, the SNP1_0912 marker is located at 21.7 cM, whereas the QTL is at 22 cM on chromosome 2 of the cowpea genome. Despite its closeness to the QTL, this marker might have been ineffective in identifying genetic variation of the phenotypic trait among the cowpea genotypes due to its non-functionality [33,57]. Similar results were observed when using CP 171F/172R, which confers resistance in SARI-2KTA-6. This suggested that, beyond the probable negative effect of the QTL-by-environment interaction and the non-functionality of the SNP marker, the substantially low level of aphid population and high plant vigour in Tswana and ER7 were causative of a probable unknown functional aphid resistance marker region for the specific aphid biotype, and that marker might be substantially expressed under high and dry temperate growth conditions.

4. Materials and Methods

4.1. Experimental Site

The experiment was conducted in a screenhouse at the Botswana University of Agriculture and Natural Resources Gardens in Sebele during the 2023/2024 growing season. This site is located at a latitude of 24°33′ S and a longitude of 25°54′ E in Sebele, Gaborone, in the southern part of Botswana. The experiment period was between 1 February and March 2024. The minimum temperature recorded during this period was 16.1 °C, the maximum was 40.6 °C, and the average was 20 °C (Meteorological Services Botswana). The average rainfall for the experimental period was 3.8 mm (Meteorological Services Botswana).

4.2. Planting Materials and Experimental Design

A total of 12 genotypes were used in the current study. This includes the local landraces B013-F, B261-B, B301, B339, and B359, as well as the farmers' preferred commercial varieties, Tswana, Blackeye, and ER7, sourced from the Botswana National Plant Genetic Resources Centre. The landraces B261-B, B301, B339, B359, Blackeye, and Tswana were selected based on their performance against aphid damages in the previous studies [28–30]. This study also considered the use of West African aphid resistance releases, which are available through the Kirkhouse Trust network of the African Cowpea Breeding program. These include aphid-resistant breeding lines IT97K-556-6 and KVX295-2-124-99 from the International Institute of Tropical Agriculture (IITA), Ibadan, Nigeria. Furthermore, aphid-resistant breeding lines SARI-21KTA-6 (Zaa/566/SARC) and SARC 1-57-2 were sourced from the Council for Scientific and Industrial Research-Savannah Agriculture Research Institute (CSIR-SARI) in Ghana.

The experiment was laid out in a completely randomized design with three replications. The aphid-resistant varieties IT97K-556-6, KVX295-2-124-99, SARI-21KTA-6 (Zaa/566/SARC), and SARC 1-57-2 were used as controls. Two seeds were sown and planted in 2.5 L polyethylene planting bags filled with soil- and later thinned to one plant per potting bag. The chemical composition of the soil was as follows: organic carbon (3.07%), phosphorus (17.21 mg/kg), calcium (19.54 Cmol/kg), magnesium (7.50 Cmol/kg), potassium (2.08 Cmol/kg), and a pH of 6.43.

4.3. Aphid Culture

Aphids were collected from a cowpea field near Gaborone and cultured on the susceptible Blackeye in an insect-proof cage. Cowpea seedlings (seven days after emergence) were infested with fourth-instar nymphs (4-to-5-day-old aphids). Five of these apterygote aphids were collected with a camel hairbrush and carefully placed on each plant (Togola et al., 2020 [14,39]). Watering was conducted carefully to prevent aphids from being washed off the plants. The pots remained in the insect-proof cages for 27 days. At the end of the experiment, all the infested plants were sprayed with Dimethoate.

4.4. Aphid Infestation and Plant Damage Severity Score

The number of live nymphs, the number of live adults, and the number of alates were recorded at 2-day intervals for 27 days after infestation. Aphids were visually inspected on the leaves, petioles, and stems using a magnifying glass, while counting was performed with the aid of a digital counter (Figure A1). At the end of the experiment, visual aphid plant damage scoring was conducted using a scale of 1 to 5 [36]. Score 1 was for plants with no aphid and no aphid damage, labelled as highly resistant; 2 was for plants with aphids and slight signs of damage (slight yellowing of lower leaves without capping), labelled as resistant; 3 was for plants showing symptoms of aphids' damage (yellowing of lower leaves and slight capping), labelled as moderately resistant; 4 was for plants with weak stems and leaves with symptoms of aphid damage (severe capping of leaves, stunted plants, yellowing of all leaves), labelled as moderately susceptible; and score 5 indicated dead plants due to aphids' damage, categorized as highly susceptible [14,36,40].

4.5. Marker-Assisted Polymorphic Tests

Local farmers' preferred commercial cowpea varieties, including Tswana, Blackeye, and ER7, were further subjected to polymorphic tests. In addition to farmers' preference and the need for improvement of these varieties against biotic stresses, the selection was also based on the phenotypic response to aphids' damage. The phenotypic resistance and susceptibility of these varieties were validated using the West African aphid-released variety IT97K-556-6. The codominant SNP1_0912 marker (Table 4), which is a flanking marker of the major aphid resistance QTL (QAc-vu7.1) in IT97K-556-6 [33] was used to confirm resistant (165 bp) and susceptible plants (150 bp). In addition, polymorphic tests were conducted on Tswana and ER7 using SARI-2KTA-6 as the control. The SARI-2KTA-6 was developed by pyramiding aphid resistance from IT97K-556-6 and SARC 1-57-2 into a susceptible Zaayura [32]. Therefore, the simple sequence repeat (SSR) marker CP 171F/172R, which flanks aphid resistance in IT97K-556-6, SARC 1-57-2, and SARI-2KTA-6 breeding lines, was used for polymorphic tests.

Table 4. Primer sequences for co-dominant SNP1_0912 marker.

Primer Name	Primer Sequence (5′-3′)	Reference
CP 171F/172R	F: GTAGGGAGTTGCCCACGAATA R: CAACCGATGTAAAAAGTGGAC	[58]
SNP1_0912	1_0912R: ACTGTTGGATCCGATTGAGG 1_0912F1: ACCATACATTACATATAACTAACTTCTCGCCATGACT 1_0912F2: ATTAATAACTAACTTCTCGCCATGAGC	[57]

Total genomic DNA (gDNA) was extracted from six replicates of liquid nitrogen-grounded leaf tissue samples using Quick-DNATM Plant/Seed Miniprep Kit Cat No; D6020 (Zymo Research Corporation, Irvine, CA, USA) as per the manufacturer's instructions. The quality and concentration of the extracted gDNA were checked on Nanodrop (Nanodrop One, Thermofisher Scientific 5225 Verona Rd. Madison, WI, USA) and a 2.5% agarose gel consisting of 2.5 g agarose, 100 mL of ×1 TAE buffer and 5 μL of ethidium bromide. The polymerase chain (PCR) reaction (PCR) was performed in a Nexus Eppendorf Master Cycler (Eppendorf AG 700W 22331 Hamburg, Germany). The 25 μL PCR reaction consisted of the following: 1X Q5® High-Fidelity reaction buffer, 200 200 μM dNTPs, 0.5 μM of each Forward and reverse primer, 0.02 U/μL Q5 High-Fidelity DNA Polymerase, 1 μg DNA template, and nuclease-free water. Thermal cycling was followed by an initial denaturing at 98 °C for 30 s, denaturing at 98 °C for 10 s, and 35 cycles of annealing at 52 °C for 30 s. The extension was performed at 72 °C for 30 s, the final extension at 72 °C for 2 min, and storage was conducted at 4 °C. The amplicons were separated by agarose gel (2.5% *w*/*v*) electrophoresis in 1X TAE (Tris base-acetic acid-EDTA) buffer at 80 V for 60 min. Gels were stained with ethidium bromide (0.002%, *v*/*v*). A 1 kilobase DNA molecular weight marker (Fast DNA ladder (New England Biolabs Inc., Ipswich, MA, USA) was used. The gels were visualized in a Molecular imager Bio-Rad Gel DocTM XR + System (Biorad Laboratories, Hercules, CA, USA).

4.6. Data Statistical Analysis

The data were subjected to one-way analysis of variance (ANOVA) using R software version 4.2.2, using the Agricolae package version 1.3-5. The Fisher's least significant difference (LSD) test was used to separate the means at a significance level of 5%.

5. Conclusions

In this study, we evaluated cowpea genotypes for aphid resistance under screehouse conditions in an insect-proof cage. Surprisingly, IT97K-556-6, SARC1-57-2, and SARI-2KTA-6 were susceptible to aphid damage. Their aphid damage susceptibility could be attributed to several factors, including the negative effect of QTL-by-environment interaction and the possible existence of different aphid biotypes. Tswana exhibited resistance to aphids, probably through the antifeedant property that suppressed the reproductive rate and multiplication of aphids, ultimately leading to their death. However, future studies should consider various mechanisms linked to antixenosis, such as the salivary-related transmission of the virus beyond the epidermis and mesophyll, aphid feeding-induced phytohormones-mediated plant defence, and aphid salivary effector-triggered immunity. The resistance and tolerance to aphids observed in Tswana, B261-B, B301, and B359 provide a basis for marker-assisted breeding in developing genotypes resistant to aphids specific to Botswana's cowpea aphid biotypes. Tswana might be used in breeding programs involving biparental lines between Tswana as the resistant donor and Blackeye as the susceptible recurrent line. Sequencing the biparental F2 progeny will be crucial in identifying novel functional markers that flank resistance to aphids. Furthermore, these findings provide an important basis for exploring the metabolic and molecular mechanisms to enhance understanding of the function of secondary metabolites and effector-triggered immunity.

Author Contributions: Conceptualization, M.P.-T. and L.L.; methodology, M.P.-T., L.L., B.T., A.B.T., S.N. and T.M.; Investigation and formal analysis, M.P.-T., L.L., B.T., A.B.T., M.M., G.G.G. and M.N.S.; writing—original draft preparation, M.M., G.G.G. and M.N.S.; writing—review and editing, M.P.-T. and L.L.; funding acquisition, M.P.-T. and L.L. All authors have read and agreed to the published version of the manuscript.

Funding: This research was funded by the Kirhouse Scio Trust, the African Cowpea Programme (ACP).

Institutional Review Board Statement: Not applicable.

Informed Consent Statement: Not applicable.

Data Availability Statement: The original contributions presented in the study are included in this article, and further inquiries can be directed to the corresponding authors.

Acknowledgments: We thank the Botswana University of Agriculture and Natural Resources (BUAN) and the National Agricultural Research and Development Institute (NARDI) for providing the research infrastructure. The technical support of the Kirkhouse Trust consultants is also highly acknowledged.

Conflicts of Interest: The authors declare no conflicts of interest.

Appendix A

Figure A1. Aphid visualisation and population counting using a magnifying glass and a digital counter, respectively.

References

1. Agbogidi, O.M. Response of six cultivars of cowpea (*Vigna unguiculata* (L.) Walp.) to spent engine oil. *Afr. J. Food Sci. Technol.* **2010**, *1*, 139–142.
2. Statistics Botswana. *Annual Agriculture Survey Report 2013*; Statistics Botswana: Gaborone, Botswana, 2015; ISBN 9789996842856.
3. Langyintuo, A.S.; Lowenberg-DeBoer, J.; Faye, M.; Lambert, D.; Ibro, G.; Moussa, B.; Kergna, A.; Kushwaha, S.; Musa, S.; Ntoukam, G. Cowpea supply and demand in West and Central Africa. *Field Crops Res.* **2003**, *82*, 215–231.
4. Abebe, B.K.; Alemayehu, M.T. A Review of the nutritional use of cowpea (*Vigna unguiculata* L. Walp.) for human and animal diets. *J. Agric. Food Res.* **2022**, *10*, 100383. [CrossRef]
5. Dube, E.; Fanadzo, M. Maximising yield benefits from dual-purpose cowpea. *Food Secur.* **2013**, *5*, 769–779. [CrossRef]
6. Kyei-Boahen, S.; Savala, E.N.; Chikoye, D.; Abaidoo, R. Growth and yield responses of cowpea to inoculation and phosphorus fertilization in different environments. *Front. Plant Sci.* **2017**, *8*, 646. [CrossRef]
7. Ayalew, T.; Yoseph, T. Cowpea (*Vigna unguiculata* L. Walp.): A choice crop for sustainability during the climate change periods. *J. Appl. Biol. Biotechnol.* **2022**, *10*, 54–162. [CrossRef]
8. Horn, L.N. Breeding Cowpea (*Vigna unguiculata* [L.] Walp.) for Improved Yield and Related Traits Using Gamma Irradiation. Ph.D. Thesis, University of KwaZulu-Natal, Pietermaritzburg, South Africa, 2016.
9. Motlhaodi, T. *NPGRC Brief*; Department of Agricultural Research: Gaborone, Botswana, 2020; Volume 21.
10. Mulila Mitti, J.; Mba, C.; Lupupa, T. Southern African countries develop the national strategy for plant genetic resources for food and agriculture. *Acta Hortic.* **2020**, *1267*, 38. [CrossRef]
11. Raina, A.; Laskar, R.A.; Tantray, Y.R.; Khursheed, S.; Wani, M.R.; Khan, S. Characterization of induced high yielding cowpea mutant lines using physiological, biochemical and molecular markers. *Sci. Rep.* **2020**, *10*, 3687. [CrossRef]
12. Jackai, L.E.N.; Daoust, R.A. Insect pests of cowpeas. *Annu. Rev. Entomol.* **1986**, *31*, 95–119. [CrossRef]

13. Obopile, M. Economic threshold and injury levels for control of cowpea aphid, *Aphis crassivora* Linnaeus (Homoptera: Aphididae), on cowpea. *Afr. Plant Prot.* **2006**, *12*, 111–115.
14. Togola, A.; Boukar, O.; Servent, A.; Chamarthi, S.; Tamò, M.; Fatokun, C. Identification of sources of resistance in cowpea mini core accessions to *Aphis craccivora* Koch (Homoptera: Aphididae) and their biochemical characterization. *Euphytica* **2020**, *216*, 88. [CrossRef] [PubMed]
15. Boukar, O.; Togola, A.; Chamarthi, S.; Belko, N.; Ishikawa, H.; Suzuki, K.; Fatokun, C. Cowpea [*Vigna unguiculata* (L.) Walp.] Breeding. In *Advances in Plant Breeding Strategies: Legumes*; Springer International Publishing: Cham, Switzerland, 2019; Volume 7, pp. 201–243, ISBN 9783030234003.
16. Mweke, A.; Akutse, K.S.; Ulrichs, C.; Fiaboe, K.K.M.; Maniania, N.K.; Ekesi, S. Integrated management of aphis craccivora in cowpea using intercropping and entomopathogenic fungi under field conditions. *J. Fungi* **2020**, *6*, 60. [CrossRef] [PubMed]
17. Blackman, R.L.; Eastop, V.F. Taxonomic Issues. In *Aphids as Crop Pests*; CABI: Wallingford, UK, 2007.
18. Ibrahim, H.; Salam, A.; Abdel-Mogib, M.; El-nagar, M.; Salem, H.; Nada, M. Survey of entomopathogenic fungi naturally infecting cowpea aphid, *Aphis craccivora*. Koch. *J. Plant Prot. Pathol.* **2011**, *2*, 1063–1070. [CrossRef]
19. Jayasinghe, W.H.; Akhter, M.S.; Nakahara, K.; Maruthi, M.N. Effect of aphid biology and morphology on plant virus transmission. *Pest Manag. Sci.* **2022**, *78*, 416–427. [CrossRef]
20. Singh, R.; Singh, G. Aphids. In *Polyphagous Pests of Crops*; Springer: Singapore, 2021; pp. 105–182, ISBN 9789811580758.
21. Zhang, R.J.; Chen, J.; Jiang, L.Y.; Qiao, G.X. The genes expression difference between winged and wingless bird cherry-oat *Aphid rhopalosiphum* Padi based on transcriptomic data. *Sci. Rep.* **2019**, *9*, 4754. [CrossRef]
22. Ofuya, T.I. Control of the cowpea aphid, *Aphis craccivora* Koch (Homoptera: Aphididae), in cowpea, *Vigna unguiculata* (L.) Walp. *Integr. Pest Manag. Rev.* **1997**, *2*, 199–207. [CrossRef]
23. Stout, M.J.; Bernaola, L.; Acevedo, F. Recent history and future trends in host-plant resistance. *Ann. Entomol. Soc. Am.* **2024**, *117*, 139–149. [CrossRef]
24. Togola, A.; Boukar, O.; Belko, N.; Chamarthi, S.K.; Fatokun, C.; Tamo, M.; Oigiangbe, N. Host plant resistance to insect pests of cowpea (*Vigna unguiculata* L. Walp.): Achievements and future prospects. *Euphytica* **2017**, *213*, 239. [CrossRef]
25. Dogimont, C.; Bendahmane, A.; Chovelon, V.; Boissot, N. Host plant resistance to aphids in cultivated crops: Genetic and molecular bases, and interactions with aphid populations. *Comptes Rendus Biol.* **2010**, *333*, 566–573. [CrossRef]
26. Hittalmani, S.; Parco, A.; Mew, T.V.; Zeigler, R.S.; Huang, N. Fine mapping and dna marker-assisted pyramiding of the three major genes for blast resistance in rice. *Theor. Appl. Genet.* **2000**, *100*, 1121–1128. [CrossRef]
27. Sarkar, S.; Ghosh, S.; Chatterjee, M.; Das, P.; Lahari, T.; Maji, A.; Mondal, N.; Pradhan, K.K.; Bhattacharyya, S. Molecular markers linked with bruchid resistance in *Vigna radiata* Var. Sublobata and their validation. *J. Plant Biochem. Biotechnol.* **2011**, *20*, 155–160. [CrossRef]
28. Obopile, M.; Ositile, B. Life table and population parameters of cowpea aphid, *Aphis craccivora* Koch (Homoptera: Aphididae) on five cowpea *Vigna unguiculata* (L. Walp.) varieties. *J. Pest Sci.* **2010**, *83*, 9–14. [CrossRef]
29. Munthali, D.C.; Lanamile, K. Evaluation of relative resistance of four cowpea (*Vigna uniguiculata* (L.) Walp.) varieties to the black eye aphid (*Aphis craccivora*) in Southeastern Botswana. *Botsw. J. Agric. Appl. Sci.* **2011**, *7*.
30. Machacha, M.; Obopile, M.; Tshegofatso, A.B.N.; Tiroesele, B.; Gwafila, C.; Ramokapane, M. Demographic parameters of cowpea aphid *Aphis craccivora* (Homoptera: Aphididae) on different botswana cowpea landraces. *Int. J. Trop. Insect Sci.* **2012**, *32*, 189–193. [CrossRef]
31. Kusi, F.; Obeng-Ofori, D.; Asante, S.; Padi, F. New sources of resistance in cowpea to the cowpea aphid (*Aphis craccivora* Koch) (Homoptera: Aphididae). *J. Ghana Sci. Assoc.* **2010**, *12*, 95–104. [CrossRef]
32. Lattanzio, V.; Arpaia, S.; Cardinali, A.; Di Venere, D.; Linsalata, V. Role of endogenous flavonoids in resistance mechanism of vigna role of to aphids. *J. Agric. Food Chem.* **2000**, *48*, 5316–5320. [CrossRef] [PubMed]
33. Huynh, B.L.; Ehlers, J.D.; Ndeve, A.; Wanamaker, S.; Lucas, M.R.; Close, T.J.; Roberts, P.A. Genetic mapping and legume synteny of aphid resistance in African cowpea (*Vigna unguiculata* L. Walp.) grown in California. *Mol. Breed.* **2015**, *35*, 36. [CrossRef] [PubMed]
34. Kusi, F.; Padi, F.K.; Obeng-Ofori, D.; Asante, S.K.; Agyare, R.Y.; Sugri, I.; Timko, M.P.; Koebner, R.; Huynh, B.L.; Santos, J.R.P.; et al. A novel aphid resistance locus in cowpea identified by combining SSR and SNP markers. *Plant Breed.* **2018**, *137*, 203–209. [CrossRef]
35. Ouedraogo, A.P.; Danquah, A.; Tignegre, J.B.; Paoda, L.S.; Batieno, J.B.; Asante, I.K.; Ouedraogo, J.T.; Ayertey, J.N.; Ofori, K. Determination of inheritance of aphid resistance in cowpea genotypes and identification of single sequence repeat markers linked to resistance genes. *Legume Sci.* **2022**, *4*, e127. [CrossRef]
36. Seram, D.; Devi, Y.K. Insect resistance screening techniques and scoring for important horticultural crops. *Plant Cell Biotechnol. Mol. Biol.* **2021**, *22*, 155–162.
37. Schreiner, I. Cowpea Aphid (*Aphis craccivora* Koch). In *Agricultural Pests of the Pacific Cowpea Aphid (Aphis craccivora Koch) 2000*; Agricultural Development in the American Pacific (AGAP): Washington, DC, USA, 2020; ISBN 1-931435-09-X.

38. Souleymane, A.; Ova, M.E.A.; Fatokun, C.A.; Alabi, O.Y. Screening for resistance to cowpea aphid (*Aphis craccivora* Koch) in wild and cultivated cowpea (*Vigna unguiculata* L. Walp.) accessions. *Int. J. Sci. Environ.* **2013**, *2*, 611–621.
39. Braendle, C.; Davis, G.K.; Brisson, J.A.; Stern, D.L. Wing dimorphism in aphids. *Heredity* **2006**, *97*, 192–199. [CrossRef] [PubMed]
40. Kityo, R.; Okdoi, J.B.; Ozimati, A.; Dramadri, I.O.; Agaba, R.; Ongom, P.O.; Nampala, P.; Edema, R.; Karungi, J.; Gibson, P.; et al. New sources and stability of resistance to aphids in cowpea. *Afr. Crop Sci. J.* **2021**, *29*, 209–228. [CrossRef]
41. Silva-Sanzana, C.; Zavala, D.; Moraga, F.; Herrera-Vásquez, A.; Blanco-Herrera, F. Oligogalacturonides enhance resistance against aphids through pattern-triggered immunity and activation of salicylic acid signaling. *Int. J. Mol. Sci.* **2022**, *23*, 9753. [CrossRef]
42. Kusi, F.; Nboyine, J.A.; Attamah, P.; Awuku, J.F.; Sugri, I.; Zakaria, M.; Lamini, S.; Mensah, G.; Larweh, V.; Owusu, R.K.; et al. Stability of sources of resistance to cowpea aphid (*Aphis craccivora* Koch, Hemiptera: Aphididae) across major cowpea production zones in Ghana. *Hindawi Int. J. Agron.* **2020**, *2020*, 8869334. [CrossRef]
43. Bala, K.; Sood, A.; Singh Pathania, V.; Thakur, S. Effect of plant nutrition in insect pest management: A Review. *J. Pharmacogn. Phytochem.* **2018**, *7*, 2737–2742.
44. Bansal, R.; Mian, M.A.R.; Michel, A. Characterizing Resistance to soybean aphid (Hemiptera: Aphididae): Antibiosis and antixenosis assessment. *J. Econ. Entomol.* **2021**, *114*, 1329–1335. [CrossRef] [PubMed]
45. MacWilliams, J.R.; Chesnais, Q.; Nabity, P.; Mauck, K.; Kaloshian, I. Cowpea aphid resistance in cowpea line CB77 functions primarily through antibiosis and eliminates phytotoxic symptoms of aphid feeding. *J. Pest Sci.* **2023**, *96*, 539–553. [CrossRef]
46. Martin, B.; Collar, J.L.; Tjallingii, W.F.; Fereres, A. Intracellular ingestion and salivation by aphids may cause the acquisition and inoculation of non-persistently transmitted plant viruses. *J. Gen. Virol.* **1997**, *78*, 2701–2705. [CrossRef]
47. Stec, K.; Kordan, B.; Sergiel, I.; Biesaga, M.; Mroczek, J.; Bocianowski, J.; Gabryś, B. Antixenosis in *Glycine max* (L.) Merr against *Acyrthosiphon pisum* (Harris). *Sci. Rep.* **2021**, *11*, 15289. [CrossRef]
48. Hu, Z.; Zhong, X.; Zhang, H.; Luo, X.; Wang, Y.; Wang, Y.; Liu, T.; Zhang, Y.; Wang, X.; An, H.; et al. *GhMYB18* confers *Aphis Gossypii* Glover resistance through regulating the synthesis of salicylic acid and flavonoids in cotton plants. *Plant Cell Rep.* **2023**, *42*, 355–369. [CrossRef] [PubMed]
49. Koch, K.G.; Palmer, N.A.; Donze-Reiner, T.; Scully, E.D.; Seravalli, J.; Amundsen, K.; Twigg, P.; Louis, J.; Bradshaw, J.D.; Heng-Moss, T.M.; et al. Aphid-responsive defense networks in hybrid switchgrass. *Front. Plant Sci.* **2020**, *11*, 1145. [CrossRef]
50. Feng, J.L.; Zhang, J.; Yang, J.; Zou, L.P.; Fang, T.T.; Xu, H.L.; Cai, Q.N. Exogenous salicylic acid improves resistance of aphid-susceptible wheat to the grain aphid, *Sitobion avenae* (F.) (Hemiptera: Aphididae). *Bull. Entomol. Res.* **2021**, *111*, 544–552. [CrossRef] [PubMed]
51. Kumaraswamy, S.; Huang, Y. Molecular interactions between plants and aphids: Recent advances and future perspectives. *Insects* **2024**, *15*, 935. [CrossRef]
52. Sombie, P.A.E.D.; Ouedraogo, I.; Tignegré, J.B.D.L.S.; Hilou, A.; Ouedraogo, T.J.; Kiendrébéogo, M. Genotypic variation of mineral elements and phytate levels of thirty cowpeas (*Vigna unguiculata* L. Walp.) varieties cultivated in Burkina Faso. *J. Food Chem. Nutr.* **2018**, *6*, 13–20. [CrossRef]
53. Omoigui, L.O.; Ugba, S.M.; Bello, L.L.; Timlo, M.P.; Gowda, B.S.; Motagi, B.N. SSR markers linked with *Alectra vogelii* resistance in cowpea *Vigna unguiculata* (L.) Walp] Omoigui. In Proceedings of the 5th International Conference on Next Generation Genomics & Integreted Breeding Crop Improvement, Patancheru, India, 18–20 February 2015; p. 158.
54. Ouédraogo, A.P.; Batieno, B.J.; Traore, F.; Tignegre, J.-B.; Huynh, B.; Roberts, P.A.; Close, T.; Ouédraogo, J.T. Screening of cowpea (*Vigna unguiculata* (L.) Walp.) lines for resistance to three aphids (*Aphis craccivora* Koch) strains in Burkina Faso. *Afr. J. Agric. Res.* **2018**, *13*, 1487–1495. [CrossRef]
55. Pholo-Tait, M.; Le Maitre, N.C.; Lloyd, J.R.; Kossmann, J.; Hills, P.N.; Schulze, W.X.; Alseekh, S.; Valentine, A.J. Proteome profiling of lumichrome-treated arabidopsis thaliana suggests that various regulatory mechanisms mediate enhanced photosynthesis and plant growth. *S. Afr. J. Bot.* **2024**, *172*, 515–528. [CrossRef]
56. Yates-Stewart, A.D.; Daron, J.; Wijeratne, S.; Shahid, S.; Edgington, H.A.; Slotkin, R.K.; Michel, A. Soybean aphids adapted to host-plant resistance by down regulating putative effectors and up regulating transposable elements. *Insect Biochem. Mol. Biol.* **2020**, *121*, 103363. [CrossRef]
57. Attamah, P.; Kusi, F.; Kena, A.W.; Awuku, F.J.; Lamini, S.; Mensah, G.; Zackaria, M.; Owusu, E.Y.; Akromah, R. Pyramiding aphid resistance genes into the elite cowpea variety, Zaayura, using marker-assisted backcrossing. *Heliyon* **2024**, *10*, e31976. [CrossRef]
58. Omoigui, L.O.; Ekeuro, G.C.; Kamara, A.Y.; Bello, L.L.; Timko, M.P.; Ogunwolu, G.O. New sources of aphids [*Aphis craccivora* (Koch)] resistance in cowpea germplasm using phenotypic and molecular marker approaches. *Euphytica* **2017**, *213*, 178. [CrossRef]

International Journal of *Molecular Sciences*

MDPI

Article

Investigating Morphological and Physiological Responses to Stress in *Begonia semperflorens*

Julian Ginori [1,2], Chi D. Nguyen [1,2,3], Sandra Wilson [4], Zhanao Deng [5] and Heqiang Huo [1,2,3,*]

1 Mid-Florida Research and Education Center, IFAS-University of Florida, Apopka, FL 32703, USA; jginori@gmail.com (J.G.); chi285@ufl.edu (C.D.N.)
2 Department of Horticulture, IFAS-University of Florida, Gainesville, FL 32611, USA
3 Crop Transformation Center, Horticulture Science Department, University of Florida, Gainesville, FL 32611, USA
4 Department of Environmental Horticulture, IFAS-University of Florida, Gainesville, FL 32611, USA; sbwilson@ufl.edu
5 Department of Environmental Horticulture, Gulf Coast Research and Education Center, IFAS-University of Florida, Wimauma, FL 33598, USA; zdeng@ufl.edu
* Correspondence: hhuo@ufl.edu

Abstract: *Begonia semperflorens*, or wax begonias, are popular ornamental plants often challenged by heat and high light stress in subtropical and tropical regions. This study examined the responses of two stress-tolerant genotypes (FB08-059 and OPGC 5104) and two stress-susceptible genotypes (Cocktail Vodka and Sprint White) under elevated temperature and light conditions. The results revealed significant genotype-dependent differences in stress responses. Stress-tolerant genotypes demonstrated distinct adaptive traits, including thicker cuticles, acute leaf folding, and elevated anthocyanin accumulation, which collectively contributed to enhanced photoprotection and mitigation of cellular damage. Notably, FB08-059 exhibited the most robust stress-adaptive responses, characterized by a 25.83% increase in cuticle thickness and a threefold increase in anthocyanin content under stress conditions. These adaptations effectively limited ROS accumulation and maintained higher Fv/Fm values, thereby sustaining photosynthetic efficiency relative to the other genotypes. In contrast, stress-susceptible genotypes exhibited increased ion leakage, reduced chlorophyll content, and impaired gas exchange rates, reflecting greater oxidative stress and cellular damage. These findings highlight cuticle thickness, anthocyanin accumulation, and leaf folding as key indicators of heat and light stress resilience. This research provides critical insights for breeding programs focused on improving the resilience of wax begonias, supporting the development of heat- and light-tolerant cultivars for sustainable production in stress-prone environments.

Keywords: *Begonia semperflorens*; abiotic stress; photoprotection; oxidative stress; stomatal regulation; cuticle thickness; leaf morphology; high light stress; high temperature stress

1. Introduction

The genus *Begonia* comprises over 1800 species, positioning it among the most taxonomically and ecologically diverse groups in the plant kingdom. This diversity is exemplified by a vast spectrum of morphological, ecological, and adaptive traits, predominantly found within tropical and subtropical ecosystems [1–3]. The evolutionary plasticity and adaptability of Begonia continue to draw significant attention from botanists, ecologists, and horticulturists. Ongoing discoveries of new species and hybrids underscore its dynamic evolutionary trajectory and emphasize its substantial horticultural and ecological potential [4]. Among the members of this genus, *Begonia semperflorens-colturum*, or wax begonia, holds considerable economic and esthetic value as one of the most widely cultivated bedding plants [5–7]. Their vibrant foliage, continuous flowering habits, and resilience to variable urban environments make them valuable in landscaping for roadsides, walkways, and garden beds. In the United States alone, the economic contribution of wax begonias surpassed USD 130 million in 2019, with Florida recognized as a central hub for their production and commercial distribution [8].

Despite their widespread cultivation and popularity, wax begonia exhibits vulnerability to abiotic stresses, particularly elevated temperatures and intense light exposure. These stressors are becoming increasingly prevalent due to global climate change, which poses significant challenges to sustainable horticultural practices. Current climate models project a global temperature increase of 1.5–4.0 °C by the end of the century, alongside a higher frequency of extreme heat events [9]. These environmental changes exacerbate the susceptibility of wax begonias to stress-induced morphological and physiological impairments, including leaf necrosis, reduced vigor, photoinhibition, and a decline in ornamental quality [10,11]. These challenges are particularly pronounced in subtropical and tropical regions, where heat and highlight intensity are pervasive, posing a considerable threat to the sustainable cultivation of wax begonias. Plants have evolved a range of adaptive strategies to mitigate and tolerate abiotic stresses, including both morphological and physiological modifications. At the morphological level, plants respond by thickening the cuticle, which reduces transpirational water loss and provides a protective barrier against excessive radiation. Additionally, leaf folding can serve as an adaptive mechanism to reduce direct light interception and minimize photodamage [12,13]. At the physiological level, both high light intensity and heat stress impose significant constraints on photosynthetic efficiency. Excessive light exposure induces photoinhibition, reducing the efficiency of photosystem II (PSII) and accelerating chlorophyll degradation, which impairs light-harvesting capacity and limits carbon assimilation [14,15]. Simultaneously, elevated temperatures destabilize chlorophyll-protein complexes and promote the breakdown of photosynthetic pigments, further diminishing the plant's ability to capture and utilize light energy [11]. To counteract these stressors, plants dynamically adjust their chlorophyll and carotenoid composition to optimize light absorption and dissipation, thereby mitigating photodamage and sustaining photosynthetic performance [16,17]. Additionally, the biosynthesis of anthocyanins and other flavonoids serves as a critical photoprotective mechanism, functioning as light filters to prevent excessive radiation from reaching chloroplasts while scavenging reactive oxygen species (ROS) [18–20].

The interplay between high light and heat stress also disrupts cellular homeostasis, triggering the overproduction of ROS, including superoxide radicals (O_2^-) and hydrogen peroxide (H_2O_2) [21–24]. When ROS accumulation surpasses cellular detoxification capacity, oxidative stress ensues, leading to membrane lipid peroxidation, protein denaturation, and organellar dysfunction, particularly in chloroplasts and mitochondria [21,25]. To counterbalance these deleterious effects, plants upregulate antioxidant defense systems, including enzymatic components such as superoxide dismutase (SOD), catalase (CAT), and peroxidases (POD), which neutralize ROS and maintain cellular redox homeostasis [22,23,26]. Moreover, heat and light stress influence carbon metabolism by altering carbohydrate partitioning and accumulation, with stressed plants often accumulating soluble sugars such as glucose, sucrose, and fructose as osmoprotectants to stabilize cellular structures and sustain metabolic balance [27]. However, prolonged exposure to these stressors disrupts sugar homeostasis, ultimately leading to impaired growth, diminished energy reserves, and compromised reproductive success [10,28,29]. Although these mechanisms have been extensively studied in major agricultural crops such as maize, wheat, and rice, they remain underexplored in ornamental species like wax begonia, creating a critical gap in our understanding of stress adaptation in these plants.

This study seeks to address this knowledge gap by systematically investigating the morphological and physiological responses of four distinct wax begonia genotypes under conditions of elevated temperature and high light intensity. Two experimental genotypes, FB08-059 and OPGC 5104, were selected based on field observations indicating superior stress tolerance. These were compared with two commercially popular cultivars, 'Cocktail Vodka' and 'Sprint White', thought to be more susceptible to abiotic stress. The research methodology involved a comprehensive assessment of key stress response parameters, including cuticle thickness, leaf folding angles, pigment composition (anthocyanin and chlorophyll content), ROS accumulation, photosynthetic efficiency (Fv/Fm), and gas exchange rates. By integrating these parameters, this study provides a holistic understanding of the interplay between structural, chemical, and physiological traits in mediating stress adaptation. The findings of this research are anticipated to yield significant insights into the genotype-specific mechanisms of stress tolerance in wax begonia. By elucidating these adaptive responses, this study aims to inform breeding programs focused on improving stress resilience, ensuring the sustainable cultivation of wax begonia in increasingly adverse climatic conditions. Moreover, this work contributes to the broader scientific discourse on abiotic stress responses in ornamental plants, offering knowledge that can drive the development of innovative breeding strategies.

2. Results

2.1. *Leaf Thickness, Fresh and Dry Weight of Four Wax Begonia Genotypes Under Shaded and Non-Shaded Conditions*

Table 1 depicts the thickness of leaves, which increased significantly when comparing the main effects of shaded to non-shaded leaves across all genotypes ($p = 2.053 \times 10^{-7}$). However, a two-way ANOVA revealed no significant interaction between genotype and condition ($p = 0.26$), indicating that the increase in total leaf thickness was consistent among the four genotypes in response to the stress condition. Despite this, FB08-059 and OPGC 5104 showed a greater increase in total leaf thickness (12.79% and 14.91%, respectively) compared to 'Cocktail Vodka' and 'Sprint White' (7.33% and 9.64%, respectively). To further understand the variation in total leaf thickness, mesophyll and cuticle layers were examined. Mesophyll thickness varied significantly by genotype ($p = 1.955 \times 10^{-7}$), but no significant effect of condition was observed ($p = 0.3649$). In contrast, a significant genotype–condition interaction was detected for cuticle thickness ($p = 0.00078$), suggesting that genotypes responded differently to elevated temperature and light under non-shaded conditions. FB08-059 exhibited the highest increase in cuticle thickness (25.83%), followed by OPGC 5104 (20.04%), while 'Cocktail Vodka' and 'Sprint White' showed lower increases (15.74% and 16.03%, respectively).

FB08-059 had the most notable cuticle contribution to total leaf thickness, accounting for 53.8% and 61.1% under shaded and non-shaded conditions, respectively. This was significantly higher than 'Cocktail Vodka' and 'Sprint White', neither of which exceeded a 46.3% cuticle-to-total leaf thickness ratio in either condition (Table 1). For OPGC 5104, the cuticle contributed 50.2% and 47.6% under non-shaded and shaded conditions, respectively. Comparisons of the tolerant genotypes (FB08-059 and OPGC 5104) to the susceptible genotypes ('Cocktail Vodka' and 'Sprint White') in the non-shaded condition revealed significant differences in both total leaf thickness ($p = 1.525 \times 10^{-10}$) and cuticle thickness ($p = 9.081 \times 10^{-12}$). These results suggest that increased cuticle thickness, contributing to total leaf thickness, may be an adaptive response to heat and light stress in the tolerant genotypes.

Table 1. Morphological and Physiological Parameters of Four Begonia semperflorens Genotypes Under Shaded and Non-Shaded Conditions.

	FB08-059		"Cocktail Vodka"		OPGC 5104		"Sprint White"	
	Shaded	Non-Shaded	Shaded	Non-Shaded	Shaded	Non-Shaded	Shaded	Non-Shaded
Total Leaf Thickness	5.35 ± 0.09^{b}	6.04 ± 0.08^{a}	4.36 ± 0.20^{d}	4.68 ± 0.12^{cd}	4.59 ± 0.09^{d}	5.22 ± 0.10^{bc}	3.15 ± 0.15^{e}	3.46 ± 0.08^{e}
Cuticle Thickness	2.89 ± 0.06^{b}	3.63 ± 0.10^{a}	1.86 ± 0.09^{d}	2.16 ± 0.05^{cd}	2.18 ± 0.06^{c}	2.61 ± 0.06^{b}	1.3 ± 0.06^{e}	1.51 ± 0.05^{e}
Mesophyll Thickness	2.47 ± 0.05	2.41 ± 0.11	2.50 ± 0.13	2.53 ± 0.09	2.41 ± 0.07	2.6 ± 0.11	1.85 ± 0.11	1.95 ± 0.08
Cuticle/Total Leaf Ratio	0.54 ± 0.01^{b}	0.61 ± 0.02^{a}	0.43 ± 0.01^{de}	0.46 ± 0.01^{cde}	0.48 ± 0.01^{cd}	0.50 ± 0.01^{bc}	0.42 ± 0.01^{e}	0.43 ± 0.02^{de}
Shoot FW (g)	356.87 ± 20.85^{a}	177.09 ± 16.60^{cd}	180.53 ± 9.55^{cd}	109.91 ± 9.74^{e}	284.11 ± 11.78^{b}	135.61 ± 7.09^{cde}	194.68 ± 15.82^{c}	119.08 ± 13.95^{de}
Shoot DW (g)	64.31 ± 1.13^{a}	33.86 ± 1.96^{c}	52.14 ± 0.49^{b}	25.07 ± 1.39^{d}	61.02 ± 0.92^{a}	31.48 ± 2.01^{c}	53.71 ± 0.57^{b}	28.70 ± 1.30^{cd}
Water Weight (g)	292.56 ± 19.77^{a}	143.23 ± 15.29^{c}	128.39 ± 9.40^{c}	84.84 ± 10.20^{c}	223.1 ± 10.99^{b}	104.13 ± 5.98^{c}	140.97 ± 15.37^{c}	90.38 ± 13.41^{c}
Root DW (g)	3.98 ± 0.43	1.83 ± 0.19	2.50 ± 0.49	1.08 ± 0.13	3.23 ± 0.34	1.56 ± 0.10	2.8 ± 0.27	1.34 ± 0.14

The table presents the mean thickness (μm, n = 12) of the cuticle and total leaf, along with the cuticle-to-leaf thickness ratio, for four genotypes under shaded and non-shaded conditions. Additional parameters include shoot fresh weight (Shoot FW), shoot dry weight (Shoot DW), root dry weight (Root DW), and water weight. Means within rows followed by different letters are significantly different according to Tukey's HSD test ($p \leq 0.05$).

All plants survived the 41-day experiment, allowing for comprehensive measurements of morphological and physiological parameters. The combined effects of heat and direct sunlight were evident in shoot and root weights (Table 1). All genotypes exhibited a consistent decrease in shoot and root dry weights with no significant genotype–condition interaction ($p = 0.17$ for shoot dry weight; $p = 0.62$ for root dry weight). Under shaded conditions, FB08-059 had the highest shoot fresh weight (357 g) and dry weight (64 g) per plant, followed by OPGC 5104 (284 g and 61 g, respectively). The decrease in water mass under non-shaded conditions was approximately 50% for FB08-059 and OPGC 5104, compared to a 35% decrease for 'Sprint White' and 'Cocktail Vodka'. Despite the higher water loss, FB08-059 maintained substantial fresh (177.09 g) and dry weights (33.86 g) under non-shaded conditions by the end of the study (Table 1). A significant genotype × condition interaction was observed for fresh weight ($p = 0.001$), indicating distinct responses to high light intensity and elevated temperature among genotypes.

2.2. *Leaf Folding and Stomatal Density of Four Wax Begonia Genotypes Under Shaded and Non-Shaded Conditions*

Leaf folding refers to the angle formed by three points: the leaf margin, the petiole-leaf blade junction, and the opposite leaf margin. All four genotypes exhibited leaf folding in response to the direct sunlight and elevated temperature conditions (non-shaded), but the degree of folding varied significantly among them (Figure 1A). A significant genotype–condition interaction effect was observed for leaf folding, indicating that the genotypes responded differently to the shaded vs. non-shaded conditions. Under non-shaded conditions or high-intensity light and elevated temperature, FB08-059 displayed the most pronounced leaf folding, with the leaf angle decreasing to 43.59°, representing a 62.30% change from the original angle (Figure 1A,B). Similarly, 'Cocktail Vodka' and OPGC 5104 also showed significant folding under non-shaded conditions, with leaf angle reductions of 49.79% (from 124.31° to 62.41°) and 52.91% (from 117.71° to 55.44°), respectively. In contrast, the 'Sprint White' genotype exhibited the least response to direct sunlight, with only a 25.97% reduction in leaf angle (from 142.56° under shaded conditions to 105.31° under non-shaded conditions).

Microscopic observations revealed significant differences in stomatal morphology and arrangement among the genotypes. All four genotypes displayed clustered stomata, a common morphological characteristic of semperflorens-type begonias, with cluster sizes ranging from one to seven stomata per cluster. Statistical analysis demonstrated significant differences in stomatal densities of new leaves between shaded and non-shaded conditions ($p = 1.91 \times 10^{-16}$) among the four genotypes (Figure 2). Under shaded conditions, FB08-059 and OPGC 5104 had significantly lower stomatal densities (74.10/mm^2 and 68.47/mm^2, respectively) compared to their commercial counterparts, 'Cocktail Vodka' (91.76/mm^2) and 'Sprint White' (90.02/mm^2; $p = 5.77 \times 10^{-13}$). This trend persisted under non-shaded conditions ($p = 3.11 \times 10^{-18}$), with FB08-059 and OPGC 5104 maintaining significantly lower stomatal densities (85.74/mm^2 and 79.58/mm^2, respectively) compared to 'Cocktail Vodka' (104.39/mm^2) and 'Sprint White' (101.95/mm^2). A two-way ANOVA revealed no significant interaction between light availability and genotype in determining stomatal density ($p = 0.97$), indicating that the increase in stomatal density from shaded to non-shaded conditions was consistent across all four genotypes. These results emphasize the uniformity of the stomatal density response to light availability despite genotype-specific differences in absolute stomatal density levels.

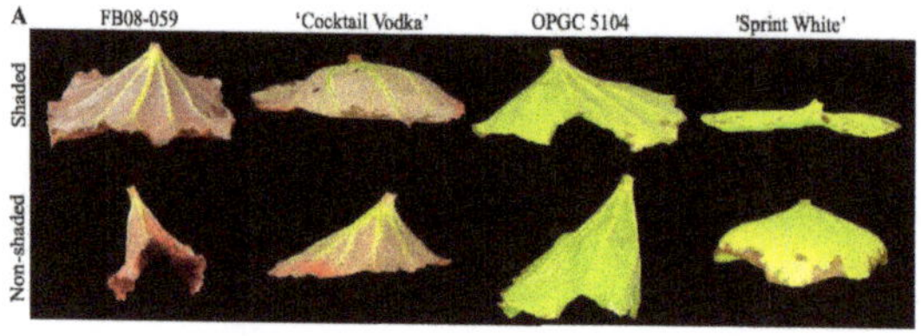

B

	FB08-059		'Cocktail Vodka'		OPGC 5104		'Sprint White'	
	Shaded	Non-shaded	Shaded	Non-shaded	Shaded	Non-shaded	Shaded	Non-shaded
Leaf Folding Angle	115.66 bc	43.59 e	124.31 b	62.41 d	117.71 b	55.44 de	142.26 a	105.31 c
% Decrease in Angle	62.30		49.79		52.91		25.97	

Figure 1. Variations in Leaf Folding Among Four Wax Begonia Genotypes Under Shaded and Non-Shaded Conditions. (**A**) Images depicting the leaf folding patterns of four genotypes in shaded (top) and non-shaded (bottom) conditions. (**B**) Table summarizing the degree (°) of leaf folding for each genotype under shaded and non-shaded conditions. Means within row marked by different letters indicate significant differences according to Tukey's HSD test ($p \leq 0.05$). Leaf angles were averaged from three leaves per plant ($n = 12$).

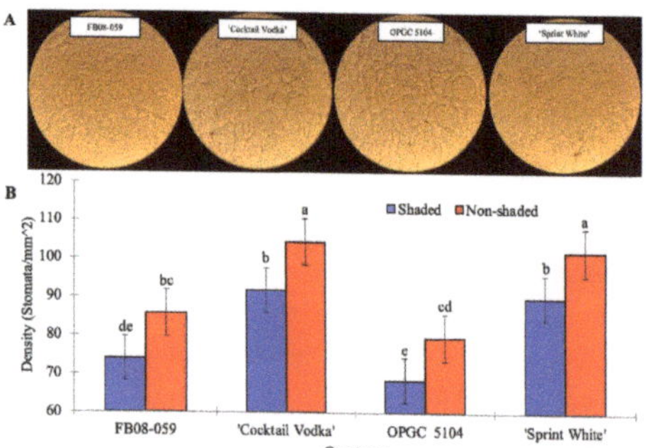

Figure 2. Stomatal Morphology and Density in Four Wax Begonia Genotypes Under Shaded and Non-Shaded Conditions. (**A**) Representative images showing clustered stomata in the four genotypes under shaded conditions. (**B**) Stomatal density for each genotype under shaded and non-shaded conditions. Bars represent the mean ± standard error (n = 12). Letters above bars indicate significant differences across all conditions and genotypes determined by Tukey's HSD test (p-value ≤ 0.05).

2.3. Photosynthetic Parameters, Chlorophyll and Anthocyanin Contents of Four Wax Begonia Genotypes Under Shaded and Non-Shaded Conditions

Stomatal conductance, an essential parameter of photosynthetic performance, exhibited significant variation among genotypes and conditions over the study period. Under shaded conditions, 'Sprint White' achieved the highest conductance rate (0.36 mol m^{-2} s^{-1}), while OPGC 5104 displayed the lowest (0.12 mol m^{-2} s^{-1}) (Figure 3A). Despite these differences, a consistent trend emerged across all genotypes in shaded conditions: stomatal conductance declined initially during the first 6–9 days, followed by gradual recovery. Commercial genotypes ('Sprint White' and 'Cocktail Vodka') consistently exhibited higher stomatal conductance rates under shaded conditions compared to the non-commercial genotypes (FB08-059 and OPGC 5104). In non-shaded conditions, FB08-059 maintained significantly higher stomatal conductance rates than 'Cocktail Vodka', OPGC 5104, and 'Sprint White' at all time points (Figure 3A). Remarkably, FB08-059 was the only genotype to achieve stomatal conductance levels in the non-shaded condition comparable to those in the shaded condition on day 41 (0.247 mol m^{-2} s^{-1} shaded; 0.234 mol m^{-2} s^{-1} non-shaded) (Figure 3A, Supplemental Figure S1A). A two-way ANOVA revealed a significant interaction between genotype and condition (p = 0.0006), underscoring the distinct physiological responses of the genotypes to the experimental conditions.

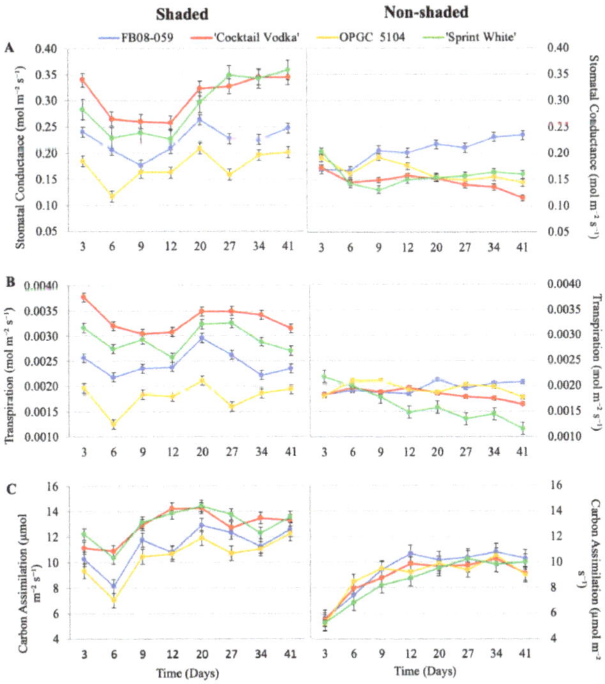

Figure 3. Stomatal Conductance, Transpiration, and Carbon Assimilation Rates of Four Wax Begonia Genotypes Under Shaded and Non-Shaded Conditions. Stomatal conductance (**A**), transpiration (**B**), and carbon assimilation (**C**) responses of wax begonia genotypes (FB08-059, OPGC 5104, 'Sprint White', and 'Cocktail Vodka') grown under shaded (**left**) and non-shaded (**right**) conditions for 41 days. Measurements were taken during peak light and heat intensity between 12 PM and 3 PM. Bars represent mean ± standard error (n = 12).

Transpiration rates mirrored the patterns observed in stomatal conductance. Under shaded conditions, the commercial genotypes ('Sprint White' and 'Cocktail Vodka') demonstrated higher rates of water loss compared to FB08-059 and OPGC 5104 throughout the study. However, in non-shaded conditions, the transpiration rates of 'Sprint White' and 'Cocktail Vodka' declined more sharply than those of FB08-059 and OPGC 5104 (Figure 3B). Notably, 'Sprint White' consistently exhibited a progressive decrease in transpiration rate under non-shaded conditions. By day 41, FB08-059 showed no significant difference in transpiration rates between shaded (0.00235 mol m^{-2} s^{-1}) and non-shaded (0.00207 mol m^{-2} s^{-1}) conditions, whereas the commercial genotypes displayed significant reductions in transpiration between conditions (Figure 3B, Supplemental Figure S1B). These results were further supported by a significant genotype–condition interaction effect observed in a two-way ANOVA ($p = 0.0025$).

Carbon assimilation rates, a direct measure of photosynthetic capacity, were consistently higher in shaded plants compared to their non-shaded counterparts, reflecting lower stress levels in shaded conditions (Figure 3C). Under non-shaded conditions, all genotypes initially exhibited suppressed carbon assimilation rates, followed by a gradual recovery over the study period (Figure 3C). On day 41, FB08-059 exhibited a significant difference in carbon assimilation between conditions (12.57 μmol m^{-2} s^{-1} shaded; 10.29 μmol m^{-2} s^{-1} non-shaded), contrasting with its consistent performance in stomatal conductance and transpiration (Figure 3C, Supplemental Figure S1C). The two-way ANOVA revealed a highly significant interaction between genotype and condition ($p = 1.53 \times 10^{-5}$), indicating substantial variability in carbon assimilation responses among genotypes. Notably, FB08-059 and OPGC 5104 experienced smaller reductions in carbon assimilation under non-shaded conditions compared to the commercial genotypes 'Cocktail Vodka' and 'Sprint White'.

Anthocyanins serve as crucial antioxidants that mitigate oxidative damage induced by abiotic stressors. Leaf pigmentation exhibited a pronounced response to direct sunlight (non-shaded conditions), with all four genotypes developing darker foliage compared to their counterparts grown under shaded conditions (Figure 4A). However, anthocyanin accumulation varied significantly among genotypes. In OPGC 5104 and 'Sprint White', both of which possess green foliage, anthocyanin content was negligible under both light conditions (Figure 4A,B). Conversely, in the red-foliaged genotypes, FB08-059 and 'Cocktail Vodka', anthocyanin levels were not significantly different under shaded conditions, as determined by Tukey's HSD test ($p \leq 0.05$). Upon exposure to direct sunlight, anthocyanin content in FB08-059 increased threefold compared to shaded conditions, whereas 'Cocktail Vodka' did not exhibit a significant change in anthocyanin accumulation (Figure 5A,B).

Chlorophyll degradation is a key physiological marker of cellular damage resulting from environmental stress. Under non-shaded conditions, 'Cocktail Vodka', 'Sprint White', and OPGC 5104 exhibited a significant reduction in total chlorophyll content compared to their shaded counterparts (Figure 4C). Notably, FB08-059 was the only genotype that maintained total chlorophyll content irrespective of light condition (Figure 4C). The observed decline in total chlorophyll content in 'Cocktail Vodka', 'Sprint White', and OPGC 5104 was primarily attributed to reductions in chlorophyll a, the dominant component of total chlorophyll (Figure 4C,D). Additionally, a significant genotype–by-condition interaction effect was detected for total chlorophyll content ($p = 3.7 \times 10^{-9}$), chlorophyll a ($p = 2.5 \times 10^{-9}$), and chlorophyll b ($p = 0.01$), indicating differential photoprotective strategies among the genotypes in response to high light intensity.

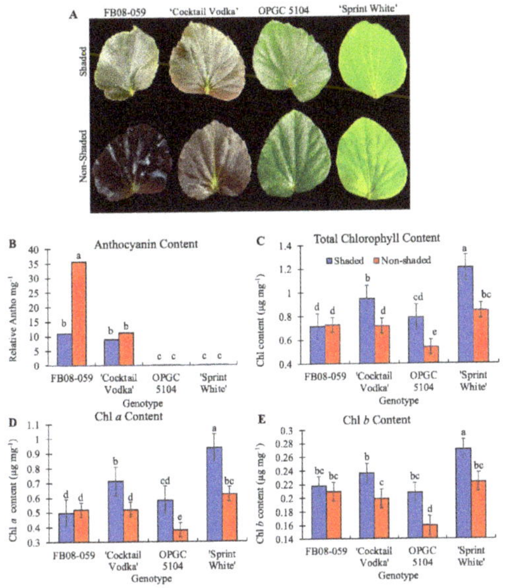

Figure 4. Anthocyanin and Chlorophyll Content in Wax Begonia Genotypes Under Shaded and Non-Shaded Conditions. (**A**) Representative leaf images from each genotype, taken 41 days after exposure to shaded and non-shaded conditions. (**B**) Relative anthocyanin content of four genotypes under shaded and non-shaded conditions. 'OPGC5104' and 'Sprint White' bars appear minimal due to their negligible anthocyanin levels (<0.1 absorbance units) but are shown to highlight differences. (**C**) Total chlorophyll content (Chl a + Chl b) in four genotypes under shaded and non-shaded conditions. (**D**) Chlorophyll a content. (**E**) Chlorophyll b content. Bars represent mean ± standard error (n = 12). Different letters above bars indicate significant differences according to Tukey's HSD test (p-value < 0.05).

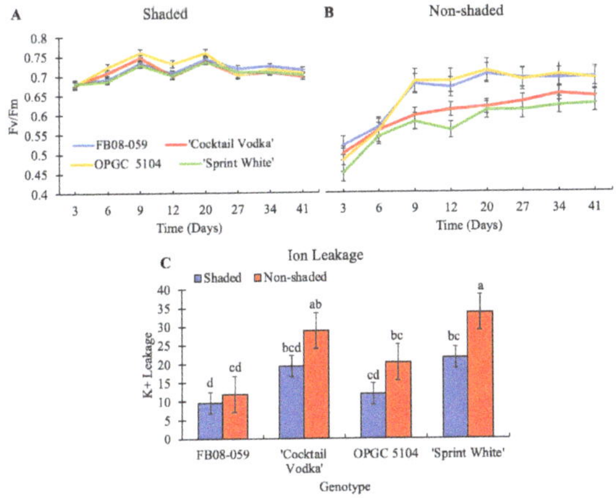

Figure 5. Effects of Shaded and Non-Shaded Conditions on Fluorescence and Electrolyte Leakage in Wax Begonia Genotypes. (**A**) Temporal dynamics of Fv/Fm under shaded conditions across four genotypes. (**B**) Temporal dynamics of Fv/Fm under non-shaded conditions. (**C**) Electrolyte leakage, expressed as the percentage of total ions released. Bars represent mean ± standard error (n = 12). Letters above bars indicate significant differences among all genotypes and conditions based on Tukey's HSD test (p-value ≤ 0.05).

2.4. Photosynthetic Efficiency and Oxidative Stress Responses of Four Wax Begonia Genotypes Under Shaded and Non-Shaded Conditions

An Fv/Fm ratio between 0.75 and 0.8 is indicative of optimal photosynthetic efficiency, reflecting minimal stress and maximal photochemical activity. Under shaded conditions, no statistically significant variation in Fv/Fm values was observed among the four genotypes (Figure 5A). All genotypes maintained relatively stable Fv/Fm values between 0.7 and 0.75 throughout the duration of the experiment, suggesting mild but non-detrimental stress effects. In contrast, exposure to direct sunlight (non-shaded conditions) resulted in a substantial decline in Fv/Fm values (0.45–0.52) by day 3 across all genotypes (Figure 5B). However, genotypic differences in exposure to high light stress became apparent as time progressed. The putatively stress-tolerant genotypes, FB08-059 and OPGC 5104, exhibited significantly higher Fv/Fm values than the susceptible genotypes, 'Cocktail Vodka' and 'Sprint White', from day 9 onward in the non-shaded condition ($p = 0.01$). Further assessment of membrane integrity through ion leakage analysis reinforced the observed genotypic disparities in stress tolerance. The non-shaded condition resulted in a markedly higher mean percentage of total ion leakage (23.70%) relative to the shaded condition (15.61%), underscoring the deleterious effects of direct sunlight and elevated temperatures. Ion leakage levels varied significantly among genotypes in both shaded ($p = 0.0002$) and non-shaded ($p = 5.85 \times 10^{-6}$) conditions. Notably, FB08-059 exhibited the lowest ion leakage, suggesting superior membrane stability compared to the commercial genotypes (Figure 5C). Under non-shaded conditions, ion leakage levels were recorded as follows: FB08-059 (11.91%), OPGC 5104 (20.34%), 'Cocktail Vodka' (28.85%), and 'Sprint White' (33.72%). Additionally, FB08-059 displayed only a 2.25% increase in ion leakage between shaded and non-shaded conditions, in stark contrast to 'Sprint White', which exhibited a 12.23% increase. 'Cocktail Vodka' showed an intermediate increase of 9.45% between conditions. These results suggest that FB08-059 demonstrates enhanced resilience to high light intensity and thermal stress compared to the commercially available genotypes.

The regulation of reactive oxygen species (ROS) homeostasis is crucial for sustaining cellular function under environmental stress. Disruptions in ROS equilibrium can lead to oxidative damage and cellular dysfunction. Two primary ROS, superoxide ($O_2{\cdot}^-$) and hydrogen peroxide (H_2O_2), were detected using nitroblue tetrazolium (NBT) and diaminobenzidine (DAB) staining, respectively (Figure 6A,B). A statistically significant increase in ROS accumulation was observed in all genotypes under non-shaded conditions, as determined by Tukey's HSD test ($p \leq 0.05$) (Figure 6C,D). FB08-059 exhibited the lowest levels of $O_2{\cdot}^-$ and H_2O_2 accumulation, followed closely by OPGC 5104. Conversely, the commercial genotypes 'Sprint White' and 'Cocktail Vodka' displayed higher $O_2{\cdot}^-$ accumulation under both shaded (52.70% and 49.84%, respectively) and non-shaded conditions (80.89% and 71.86%, respectively) (Figure 6C). A similar pattern was evident in H_2O_2 accumulation, with 'Sprint White' and 'Cocktail Vodka' exhibiting higher values under both shaded (44.07% and 39.81%, respectively) and non-shaded conditions (61.85% and 60.21%, respectively) (Figure 6D).

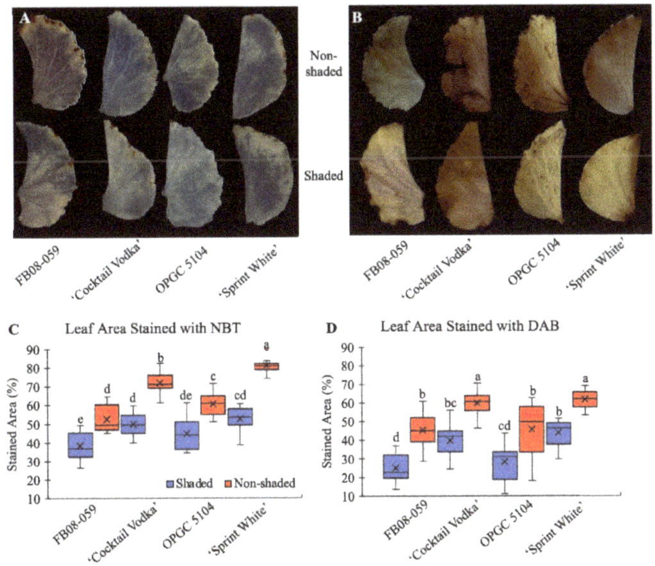

Figure 6. Oxidative Stress Responses in Wax Begonia Genotypes Under Shaded and Non-Shaded Conditions (**A**) NBT staining for superoxide ($O_2\cdot^-$) accumulation in leaf tissue of four genotypes under shaded and non-shaded conditions. (**B**) DAB staining for hydrogen peroxide (H_2O_2) accumulation in leaf tissue under both conditions. (**C**) Quantification of $O_2\cdot^-$ accumulation. (**D**) Quantification of H_2O_2 accumulation. Bars represent mean ± standard error (n = 12). Different letters indicate significant differences among genotypes as determined by Tukey's HSD test (p-value $\leq$ 0.05).

3. Discussion

3.1. Stomatal Density and Gas Exchange Regulation in Wax Begonia

One of the primary responses observed was an increase in stomatal density under non-shaded conditions across all genotypes, consistent with previous research indicating that high light intensity stimulates stomatal development [30,31]. Stomata play a crucial role in regulating gas exchange by facilitating CO_2 uptake for photosynthesis while minimizing water loss through transpiration. However, despite the increase in stomatal density, CO_2 assimilation rates declined under high light and temperature conditions, particularly in the stress-susceptible genotypes. This suggests that stomatal density alone does not determine photosynthetic efficiency; rather, other factors such as cuticle thickness, leaf folding, and oxidative stress regulation play a more substantial role in stress adaptation. A similar increase in stomatal density under high light conditions has been documented in Arabidopsis thaliana, where plants grown under elevated light intensities exhibited greater stomatal development; however, their water-use efficiency varied depending on the genotype's ability to regulate stomatal aperture [32]. In wheat (*Triticum aestivum*), an increase in stomatal density has been correlated with enhanced photosynthetic performance under moderate stress conditions but was detrimental under prolonged drought due to excessive water loss [33]. Additionally, research in grapevines (*Vitis vinifera*) has demonstrated that cultivars adapted to hot and arid climates maintain lower stomatal density but compensate with increased stomatal size, thereby optimizing water use efficiency [34]. These examples highlight that stomatal density alone does not dictate overall plant performance, as its interaction with stomatal conductance, cuticle properties, and internal water regulation determines stress adaptation.

Plants have evolved trade-offs in stomatal regulation under environmental stress. Increased stomatal density can enhance photosynthetic rates in favorable conditions but may lead to excessive water loss under drought and heat stress [35]. In our study, 'Cocktail Vodka' and 'Sprint White' exhibited significantly lower stomatal conductance under non-shaded conditions than FB08-059, indicating greater stomatal closure as a water conservation strategy. This may explain their reduced transpiration and carbon assimilation rates compared to the tolerant genotypes, which maintained higher stomatal conductance. This pattern is also evident in corn (*Zea mays*), where drought-tolerant hybrids maintain stomatal conductance despite lower stomatal density, optimizing carbon assimilation while preventing excessive water loss [36]. Similarly, in rice (*Oryza sativa*), cultivars with greater stomatal density exhibited higher transpiration rates, but their photosynthetic efficiency was largely dependent on water availability [37].

Transpiration rates also varied significantly among genotypes. While 'Cocktail Vodka' and 'Sprint White' exhibited a sharp decline in transpiration under non-shaded conditions, FB08-059 maintained relatively stable transpiration rates, indicating an improved capacity for evaporative cooling. This balance between transpiration and stomatal conductance may contribute to the superior stress tolerance of FB08-059 [38]. This relationship has been observed in sunflowers (*Helianthus annuus*), where drought-resistant varieties show lower transpiration rates but maintain sufficient stomatal opening for sustained photosynthesis, allowing them to endure extended periods of heat stress [39]. In contrast, more sensitive varieties experience rapid stomatal closure under high temperatures, leading to reductions in gas exchange and growth rates. Similarly, in cotton (*Gossypium hirsutum*), heat-tolerant genotypes exhibit higher transpiration rates, enabling them to dissipate excess heat and maintain lower leaf temperatures [40].

3.2. Structural Adaptations: Leaf Folding and Cuticle Thickening in Wax Begonia

Leaf folding is an adaptive mechanism that reduces light interception and minimizes photodamage under excessive radiation [41]. In this study, FB08-059 displayed the most pronounced leaf folding under non-shaded conditions, followed by OPGC 5104. Leaf folding limits direct light penetration, thereby reducing the risk of photoinhibition and excessive heat buildup [42]. In contrast, 'Sprint White' exhibited minimal leaf folding, which may explain its greater susceptibility to photoinhibition and oxidative stress. Leaf folding has been reported as a photoprotective response in various plant species. In cotton (*Gossypium hirsutum*), leaves subjected to high light intensity exhibit enhanced folding to limit radiation absorption and maintain optimal photosynthetic rates [40]. In tropical grasses, leaf rolling, a more extreme form of leaf folding, is a well-documented response to heat and drought stress, helping to conserve moisture and reduce excessive transpiration [43]. These findings suggest that structural modifications, such as leaf folding, contribute significantly to stress mitigation across diverse plant taxa.

Cuticle thickening was another significant morphological response observed in this study. FB08-059 exhibited the highest increase in cuticle thickness under non-shaded conditions, followed by OPGC 5104. A thicker cuticle serves multiple protective functions, including reducing transpirational water loss, enhancing leaf reflectance, and providing a physical barrier against excessive light penetration [44]. Previous studies have also shown that cuticle thickness is correlated with drought and heat tolerance in multiple plant species, including corn and Arabidopsis [12,13]. The greater cuticle thickness in FB08-059 and OPGC 5104 likely contributed to their ability to maintain higher photosynthetic efficiency under stress conditions. Similar responses have been documented in numerous plant species. In wheat, genotypes with thicker cuticles exhibit greater resistance to heat-induced oxidative stress and maintain better water retention than those with thinner cuticles [45]. In rice, increased cuticle deposition has been linked to improved drought tolerance, reducing non-stomatal water loss while simultaneously protecting against leaf desiccation [46]. These findings further reinforce the importance of cuticle thickening as a conserved trait among plant species adapting to extreme environmental conditions.

The combination of leaf folding and cuticle thickening provides an integrated strategy for mitigating abiotic stress. While leaf folding reduces direct light exposure, cuticle thickening enhances overall structural integrity, allowing plants to tolerate prolonged periods of heat and light stress. The differential responses observed among the genotypes in this study indicate that FB08-059 and OPGC 5104 possess superior adaptive traits, which could be valuable for breeding programs aimed at improving stress resilience in ornamental and crop species.

3.3. Anthocyanin Accumulation and Photoprotection in Wax Begonia

Anthocyanins have been widely recognized for their role in photoprotection and oxidative stress mitigation. The results of this study demonstrated a significant increase in anthocyanin accumulation in FB08-059 under non-shaded conditions, whereas 'Cocktail Vodka' did not exhibit a comparable increase. This suggests that the genetic capacity to upregulate anthocyanin production in response to environmental stress is a key determinant of heat and light tolerance [47,48]. The protective role of anthocyanins is twofold: (1) they act as light filters that absorb excess radiation, thereby reducing potential damage to the photosynthetic apparatus, and (2) they function as antioxidants that neutralize reactive oxygen species (ROS) [18]. The elevated anthocyanin levels in FB08-059 corresponded with lower ROS accumulation and reduced ion leakage, further supporting the hypothesis that anthocyanins play a crucial role in stress tolerance. Anthocyanin accumulation has been documented as a protective mechanism in multiple plant species exposed to high light intensity, reinforcing the role of this pigment in photoprotection [49].

Anthocyanin biosynthesis has been shown to be a critical photoprotective strategy in a variety of plant species. In corn, high light intensity induces anthocyanin accumulation in leaf tissue, leading to improved tolerance against photodamage and heat stress [50]. Similarly, in apples (*Malus domestica*), anthocyanin accumulation in response to high light conditions enhances both photosynthetic efficiency and ROS scavenging, reducing oxidative damage to cellular structures [51]. In Arabidopsis, mutants deficient in anthocyanin biosynthesis exhibit increased susceptibility to high light stress, further emphasizing the importance of this pigment in photoprotection [52]. Additionally, anthocyanin accumulation has been linked to improved drought tolerance in crops such as wheat and rice. Under limited water availability, anthocyanins not only protect against excessive light damage but also enhance osmotic regulation, allowing plants to maintain cellular hydration [53,54]. The findings in FB08-059 align with these studies, suggesting that anthocyanins confer a multifaceted advantage by protecting against both high light and oxidative stress while maintaining physiological stability. Interestingly, in 'Cocktail Vodka', the absence of a significant anthocyanin increase suggests that commercial breeding selections may prioritize esthetic consistency over environmental responsiveness. This highlights the potential trade-off between ornamental value and stress resilience, suggesting that future breeding efforts should consider incorporating genetic diversity in anthocyanin regulation to enhance stress tolerance.

3.4. Oxidative Stress and Membrane Stability in Wax Begonia

Exposure to high light and temperature conditions induces oxidative stress, leading to ROS accumulation and membrane damage. In this study, FB08-059 and OPGC

5104 exhibited significantly lower ROS accumulation and ion leakage compared to 'Cocktail Vodka' and 'Sprint White'. This indicates that these genotypes possess more efficient ROS scavenging mechanisms, which help maintain membrane stability and overall plant health under stress conditions [24]. The detrimental effects of ROS accumulation have been widely studied in various plant species. In wheat, genotypes with higher levels of enzymatic antioxidants, such as superoxide dismutase (SOD) and catalase (CAT), exhibit greater resilience to heat stress by mitigating oxidative damage [55]. Similarly, in rice, plants with upregulated ascorbate peroxidase (APX) activity maintain superior membrane stability and reduced ion leakage when subjected to high-temperature stress [26]. The ability to efficiently detoxify ROS is, therefore, a crucial determinant of stress tolerance across diverse plant taxa.

Chlorophyll fluorescence (Fv/Fm) measurements further reinforced these findings. While all genotypes experienced a decline in Fv/Fm under non-shaded conditions, FB08-059 and OPGC 5104 maintained significantly higher values than the stress-susceptible genotypes, suggesting a lower degree of photoinhibition. Similar trends have been reported in sunflower (*Helianthus annuus*), where drought-tolerant cultivars exhibit higher Fv/Fm values under water stress due to enhanced protective mechanisms in the photosynthetic apparatus [14]. In corn, genotypes with improved ROS scavenging capacity show sustained Fv/Fm ratios under combined drought and heat stress, allowing for greater retention of photosynthetic efficiency [56]. Beyond enzymatic ROS detoxification, non-enzymatic antioxidants such as flavonoids, carotenoids, and tocopherols also play an essential role in oxidative stress mitigation. In Arabidopsis, flavonoid-deficient mutants display increased susceptibility to high light stress, underscoring the importance of these compounds in photoprotection [54]. Similarly, in grapevines, cultivars with higher carotenoid content exhibit reduced oxidative damage and improved physiological stability under prolonged heat stress [57]. The significant differences in ROS accumulation and ion leakage observed in this study suggest that FB08-059 and OPGC 5104 may possess enhanced antioxidant systems, contributing to their superior stress tolerance.

The interplay between oxidative stress regulation and membrane stability is a key determinant of plant resilience. The observed differences in ROS accumulation, ion leakage, and Fv/Fm values among the genotypes indicate that FB08-059 and OPGC 5104 employ more robust protective mechanisms compared to 'Cocktail Vodka' and 'Sprint White'. These findings align with previous research in multiple crop and ornamental species, highlighting the importance of integrating ROS detoxification pathways into breeding programs aimed at improving stress tolerance.

4. Materials and Methods

4.1. Plant Material and Growing Conditions

Four wax begonia (*Begonia semperflorens-coltorum*) genotypes were selected for this study: two bronze-foliaged genotypes (FB08-059 and 'Cocktail Vodka') and two green-foliaged genotypes (OPGC 5104 and 'Sprint White'). The plants were propagated from cuttings and cultivated for ten weeks in a low-light greenhouse environment (photosynthetically active radiation, PAR = 750 μmol m^{-2} s^{-1}) under a controlled temperature regime of 28/23 °C (day/night). FB08-059, previously identified as heat-tolerant [58], exhibits dark green to red foliage in shaded environments. OPGC 5104 is a wild-type begonia with bright green foliage, originally collected from Hawaii. 'Sprint White' and 'Cocktail Vodka' are commercially available genotypes with green and red foliage, respectively (Figure 7). Two weeks prior to initiating the experimental conditions, uniform plugs were transplanted into 1-gallon pots containing a bark- and peat-based soilless medium (Premium Nursery and Veg Mix, Reliable Peat Company, Leesburg,

FL, USA), supplemented with 5 g/L of 14N-14P-14K slow-release fertilizer (Osmocote, The Scotts Company, Marysville, OH, USA). Plants were then randomly assigned to either a shaded or non-shaded condition under natural sunlight at the Mid-Florida Research and Education Center in Apopka, FL, USA. The average day/night temperatures and humidity are presented in Supplementary Figure S2. The non-shaded condition was characterized by peak PAR values of 2100 μmol m^{-2} s^{-1}, with day/night temperatures of 35/22.5 °C. Shading was achieved using a black shade net (Agfabric, Home Depot, Atlanta, GA, USA), which maintained a peak PAR of 750 μmol m^{-2} s^{-1} and temperatures of 30/22.5 °C. Relative humidity levels fluctuated between 65% and 100% in both conditions. To ensure that drought stress did not confound the results, plants were regularly irrigated via rainfall or an overhead sprinkler system.

Figure 7. Representative top and side views of four wax begonia genotypes used in this study. FB08-059 (**A**,**B**), 'Cocktail Vodka' (**C**,**D**), OPGC 5104 (**E**,**F**), and 'Sprint White' (**G**,**H**). Images were captured ten weeks post-cutting propagation, prior to exposure to experimental light conditions. The study examined stress tolerance under shaded and non-shaded conditions.

4.2. Morphological Measurements

Morphological assessments were conducted at the conclusion of the 41-day experiment. Measurements were taken from the first, second, and third fully expanded leaves from the apical meristem. Stomatal density and cluster formation were determined using the clear nail varnish impression method [59]. Stomatal impressions were obtained from the abaxial leaf surface and examined under a light microscope. Stomatal density was calculated by averaging counts from three leaves per plant and normalizing per unit area. To evaluate anatomical differences, leaf cross-sections were obtained using a microtome blade. Sections were examined under a calibrated light microscope to determine total leaf thickness and cuticle thickness. Measurements were restricted to the lamina, excluding primary and secondary veins.

Leaf folding was quantified using ImageJ Version 1.51 software (NIH, Bethesda, MD, USA). Leaves were laid flat with the petiole oriented upwards. Images were captured from a standardized distance, and the angle between the leaf margins and petiole was measured. Each measurement was replicated three times per sample. Shoot and root biomass were measured following the 41-day stress conditions. Above-ground fresh weights were recorded immediately after excising plants at the base of the stem. Dry weights for shoot and root tissues were determined after oven drying at 50 °C for 72 h.

4.3. Physiological Measurements

Photosynthetic parameters, including carbon assimilation, stomatal conductance, and transpiration, were recorded throughout the 41-day experiment using a LI-COR 6800 Portable Photosynthesis System (LI-COR Biosciences, Lincoln, NE, USA). Measurements were taken during peak daylight hours (12:00–16:00 h) under the following chamber conditions: relative humidity 40–50%, photosynthetic photon flux density (PPFD) 1000 μmol m^{-2} s^{-1}, reference CO_2 concentration 400 ppm, flow rate 500 μmol s^{-1}, and fan speed 10,000 rpm. Following stabilization, three consecutive measurements were recorded for each plant at 30 s intervals.

Anthocyanin and chlorophyll contents were quantified using the protocols described by Taghavi and Porra et al., respectively [60,61]. Leaf samples were collected from each plant at the conclusion of the experiment. Three lamina disks were excised per sample, ensuring the exclusion of major veins. Samples were weighed and processed for absorbance readings using a Synergy H1 microplate spectrophotometer (BioTek, Winooski, VT, USA). Pigment concentrations were calculated using the following equations:

$$Anthocyanin\ content = \frac{(A_{530} - A_{657}) \times 1000}{mg\ tissue}$$

$$Chlorophyll\ a\ and\ b\ Content = \frac{((22.12 \times A_{652}) + (2.71 \times A_{665})) \times 0.5}{mg\ tissue}$$

4.4. Measurements of Stress Responses

Chlorophyll fluorescence (Fv/Fm) was measured using an OS30p chlorophyll fluorometer (Opti-Sciences, Hudson, NH, USA). To ensure dark adaptation, measurements were taken two hours before dawn on three fully expanded leaves per plant. Readings were averaged to determine mean Fv/Fm values.

Electrolyte leakage was quantified as an indicator of membrane stability. Three leaf disks from each sample were submerged in 5 mL deionized water for four hours, and initial conductivity readings were taken using an Orion Star A215 conductivity meter (Thermo Fisher Scientific, Waltham, MA, USA). Samples were then autoclaved at 121 °C for 20 min, allowed to cool for 15 min, and final conductivity readings were obtained. Electrolyte leakage was calculated as: (initial conductivity/final conductivity) × 100. Reactive oxygen species (ROS) accumulation was assessed using diaminobenzidine (DAB) and nitroblue tetrazolium (NBT) staining [62]. Leaves were collected at the end of the study and bisected along the midrib. One half was stained with NBT to detect superoxide radicals ($O_2 \cdot^-$), while the other half was stained with DAB to visualize hydrogen peroxide (H_2O_2) accumulation. High-anthocyanin samples (FB08-059 and 'Cocktail Vodka') required ethanol washes for accurate staining observation.

4.5. Statistical Analysis

The experimental design consisted of a completely randomized block arrangement with three replications per block and four plants per genotype in each condition. Trait means and standard errors (n = 12) were calculated at each time interval. At the conclusion of the experiment (day 41), a two-way analysis of variance (ANOVA) was performed to evaluate the effects of genotype and light condition, and their interaction. Significant differences were determined using Tukey's honestly significant difference (HSD) test (p-value $\leq$ 0.05) in R Studio Version 3.6.0 (Boston, MA, USA) with the Agricolae package. Paired Student's t-tests were conducted to assess significant differences between putatively tolerant and susceptible genotypes. Leaf folding images were captured with a Nikon digital camera and analyzed using ImageJ public domain software. ROS-stained leaf images were processed with GIMP (version 2.10) before quantification in ImageJ.

5. Conclusions

The findings of this study showcase key morphological and physiological traits associated with heat and high light stress tolerance in wax begonia. Specifically, increased cuticle thickness, enhanced anthocyanin accumulation, and acute leaf folding were identified as critical adaptive traits. These insights can be directly applied to breeding programs aimed at developing more resilient ornamental cultivars for commercial and landscaping applications. FB08-059, in particular, demonstrated the most robust stress-adaptive responses, making it a valuable candidate for breeding initiatives focused on enhancing heat and light tolerance in wax begonias. Future research should explore the genetic regulation of these adaptive traits and evaluate their heritability to facilitate targeted breeding strategies.

Supplementary Materials: The following supporting information can be downloaded at: https://www.mdpi.com/article/10.3390/ijms26083514/s1.

Author Contributions: Conceptualization, J.G., C.D.N. and H.H.; methodology, J.G.; formal analysis, J.G.; data curation, J.G.; writing—original draft preparation, J.G. and C.D.N.; writing—review and editing, S.W., Z.D. and H.H.; supervision, H.H. All authors have read and agreed to the published version of the manuscript.

Funding: This research was funded by the Florida Specialty Crop Block Grant Program (AWD11668) and the HTACH Grant (FLA-MFC-006387).

Informed Consent Statement: Not applicable.

Data Availability Statement: All data is presented in the manuscript and Supplementary Materials.

Acknowledgments: The authors would like to acknowledge Matthew Creech, Biological Scientist, for his technical support and assistance throughout the study.

Conflicts of Interest: The authors declare no conflicts of interest.

References

1. Kidner, C.; Neale, S.; Goodall-Copestake, W.; Kidner, C.A. *The Evolution of Diversity in Begonia*; Global Science Books: Bexhill-On-Sea, UK, 2014.
2. Moonlight, P.W.; Richardson, J.E.; Tebbitt, M.C.; Thomas, D.C.; Hollands, R.; Peng, C.I.; Hughes, M. Continental-Scale Diversification Patterns in a Megadiverse Genus: The Biogeography of Neotropical Begonia. *J. Biogeogr.* **2015**, *42*, 1137–1149. [CrossRef]
3. Moonlight, P.W.; Ardi, W.H.; Padilla, L.A.; Chung, K.F.; Fuller, D.; Girmansyah, D.; Hollands, R.; Jara-Muñoz, A.; Kiew, R.; Leong, W.C.; et al. Dividing and Conquering the Fastest-Growing Genus: Towards a Natural Sectional Classification of the Mega-Diverse Genus Begonia (*Begoniaceae*). *Taxon* **2018**, *67*, 267–323. [CrossRef]
4. Ardi, W.H.; Thomas, D. Three new species of begonia from the outer islands of southeastern sulawesi. *Edinb. J. Bot.* **2023**, *80*, 1–17. [CrossRef]
5. Hughes, M.; Girmansyah, D.; Ardi, W.H. Further Discoveries in the Ever-Expanding Genus Begonia (*Begoniaceae*): Fifteen New Species from Sumatra. *Eur. J. Taxon.* **2015**, *2015*, 1–140. [CrossRef]
6. Kang, J.-G.; Van Iersel, M.W. Managing Fertilization of Bedding Plants: A Comparison of Constant Fertilizer Concentrations versus Constant Leachate Electrical Conductivity. *HortScience* **2009**, *44*, 151–156. [CrossRef]
7. Broschat, T.K.; Moore, K.A. Influence of Substrate and Fertilizer Analysis and Rate on Growth and Quality of Five Species of Bedding Plants. *Horttechnology* **2001**, *11*, 434–437. [CrossRef]
8. *Floriculture in 2022*. Commercial Floriculture Survey. January 2023 (for 2022 Production). Available online: https://www.nass.usda.gov/Publications/Methodology_and_Data_Quality/Floriculture/07-2023/Floriculture_Questionnaire_CA_Jan2023.pdf (accessed on 14 February 2025).
9. Christensen, J.H.; Christensen, O.B. A Summary of the PRUDENCE Model Projections of Changes in European Climate by the End of This Century. *Clim. Change* **2007**, *81*, 7–30. [CrossRef]
10. Hasanuzzaman, M.; Nahar, K.; Alam, M.M.; Roychowdhury, R.; Fujita, M. Physiological, Biochemical, and Molecular Mechanisms of Heat Stress Tolerance in Plants. *Int. J. Mol. Sci.* **2013**, *14*, 9643–9684. [CrossRef]
11. Wahid, A.; Gelani, S.; Ashraf, M.; Foolad, M.R. Heat Tolerance in Plants: An Overview. *Environ. Exp. Bot.* **2007**, *61*, 199–223. [CrossRef]

12. Bourgault, R.; Matschi, S.; Vasquez, M.; Qiao, P.; Sonntag, A.; Charlebois, C.; Mohammadi, M.; Scanlon, M.J.; Smith, L.G.; Molina, I. Constructing Functional Cuticles: Analysis of Relationships between Cuticle Lipid Composition, Ultrastructure and Water Barrier Function in Developing Adult Maize Leaves. *Ann. Bot.* **2020**, *125*, 79–91. [CrossRef]
13. Kosma, D.K.; Bourdenx, B.; Bernard, A.; Parsons, E.P.; Lü, S.; Joubès, J.; Jenks, M.A. The Impact of Water Deficiency on Leaf Cuticle Lipids of Arabidopsis. *Plant Physiol.* **2009**, *151*, 1918–1929. [CrossRef] [PubMed]
14. Demmig-Adams, B.; Adams, W.W. Photoprotection in an Ecological Context: The Remarkable Complexity of Thermal Energy Dissipation. *New Phytol.* **2006**, *172*, 11–21. [CrossRef]
15. Nemali, K.S.; Van Iersel, M.W. Acclimation of Wax Begonia to Light Intensity: Changes in Photosynthesis, Respiration, and Chlorophyll Concentration. *J. Am. Soc. Hortic. Sci.* **2004**, *129*, 745–751. [CrossRef]
16. Baker, N.R. Chlorophyll Fluorescence: A Probe of Photosynthesis in Vivo. *Annu. Rev. Plant Biol.* **2008**, *59*, 89–113. [CrossRef] [PubMed]
17. Niyogi, K.K. PHOTOPROTECTION REVISITED: Genetic and Molecular Approaches. *Annu. Rev. Plant Biol.* **2025**, *50*, 333–359. [CrossRef]
18. Gould, K.S.; Neill, S.O.; Vogelmann, T.C. A Unified Explanation for Anthocyanins in Leaves? *Adv. Bot. Res.* **2002**, *37*, 167–192. [CrossRef]
19. Gould, K.S.; Jay-Allemand, C.; Logan, B.A.; Baissac, Y.; Bidel, L.P.R. When Are Foliar Anthocyanins Useful to Plants? Re-Evaluation of the Photoprotection Hypothesis Using Arabidopsis Thaliana Mutants That Differ in Anthocyanin Accumulation. *Environ. Exp. Bot.* **2018**, *154*, 11–22. [CrossRef]
20. Agati, G.; Azzarello, E.; Pollastri, S.; Tattini, M. Flavonoids as Antioxidants in Plants: Location and Functional Significance. *Plant Sci.* **2012**, *196*, 67–76. [CrossRef]
21. Miller, G.; Suzuki, N.; Ciftci-Yilmaz, S.; Mittler, R. Reactive Oxygen Species Homeostasis and Signalling during Drought and Salinity Stresses. *Plant Cell Environ.* **2010**, *33*, 453–467. [CrossRef]
22. Sies, H.; Jones, D.P. Reactive Oxygen Species (ROS) as Pleiotropic Physiological Signalling Agents. *Nat. Rev. Mol. Cell Biol.* **2020**, *21*, 363–383. [CrossRef]
23. Choudhury, F.K.; Rivero, R.M.; Blumwald, E.; Mittler, R. Reactive Oxygen Species, Abiotic Stress and Stress Combination. *Plant J.* **2017**, *90*, 856–867. [CrossRef]
24. Gill, S.S.; Tuteja, N. Reactive Oxygen Species and Antioxidant Machinery in Abiotic Stress Tolerance in Crop Plants. *Plant Physiol. Biochem.* **2010**, *48*, 909–930. [CrossRef] [PubMed]
25. Apel, K.; Hirt, H. Reactive Oxygen Species: Metabolism, Oxidative Stress, and Signal Transduction. *Annu. Rev. Plant Biol.* **2004**, *55*, 373–399. [CrossRef]
26. Foyer, C.H.; Noctor, G. Ascorbate and Glutathione: The Heart of the Redox Hub. *Plant Physiol.* **2011**, *155*, 2–18. [CrossRef] [PubMed]
27. Krasensky, J.; Jonak, C. Drought, Salt, and Temperature Stress-Induced Metabolic Rearrangements and Regulatory Networks. *J. Exp. Bot.* **2012**, *63*, 1593–1608. [CrossRef]
28. Rahmati Ishka, M.; Julkowska, M. Tapping into the Plasticity of Plant Architecture for Increased Stress Resilience. *F1000Res* **2023**, *12*, 1257. [CrossRef]
29. Jeandet, P.; Formela-Luboińska, M.; Labudda, M.; Morkunas, I. The Role of Sugars in Plant Responses to Stress and Their Regulatory Function during Development. *Int. J. Mol. Sci.* **2022**, *23*, 5161. [CrossRef] [PubMed]
30. Casson, S.; Gray, J.E. Influence of Environmental Factors on Stomatal Development. *New Phytol.* **2008**, *178*, 9–23. [CrossRef]
31. Hetherington, A.M.; Woodward, F.I. The Role of Stomata in Sensing and Driving Environmental Change. *Nature* **2003**, *424*, 901–908. [CrossRef]
32. Tanaka, Y.; Sugano, S.S.; Shimada, T.; Hara-Nishimura, I. Enhancement of Leaf Photosynthetic Capacity through Increased Stomatal Density in Arabidopsis. *New Phytol.* **2013**, *198*, 757–764. [CrossRef]
33. Haworth, M.; Elliott-Kingston, C.; McElwain, J.C. Stomatal Control as a Driver of Plant Evolution. *J. Exp. Bot.* **2011**, *62*, 2419–2423. [CrossRef] [PubMed]
34. Pagay, V.; Zufferey, V.; Lakso, A.N. The Influence of Water Stress on Grapevine (*Vitis vinifera* L.) Shoots in a Cool, Humid Climate: Growth, Gas Exchange and Hydraulics. *Funct. Plant Biol.* **2016**, *43*, 827–837. [CrossRef]
35. Hepworth, C.; Caine, R.S.; Harrison, E.L.; Sloan, J.; Gray, J.E. Stomatal Development: Focusing on the Grasses. *Curr. Opin. Plant Biol.* **2018**, *41*, 1–7. [CrossRef]
36. Galmés, J.; Aranjuelo, I.; Medrano, H.; Flexas, J. Variation in Rubisco Content and Activity under Variable Climatic Factors. *Photosynth. Res.* **2013**, *117*, 73–90. [CrossRef]
37. García-Mata, C.; Lamattina, L. Abscisic Acid, Nitric Oxide and Stomatal Closure-Is Nitrate Reductase One of the Missing Links? *Trends Plant Sci.* **2003**, *8*, 20–26. [CrossRef]

38. Lawson, T.; Blatt, M.R. Stomatal Size, Speed, and Responsiveness Impact on Photosynthesis and Water Use Efficiency. *Plant Physiol.* **2014**, *164*, 1556–1570. [CrossRef] [PubMed]
39. Roche, D. Stomatal Conductance Is Essential for Higher Yield Potential of C3 Crops. *CRC Crit. Rev. Plant Sci.* **2015**, *34*, 429–453. [CrossRef]
40. Mahmood, T.; Khalid, S.; Abdullah, M.; Ahmed, Z.; Shah, M.K.N.; Ghafoor, A.; Du, X. Insights into Drought Stress Signaling in Plants and the Molecular Genetic Basis of Cotton Drought Tolerance. *Cells* **2020**, *9*, 105. [CrossRef]
41. Givnish, T.J. Comparative studies of leaf form: Assessing the relative roles of selective pressures and phylogenetic constraints. *New Phytol.* **1987**, *106*, 131–160. [CrossRef]
42. Vogel, S. Leaves in the Lowest and Highest Winds: Temperature, Force and Shape: Tansley Review. *New Phytol.* **2009**, *183*, 13–26. [CrossRef]
43. O'toole, J.C.; Cruz, R.T. Response of Leaf Water Potential, Stomatal Resistance, and Leaf Rolling to Water Stress. *Plant Physiol.* **1980**, *65*, 428–432. [CrossRef]
44. Shepherd, T.; Griffiths, D.W. The Effects of Stress on Plant Cuticular Waxes. *New Phytol.* **2006**, *171*, 469–499. [CrossRef]
45. Mondal, S.; Singh, R.P.; Huerta-Espino, J.; Kehel, Z.; Autrique, E. Characterization of Heat-and Drought-Stress Tolerance in High-Yielding Spring Wheat. *Crop Sci.* **2015**, *55*, 1552–1562. [CrossRef]
46. Zhu, X.; Xiong, L. Putative Megaenzyme DWA1 Plays Essential Roles in Drought Resistance by Regulating Stress-Induced Wax Deposition in Rice. *Proc. Natl. Acad. Sci. USA* **2013**, *110*, 17790–17795. [CrossRef]
47. Gould, K.S. Nature's Swiss Army Knife: The Diverse Protective Roles of Anthocyanins in Leaves. *BioMed Res. Int.* **2004**, *2004*, 314–320. [CrossRef]
48. Steyn, W.J.; Wand, S.J.E.; Holcroft, D.M.; Jacobs, G. Anthocyanins in Vegetative Tissues: A Proposed Unified Function in Photoprotection. *New Phytol.* **2002**, *155*, 349–361. [CrossRef] [PubMed]
49. Close, D.C.; Beaoce, C.L. The Ecophysiology of Foliar Anthocyanin. *Bot. Rev.* **2003**, *69*, 149–161. [CrossRef]
50. Cai, T.; Ge-Zhang, S.; Song, M. Anthocyanins in Metabolites of Purple Corn. *Front. Plant Sci.* **2023**, *14*, 1154535. [CrossRef] [PubMed]
51. Zhao, S.; Blum, J.A.; Ma, F.; Wang, Y.; Borejsza-Wysocka, E.; Ma, F.; Cheng, L.; Li, P. Anthocyanin Accumulation Provides Protection against High Light Stress While Reducing Photosynthesis in Apple Leaves. *Int. J. Mol. Sci.* **2022**, *23*, 12616. [CrossRef]
52. Zheng, X.T.; Yu, Z.C.; Tang, J.W.; Cai, M.L.; Chen, Y.L.; Yang, C.W.; Chow, W.S.; Peng, C.L. The Major Photoprotective Role of Anthocyanins in Leaves of Arabidopsis Thaliana under Long-Term High Light Treatment: Antioxidant or Light Attenuator? *Photosynth. Res.* **2021**, *149*, 25–40. [CrossRef]
53. Agati, G.; Biricolti, S.; Guidi, L.; Ferrini, F.; Fini, A.; Tattini, M. The Biosynthesis of Flavonoids Is Enhanced Similarly by UV Radiation and Root Zone Salinity in L. Vulgare Leaves. *J. Plant Physiol.* **2011**, *168*, 204–212. [CrossRef] [PubMed]
54. Hernández, I.; Alegre, L.; Van Breusegem, F.; Munné-Bosch, S. How Relevant Are Flavonoids as Antioxidants in Plants? *Trends Plant Sci.* **2009**, *14*, 125–132. [CrossRef] [PubMed]
55. Balla, K.; Bencze, S.; Janda, T.; Veisz, O. Analysis of Heat Stress Tolerance in Winter Wheat. *Acta Agron. Hung.* **2009**, *57*, 437–444. [CrossRef]
56. Murchie, E.H.; Lawson, T. Chlorophyll Fluorescence Analysis: A Guide to Good Practice and Understanding Some New Applications. *J. Exp. Bot.* **2013**, *64*, 3983–3998. [CrossRef]
57. Gashu, K., Song, C.; Dubey, A.K.; Acuña, T.; Sagi, M.; Agam, N.; Bustan, A.; Fait, A. The Effect of Topo-Climate Variation on the Secondary Metabolism of Berries in White Grapevine Varieties (*Vitis vinifera*). *Front. Plant Sci.* **2022**, *13*, 847268. [CrossRef]
58. Ginori, J.; Huo, H.; Wilson, S. Physiological Response of Wax Begonia to Heat and Light Stress. *Int. Plant Propagator's Soc.* **2022**, *72*, 94–101.
59. Weyers, J.D.B.; Johansen, L.G. Accurate estimation of stomatal aperture from silicone rubber impressions. *New Phytol.* **1985**, *101*, 109–115. [CrossRef]
60. Porra, R.J.; Thompson, W.A.; Kriedemann, P.E. Determination of accurate extinction coefficients and simultaneous equations for assaying chlorophylls *a* and *b* extracted with four different solvents: Verification of the concentration of chlorophyll standards by atomic absorption spectroscopy. *Biochim. Biophys. Acta* **1989**, *975*, 384–394. [CrossRef]
61. Taghavi, T.; Patel, H.; Rafie, R. Anthocyanin Extraction Method and Sample Preparation Affect Anthocyanin Yield of Strawberries. *Nat. Prod. Commun.* **2022**, *17*, 1934578X221099970. [CrossRef]
62. Hincha, D.K.; Zuther, E. *Plant Cold Acclimation Methods and Protocols Methods in Molecular Biology*; Springer: Clifton, NJ, USA, 2020.

International Journal of
Molecular Sciences

MDPI

Article

Genome-Wide Identification of the BTB Domain-Containing Protein Gene Family in Pepper (*Capsicum annuum* L.)

Qiaoling Yuan [1,2], Jin Wang [1,2], Feng Liu [1,2], Xiongze Dai [1,2], Fan Zhu [1,2], Xuexiao Zou [1,2,*] and Cheng Xiong [1,2,*]

[1] Key Laboratory for Vegetable Biology of Hunan Province, Engineering Research Center for Horticultural Crop Germplasm Creation and New Variety Breeding, Ministry of Education, College of Horticulture, Hunan Agricultural University, Changsha 410125, China; qiaoling_y1993@163.com (Q.Y.); jinwang214@163.com (J.W.); jwszjx@hunau.edu.cn (F.L.); xiongzedai@126.com (X.D.); zf1968001@hunau.edu.cn (F.Z.)

[2] Yuelushan Lab, Changsha 410128, China

* Correspondence: zouxuexiao@hunau.edu.cn (X.Z.); xiongchenghzau@163.com (C.X.)

Abstract: Pepper (*Capsicum annuum* L.), recognized as a globally preeminent vegetable, holds substantial economic and nutritional value. The BTB (broad-complex, tramtrack, and bric-a-brac) family of proteins, characterized by a highly conserved BTB domain, also denoted as the POZ domain, are intricately involved in a diverse array of biological processes. However, the existing corpus of research regarding pepper *BTB* genes remains relatively meager. In this study, a total of 72 *CaBTB* gene members were meticulously identified from the entire genome of pepper. Phylogenetic analysis illuminated the presence of conspicuous collinear relationships between the *CaBTB* genes and those of its closely affiliated species. Gene expression profiling and RT-qPCR analysis revealed that multiple *CaBTB* genes exhibited pronounced differential expression under diverse treatment regimens. Expression pattern analysis unveiled that *CaBTB25* manifested a remarkably elevated abundance in leaves. Moreover, its promoters were replete with an abundance of light-responsive *cis*-elements. Our comprehensive and in-depth explorations into subcellular localization revealed that CaBTB25 was predominantly detected to localize within the nucleus and lacked transcriptional activation. This research provides a crucial theoretical edifice, enabling a more profound understanding of the biological functions of the *BTB* gene family in pepper, thereby underscoring its potential significance within the intricate network of gene–environment interactions.

Keywords: *Capsicum annuum* L.; *CaBTB* gene family; abiotic stresses; expression patterns; RT-qPCR

1. Introduction

The BTB (broad-complex, tramtrack, and bric-a-brac) protein was initially discovered in the fruit fly (*Drosophila melanogaster*). It encompasses a highly conserved domain consisting of approximately 115 amino acids [1]. Structurally, its core is composed of five α-helices (designated from A1 to A5) and three β-sheets (from B1 to B3). The A1/2 and A4/5 segments form an α-helical hairpin configuration, while the B1/B2/A1/A2/B3 region is connected to the A4/A5 region via helix A3 and a variable linker region [2]. Besides the core BTB domain, the diverse BTB subclasses feature extended regions at the amino-terminal (N-terminal) and carboxyl-terminal (C-terminal). These extensions are pivotal in endowing the protein with its specific functionality [2,3]. Based on the additional regions associated with the BTB domain, it can be classified into nine distinct subfamilies: BTB-only,

BTB-BACK, BTB-Kelch, BTB-back-Kelch, BTB-Arm, BTB-TAZ, BTB-ANK, BTB-PHR, and Rho-BTB [4]. Currently, within the BTB protein family, research on the NPH3 subfamily is relatively inadequate. Previous studies suggest that NPH3 subfamily proteins may be involved in plant growth and development. In maize, the *BTB* family genes exhibit distinct tissue-specific expression patterns. Although the precise function of the NPH3 subfamily remains ill defined, it can be surmised that, akin to other subfamilies, they may partake in specific physiological processes during maize growth and development [5]. Moreover, studies have demonstrated that BTB-NPH3 is associated with phototropism, root geotropism, and auxin-mediated plant development [6]. The MATH subfamily, frequently encountered in proteins engaged in cytoplasmic signal transduction, and *BTB* genes bearing certain other functional domains, may potentially influence processes such as early flowering, branching, leaf development, and long-horned fruit development [7]. The BTB Armadillo (ARM) protein, characterized by an N-terminal ARM repeat sequence and a C-terminal BTB domain, may be implicated in the growth, development, and stress resistance of sugar beet [8]. However, its exact function necessitates further in-depth exploration. The BTB-TPR protein is distinguished by an N-terminal BTB domain and six TPR motifs, along with a coiled-coil motif at the C-terminal. This protein is exclusive to the plant kingdom. In *Arabidopsis thaliana*, it is represented by three members, namely *ETO1* and *EOL1/2*, which are involved in ethylene biosynthesis [9]. Proteins containing the BACK (BTB and C-terminal Kelch) domain comprise a BTB domain and a C-terminal Kelch repeat sequence. The BTB Ankyrin protein family is present in metazoans and higher plants. Its members possess an N-terminal BTB domain, a C-terminal ankyrin protein repeat domain, and a nuclear localization sequence [10].

BTB proteins have been extensively studied in the context of plant growth and development, stress resistance, protein ubiquitination and degradation, cytoskeleton organization, ion channel regulation, and cell cycle control [11,12]. Besides, they can interact with E3 ligase CUL3 to form an E3 ubiquitin ligase complex, thereby participating in the ubiquitination process of target proteins [13]. For instance, an interaction between OsMBTB32, a protein encompassing a MATH BTB domain, and OsCUL1s has been detected to modulate salt tolerance in *Oryza sativa* [14]. The BTB-BACK-TAZ domain-containing protein MdBT2 negatively modulates the drought tolerance of apple plantlets by attenuating the positive regulatory effect of MdHDZ27 through ubiquitination [15]. Four *CmBTs* genes negatively regulate root development in chrysanthemum by mediating nitrate assimilation, amino acid biosynthesis, and the auxin and jasmonic acid (JA) signaling pathways [16]. Additionally, OsBTB97 harbors two BBX proteins, and the OsBBX19-OsBTB97/OsBBX11 module may be involved in spikelet development and seed production of rice [17].

Multiple members of the BTB protein family are involved in regulating the reaction of the plant-related signal network system mediated by hormones such as SA, JA, ABA, GA_3, etc. Subsequently, they regulate a series of physiological and biochemical processes, including plant germination, metabolism, growth, and development [3,18–20]. Yeast two-hybrid (Y2H) and bimolecular fluorescence complementation (BiFC) assays reveal that the IbBT4 protein interacts with BR-ENHANCED EXPRESSION 2 (BEE2) to enhance the BR signaling pathway and proline biosynthesis and further activate the ROS-scavenging system. Ultimately, the drought resistance of transgenic *Arabidopsis thaliana* is augmented [21]. A comprehensive RT-qPCR analysis of 11 poplar *BTB* genes in leaf, root, and xylem tissues indicates their expression patterns in response to different hormonal and biotic/abiotic stress conditions, with varying degrees of regulation manifested in the results [22]. In soybean, multiple *GmBTBs* are verified by RT-qPCR to be significantly upregulated under salinity, drought, and nitrate stress conditions [23]. Studies have shown that *CaBPM4*-silenced plants exhibit markedly increased peroxidase activity and decreased malondialdehyde concentrations, suggesting that *CaBPM4* may enhance resistance to salt and drought stress in pepper [24]. In summary, BTB proteins play an essential role in regulating plant stress responses.

Pepper (*Capsicum annuum* L.) represents a vegetable crop of substantial economic importance on a global scale. Nevertheless, the extant literature regarding the members of the *BTB* genes has been rather scarce heretofore. Consequently, in this study, a genome-wide exploration was undertaken, entailing comprehensive bioinformatics and expression analyses of *CaBTBs*, with the aim of attaining a more profound comprehension of the pepper *CaBTB* genes. Specifically, a total of 72 members of the *CaBTB* family were identified. Subsequently, their phylogenetic relationships, chromosomal localizations, gene architectures, and conserved motifs were meticulously elucidated. The expression profiles in diverse tissues and under a plethora of stress conditions were also scrutinized. Notably, *CaBTB25* was significantly enriched in leaves and harbored an abundance of light-responsive *cis*-elements within its promoters, strongly suggesting its involvement in the process of photomorphogenesis. In summary, this research expounds upon the evolutionary trajectory and functional divergence of the *BTB* gene family in pepper. Moreover, it proffers potential gene resources for the genetic enhancement of pepper, particularly with respect to tolerance to light and stress.

2. Results

2.1. Identification of BTB Genes in Pepper

In reference to the sequences of BTB proteins in *Arabidopsis thaliana* and *Oryza sativa*, a total of 72 *CaBTB* genes were identified within the 'Zunla-1' genome databases. These genes were renamed as *CaBTB1* to *CaBTB72*, contingent upon their distributions and relative linear orders along their respective chromosomes (Table S1). The analysis outcomes disclosed that the divergent genomic characteristics, coding sequence lengths, and amino acid sequences culminated in distinct relative molecular masses (MW) and isoelectric points (pI) among the identified *CaBTB* candidates. The lengths of the 72 putative proteins spanned from 126 (*CaBTB15*) to 966 (*CaBTB53*) amino acid residues, with predicted molecular weights (MW) ranging from 14.1 to 110.1 kDa. Moreover, the isoelectric points (pI) of the *CaBTB* gene family exhibited a spectrum from 3.88 to 9.87, with an average value of 6.44. A multitude of CaBTB proteins were localized in the nucleus, endomembrane system, and plasma membrane, while a small number were found in the chloroplast and other subcellular compartments.

2.2. Phylogenetic Analysis of BTB Family

To further delineate the evolutionary pathway of *BTB* genes across diverse species, a phylogenetic tree was constructed based on 72 pepper BTB amino acid sequences, along with those from tomato and *Arabidopsis* (Figure 1). The bootstrap values, which ranged predominantly from 0.88 to 1, demonstrated a reliable phylogenetic relationship. The graphical illustration revealed that all the identified BTB proteins were classified into seven subfamilies, comprising NPR1 (nonexpressor of pathogenesis-related proteins 1), MATH (meprin and TRAF homology), C-terminal Kelch-related, TAZ zinc finger, ARM (Armadillo/beta-catenin-like repeat), SKP1 (S-phase kinase-associated protein 1), and NPH3 (Nonphototropic hypocotyl 3), respectively. Notably, the NPH3 domain was the most prevalent putative domain. It was specifically present in pepper and *Arabidopsis*, with the exception of SlBTB13 of tomato.

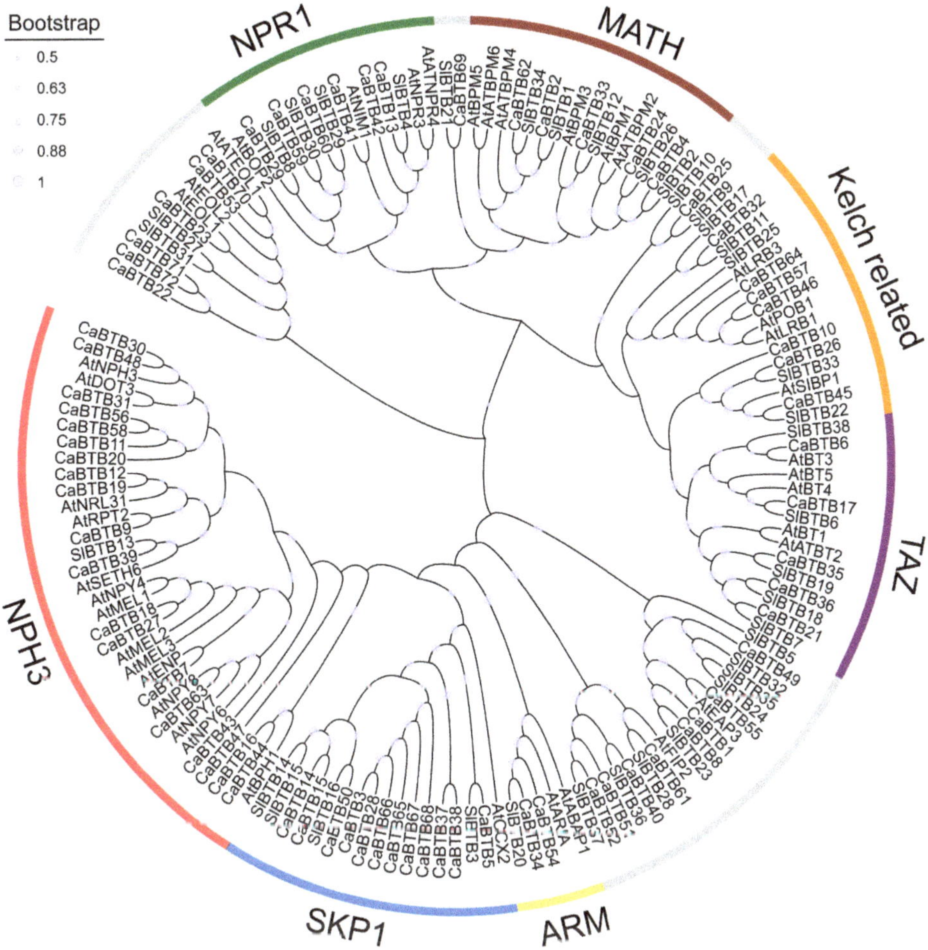

Figure 1. Phylogenetic analysis of the *BTB* gene family in *Capsicum annuum*, *Solanum lycopersicum*, and *Arabidopsis thaliana* was conducted. The full-length sequences of BTB proteins were meticulously analyzed to construct an evolutionary tree. To this end, the MEGA5.2 software was employed, and the neighbor-joining algorithm was implemented with 1000 bootstrap replications to ensure the robustness and reliability of the phylogenetic inferences.

2.3. Structure Characterization and Motif Composition of CaBTB Genes

To comprehensively understand the structural diversity, similarity, and evolutionary relationships of *CaBTB* genes, an analysis was conducted on conserved motifs, protein structures, and exon–intron arrangements. Using the MEME suite, ten conserved motifs were identified among members of the *CaBTB* gene family. Each gene contained 1–7 motifs (Figure 2a,b). The sequence logos of these motifs, presented in Figure S1, had a minimum length of 21 amino acids (motif 3) and a maximum length of 80 amino acids (motif 1). Given that members of the same subfamily shared similar motifs, the 72 *CaBTB* genes were classified into four clades. Notably, except for *CaBTB47*, motif 8 was unique to clade I at the C-terminal and absent in other clades, suggesting its significance in conferring specialized genetic functions. CaBTB proteins were categorized into several subfamilies based on the domains at the N- and C-terminal, such as the BTB-NPH3 subfamily (Figure 2c). The gene structure of *CaBTB* gene family members was characterized by the exon–intron pattern (Figure 2d). The results indicated that within the same subfamily, almost all *CaBTB* gene structures were highly conserved. However, the variation in exon–intron structures among the four distinct subgroups contributed to the diversification of the functional range of the *CaBTB* gene family during evolution.

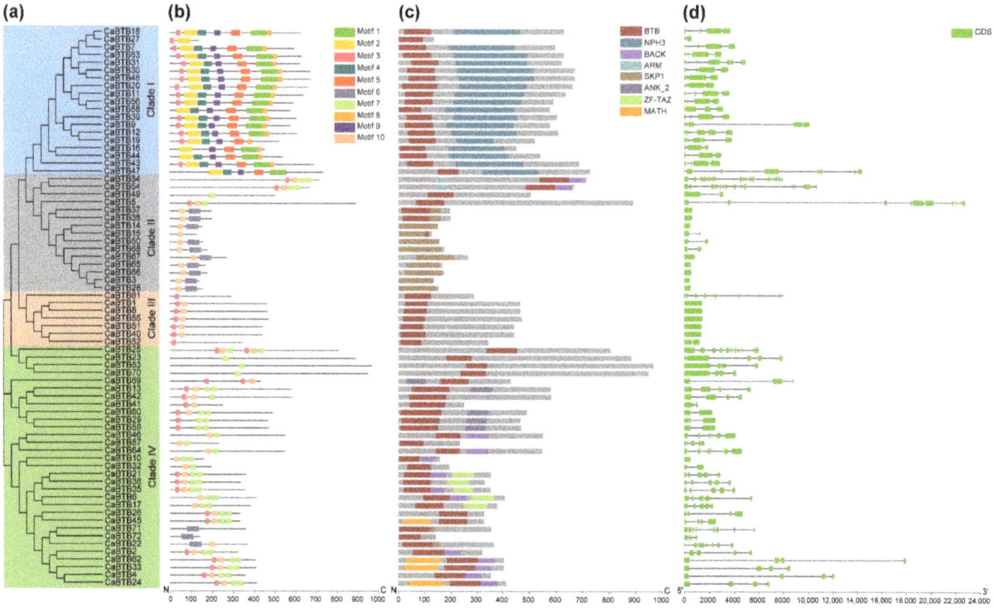

Figure 2. Conserved motif and gene structure analysis of *CaBTBs*. (**a**) Phylogenetic relationship of *CaBTB* gene family. The *CaBTB* gene family in pepper was classified into four distinct clades according to its evolutionary relationship. (**b**) Gene motifs analysis. A total of ten conserved motif structures were identified in CaBTB proteins. These motifs are represented by diverse-colored rectangles, which visually display the distribution of motifs among different CaBTB proteins. (**c**) Structure of the pepper BTB proteins. The *CaBTB* gene family was discriminated based on the functional domains located at the N- and C-terminal of the BTB domain (**d**) Gene structure. The green block represents the CDS, and the solid line indicates the intron.

2.4. *Chromosomal Localization and Duplication Analysis of CaBTB Genes*

Chromosome mapping analysis revealed that approximately 94% of *CaBTB* family genes were heterogeneously distributed across 12 chromosomes. However, four genes (*CaBTB69*, *CaBTB70*, *CaBTB71*, and *CaBTB72*) were located on as-yet-unmapped scaffolds (Figure 3). Among the chromosomes, chromosome 2 harbored the largest number of *CaBTB* genes, followed by chromosomes 1, 6, 7, and 9, respectively. Chromosome 11 had the fewest *CaBTB* genes. Interestingly, a significant number of genes were clustered at the chromosome ends, while the middle regions contained a relatively small number of genes. Segmental duplications analysis, following MCScanX searching and manual screening, confirmed the existence of *CaBTB* gene pairs. Moreover, most segmental duplications occurred between different pepper chromosomes, with fewer within the same chromosome. These findings were highly consistent with the evolutionary grouping tree, as each pair of segmentally duplicated genes was located within the same phylogenetic clade.

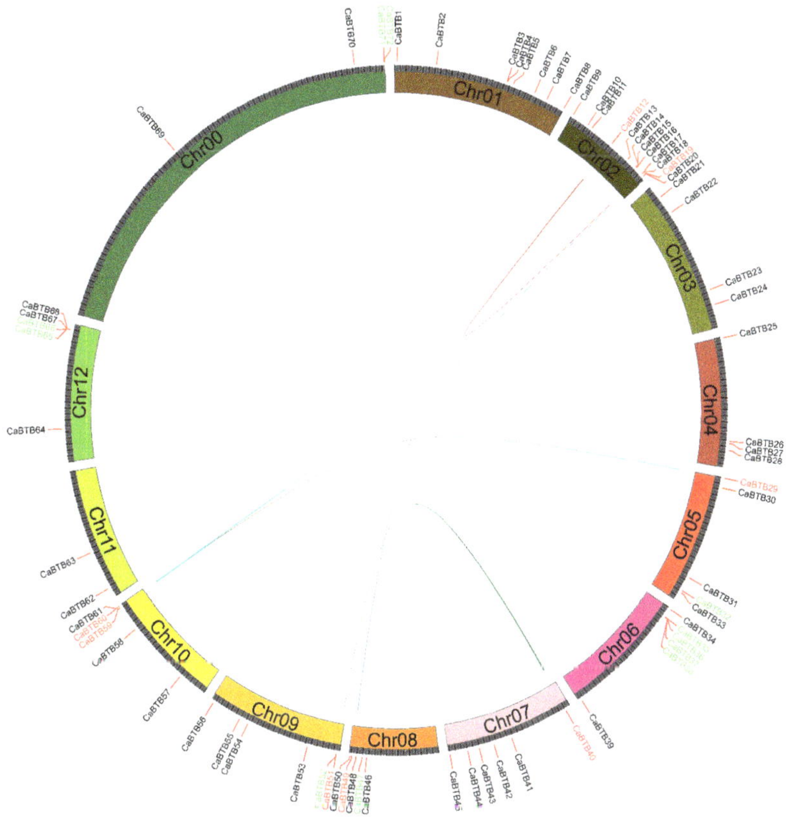

Figure 3. Chromosomal localization and duplication events of *BTB* genes in pepper. The decentralized rings denote the distribution of consecutive chromosomes and the interchromosomal linkages. Multicolored solid-dotted lines are applied to signify the duplicating gene pairs of *CaBTBs*.

2.5. Cis-Acting Element Prediction of CaBTB Promoters

Investigating the putative *cis*-acting elements is crucial for a deeper familiarization of the *CaBTB* gene expression level. Additionally, it helps elucidate the potential functions and molecular mechanisms under the regulation of their corresponding trans-regulatory factors. Therefore, the promoter sequences located 2000 bp upstream of *CaBTB* genes were retrieved using TBtools and analyzed via the PlantCARE database for *cis*-acting element identification. The data revealed the presence of a total of 47 *cis*-acting elements. Invariably, the core promoter elements TATA-box and CAAT-box (not shown in the legend due to their large numbers) were present in all *CaBTB* promoters. The remaining *cis*-acting elements were mainly classified into four categories (Figures 4 and S2). Light-responsive elements were the most prevalent, amounting to 16 (34%), indicating that *CaBTB* gene family members are highly induced by light. Among them, another category consisted of 12 elements that functioned as abiotic and biotic stress-responsive elements, including box-S, CARE, CCAAT-box, LTR, MBS, MYB, MYC, STRE, TC-rich repeats, W box, WRE3, and WUN-motif. Specifically, 10 elements were associated with the adaptation to endogenous hormone changes: ABRE, as-1, AuxRR-core, CGTCA-motif, DRE, ERE, GARE-motif, TCA-element, TGACG-motif, and TGA-element. Moreover, the remaining nine *cis*-acting elements, accounting for 19% of the total number of elements, were involved in plant growth and development. Collectively, the presence of diverse response components provides empirical support for the regulatory role of *CaBTB* genes in coordinating plant biotic and abiotic responses through hormone signaling pathways.

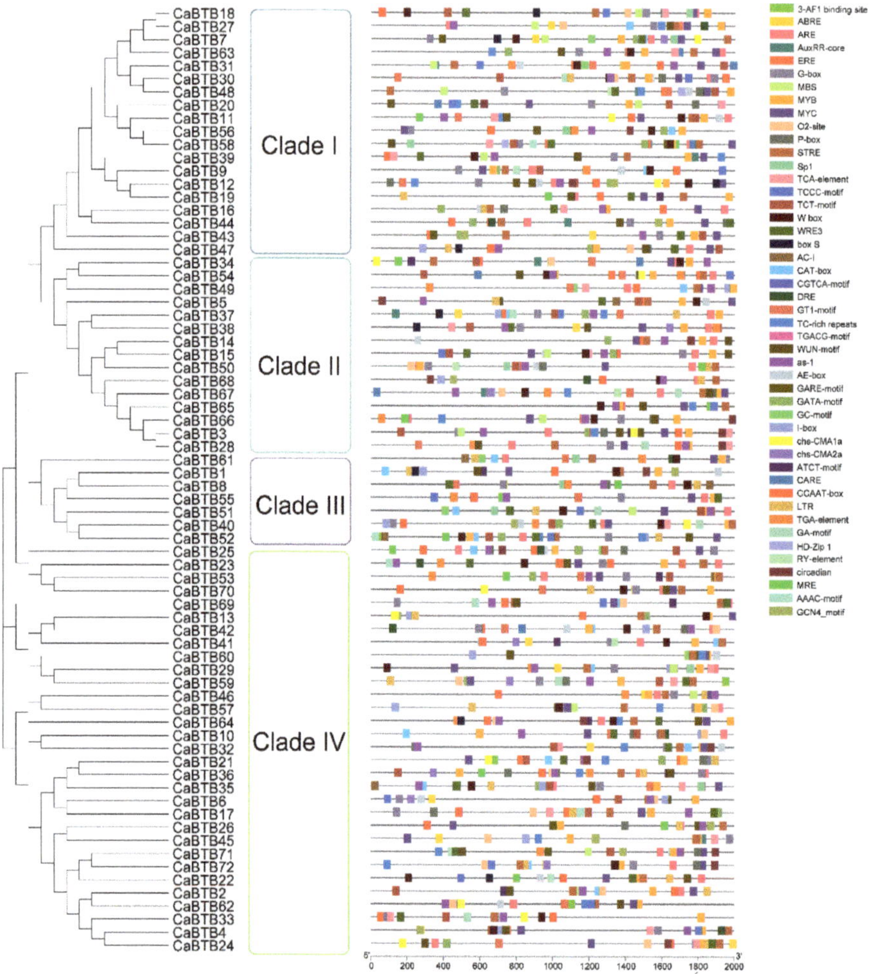

Figure 4. Analysis of *cis*-acting element in the *BTB* gene family of pepper.

2.6. Expression Pattern Analysis of CaBTB Genes in Various Tissues

To precisely determine the transcription levels of 72 pepper *BTB* genes across 14 tissues, namely root, stem, leaf, bud, flower, and nine stages of fruit development, a hierarchical clustering heatmap was constructed based on the previously obtained 'Zunla-1' transcriptome data. In view of the aforementioned *cis*-element analysis findings, the majority of *CaBTB* genes are presumably involved in the photoperiodic response. Given that the principal site of photoperiod perception in plants is the leaf, especially the mature leaf, six *CaBTB* genes (*CaBTB1*, *CaBTB21*, *CaBTB25*, *CaBTB29*, *CaBTB34*, and *CaBTB42*) exhibiting relatively high transcriptional levels in leaves were selected for RT-qPCR validation to corroborate the accuracy of the data. The RT-qPCR results were congruent with those of RNA-sequencing (Figure 5b–g). Among the identified *CaBTB* genes, *CaBTB3*, *CaBTB14*, and *CaBTB72* were clearly not expressed. Conversely, the vast majority of genes were expressed across nearly all tissues and developmental stages. Nevertheless, several genes demonstrated tissue-specific and stage-dependent expression patterns (Figure 5a). For instance, *CaBTB32*, *CaBTB60*, and *CaBTB70*, members of clade IV, exhibited pronounced enrichment in buds and floral tissues. This expression profile strongly suggests that these genes play pivotal roles in the morphogenesis and ontogeny of floral organs.

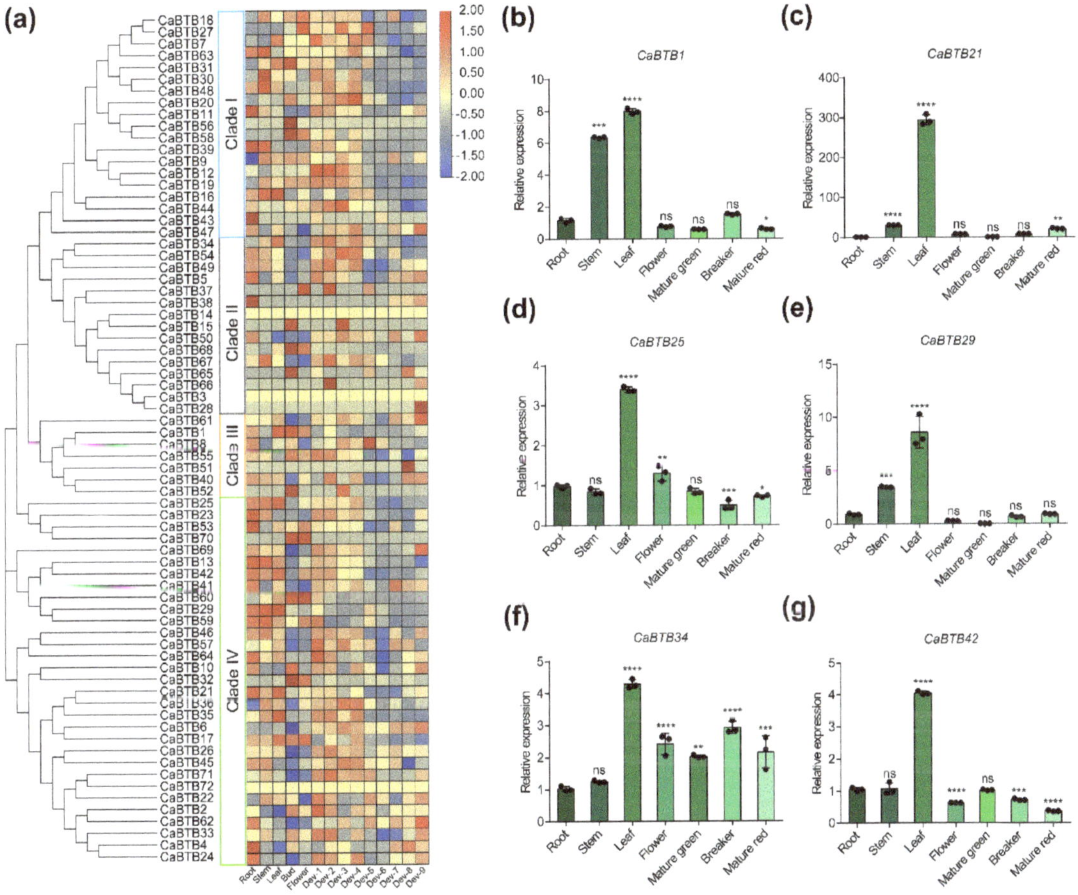

Figure 5. Expression pattern of *CaBTB* genes in pepper. (**a**) By integrating the outcomes of phylogenetic analysis, we harnessed transcriptome datasets to comprehensively profile the tissue-specific expression patterns of the *CaBTB* gene family across diverse tissues and organs. The tissues include the following: root (developing roots), stem (developing stems), leaf (mature leaves), bud (closed flower buds), flower (blooming flowers), Dev1 (fruits with a length of 0–1 cm), Dev2 (fruits with a length of 1–3 cm), Dev3 (fruits with a length of 3–4 cm), Dev4 (fruits with a length of 4–5 cm), Dev5 (mature green fruits), Dev6 (fruits at breaker stage), Dev7 (fruits 3 days after breaker stage), Dev8 (fruits 5 days after breaker stage), and Dev9 (fruits 7 days after breaker stage). (**b**–**g**) The relative expression levels of six *CaBTB* genes in various tissues were measured. Values are presented as means ± SD (n = 3 biological replicates). Significant differences were determined by one-way ANOVA analysis (* $p < 0.05$; ** $p < 0.01$; *** $p < 0.001$; **** $p < 0.0001$; ns, no significant difference).

2.7. Expression Profiles of CaBTB Genes in Response to Exogenous Hormones

The *CaBTB* gene family members likely partake in multiple biological processes, contingent upon the profusion of *cis*-acting elements enriched within the promoter analysis outcomes. These elements are associated with phytohormone responses and abiotic stress. To explore the potential functions of *CaBTB* genes in their response to phytohormones, heat map visualizations were employed to analyze the expression profiles of all *CaBTB* genes in either pepper leaf or root tissues. These tissues were subjected to five phytohormone treatments—ABA, GA_3, IAA, JA, and SA—and were contrasted with untreated control samples. The results revealed that the expression levels of the majority of *CaBTB* genes were markedly influenced by diverse treatments, manifesting as varying degrees of upregulation or downregulation. Nevertheless, *CaBTB3, CaBTB14, CaBTB56,* and *CaBTB66* were not detectable in terms of expression, thereby elucidating the co-existence of both correlations and disparities in the expression patterns among these *CaBTB* members.

Figure 6a demonstrated that upon treatment with abscisic acid (ABA, 30 μM), the expression levels of *CaBTB7, CaBTB30, CaBTB36, CaBTB48,* and *CaBTB72* were upregulated in leaves at stage AL4. Notably, *CaBTB28* was distinctly upregulated only at stage AL2. Conversely, *CaBTB10, CaBTB18, CaBTB69,* and *CaBTB71* showed pronounced downregulation in expression at stage AL3. In contrast, the expression of three other *CaBTB* genes, namely clustered *CaBTB32, CaBTB51,* and *CaBTB70,* was significantly upregulated in roots. When treated with gibberellic acid (GA_3, 2 μM), there was a substantial difference in expression between leaves and roots. Specifically, *CaBTB27, CaBTB36, CaBTB37, CaBTB51,* and *CaBTB67* were significantly upregulated in leaves, while the expressions of *CaBTB49* and *CaBTB63* were inhibited by GA_3 treatment (Figure 6b). In response to indole acetic acid (IAA, 2 μM) treatment, the majority of *CaBTB* gene expressions were downregulated in leaves, with a particularly significant decrease in the transcriptional abundance of the *CaBTB6* gene. Notably, *CaBTB50* and *CaBTB65* showed markedly higher upregulation when exposed to IAA treatment in roots (Figure 6c). Similarly, after jasmonic acid (JA, 10 μM) treatment, numerous gene expressions were downregulated. For instance, *CaBTB2, CaBTB8, CaBTB21,* and *CaBTB49* genes underwent significant changes (Figure 6d). However, *CaBTB36, CaBTB37,* and *CaBTB58* were upregulated in leaves. Simultaneously, most pepper *BTB* genes were upregulated in roots under JA treatment. During the salicylic acid (SA, 2 mM) treatment, the expression levels of *CaBTBs* showed a significant response in roots, presenting an opposite pattern to that in leaves (Figure 6e).

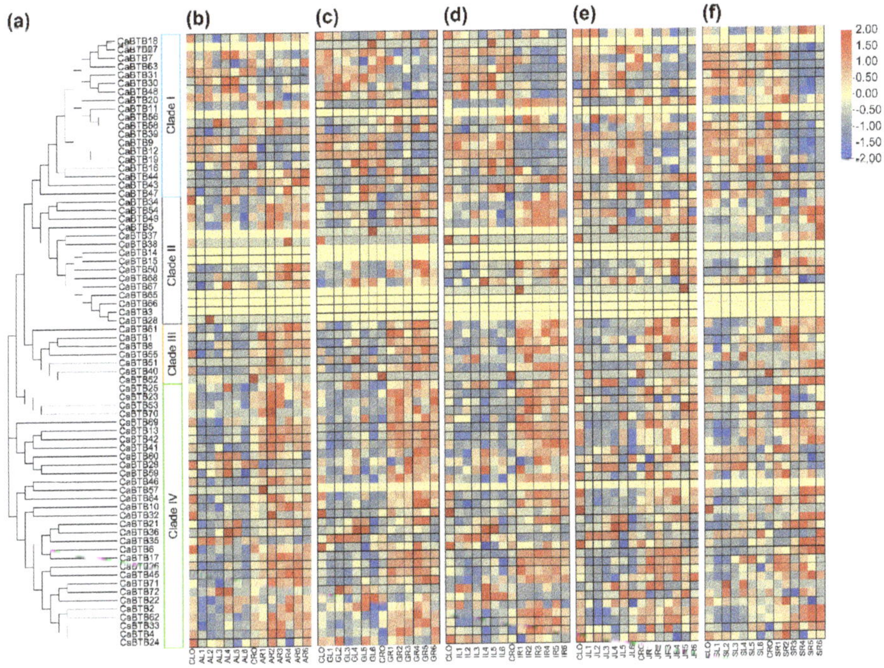

Figure 6. Expression profiles of *CaBTB* genes under exogenous hormones. Red blocks denote genes with high expression levels, while blue blocks represent genes with low expression levels. (**a**) Phylogenetic analysis. (**b**) ABA treatment. (**c**) GA_3 treatment. (**d**) IAA treatment. (**e**) JA treatment. (**f**) SA treatment. CL: control leaf; CR: control root; AL: ABA-treated leaf; AR: ABA-treated root; GL: GA_3-treated leaf; GR: GA_3-treated root; IL: IAA-treated leaf; IR: IAA-treated root; JL: JA-treated leaf; JR: JA-treated root; SL: SA-treated leaf; SR: SA-treated root. All samples were collected from leaves or roots, with three biological replicates for each treatment.

2.8. Expression Profiles of CaBTB Genes Under Multiple Abiotic Stresses

To further explore the potential role of pepper *BTB* genes under multiple abiotic stresses, we analyzed their expression patterns. Pepper leaves and roots were treated with low-temperature stress (cold), oxidative stress (H_2O_2), high-temperature stress (heat), dehydration stress (mannitol), and salt stress (NaCl). Low-temperature treatment significantly induced the upregulation of *CaBTB57* and *CaBTB65* in leaves (Figure 7a) while downregulating *CaBTB24*, *CaBTB39*, *CaBTB49*, and *CaBTB63*. The expression patterns of some genes (*CaBTB3*, *CaBTB14*, *CaBTB15*, *CaBTB28*, *CaBTB56*, and *CaBTB66*) were undetectable. In roots, cold stress led to a significant increase in the expression levels of *CaBTB1*, *CaBTB27*, *CaBTB38*, *CaBTB57*, and *CaBTB58*, while *CaBTB2* and *CaBTB48* were markedly suppressed. The gene expression changes under H_2O_2 treatment were found to be similar to those under low-temperature stress (Figure 7b). Notably, *CaBTB27* and *CaBTB57* showed a more pronounced increase in leaves with enhanced oxidation effects. Elevated temperature mainly upregulated the expression of *CaBTB51* and *CaBTB72* in both leaves and roots (Figure 7c). Conversely, *CaBTB4* and *CaBTB42* were sharply downregulated by high temperature. Almost all *CaBTB* genes showed opposite expression patterns in leaves and roots under drought stress (Figure 7d). For example, *CaBTB47* was upregulated in leaves but downregulated in roots. In leaves treated with sodium chloride stress (Figure 7e), *CaBTB2*, *CaBTB10*, *CaBTB18*, *CaBTB53*, and *CaBTB71* were transcribed at relatively low levels, while *CaBTB7*, *CaBTB16*, *CaBTB30*, *CaBTB36*, *CaBTB37*, *CaBTB47*, and *CaBTB58* were evidently enriched. In roots, except for the increased expression of *CaBTB51* and *CaBTB65*, the expression changes of other genes were not significant.

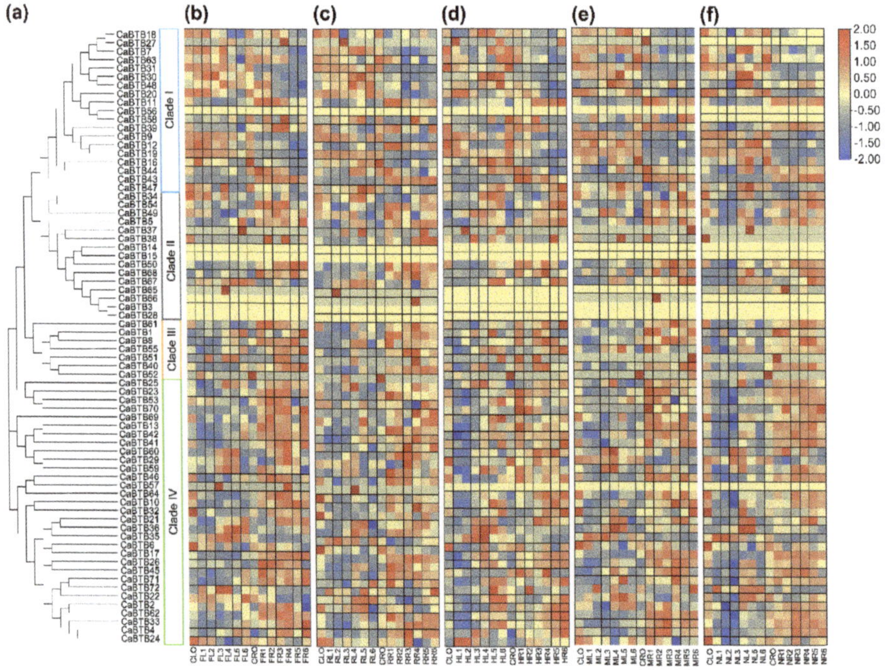

Figure 7. Expression profiles of *CaBTB* genes under different abiotic stresses. Colorful squares indicate $\log_2$ expression value. (**a**) Phylogenetic analysis. (**b**) Cold treatment. (**c**) H_2O_2 treatment. (**d**) Heat treatment. (**e**) Mannitol treatment. (**f**) NaCl treatment. CL: control leaf; CR: control root; FL: cold-treated leaf; FR: cold-treated root; RL: H_2O_2-treated leaf; RR: H_2O_2-treated root; HL: heat-treated leaf; HR: heat-treated root; ML: mannitol-treated leaf; MR: mannitol-treated root; NL: NaCl-treated leaf; NR: NaCl-treated root. All samples were collected from leaves or roots, with three biological replicates for each treatment.

2.9. Homology, Transcription Activation, and Subcellular Localization of CaBTB25

In this study, our focus was directed towards the *CaBTB25* gene. Given its particularly high expression in leaves and the presence of a greater number of light-responsive *cis*-elements in its promoter regions, it was hypothesized that *CaBTB25* might play a pivotal role in photomorphogenesis. To further explore this, a phylogenetic tree was constructed by comparing the amino acid sequences of CaBTB25 with its homologous sequences from 29 other species. As depicted in Figure 8a, CaBTB25 exhibited the closest relationship with tobacco (*Nicotiana tabacum* L.) due to their membership of the same Solanaceae family. Moreover, the growth status of yeast cells was examined to assess the transcriptional activity of CaBTB25. Using the combinations BD-Lam + AD-T and BD + AD as negative controls and the combination BD-53 + AD-T as the positive control, it was revealed that the full-length CaBTB25 lacked transcriptional activity (Figure 8b). Subcellular localization analysis of the CaBTB25-eGFP fusion protein, carried out in *N. benthamiana* leaves with mCherry serving as a marker for nuclear proteins, demonstrated that CaBTB25 was localized within the nucleus (Figure 8c).

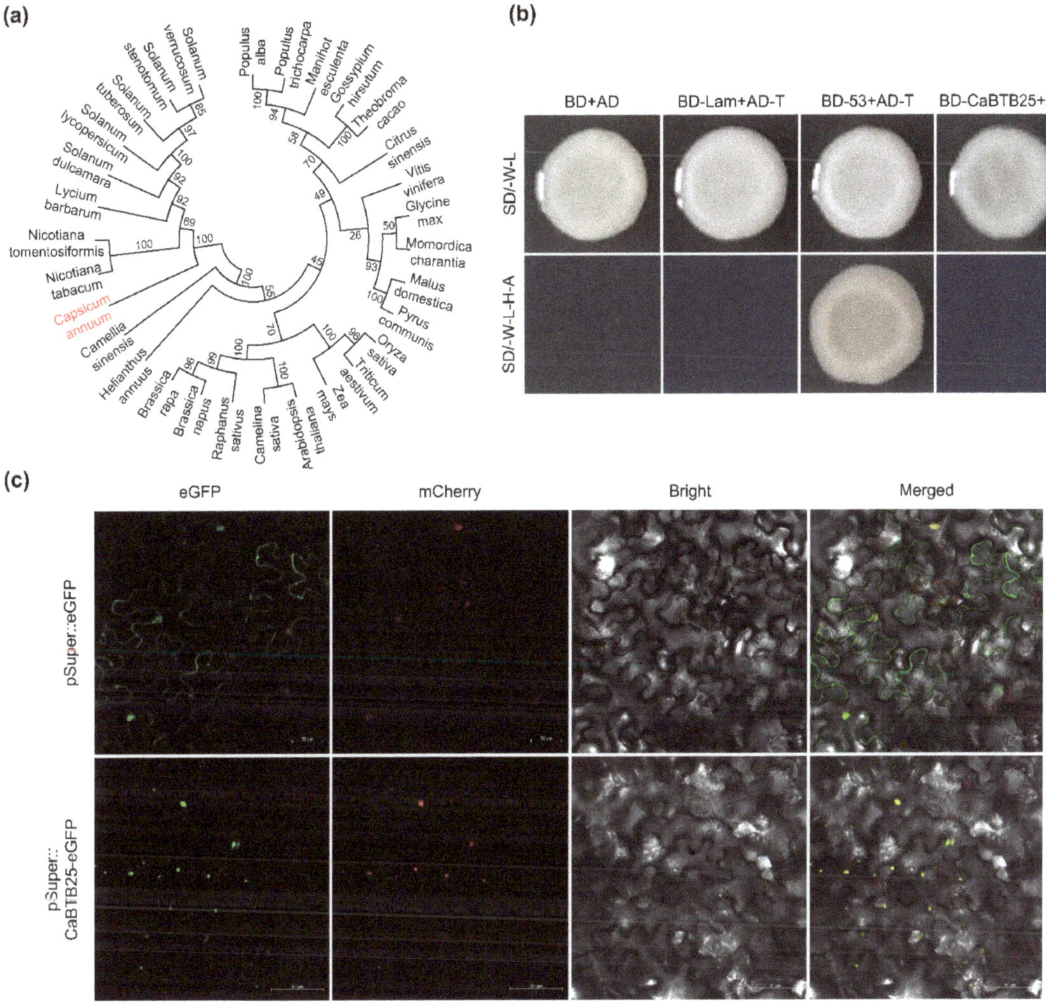

Figure 8. Gene basic information of *CaBTB25*. (**a**) Phyloenetic analysis of CaBTB25 with its homolog proteins in other species. (**b**) Transcriptional activation of CaBTB25 in yeast cells. BD + AD indicates pGBKT7 + pGADT7; BD-Lam + AD-T indicates pGBKT7-Lam + pGADT7-T; BD-53 + AD-T indicates pGBKT7-53 + pGADT7-T; BD-CaBTB25 + AD indicates pGBKT7-CaBTB25 + pGADT7. SD/-W-L, selective medium lacking tryptophan and leucine; SD/-W-L-H-A, selective medium lacking tryptophan, leucine, histidine, and adenine. (**c**) Subcellular localization analysis of CaBTB25-eGFP recombinant protein. Scale bar, 50 μm.

3. Discussion

As previously documented, proteins belonging to the BTB/POZ family play pivotal roles in a plethora of functions within plants, encompassing growth and development, cytoskeletal modulation, chromatin remodeling, transcriptional regulation, and proteolytic degradation [7]. Extensive research has been carried out on BTB proteins in animals [25–27]. In recent times, BTB proteins have been identified in a wide array of plant species, with 158 members in rice [28], 38 members in tomato [29], 95 members in poplar [22], 122 members in soybean [23], 69 members in grape [30], and 62 members in *paulownia* [31]. In this study, we executed a comprehensive and systematic bioinformatics-based analysis of the BTB family members in chili pepper. The number of *CaBTB* genes was determined to be 72, which reveals the distinct genomic sizes across diverse plant species, suggesting that *BTB* genes have undergone evolutionary alterations. In this investigation, based on the BTB domain in conjunction with other conserved domains, 72 *CaBTB* members were categorized into 7 subfamilies. However, in comparison to the tomato *BTB* subfamily classification, pepper lacks the KCTD subfamily (Figure 1). This phenomenon may be attributed to the specific gene contraction in pepper, which has adapted to specific ecological environments during long-term evolution. Additionally, the completeness and accuracy of genome annotation have, to some degree, influenced the results.

Based on the phylogenetic relationships, the *CaBTB* genes in pepper were classified into four distinct clades comprising seven subfamilies. Members within these subfamilies exhibited similarities in terms of their motifs, protein structures, and exon–intron organizations. In peach, four novel subgroups of BTB proteins—BTBAND, BTBBL, BTBP, and BTBAN—were identified for the first time [32]. Notably, the newly discovered BTBAND subgroup was demonstrated to be a composite of ANK, NPR, and DUF domains. These domains, which are components of these new subgroups, have been detected in previous investigations [2,4,29], suggesting that these genes may execute novel or hitherto uncharacterized functions during plant growth and development. These studies imply that within the diverse subfamilies of BTB proteins, each subfamily member likely performs analogous functions in plant-specific biological processes. There exist intricate interactions between proteins and other molecules within different subcellular compartments. Elucidating the subcellular localization of proteins is essential for uncovering the underlying mechanisms of these molecular interactions. The expression patterns and subcellular localization of BPM proteins in *Arabidopsis* display distinct characteristics across various tissues and under abiotic stress conditions [33]. By investigating gene structure and motif distribution, the composition, organization, and evolution of genes in different organisms can be ascertained. This knowledge forms the cornerstone for comprehending gene function and regulatory mechanisms. We also observed a high degree of conservation among BTB family members in pepper, resulting in genes within the same subfamily sharing identical or similar exon–intron structures and motif types (Figure 2). Evidently, genes with multiple introns may experience transcriptional retardation, and their gene expression levels may be inhibited in response to stressful environments [34,35]. The motifs within the BTB domain are ubiquitously present across eukaryotes and are significant for their interactions with other proteins, transcriptional activity, and DNA binding [36]. In pepper *BTB* genes, we identified 10 motifs, which may aid in the identification of potential transcription factor binding sites, thus facilitating our understanding of the diverse gene regulatory mechanisms (Figure S1).

The *CaBTB* genes exhibit an uneven distribution across the twelve chromosomes of the pepper genome. The overwhelming majority of these genes are localized at the termini of the chromosomes and predominantly exist in tandem duplications (Figure 3). Gene duplication represents one of the primary driving forces in genome evolution. In pepper, gene duplication events are likely responsible for the expansion of the *BTB* gene family. A substantial number of gene duplications provide pepper with supplementary genetic material and augment gene diversity [37]. For example, in other plant species, gene duplication events frequently precipitate the expansion of gene families, endowing plants with the capacity to adapt to diverse environmental conditions. As a crucial economic crop, the expansion of the *BTB* gene family in pepper may facilitate its survival and reproduction in various growth environments. In grasses, the MATH-BTB protein functions as a substrate-specific adaptor within the context of ubiquitin E3 ligase and exerts a regulatory effect during the transition from meiosis to mitosis. In contrast to *Arabidopsis*, which possesses only six homologous genes, this family has undergone extensive amplification in grasses. Research has shown that this amplification is predominantly attributed to local gene duplication [38].

Analyzing the spatiotemporal expression patterns of genes, along with the expression levels of regulated genes and their responses to environmental and endogenous signals, can reveal promoter analysis [39,40]. This offers novel insights and methodologies for researching the molecular mechanisms of plant growth and development, breeding superior varieties, and studying plant development. In sweet potato, several stress-responsive *cis*-acting regulatory elements, including the GARE-motif, O2-site, LTR, TC-rich repeats, TCA-element, MBS, ARE, and MBSI, were detected in the promoter regions of *IbBT4* [21]. In *GmBTBs* promoters, various light-, hormone-, and stress-responsiveness *cis*-elements were identified, suggesting that the associated proteins may be involved in soybean growth and stress responses [23]. Regarding the *cis*-elements analysis in *CaBTB* gene promoters, we obtained a series of abiotic and biotic stress-responsive elements, light-responsive elements, phytohormone-responsive elements, and plant growth and development-responsive elements (Figure 4 and Supplemental Figure 2). Furthermore, light-responsive elements are ubiquitously present and are the most abundant, which is consistent with the findings in tomato [29], sugar beet [8], grape [30], and *paulownias* [31]. This implies that BTB family members are likely influenced by light in most plant species. It is essential to highlight that the ubiquitin ligase CONSTITUTIVELY PHOTOMORPHOGENIC 1 (COP1) assumes a pivotal role in light-induced anthocyanin biosynthesis. In the context of apple, MdBT2, a BTB protein, has been demonstrated to directly interact with and stabilize MdCOP1 through the inhibition of self-ubiquitination [41]. This process is closely intertwined with the photoperiodic response, as light perception and subsequent signaling pathways are fundamental to regulating anthocyanin production. The potential connection between *CaBTB* genes and the COP1-mediated, light-induced anthocyanin biosynthesis pathway warrants further investigation to elucidate the complex molecular mechanisms underlying photoperiod-related physiological processes in plants. Within the purview of this study, *CaBTB25*, the CaBTB25-eGFP fusion protein of which is localized to the nucleus and devoid of transcriptional activity, manifested the most conspicuous level of expression in leaves, and its promoter was replete with an ample quality of light-responsive *cis*-acting elements (Figure 8). It is postulated that this phenomenon is associated with photomorphogenesis. Meanwhile, stress-responsive elements such as MBS, MYB, and W box were also identified in the promoters of *CaBTB* genes. Previous research on pepper has demonstrated that the transcript level of *CaBPM4* (*CaBTB62*) increases when the plant is exposed to abiotic stresses such as salt, cold, and drought, owing to the presence of corresponding response elements in the promoter site [24]. In conclusion, the analysis of pepper *cis*-elements is of great significance for uncovering gene expression regulation mechanisms, studying pepper's response to adversity, guiding pepper variety improvement, and exploring the functions of the pepper gene family.

Based on the expression patterns and RT-qPCR results of *CaBTB* genes in diverse tissues, our findings demonstrated that three genes, namely *CaBTB3*, *CaBTB14*, and *CaBTB72*, were not expressed. Conversely, the remaining genes exhibited high expression levels across nearly all the examined pepper tissues (Figure 5a). Certain specific *CaBTB* genes, such as *CaBTB1*, *CaBTB21*, *CaBTB25*, *CaBTB29*, *CaBTB34*, and *CaBTB42*, manifested high expression levels in the leaves, implying their potential involvement in leaf development (Figure 5b–g). TMF (TERMINATING FLOWER) is recognized as a transcription factor during the formation of multiflowered inflorescences. In tomato, BOP (BLADE-ON-PETIOLE) transcriptional cofactors, characterized by the conserved BTB/POZ domain, were found to mediate the regulation of inflorescence architecture through interaction with TMF [42]. Intriguingly, *CaBTB60*, which is a homologous gene of *SlBOP3*, is highly transcribed in pepper buds and flowers. This suggests that *BTB* genes may play a role in flower development, although their specific functions await further investigation.

Previously, studies have elucidated that the BTB protein assumes a critical role in plants' defense mechanisms against both abiotic and biotic stresses. A multitude of plant BTB/POZ protein family members are implicated in governing hormone-mediated processes and phytologically relevant signal network responses [31]. These proteins serve as pivotal regulators, precisely modulating the intricate signaling cascades initiated by diverse hormones, including JA [43], SA [44,45], ABA [46], and GA_3. *OsBTBZ1*, a salt-tolerant gene operating within the ABA-dependent pathways, not only restores the phenotypes of the *bt3* mutant line but also augments the growth of wild-type (WT) *Arabidopsis* under stress conditions when ectopically overexpressed [46]. Additionally, in the ethylene (ET)/jasmonic acid (JA) pathway of *Arabidopsis thaliana*, the expression of *AtBT4* is markedly upregulated, thereby promoting the plant's resistance to *Botrytis griseus* [47]. The DELLA protein functions as a central regulatory factor that interconnects the gibberellin (GA) signaling pathway with other signaling pathways. It accomplishes this by directly interacting with key transcription factors or regulators, thus playing an indispensable role in integrating and coordinating various signaling processes within the plant [48,49]. Research findings demonstrate that MdBT2 is involved in regulating nitrate-induced plant growth through its interaction with the DELLA protein MdRGL3a. Furthermore, compared to WT plants, the OE-MdBT2 lines exhibit a substantial increase in plant height and biomass [50]. In our study, to explore the potential response capabilities of *CaBTB* genes to abiotic and biotic stresses, we harnessed RNA-seq data to analyze their expression patterns under cold, H_2O_2, heat, mannitol, and NaCl conditions, as well as various hormone treatments. Obviously, the expression level of *CaBTB36*, which is homologous to *MdBT2* in apple, was remarkably upregulated upon treatment with ABA, GA_3, and JA. These results imply that *CaBTB36* may play significant roles in plant responses to different environmental stimuli and hormonal regulations. Further research on its function could offer deeper insights into the molecular mechanisms underlying plant stress responses and growth regulations. By modulating these processes, BTB proteins can exert an influence on plant growth, development, and responses to environmental stimuli. Their functions in signal network responses are also of utmost importance for plants to coordinate different physiological processes, ensuring proper growth and survival in changing environments. This further underscores the central and diverse roles of the BTB protein family in plant biology.

4. Materials and Methods

4.1. Plant Materials and Growth Conditions

In this study, the wild-type (WT) parent 'Zhangshugang' line, a widely employed backbone and breeding parent in China's pepper breeding program and an advanced inbred line, was sourced from the Pepper Research Laboratory, College of Horticulture, Hunan Agricultural University (latitude 28°11′49″ N, longitude 112°58′429″ E, Changsha, China). Prior to cultivation, seeds were germinated at 55 °C for 15 min. Subsequently, all plants were cultivated in growth chambers under a 16-h light (25 °C)/8-h dark (18 °C) photoperiod cycle, with a relative humidity of 65% and a light intensity of 300 $\mu mol \cdot m^{-2} \cdot s^{-1}$. In the context of *Nicotiana tabacum* L., nuclear-localized tobacco proteins were tagged with a nuclear-localization marker. Concurrently, tobacco plants were cultured in an incubator maintained at a constant temperature of 22 °C. The photoperiod was strictly regulated to a

16-hour light/8-hour dark cycle. At the 8-to-10-leaf developmental stage, leaves that were fully expanded and exhibited optimal physiological vigor were meticulously selected for subsequent biochemical analyses.

4.2. Identification of Pepper Genes Encoding BTB Domain Proteins

The CaBTB protein sequences were retrieved from the high-quality 'Zunla 1' Pepper Genomic Database website (https://solgenomics.net/organism/Capsicum_annuum/genome, accessed on 13 January 2025). The *Solanum lycopersicum* L. genome (SGN: http://solgenomics.net/, accessed on 5 January 2025) and *Arabidopsis thaliana* genome (TAIR: https://www.Arabidopsis.org/, accessed on 5 January 2025) were referenced. Moreover, the HMM file (PF00651) of the conserved domain was obtained from the Pfam database website (http://pfam-legacy.xfam.org/, accessed on 13 January 2025). The prediction of the BTB domain was validated using the NCBI-CDD database (http://www.ncbi.nlm.nih.gov/cdd/, accessed on 13 January 2025) and SMART database (http://smart.embl-heidelberg.de, accessed on 13 January 2025), with redundant sequences being filtered out. The physicochemical properties, such as the relative molecular weight (MW) and isoelectric point (pI) of the final CaBTB proteins, were analyzed via the ExPASy website (http://www.expasy.org, accessed on 13 January 2025). Subcellular localization prediction was conducted using the BUSCA website (https://busca.biocomp.unibo.it/, accessed on 13 January 2025) and CELLO online analysis tool (http://cello.life.nctu.edu.tw/, accessed on 13 January 2025).

4.3. Phylogenetic Analysis of the BTB Proteins in Pepper

In the context of analyzing the amino acid sequences of tomato (*Solanum lycopersicum*) and *Arabidopsis thaliana*, in addition to those of pepper (*Capsicum annuum*) BTB family members, the MEGA (Molecular Evolutionary Genetics Analysis, available at (https://www.megasoftware.net/, Version 11.0)) software was utilized to construct a phylogenetic tree. The neighbor-joining (NJ) algorithm was implemented, with the *p*-distance model adopted as the substitution model and 1000 bootstrap replicates performed for statistical robustness. As a result, the *CaBTB* gene family in pepper was partitioned into four distinct subfamilies.

4.4. Gene Structure and Significant Motifs Analysis

TBtools (Version 2.152) was harnessed to delineate the gene structure maps of *CaBTB* gene family members [51]. Concurrently, the exon–intron distribution was annotated based on the coding DNA sequence (CDS) and genome sequences of the *CaBTB* genes, leveraging the GSDS (http://gsds.cbi.pku.edu.cn/, accessed on 13 January 2025) visualization platform. For the prediction of conserved motifs, the Multiple Expectation Maximization for Motif Elucidation (MEME) tool (http://meme-suite.org/tools/meme, accessed on 13 January 2025) was utilized. The search mode was configured as 'any number of repetitions', the number of motifs was set to 10, the minimum width was specified as 6 amino acid residues, and the maximum width was set at 100 amino acid residues.

4.5. Chromosomal Distribution and Detection of Gene Duplication

The chromosomal localization of each *CaBTB* gene was visualized by discerning its position on chromosomes, referring to the online Sol Genomics Network (SGN) database. The MCScanX (Version 2019) software was utilized to detect segmental duplication alignments and identify intraspecific gene pairs within the pepper genome. Subsequently, the chromosome mapping and gene duplication results of the *CaBTB* genes were visualized using TBtools software, enabling a comprehensive and detailed display of the genetic relationships and distribution patterns.

4.6. Prediction of the Cis-Acting Elements in the CaBTB Promoters

The 2-kilobase (kb) upstream regions of each CDS within the *CaBTB* gene family were extracted using TBtools software, which were designated as the promoter regions. Subsequently, all the *BTB* genes under investigation were submitted to the PlantCARE online database (http://bioinformatics.psb.ugent.be/webtools/plantcare/html/, accessed on 15 January 2025) for the prediction of primary stress-response *cis*-elements. Finally, all the identified *cis*-acting elements were visualized with the assistance of TBtools software, facilitating a comprehensive and intuitive understanding of their distribution patterns.

4.7. Expression Pattern Analysis of Pepper CaBTB Genes

The gene expression and RNA-seq data were downloaded from the National Center for Biotechnology Information (NCBI) under the accession number GSE45037 and the Pepper Information Data Center (http://lifenglab.hzau.edu.cn/PepperHub/index.php, accessed on 15 January 2025) [52,53]. The 'Zunla-1' genome was utilized as the reference genome. Experimental treatment conditions and statistical analysis methods were implemented in accordance with those described in previous studies [53]. Subsequently, the gene expression results were calculated, and heatmaps were generated using TBtools software, enabling a comprehensive visualization and interpretation of the gene expression profiles.

4.8. RNA Isolation and Reverse Transcription Quantitative PCR (RT-qPCR)

Total RNA was extracted from the leaves of WT pepper 'Zhangshugang' using the FastPure® Universal Plant Total RNA Isolation Kit (Vazyme Biotech Co., Ltd., Piscataway, NJ, USA). Each sample group was prepared with three biological replicates. Subsequently, 1 µg of the isolated total RNA was reverse-transcribed into complementary DNA (cDNA) using the HiScript® IIQ RT SuperMix (+ gDNAwiper) kit (Vazyme Biotech) in a 20 µL reaction system. Real-time quantitative polymerase chain reaction (RT-qPCR) was carried out on a LightCycler® 96 Real-Time PCR System (Roche, Basel, Switzerland) with a 20 µL reaction system, employing the ChamQ Universal SYBR qPCR Master Mix (Vazyme Biotech Co., Ltd., Piscataway, NJ, USA). The relative expression levels were calculated using the $2^{-\Delta\Delta Ct}$ method [54], with *CaActin7* (*Caz04g16060*) serving as the internal reference gene for normalization. The specific primers utilized in this study are listed in Table S2. These primers were designed using the GenScript Online PCR Primers Designs Tool (https://www.genscript.com/tools/pcr-primers-designer, accessed on 15 January 2025).

4.9. Subcellular Localization

The full-length open reading frames (ORFs) of *CaBTB25*, excluding the stop codon, were inserted into the pSuper1300-eGFP plasmid via homologous recombination, thereby generating the CaBTB25-eGFP fusion protein construct. Subsequently, this recombinant plasmid was introduced into *Agrobacterium* competent GV3101 (Weidi Biotechnology Co., Ltd., Shanghai, China). The correct bacterial solution was first cultured in LB medium supplemented with kanamycin and rifampicin antibiotics overnight for 16 h. Then, 1 mL of the sample was transferred and further cultured in 20 mL of induced medium (comprising 50 $\mu g \cdot mL^{-1}$ kanamycin, 50 $\mu g \cdot mL^{-1}$ rifampicin, 10 mM MES, and 20 µM acetosyringone [AS]) for an additional 12 h. The bacterial liquid was then collected by centrifugation at 5000 rpm for 10 min. The supernatant was discarded, and an appropriate volume of infection buffer was added for resuspension. The infection buffer consisted of 10 mM

MES and 10 mM $MgCl_2$, with the pH adjusted to 5.6. Subsequently, AS was added to reach a final concentration of 200 μM. Finally, the optical density at 600 nm (OD_{600}) of the resuspended bacterial solution was adjusted to 1.0. This prepared bacterial suspension was then inoculated into the nuclear localization *Nicotiana benthamiana* leaves. The inoculated plants were incubated at 28 °C in the dark for at least 3 h before injection. Then, 2–3 days post-injection, the optimal time for observation, the enhanced green fluorescence protein was visualized using a laser confocal microscope (Zeiss LSM510 META, Oberkochen, Germany) and photographed. The primer sequences used in this experimental procedure are presented in Table S2.

4.10. Transcriptional Activation Analysis

The full-length coding sequence (CDS) of *CaBTB25* was amplified from the cDNA extracted from the young and tender leaves of the 'Zhangshugang' pepper variety. The amplified product was then integrated into a plasmid vector, and the reconstructed plasmid was subjected to sequencing to confirm its integrity and accuracy. Subsequently, in accordance with the corresponding instruction manual, the plasmid was transformed into Y2HGold chemically competent cells (Weidi Biotechnology Co., Ltd., Shanghai, China). Following transformation, the bacterial strain was cultured on SD/-Trp-Leu (synthetic dropout medium lacking tryptophan and leucine) at 30 °C for a duration of three days. This selective medium was used to screen for cells that had successfully taken up the plasmid. Positive clones were then selected and further cultivated on SD/-Trp-Leu-His-Ade (synthetic dropout medium lacking tryptophan, leucine, histidine, and adenine) for a period of 3–5 days, with appropriate controls being maintained concurrently. The primer sequences utilized in this experimental procedure are detailed in Table S2.

5. Conclusions

In our research, an exhaustive exploration of the pepper (*Capsicum annuum* L.) genome culminated in the successful discernment of 72 *CaBTB* genes, which were subsequently subjected to an intensive analysis. This analysis encompassed diverse facets, such as their physicochemical attributes, phylogenetic interrelationships, subcellular localization prognoses, gene architectures, duplication occurrences, and promoter *cis*-regulatory elements The BTB domain was phylogenetically categorized into seven discrete subfamilies. A substantial number of *CaBTB* genes manifested tissue-specific expression profiles across diverse tissues. Furthermore, they evinced responsiveness to five phytohormone regimens, specifically abscisic acid (ABA), gibberellic acid (GA_3), indole-3-acetic acid (IAA), jasmonic acid (JA), and salicylic acid (SA), as well as five abiotic stress treatments, namely cold stress, hydrogen peroxide (H_2O_2) stress, heat stress, mannitol-induced osmotic stress, and sodium chloride (NaCl)-induced salinity stress. The results from RT-qPCR incontrovertibly demonstrated substantial upregulation or downregulation of numerous *CaBTB* genes in response to specific hormonal and stress stimuli, thereby furnishing robust experimental substantiation for the responsiveness of the *CaBTB* gene family to these incitements. Notably, the *CaBTB25* gene was screened as a candidate gene that could be regulated by light. In summary, the genome-wide identification and characterization of the *BTB* gene family in pepper are pivotal in unearthing the potential applications of these research outcomes in pepper breeding initiatives and agricultural practices.

Supplementary Materials: The following supporting information can be downloaded at: https://www.mdpi.com/article/10.3390/ijms26073429/s1.

Author Contributions: Validation, Q.Y.; data curation, Q.Y.; writing—review and editing, Q.Y.; investigation, Q.Y., J.W., F.Z., X.D., and F.L.; funding acquisition, X.Z., C.X., and X.D. All authors have read and agreed to the published version of the manuscript.

Funding: This work was funded by the National Natural Science Foundation of China (32130097, 32341047, and U22A20497) and the Postgraduate Scientific Research Innovation Project of Hunan Province (CX20230702).

Institutional Review Board Statement: Not applicable.

Informed Consent Statement: Not applicable.

Data Availability Statement: All data are contained within the article and Supplementary Materials.

Conflicts of Interest: The authors declare no conflicts of interest.

References

1. Chaharbakhshi, E.; Jemc, J.C. Broad-complex, tramtrack, and bric-a-brac (BTB) proteins: Critical regulators of development. *Genesis* **2016**, *54*, 505–518. [CrossRef] [PubMed]
2. Stogios, P.; Downs, G.; Jauhal, J.; Nandra, S.; Privé, G. Sequence and structural analysis of BTB domain proteins. *Genome Biol.* **2005**, *6*, R82. [CrossRef]
3. Weber, H.; Bernhardt, A.; Dieterle, M.; Hano, P.; Mutlu, A.; Estelle, M.; Genschik, P.; Hanjo Hellmann, H. *Arabidopsis* AtCUL3a and AtCUL3b form complexes with members of the BTB/POZ-MATH protein family. *Plant Physiol.* **2005**, *137*, 83–93. [CrossRef] [PubMed]
4. Cheng, D.; Qian, W.; Meng, M.; Wang, Y.; Peng, J.; Xia, Q. Identification and expression profiling of the BTB domain-containing protein gene family in the silkworm, *Bombyx mori*. *Int. J. Genom.* **2014**, *2014*, 865065. [CrossRef] [PubMed]
5. Liu, P.F.; Zhang, K.; Li, Y.Q.; Zhou, F.; Bai, H.; Zang, J.P.; Cao, H.Z.; Xing, J.H.; Dong, J.G. Identification and expression analysis of *BTB* family genes in *Zea mays*. *Acta Phytopathol. Sin.* **2021**, *51*, 268–281.
6. Cheng, Y.F.; Qin, G.J.; Dai, X.H.; Zhao, Y.D. NPY1, a BTB-NPH3-like protein, plays a critical role in auxin-regulated organogenesis in *Arabidopsis*. *Proc. Natl. Acad. Sci. USA* **2007**, *104*, 18825–18829. [CrossRef]
7. Shalman, A.; Huang, Y.B.; Chen, Y.B.; Muhammad, I.; Li, B.B.; Ullah, U.; Jing, X.Q.; Bhanbhro, N.; Liu, W.T.; Li, W.Q.; et al. The highly interactive BTB domain targeting other functional domains to diversify the function of BTB proteins in rice growth and development. *Int. J. Biol. Macromol.* **2021**, *192*, 1311–1324. [CrossRef]
8. Liu, Y.; Yang, Q.; Wang, X.D.; Zhao, C.L.; Wang, X.; Li, Y.L.; Jia, H.L.; Ding, G.Z.; Chen, L. Mining and bioinformatics analysis of BTB protein family in Sugar beet. *Sugar Crops China* **2021**, *43*, 9–16.
9. Gu, S.Y.; Lo, W.S.; Wu, S.J.; Wang, L.C. Dimerization of the ETO1 family proteins plays a crucial role in regulating ethylene biosynthesis in *Arabidopsis thaliana*. *Plant J.* **2021**, *105*, 1293–1308. [CrossRef]
10. Mosavi, L.; Cammett, T.; Desrosiers, D.; Peng, Z.Y. The ankyrin repeat as molecular architecture for protein recognition. *Protein Sci.* **2004**, *13*, 1435–1448. [CrossRef]
11. Bomont, P.; Cavalier, L.; Blondeau, F.; Ben, H.C.; Belal, S.; Tazir, M.; Demir, E.; Topaloglu, H.; Korinthenberg, R.; Tüysüz, B.; et al. The gene encoding gigaxonin, a new member of the cytoskeletal BTB/kelch repeat family, is mutated in giant axonal neuropathy. *Nat. Genet.* **2000**, *26*, 370–374. [PubMed]
12. Collins, T.; Stone, J.R.; Williams, A.J. All in the family: The BTB/POZ, KRAB, and SCAN domains. *Mol. Cell. Biol.* **2001**, *21*, 3609–3915. [PubMed]
13. Ban, Z.; Estelle, M. CUL3 E3 ligases in plant development and environmental response. *Nat. Plants* **2021**, *7*, 6–16. [CrossRef] [PubMed]
14. Ullah, U.; Mao, W.L.; Abbas, W.; Alharthi, B.; Bhanbhro, N.; Xiong, M.; Gul, N.; Shalmani, A. OsMBTB32, a MATH-BTB domain-containing protein that interacts with OsCUL1s to regulate salt tolerance in rice. *Funct. Integr. Genom.* **2023**, *23*, 139.
15. Zhang, Q.Y.; Ma, C.N.; Gu, K.D.; Wang, J.H.; Yu, J.Q.; Liu, B.; Wang, Y.; He, J.X.; Hu, D.G.; Sun, Q. The BTB-BACK-TAZ domain protein MdBT2 reduces drought resistance by weakening the positive regulatory effect of MdHDZ27 on apple drought tolerance via ubiquitination. *Plant J.* **2024**, *119*, 283–299.
16. Du, L.; Guan, Z.J.; Liu, Y.L.; Hu, D.G.; Gao, J.P.; Sun, C.H. Scaffold protein BTB/TAZ domain-containing genes (*CmBTs*) play a negative role in root development of chrysanthemum. *Plant Sci.* **2024**, *341*, 111997.
17. Shalmani, A.; Ullah, U.; Tai, L.; Zhang, R.; Jing, X.Q.; Muhammad, I.; Bhanbhro, N.; Liu, W.T.; Li, W.Q.; Chen, K.M. OsBBX19-OsBTB97/OsBBX11 module regulates spikelet development and yield production in rice. *Plant Sci.* **2023**, *334*, 111779.

18. Fincher, G.B. Molecular and cellular biology associated with endo sperm mobilization in germinating cereal grains. *Annu. Rev. Plant Biol.* **2003**, *40*, 305–346. [CrossRef]
19. Woodger, F.J.; Jacobsen, J.V.; Gubler, F. GMPOZ, a BTB/POZ domain nuclear protein, is a regulator of hormone responsive gene expression in barley aleurone. *Plant Cell Physiol.* **2004**, *45*, 945–950.
20. Kim, S.H.; Seo, D.H.; Chung, S.L.; Kim, S.W.; Lee, J.S.; Kim, W.T.; Lee, J.H. ABA-HYPERSENSITIVE BTB/POZ PROTEIN 1 functions as a negative regulator in ABA-mediated inhibition of germination in *Arabidopsis*. *Plant Mol. Biol.* **2016**, *90*, 303–315. [CrossRef]
21. Zhou, Y.Y.; Zhai, H.; He, S.Z.; Zhu, H.; Gao, S.P.; Xing, S.H.; Wei, Z.H.; Zhao, N.; Liu, Q.C. The sweet potato BTB-TAZ protein gene, *IbBT4*, enhances drought tolerance in transgenic *Arabidopsis*. *Front. Plant Sci.* **2020**, *11*, 877. [CrossRef]
22. Yue, J.; Dai, X.R.; Li, Q.Z.; Wei, M.K. Genome-wide characterization of the *BTB* gene family in poplar and expression analysis in response to hormones and biotic/abiotic stresses. *Int. J. Mol. Sci.* **2024**, *25*, 9048. [CrossRef] [PubMed]
23. Elsanosi, H.A.; Zhang, J.H.; Mostafa, S.; Geng, X.Y.; Zhou, G.S.; Awdelseid, A.H.M.; Song, L. Genome-wide identification, structural and gene expression analysis of *BTB* gene family in soybean. *BMC Plant Biol.* **2024**, *24*, 663.
24. He, Y.M.; Liu, K.K.; Zhang, H.X.; Cheng, G.X.; Ali, M.; Haq, S.U.; Wei, A.M.; Gong, Z.H. Contribution of *CaBPM4*, a BTB domain-containing gene, to the response of pepper to *phytophthora capsici* infection and abiotic stresses. *Agronomy* **2019**, *9*, 417. [CrossRef]
25. Lu, M.C.; Ji, J.N.; Lv, Y.F.; Zhao, J.; Liu, Y.T.; Jiao, Q.; Liu, T.; Mou, Y.; You, Q.D.; Jiang, Z.Y. Bivalent inhibitors of the BTB E3 ligase KEAP1 enable instant NRF2 activation to suppress acute inflammatory response. *Cell Chem. Biol.* **2024**, *31*, 1188–1202.
26. Li, P.F.; Liu, P.; Zang, D.S.; Li, C.C.; Wang, C.; Zhu, Y.Z.; Liu, M.Q.; Lu, L.L.; Wu, X.B.; Nie, H.T. Genome-wide identification and expression analysis of the *BTB* gene superfamily provides insight into sex determination and early gonadal development of *alligator sinensis*. *Int. J. Mol. Sci.* **2024**, *25*, 10771. [CrossRef] [PubMed]
27. Liu, L.; Zhou, N.; Fu, S.N.; Wang, L.L.; Liu, Y.D.; Fu, C.F.; Xu, F.; Guo, W.Y.; Wu, Y.H.; Cheng, J.; et al. Endothelial cell-derived exosomes trigger a positive feedback loop in osteogenesis-angiogenesis coupling via up-regulating zinc finger and BTB domain containing 16 in bone marrow mesenchymal stem cell. *J. Nanobiotechnol.* **2024**, *22*, 721. [CrossRef]
28. Gingerich, D.J.; Hanada, K.; Shiu, S.H.; Vierstra, R.D. Large-scale, lineage-specific expansion of a bric-a-brac/tramtrack/broad complex ubiquitin-ligase gene family in rice. *Plant Cell* **2007**, *19*, 2329–2348.
29. Li, J.; Su, X.; Wang, Y.; Yang, W.; Pan, Y.; Su, C.; Zhang, X. Genome-wide identification and expression analysis of the BTB domain-containing protein gene family in tomato. *Genes Genom.* **2018**, *40*, 1–15.
30. Goyal, N.; Bhuria, M.; Verma, D.; Garewal, N.; Singh, K. Genome-wide identification of BTB domain-containing gene family in grapevine (*Vitis vinifera* L.). *Agriculture* **2023**, *13*, 252. [CrossRef]
31. Zhu, P.P.; Fan, Y.J.; Xu, P.L.; Fan, G.Q. Bioinformatic analysis of the *BTB* gene family in *paulownia fortunei* and functional characterization in response to abiotic and biotic stresses. *Plants* **2023**, *12*, 4144. [CrossRef] [PubMed]
32. Tan, B.; Lian, X.D.; Cheng, J.; Zeng, W.F.; Zheng, X.B.; Wang, W.; Ye, X.; Li, J.D.; Li, Z.Q.; Zhang, L.L.; et al. Genome-wide identification and transcriptome profiling reveal that E3 ubiquitin ligase genes relevant to ethylene, auxin and abscisic acid are differentially expressed in the fruits of melting flesh and stony hard peach varieties. *BMC Genom.* **2019**, *20*, 892.
33. Weber, H.; Hellmann, H. *Arabidopsis thaliana* BTB/POZ-MATH proteins interact with members of the ERF/AP2 transcription factor family. *FEBS J.* **2009**, *276*, 6624–6635. [CrossRef]
34. Jeffares, D.C.; Penkett, C.J.; Bähler, J. Rapidly regulated genes are intron poor. *Trends Genet.* **2008**, *24*, 375–378.
35. Heyn, P.; Kalinka, A.T.; Tomancak, P.; Neugebauer, K.M. Introns and gene expression: Cellular constraints, transcriptional regulation, and evolutionary consequences. *Bioessays* **2015**, *37*, 148–154. [CrossRef] [PubMed]
36. Perez-Torrado, R.; Yamada, D.; Defossez, P.A. Born to bind: The BTB protein-protein interaction domain. *Bioessays* **2006**, *28*, 1194–1202.
37. Teufel, A.I.; Johnson, M.M.; Laurent, J.M.; Kachroo, A.H.; Marcotte, E.M.; Wilke, C.O. The many nuanced evolutionary consequences of duplicated genes. *Mol. Biol. Evol.* **2019**, *36*, 304–314. [PubMed]
38. Juranić, M.; Dresselhaus, T. Phylogenetic analysis of the expansion of the MATH-BTB gene family in the grasses. *Plant Signal. Behav.* **2014**, *9*, e28242.
39. Villao-Uzho, L.; Chávez-Navarrete, T.; Pacheco-Coello, R.; Eduardo Sánchez-Timm, E.; Santos-Ordóñez, E. Plant promoters: Their identification, characterization, and role in gene regulation. *Genes* **2023**, *14*, 1226. [CrossRef]
40. Nanjareddy, K.; Arthikala, M.K.; Aguirre, A.L.; Gómez, B.M.; Lara, M. Plant promoter analysis: Identification and characterization of root nodule specific promoter in the common bean. *J. Vis. Exp.* **2017**, *130*, 56140.
41. Kang, H.; Zhang, T.T.; Li, Y.Y.; Wang, K.L.; Espley, R.V.; Du, Y.P.; Guan, Q.M.; Ma, F.W.; Hao, Y.J.; You, C.X.; et al. The apple BTB protein MdBT2 positively regulates MdCOP1 abundance to repress anthocyanin biosynthesis. *Plant Physiol.* **2022**, *190*, 305–318. [PubMed]

42. Xu, G.; Park, S.J.; Eck, J.V.; Lippman, Z.B. Control of inflorescence architecture in tomato by BTB/POZ transcriptional regulators. *Genes Dev.* **2016**, *30*, 2048–2061. [PubMed]
43. An, J.P.; Wang, X.F.; Zhang, X.W.; You, C.X.; Yu-Jin Hao, Y.J. Apple BT2 protein negatively regulates jasmonic acid-triggered leaf senescence by modulating the stability of MYC2 and JAZ2. *Plant Cell Environ.* **2021**, *44*, 216–233.
44. Ding, Y.; Sun, T.J.; Ao, K.; Peng, Y.J.; Zhang, Y.X.; Li, X.; Zhang, Y.L. Opposite roles of salicylic acid receptors NPR1 and NPR3/NPR4 in transcriptional regulation of plant immunity. *Cell* **2018**, *173*, 1454–1467.e15. [CrossRef]
45. Silva, K.J.P.; Brunings, A.; Peres, N.A.; Mou, Z.L.; Folta, K.M. The *Arabidopsis* NPR1 gene confers broad-spectrum disease resistance in strawberry. *Transgenic Res.* **2015**, *24*, 693–704.
46. Saputro, T.B.; Jakada, B.H.; Chutimanukul, P.; Comai, L.; Buaboocha, T.; Chadchawan, S. *OsBTBZ1* confers salt stress tolerance in *Arabidopsis thaliana*. *Int. J. Mol. Sci.* **2023**, *24*, 14483. [CrossRef] [PubMed]
47. Zhou, T.T.; Lin, Y.; Lin, Y.; Luo, J.Z.; Ding, J.H. Regeneration and *Agrobacterium*-mediated genetic transformation of twelve *Eucalyptus* species. *For. Res.* **2022**, *2*, 15.
48. Xu, H.; Liu, Q.; Yao, T.; Fu, X.D. Shedding light on integrative GA signaling. *Curr. Opin. Plant Biol.* **2014**, *21*, 89–95. [PubMed]
49. Davière, J.M.; Achard, P. A pivotal role of DELLAs in regulating multiple hormone signals. *Mol. Plant* **2016**, *9*, 10–20.
50. Ren, Y.R.; Zhao, Q.; Yang, Y.Y.; Zhang, R.; Wang, X.F.; Zhang, T.E.; You, C.X.L.; Huo, H.Q.; Hao, Y.J. Interaction of BTB-TAZ protein MdBT2 and DELLA protein MdRGL3a regulates nitrate-mediated plant growth. *Plant Physiol.* **2021**, *6*, 750–766.
51. Chen, C.J.; Chen, H.; Zhang, Y.; Thomas, H.R.; Frank, M.H.; He, Y.H.; Xia, R. TBtools: An integrative toolkit developed for interactive analyses of big biological data. *Mol. Plant* **2020**, *13*, 1194–1202. [CrossRef] [PubMed]
52. Qin, C.; Yu, C.S.; Shen, Y.; Fang, X.D.; Chen, L.; Min, J.M.; Cheng, J.W.; Zhao, S.C.; Xu, M.; Luo, Y.; et al. Whole-genome sequencing of cultivated and wild peppers provides insights into *Capsicum* domestication and specialization. *Proc. Natl. Acad. Sci. USA* **2014**, *111*, 5135–5140. [CrossRef] [PubMed]
53. Liu, F.; Yu, H.Y.; Deng, Y.T.; Zheng, J.Y.; Liu, M.L.; Ou, L.J.; Yang, B.Z.; Dai, X.Z.; Ma, Y.Q.; Feng, S.Y.; et al. PepperHub, an informatics hub for the chili pepper research community. *Mol. Plant* **2017**, *10*, 1129–1132. [CrossRef] [PubMed]
54. Livak, K.J.; Schmittgen, T.D. Analysis of relative gene expression data using real-time quantitative PCR and the 2(-Delta Delta C(T)) Method. *Methods* **2001**, *25*, 402–408. [CrossRef]

International Journal of
Molecular Sciences

MDPI

Article

Decoding PHR-Orchestrated Stress Adaptation: A Genome-Wide Integrative Analysis of Transcriptional Regulation Under Abiotic Stress in *Eucalyptus grandis*

Huiming Xu [1,2], Yifan Xing [2], Guangyou Li [1], Xin Wang [1,2], Xu Zhou [2], Zhaohua Lu [1], Liuyin Ma [2,*] and Deming Yang [1,*]

1 Research Institute of Tropical Forestry, Chinese Academy of Forestry, Guangzhou 510520, China; seimei2020@163.com (H.X.); lgy@caf.ac.cn (G.L.); 18735788993@163.com (X.W.); luzhaohua@caf.ac.cn (Z.L.)

2 Center for Genomics, Fujian Provincial Key Laboratory of Haixia Applied Plant Systems Biology, School of Future Technology, College of Forestry, Fujian Agriculture and Forestry University, Fuzhou 350002, China; 13675008050@163.com (Y.X.); z18584014241@163.com (X.Z.)

* Correspondence: lma223@fafu.edu.cn (L.M.); dmyang@caf.ac.cn (D.Y.)

Abstract: The phosphate starvation response (PHR) transcription factor family play central regulatory roles in nutrient signaling, but its relationship with other abiotic stress remains elusive. In the woody plant *Eucalyptus grandis*, we characterized 12 EgPHRs, which were phylogenetically divided into three groups, with group I exhibiting conserved structural features (e.g., unique motif composition and exon number). Notably, a protein–protein interaction network analysis revealed that EgPHR had a species-specific protein–protein interaction network: EgPHR6 interacted with SPX proteins of multiple species, while *Eucalyptus* and poplar PHR uniquely bound to TRARAC-kinesin ATPase, suggesting functional differences between woody and herbaceous plants. A promoter sequence analysis revealed a regulatory network of 59 transcription factors (TFs, e.g., BPC, MYBs, ERFs and WUS), mainly associated with tissue differentiation, abiotic stress, and hormonal responses that regulated *EgPHRs'* expression. Transcriptomics and RT-qPCR gene expression analyses showed that all *EgPHRs* dynamically responded to phosphate (Pi) starvation, with the expression of *EgPHR2* and *EgPHR6* exhibiting sustained induction, and were also regulated by salt, cold, jasmonic acid, and boron deficiency. Strikingly, nitrogen starvation suppressed most *EgPHRs*, highlighting crosstalk between nutrient signaling pathways. These findings revealed the multifaceted regulatory role of *EgPHRs* in adaptation to abiotic stresses and provided insights into their unique evolutionary and functional characteristics in woody plants.

Keywords: PHR; transcription factor; gene expression; cold stress; salt stress; phosphate starvation; nitrogen starvation; boron deficiency; *Eucalyptus grandis*

1. Introduction

Phosphorus (P) is one of the main mineral nutrients required for plant growth and development [1,2]. Phosphorus is an important component of macromolecules such as adenosine triphosphate (ATP), phospholipids, DNA, and RNA and is required for phosphorylation [3,4]. Therefore, phosphorus is essential for many plant life processes such as energy metabolism, photosynthesis, respiration, cell membrane stabilization, and genetic information transmission [5,6]. The phosphorus that can be absorbed and utilized by plants is soluble inorganic phosphate (Pi), but Pi is easily fixed by cation such as iron and aluminum in the soil [7,8]. As a result, P has become the second largest environmental factor limiting plant growth and development after nitrogen in many ecosystems.

Plants have evolved complex low-Pi signaling systems that adapt to soil phosphate limitations through molecular and physiological regulatory mechanisms [6]. These processes include Root system architecture (RSA) remodeling and activation of phosphate transporter (PHT) expression to improve phosphate uptake efficiency, as well as the modulation of growth and development processes in response to phosphate availability [9]. Studies in model plants have shown that phosphate starvation responses (PHRs) are the core regulatory transcription factors of the plant systematic low-Pi signaling network [10]. Under low Pi conditions, cellular Pi and ATP levels are reduced in Arabidopsis and rice, inducing PPIP5K-mediated conversion of inositol polyphosphates (InsPs) from InsP7 to InsP6 instead of InsP8 [11]. As InsP8 can stabilize the SPXs–PHRs interaction and inhibit the release of PHR from its negative regulators, SPXs, the decrease in InsP8 levels at low Pi further destabilizes the PHRs–SPXs interaction, so PHRs can be freely released from the SPXs–PHRs interaction [12,13]. PHR then shuttles from the cytoplasm into the nucleus and activates the transcription of downstream phosphate starvation-inducing genes (PSIs) [12,13]. PHR binds to the conserved *cis*-elements P1BS (*GNATATNC*) in the promoter of the PSI gene and activates or inhibits gene transcription of ~3800 PSI genes in Arabidopsis [14]. These PHR-dependent PSI genes include phosphate transporters *PHT1;1* and *PHT1;4*, microRNAs *microRNA399* and *microRNA827*, and transcription factors such as *NIGT1* [15,16]. Conversely, PPIPK5 converts InsP7 to InsP8 and then stabilizes the SPXs–PHRs interaction under conditions of sufficient Pi [12,13]. As a result, the function of PHR is inhibited, and low-Pi signaling is disrupted [12,13]. Therefore, PHRs play an important role in regulating low-Pi signaling in plants. After the discovery of AtPHR1 in low-Pi signaling in 2001, a total of 15, 12, 20, 14, 18, 22, 23, 41, and 42 PHRs were characterized in Arabidopsis, rice, brachypodium, sorghum, *Zea mays*, *G. arboreum*, *G. raimondii*, *G. hirsutum* and *G. barbadense* [17–19]. Currently, 21 PHRs have been characterized in tea plants, and 2 PHRs from apple (MdPHR1) or 1 from poplar (PtoPHR1) have been functionally validated in response to low-Pi stress in woody plants [20–22]. However, research on PHR in many woody plants lags behind to a large extent compared to model plants and crops. In addition, research on PHR has focused on low-Pi signaling, and the function of PHR in other abiotic stresses remains largely unexplored.

Eucalyptus are perennial tall trees (rarely shrubs) from three genera: *Eucalyptus*, *Angophora*, and *Corymbia* in the family *Myrtaceae* of dicot plants, with a total of 1239 species (869 species and 370 subspecies), which are naturally distributed in Australia, Papua New Guinea, Indonesia, and the Philippines [23]. *Eucalyptus* is known for its rapid growth, with a maximum growth rate of ~10 m per year, superior wood performance, and strong environmental adaptability, resulting in *Eucalyptus* having been planted in more than 100 countries and regions, accounting for ~23% of the world's forest plantations [24]. In China, *Eucalyptus* provides more than one-third of commercial wood production and is well adapted to the acidic soils of southern China, where effective free-Pi levels are known to be low due to the immobilization of free phosphorus by abundant cations such as aluminum and iron [7,8]. Therefore, understanding how *Eucalyptus* adapts and responds to low-Pi deficiency will be an interesting scientific question. However, to date, *PHRs* have not been characterized in *Eucalyptus*, and the expression patterns of *PHRs* under environmental stresses remain unexplored.

In this study, we performed a gene family analysis and characterized all 12 EgPHRs from *Eucalyptus grandis*, a model species in *Eucalyptus* spp. More importantly, we performed a systematic gene expression analysis of *EgPHRs* with different tissues, developmental stages, or environmental stresses. Our results provide useful preliminary results for further mechanistic studies of *EgPHRs*. It also unveils how *PHR* responds to different environmental stresses in woody plants and provides genetic information for the future design of high-phosphorus use efficiency (PUE) woody plants.

2. Results

2.1. Genome-Wide Characterization of PHRs in Eucalyptus grandis

A total of 12 PHR transcription factors in *Eucalyptus grandis* (*E. grandis*) were named from EgPHR1 to EgPHR12. The amino acid and CDS sequences of *EgPHRs* are listed in Table S1. Twenty-two PHR transcription factors were also characterized from *Populus trichocarpa* for the subsequent phylogenetic analysis (Table S1). To understand the biochemical properties of EgPHRs, we calculated the physicochemical properties of EgPHRs using Expasy (Table 1). In short, the amino acid length (<424.34 AAs) of all EgPHRs except EgPHR1 and EgPHR2 was shorter than the average length of plant proteins [25]. With the exception of EgPHR6, all EgPHRs showed higher isoelectric points (PI > 5.62) than the average PI of plant proteins [25]. As the instability index of all EgPHRs is above 40, the results suggested that EgPHRs might be unstable proteins [26]. All EgPHRs had a negative grand average of hydropathicity indicating that EgPHRs might be hydrophilic proteins [27]. Consistent with the results of many other plants, subcellular localization prediction showed that EgPHRs might be expressed in the nucleus.

Table 1. PHR proteins in *Eucalyptus grandis*.

Name	Sequence ID	Size (AA)	MW (KDa)	PI	Instability Index	GRAVY	Subcellular Localization	Group
EgPHR1	LOC104436960	581	63.21	5.98	67.24	−0.80	Nucleus	I
EgPHR2	LOC104450718	498	55.00	5.68	71.40	−0.72	Nucleus	I
EgPHR3	LOC104425247	414	45.41	6.79	42.30	−0.63	Nucleus	II
EgPHR4	LOC104445780	384	43.14	7.62	71.83	−0.76	Nucleus	I
EgPHR5	LOC104442716	378	41.88	6.82	46.29	−0.73	Nucleus	II
EgPHR6	LOC104432550	371	40.93	5.20	53.12	−0.76	Nucleus	I
EgPHR7	LOC104453912	353	39.24	7.14	59.45	−0.75	Nucleus	II
EgPHR8	LOC104451042	344	37.97	7.90	46.69	−0.75	Nucleus	III
EgPHR9	LOC104447696	330	36.09	5.69	53.79	−0.78	Nucleus	I
EgPHR10	LOC104441194	330	35.67	6.39	44.07	−0.70	Nucleus	I
EgPHR11	LOC104456029	316	34.12	6.08	45.25	−0.34	Nucleus	III
EgPHR12	LOC104442369	294	32.13	9.04	50.46	−0.69	Nucleus	III

AA: amino acids, MW: molecular weight, KDa: kilodalton, PI: isoelectric point, GRAVY: grand average of hydropathicity, Group I–III: classification according to the phylogenetic analysis.

2.2. Phylogenetic Analysis of Eucalyptus PHR Proteins

To understand the evolutionary characteristics of EgPHRs, phylogenetic analyses were performed on protein sequences of 61 PHRs from *Arabidopsis thaliana* (15), rice (12), *Populus trichocarpa* (*Poplar*, 22), and *Eucalyptus* (12). The results showed that the PHR proteins were divided into three groups (groups I-III, Figure 1, Table S1). Group I was the most dominant clade (29 PHRs), and the most well-studied PHRs for Pi signaling regulation including AtPHR1, AtPHL1, OsPHR2, OsPHR3, and OsPHR4 belonged to that group [28–30]. Half of the *Eucalyptus* PHR proteins (EgPHR1, EgPHR2, EgPHR4, EgPHR6, EgPHR9, and EgPHR10) were clustered with six Arabidopsis, six rice, and ten poplar PHR proteins in group I (Figure 1). The number of group I PHRs was almost the same in all four species, suggesting that group I PHRs had a potentially conserved function in different species (Figure 1). Group II had the lowest number of PHR proteins (12 PHRs), with three EgPHRs (EgPHR3, EgPHR5, and EgPHR7), two Arabidopsis, two rice, and four poplar PHRs (Figure 1). Group III was the second largest clade (20 PHRs), but only three *Eucalyptus* PHRs (EgPHR8, EgPHR11, EgPHR12) were characterized, which was significantly lower than the PHR proteins of the other three species: five Arabidopsis, four rice, eight poplar PHR proteins (Figure 1).

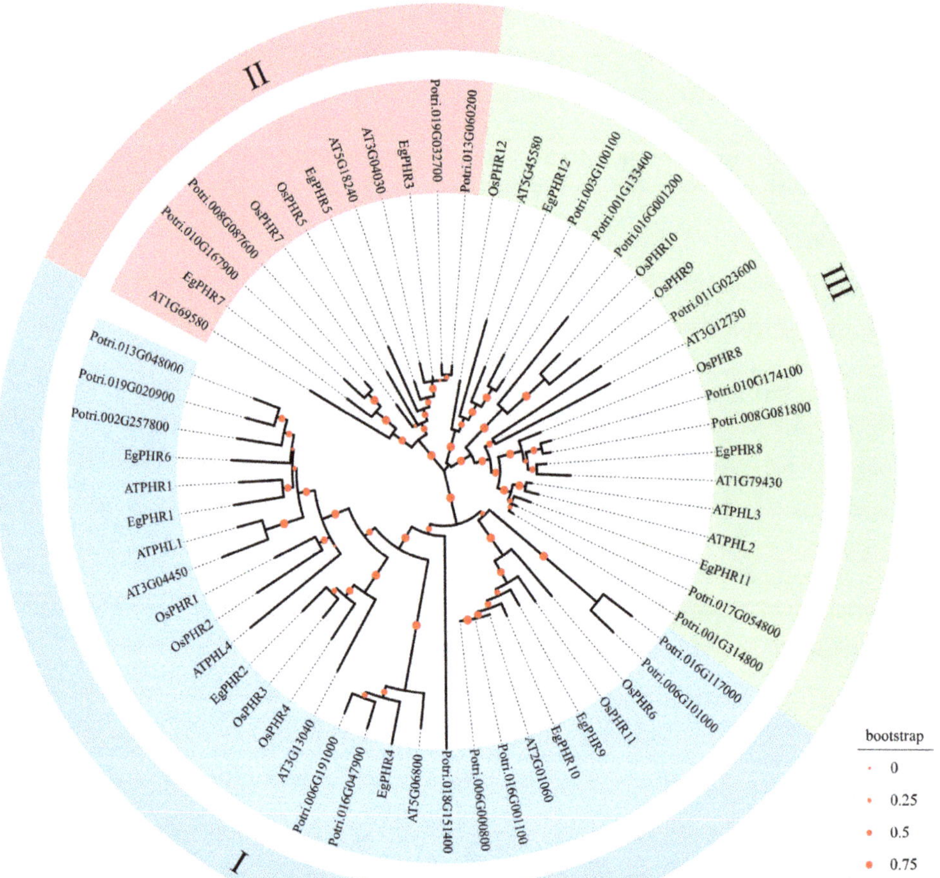

Figure 1. Phylogenetic analysis of the *PHR* gene family of *Eucalyptus grandis* with *Arabidopsis thaliana*, *Oryza sativa*, and *Populus trichocarpa*. Red dots indicate bootstrap values/metadata. The size of the circle corresponds to the level of bootstrap support for the branch. Circle diameters scale proportionally with bootstrap support values from 0 to 1, where larger diameters indicate stronger statistical confidence. Values exceeding 0.5 were generally considered reliable. The phylogenetic tree was constructed with MEGA7.0 using the maximum likelihood method with 1000 bootstraps. The tree uses three different colors to indicate the three evolutionary branches (**I–III**). The names of *PHRs* in *Arabidopsis*, *Oryza sativa*, *Populus trichocarpa*, and *Eucalyptus grandis* begin with "At", "Os", "Potri", and "Eg", respectively, at the beginning. The protein sequences of *Arabidopsis*, *Oryza sativa*, *Populus trichocarpa*, and *Eucalyptus grandis* PHRs are listed in Table S1.

2.3. Distinct Motifs Were Present or Absent Among Different Groups of PHRs

To further understand the function and conserved nature of EgPHR, a multi-sequence alignment was performed for the amino acid sequences of all 12 EgPHRs (Figure S1). Similar to PHRs from other plants, all EgPHRs contained two conserved domains: a Myb-like DNA-binding domain (Pfam ID: PF00249) and a Myb_CC_LHEQLE motif (Pfam ID: PF14379) (Figure S1). The results showed that EgPHR had a highly conserved PHR domain.

To explore the function of EgPHR, the top ten enrichment motifs were characterized by MEMEs with all EgPHR amino acid sequences (Figure 2). Notably, all EgPHR genes had three conserved motifs: motif1, motif2, and motif3 (Figure 2A, Table S2). Two of them were the conserved Myb_DNA-binding domains (motif1) and Myb_CC_LHEQLE domains (motif3, Table S2). However, an additional motif (motif2, VQRHLQLRIEAQGKYLQKILEKAQKTL) was also ubiquitous in all EgPHR proteins. Interestingly, motif 4 (GDSGLVLTTDPKPRL) was absent in group I EgPHRs, but motif 5 (LAKYMPDSSE) was only present in group I EgPHRs. Motif 7 (MYNHSQYLGKNISPSSRMSIPSERH) occurred only in EgPHR3 and EgPHR5. As shown in Figure 2C, the gene structure analysis revealed that each group I EgPHRs had eight exons, while each group II or III EgPHRs contained only six exons (Table S3). Overall, the results suggested that group I EgPHRs might differ from the other two groups of EgPHRs in terms of conserved motifs and gene structure.

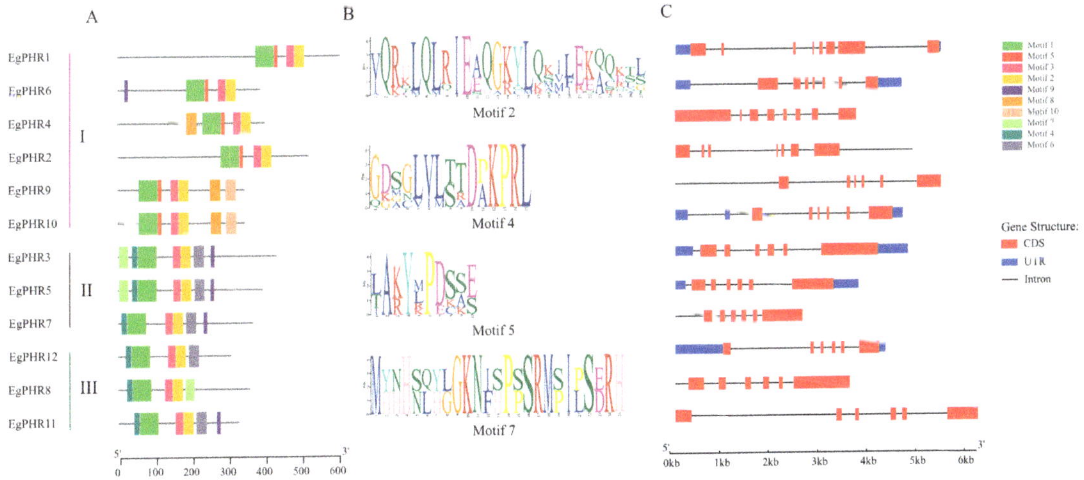

Figure 2. Conserved motifs and gene structure of the *Eucalyptus grandis* PHRs. (**A**) Conserved motifs of EgPHR; (**B**) sequence identification of several special EgPHR motifs; (**C**) genetic structure of *EgPHRs*, including CDS, UTR, and introns. The position of the sequence motifs, the domains, and the size of the exons or UTRs are estimated by the scale at the bottom.

2.4. Gene Duplication and Syntenic Analyses of EgPHRs

To understand the origin of the *EgPHR* gene family, chromosomal distribution, gene duplication, and syntenic analyses of the *EgPHRs* were performed (Figures S2 and 3). Gene duplication and syntenic analyses of *EgPHRs* showed that there were no tandem duplications within *EgPHRs*, while *EgPHR3* and *EgPHR5* were identified as a pair of segmental duplications (Figure S2). The syntenic analysis showed that *EgPHRs* formed orthologous gene pairs with ten Arabidopsis, five rice, and seventeen poplar *PHRs* (Figure 3, Table S4). The results suggested that *EgPHRs* were a highly conserved gene family, with the *EgPHRs* evolutionarily closer to poplar and Arabidopsis than rice *PHRs*.

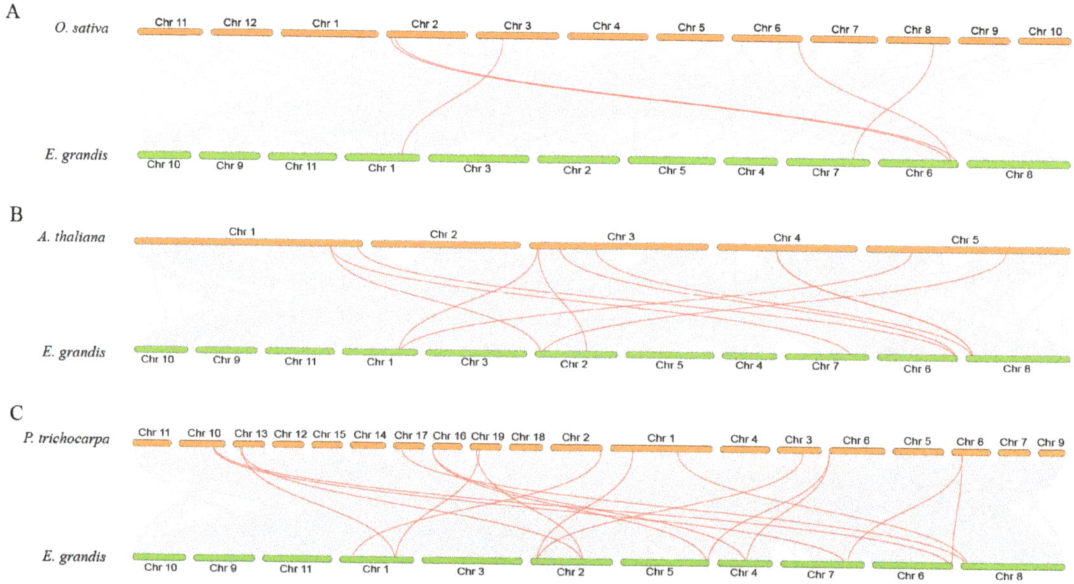

Figure 3. Analysis of covariance between *Eucalyptus grandis PHRs* and *Oryza sativa* (**A**), *Arabidopsis thaliana* (**B**), and *Populus trichocarpa* (**C**). Orange represents the chromosomes of *Eucalyptus grandis*, green represents the chromosomes of *Oryza sativa*, *Arabidopsis thaliana*, and *Populus trichocarpa* and the red line highlights *PHR* gene pairs with covariance.

2.5. The Potential Upstream Regulatory Transcription Factors of EgPHRs

To further identify the potential transcription factors (TFs) that regulated gene expression in *EgPHRs*, the PlantRegMap database was introduced to characterize the regulatory transcription factors of *EgPHRs* (Figure 4, Table S5). A total of 59 transcription factors might regulate the gene expression of *EgPHRs*, and the potential transcription factors and *EgPHRs* formed a regulatory network chord (Figure 4A), and the binding sites of these TFs were located in the promoter region of *EgPHRs*. These 59 TFs were mainly composed of 9 MYBs, 4 ERFs (ERF035, LEP, ERF021, ERF003), 9 ZF-DoF (DOF2.4, DOF3.4, CDF2), 5 MADs-Box (SVP, AGL13), 5 homology boxes (SIP1, KNOX1, WUS, WOX13), 3 GATA (GATA1, GATA15, GATA24), 2 RAX1, 2 AP2, 3 DREB (DREB1A, DREB2A, DREB2F), 2 TCP (TCP18), 2 PBP (PBP6, PBP1), 1 REF (REF6), and 1 NLP (NLP6) (Figure 4B). Among all the TFs that regulated *EgPHRs*, BPC was the most abundant (312), followed by HOMEOBOX (156) and the Dof domain, ZF-Dof (121) (Figure 4B,C). *EgPHR6* was found to be the most transcriptionally regulated gene (182), followed by *EgPHR12* (156) and *EgPHR11* (145) (Figure 4B). Notably, *EgPHR12* and *EgPHR11* were predominantly regulated by BPC. Most of the BPC-binding motifs in the promoter region of the *EgPHR* genes had significant bias.

These 59 TFs could also be divided into seven categories, with tissue differentiation, abiotic stress, growth and development, and hormones being the top-four-abundance functional categories (Figure 4B,D). These results suggest that the expression of *EgPHRs* is highly regulated by transcription factors associated with tissue differentiation, abiotic stress, growth and development, and hormonal response.

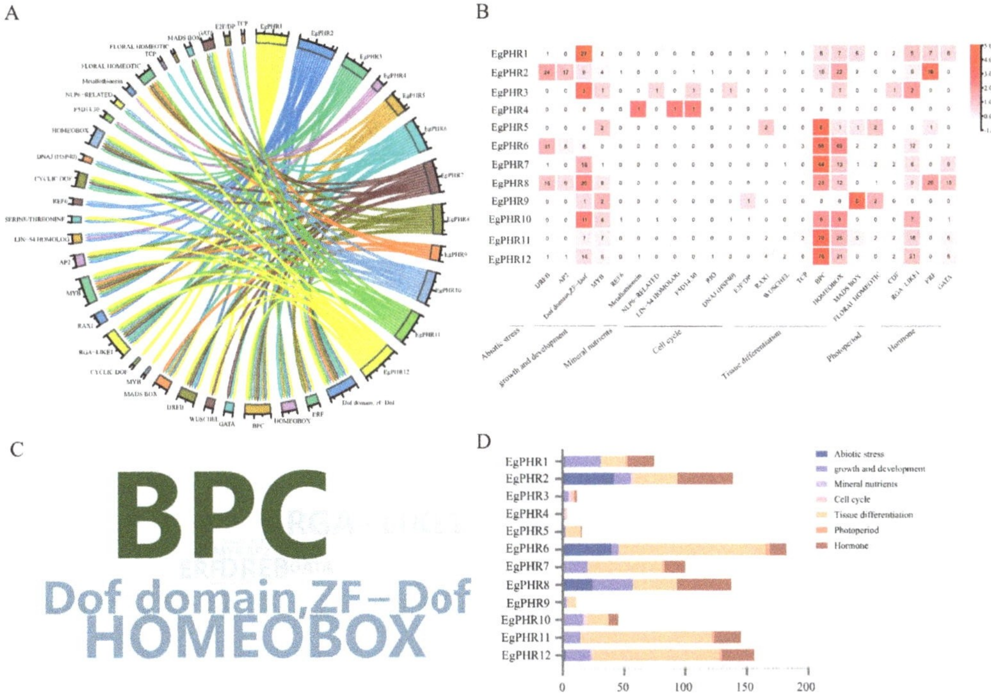

Figure 4. Prediction of potential regulatory transcription factors in the *Eucalyptus grandis PHR* promoters. (**A**) Network chords of predicted transcription factors targeting the *EgPHR* genes. The direction of the arrow is from *EgPHRs* to the transcription factors. (**B**) Statistics on the type of potential regulatory transcription factors in the promoter region of each gene. (**C**) The word cloud and font size of the transcription factors are positively correlated with the number of corresponding transcription factors. (**D**) Horizontal bar graph representing the number of transcription factors.

2.6. Protein–Protein Interaction Network of EgPHRs

To further elucidate the function of EgPHR, a protein–protein interaction (PPI) analysis of EgPHRs was performed with STRING (screening threshold: minimum required interaction score: medium confidence 0.4, maximum interaction shown: 30) (Figures 5A and S3, Table S6). With the exception of EgPHR7 and EgPHR8, all EgPHRs had interacting proteins in *Eucalyptus*. SPXs were negative regulators of PHR and inhibited the Pi starvation response by interacting with and inhibiting the release of PHR [1]. EgPHR1 and EgPHR6 were the only two EgPHRs that interacted with EgSPXs in *Eucalyptus*. In addition to EgPHR1 and EgPHR6, EgPHR10 also interacted with PtrSPXs in poplars (Figure S3A, Table S6). However, only two EgPHRs, EgPHR11 and EgPHR6, interacted with AtSPXs in Arabidopsis (Figure 5B, Table S6). As a result, only EgPHR6 interacted with SPX proteins from all three species, and the protein–protein interaction network prediction showed that there were differences in the interaction between SPX and EgPHR in different species. PHT1 is a Pi transporter, and purple acid phosphatase (PAP) releases Pi from immobilized forms of P such as Al-P or Fe-P oxides [1]. Although no PHTs or PAPs interacted with EgPHRs in *Eucalyptus* from current data, EgPHR11 and EgPHR6 indeed interacted with PHT1 and PAP10-2 in Arabidopsis. The cluster enrichment analysis further supported the hypothesis that Pi signaling-related terms were enriched in EgPHR-interacting proteins from *Euca-*

lyptus, poplar, and Arabidopsis, and therefore, EgPHRs might play a role in the Pi starvation response.

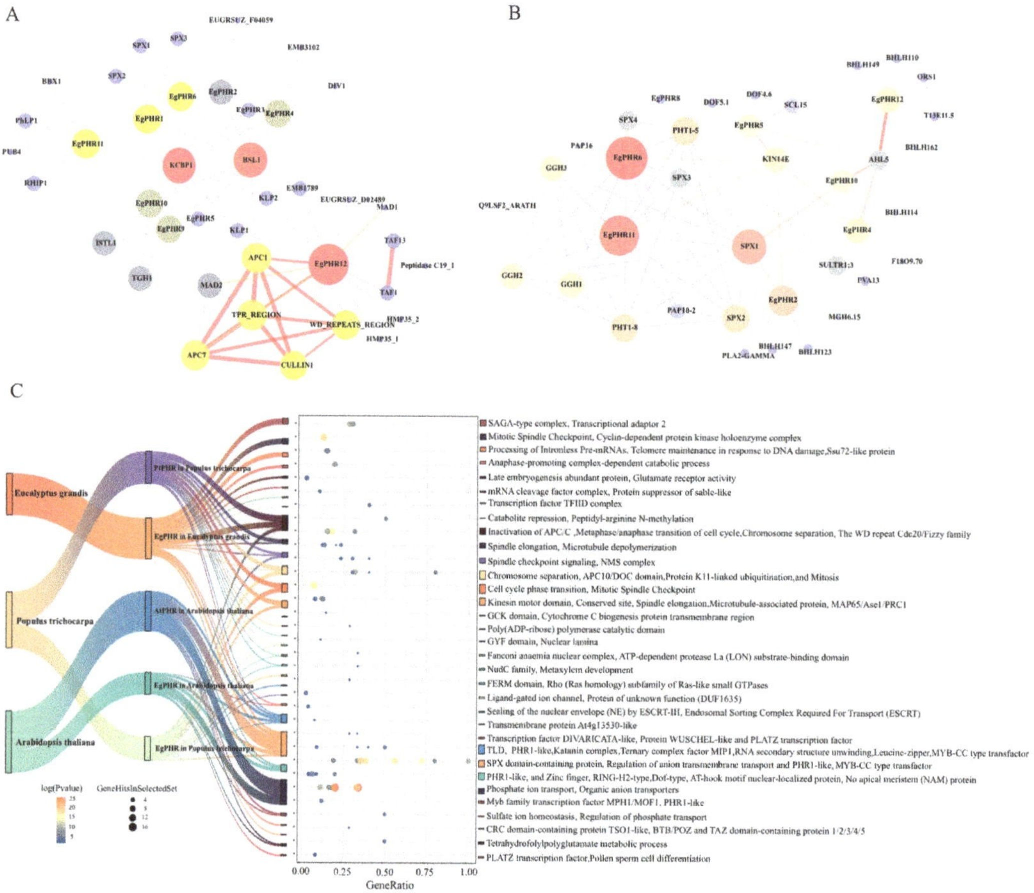

Figure 5. Protein–protein interaction network analysis and interaction protein enrichment analysis of PHR proteins in *Eucalyptus grandis*, *Populus trichocarpa*, and Arabidopsis. (**A**) Interaction network between EgPHR proteins and other *Eucalyptus grandis* proteins. (**B**) Interaction network between EgPHR proteins and Arabidopsis proteins. A network diagram consists of nodes and edges, each representing a protein; node-to-node connections (Edges) represent the interactions between these nodes; the connection's line color, shading, and thickness indicate the degree of interaction, and the node color shading and node size indicate the thickness of the node (red: large, blue: small; thick: large, thin: small). (**C**) Cluster enrichment of EgPHR-, PtrPHR-, and AtPHR-interacting proteins in *Eucalyptus grandis*, *Populus trichocarpa*, and Arabidopsis.

Ten EgPHRs (EgPHR1, EgPHR2, EgPHR3, EgPHR4, EgPHR5, EgPHR6, EgPHR9, EgPHR10, EgPHR11, and EgPHR12) interacted with at least one of the TRAFAC myosin-kinesin ATPase superfamily proteins represented by BSL1, KCBP1, KLP1, and KLP2 in *Eucalyptus* (Figure 5A). The protein–protein interaction cluster enrichment analysis of *Eucalyptus* and poplar was enriched for mitotic spindle checkpoint signaling, positive regulation of chromosome segregation, spindle elongation, and microtubule depolymerization (Figure 5C, Table S7). Notably, PtrPHRs also interacted with the TRARAC-kinesin ATPase superfamily proteins in poplar (Figure 5C, Table S7), whereas EgPHR and AtPHRs did not interact with these proteins in Arabidopsis (Figures 5B and S3C, Table S6). Therefore, PHR from two woody plants and PHR from Arabidopsis might have different functions.

2.7. Gene Expression Patterns of EgPHRs in the Development of Different Tissues

To explore the potential role of *EgPHRs*, we analyzed the gene expression of *EgPHRs* in various tissues using transcriptome data from *Eucalyptus grandis* (Figure 6A, Table S8). The gene expression analysis showed that the expression of *EgPHRs* in different tissues such as seedlings, roots, stems, leaves, flowers, lateral branches, shoot apex, 3rd, 5th, 7th, 9th, and 11th internodes showed dynamic changes.

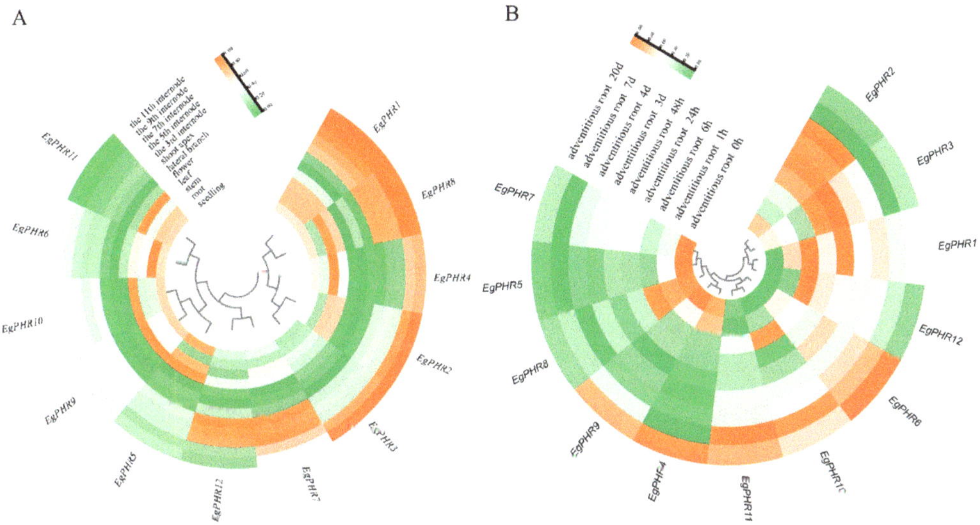

Figure 6. Gene expression analysis of *EgPHRs* in different *Eucalyptus* tissues and adventitious root development. Expression levels of *EgPHR*s in 12 different tissues (**A**) and 8 adventitious root induction states (**B**) in *Eucalyptus*. All samples were normalized using R software (v4.3.1) DEseq2 to obtain the final gene expression matrix. Heatmaps were plotted using R's pheatmap and ggplot2 packages. The clustering is by rows, with orange representing high expression and green representing low expression.

Generally, *EgPHR1* and *EgPHR8* clustered with each other and had a higher expression in nearly all tissues except for the flower, lateral branch, and shoot apex (Figure 6A). Notably, they were the only two *EgPHR*s that exhibited higher expression in all tested internodes. The slight difference between *EgPHR1* and *EgPHR8* was that *EgPHR8* showed a lower expression in the root but expressed higher in the stem (Figure 6A). The *EgPHR2*, *EgPHR3* and *EgPHR4* all showed a higher expression in the 9th and 11th internodes (Figure 6A). The difference among these three *EgPHR*s was that *EgPHR4* had the highest expression in the stem. *EgPHR7* and *EgPHR12* exhibited a higher expression in the 7th, 5th, and 3rd internodes. *EgPHR5*, *EgPHR9* and *EgPHR10* exhibited a relatively higher expression in the seedling and leaf. The slight difference was that *EgPHR5* also showed a relatively higher expression in the stem. *EgPHR6* and *EgPHR11* showed a relatively higher expression in the seedling, root and leaf.

From each tissue aspect, *EgPHR5* showed the highest expression, while *EgPHR3* had the lowest expression in seedlings (Figure 6A). In roots, *EgPHR6* had the highest expression level, and the other two *EgPHR*s (*EgPHR1* and *EgPHR11*) also showed a relatively higher expression level (Figure 6A). Taken together with the fact that *EgPHR6* and *EgPHR1* were predicted to interact with *EgSPX*s, we speculate that *EgPHR6* and *EgPHR1* might play central roles in the Pi deficiency signaling. Three *EgPHR*s (*EgPHR4*, *EgPHR5*, and *EgPHR8*) increased their expression in stem, suggesting that these *EgPHR*s might be involved in regulating stem development (Figure 6A). Many *EgPHR*s such as *EgPHR5*, *EgPHR9*, *EgPHR10*, and *EgPHR11* showed a higher expression in the leaf, indicating that *EgPHR*s might have important function in regulating leaf development (Figure 6A). All *EgPHR*s had a relatively lower expression in the lateral root and shoot apex, suggesting the *EgPHR*s might have limited roles in regulating the lateral root and shoot apex under normal conditions (Figure 6A). Seven *EgPHR*s (*EgPHR1*, *EgPHR2*, *EgPHR3*, *EgPHR4*, *EgPHR7*, *EgPHR8*, and *EgPHR12*) showed a higher expression in the development of different internodes. Overall, these results suggest that *EgPHR*s might play important roles in regulating the development of many *Eucalyptus* tissues.

2.8. Gene Expression Patterns of EgPHRs During the Adventitious Root Development

The proliferation of *Eucalyptus* depends on tissue culture, and the development of adventitious roots is one of the important characteristics of *Eucalyptus* production. To explore the potential function of *EgPHRs* in adventitious root (AR) development, we performed a gene expression analysis of *EgPHRs* using time-course (0 h, 1 h, 6 h, 24 h, 48 h, 3 days, 4 days, 7 days, and 20 days) transcriptome data during adventitious root development (Figure 6B, Table S9).

*EgPHR*s could be classified into three groups according to their gene expression pattern. The first group had five *EgPHR*s (*EgPHR4*, *EgPHR5*, *EgPHR7*, *EgPHR8*, and *EgPHR9*) (Figure 6B). They all had a higher expression in the control group, while showing a lower expression in many AR developmental stages (Figure 6B). There were several exceptions for these five *EgPHR*s. The first exception was *EgPHR8*, which had a higher expression at the 1 h and 6 h AR developmental stages (Figure 6B). The second exception was *EgPHR9*, which showed an increased expression at the 20-day AR developmental stage. The third exception was *EgPHR4*, which exhibited a higher expression at the 1 h and 20 days of AR developmental stage. The second group of *EgPHR* had three members (*EgPHR6*, *EgPHR10*, and *EgPHR11*) (Figure 6B). These three *EgPHR*s all had a higher expression at the 7-day and 20-day AR developmental stages, and with *EgPHR10* also having an increased expression at the 6 h AR developmental stage (Figure 6B). The last group of *EgPHR*s were *EgPHR1*, *EgPHR2*, *EgPHR3*, and *EgPHR12*. These four *EgPHR*s showed a higher expression at the 6 h (*EgPHR12*), 24 h (*EgPHR2* and *EgPHR3*), 48 h (*EgPHR1*, *EgPHR2*, and *EgPHR3*), 3-day (*EgPHR2*) and 4-day (*EgPHR2*) AR developmental stages (Figure 6B). Overall, these results suggest that *EgPHR*s might play an important role in AR development in *Eucalyptus grandis*.

2.9. Gene Expression Pattern of EgPHRs Under SA and JA Treatment

Pathogens threaten *Eucalyptus* growth and reduce the productivity of *Eucalyptus* plantations [31]. Two plant hormones, salicylic acid (SA) and jasmonic acid (JA), trigger plant immunity to defend against pathogen infection [31]. To understand whether *EgPHR*s were also involved in the regulation of *Eucalyptus* immunity, a gene expression analysis of *EgPHRs* in response to SA (leaf mock, 1 h, 6 h, and 7 d) and JA (leaf mock, 1 h, 6 h, and 7 d) was conducted with transcriptome data, respectively (Figure 7A, Table S10, Figure 7B, Table S11).

*EgPHR*s could be classified into eight groups according to their expression pattern under the SA treatment (Figure 7A). The first group contained four *EgPHRs* (*EgPHR1*,

EgPHR8, *EgPHR9*, and *EgPHR10*), and they showed a higher expression under all SA treatment conditions (Figure 7A). The second group had *EgPHR5* and *EgPHR6*, and their expression was highly induced after 6 h of SA treatment (Figure 7A). The *EgPHR3* and *EgPHR4* constituted the fourth group, and they also had a higher expression after 6 h of SA treatment, while having a lower expression after 7 days of SA treatment (Figure 7A). *EgPHR2* was the only member in the fifth group, and it showed the highest expression after 1 h of SA treatment (Figure 7A). The sixth group only had *EgPHR12*, and its expression was induced after 6 h and 7 days of SA treatment. *EgPHR7* and *EgPHR11* were the only members in the seventh and eighth group, respectively (Figure 7A). They all showed a high expression after 7 days of SA treatment, while *EgPHR11* also showed a decreased expression after 1 day of SA treatment. Overall, these results suggest that all *EgPHR*s might be involved in response to SA in *Eucalyptus*.

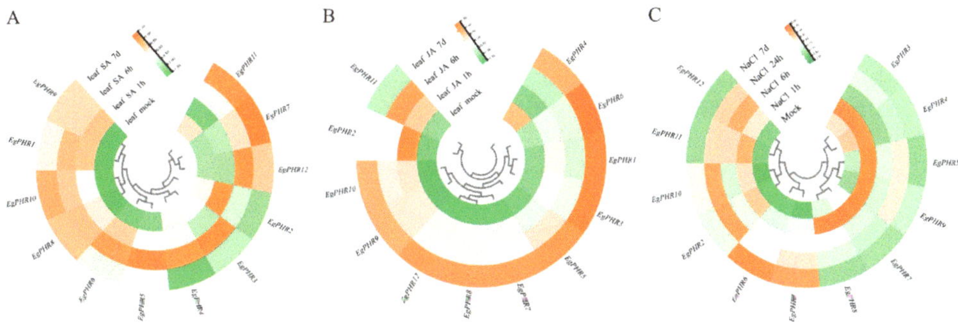

Figure 7. Gene expression analysis of *EgPHRs* under JA, SA, and salt stress treatments. The heatmap displays the expression levels of *EgPHRs* following treatment with SA (**A**) and JA (**B**) hormones, in the SA and JA treatment within 1 h, 6 h, and 7 h. (**C**) Expression levels of *EgPHRs* under salt stress, in the 200 mM NaCl treatment within 0 h, 1 h, 6 h, 24 h and 7d. All samples were normalized using R software DEseq2 to obtain the final gene expression matrix. Heatmaps were plotted using R's pheatmap and ggplot2 packages. Clustering is by rows, orange represents high expression, and green represents low expression.

Similarly, *EgPHR*s could be classified into four groups according to their expression pattern under the JA treatment (Figure 7B). The first group only had one member, *EgPHR11*, which showed the highest expression after 6 h of JA treatment. *EgPHR2* was the only member of the second group, and it showed the highest expression after 1 h of JA treatment (Figure 7B). The third group presented nine *EgPHR*s (*EgPHR1*, *EgPHR3*, *EgPHR5*, *EgPHR6*, *EgPHR7*, *EgPHR8*, *EgPHR9*, *EgPHR10*, and *EgPHR12*), and their expression was significantly induced after 7 days of JA treatment (Figure 7B). EgPHR4 was the only member of the last group, and it showed a higher expression under control and *EgPHR4* (Figure 7B). Overall, these results suggest that all *EgPHR*s might be involved in response to JA, especially under long-term JA treatment in *Eucalyptus*.

2.10. Gene Expression Pattern of EgPHRs Under Salt Stress

Salt stress is one of the major environmental stresses that limit plant growth and productivity [32,33]. To understand how *EgPHRs* respond to salt stress, the gene expression of *EgPHRs* was analyzed using transcriptome data in time-course salt-stress experiments (mock, 200 mM NaCl treatment for 1 h, 6 h, 24 h, and 7 days) (Figure 7C).

*EgPHR*s could be classified into four groups according to their expression pattern under salt stress. Four *EgPHR*s (*EgPHR2*, *EgPHR10*, *EgPHR11* and *EgPHR12*) showed a relatively higher expression after 1 h (*EgPHR2, EgPHR10, EgPHR11* and *EgPHR12*), 6 h (*EgPHR11* and *EgPHR12*), or 24 h (*EgPHR2, EgPHR10, EgPHR11* and *EgPHR12*) of salt stress (Figure 7C). The *EgPHR1* and *EgPHR6* together formed the second group; their expression was highly induced under 7 days of salt stress (Figure 7C). The third group comprised four *EgPHR*s (*EgPHR5, EgPHR7, EgPHR8* and *EgPHR9*), and they were highly induced under 1 h of salt stress (Figure 7C). The last group only contained *EgPHR3* and *EgPHR4*, and they were highly expressed under control and 1 h of salt stress (Figure 7C). Overall, these results suggest that all of *EgPHR*s might respond to salt stress at the transcriptional level in *Eucalyptus*.

2.11. Gene Expression Pattern of EgPHRs Under Cold Stress

Eucalyptus plants are sensitive to cold damage, and cold stress is one of the main environmental stresses that permanently reduce the development and productivity of *Eucalyptus* plantations [34]. To test whether *EgPHRs* are involved in cold tolerance in *Eucalyptus*, gene expression analyses of *EgPHR*s were performed using RT-qPCR experiments with the leaves from seedlings under 24 h at 4 °C (cold stress) or 25 °C (control) conditions in *Eucalyptus grandis* (Figure 8A–L). Briefly, eight *EgPHR*s (*EgPHR1, EgPHR2, EgPHR3, EgPHR4, EgPHR5, EgPHR6, EgPHR9,* and *EgPHR11*) showed significant ($p < 0.05$) gene expression changes under cold stress (Figure 8A–F,I–K). Notably, cold stress significantly ($p < 0.05$) induced mRNA accumulation level of six *EgPHRs* (*EgPHR1, EgPHR2, EgPHR3, EgPHR4, EgPHR5,* and *EgPHR6*) (Figure 8A–F). Conversely, two *EgPHR*s, *EgPHR9* and *EgPHR11*, significantly ($p < 0.05$) reduced the mRNA accumulation level under salt stress (Figure 8I,K). Therefore, cold stress indeed affected the gene expression of many *EgPHRs* at the transcriptional level.

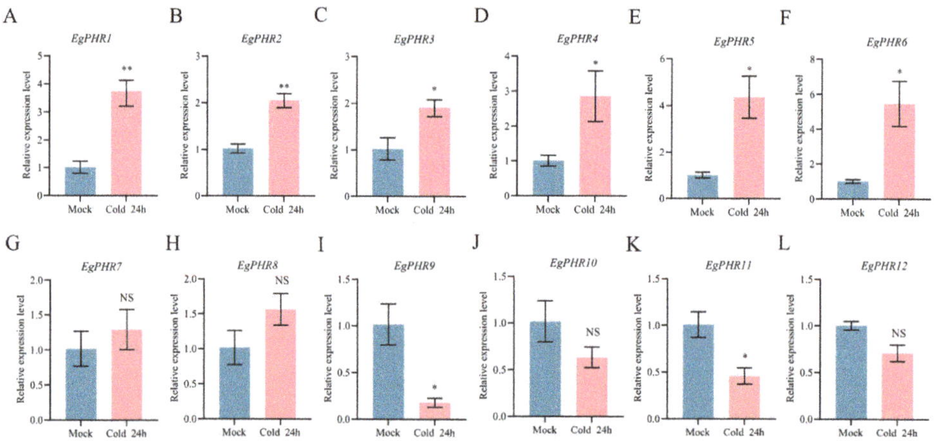

Figure 8. Gene expression analysis of *EgPHR*s in response to cold stress. **(A–L)** Expression of the *EgPHRs* in mock and cold treatment in the leaf. Seedlings were growth at 4 °C (cold stress) or 25 °C (control) for 24 h. *t*-test: "NS" represents no significance, "*" represents $p < 0.05$, "**" represents $p < 0.01$.

2.12. Gene Expression Pattern of EgPHRs Under Low-Phosphate Starvation

To investigate the role of *EgPHR*s in response to Pi starvation, a gene expression analysis of *EgPHRs* was performed using RT-qPCR experiments using the roots under different time-courses of Pi starvation (control: 0.5 mM KH_2PO_4; Pi starvation, LP: 0.005 mM KH_2PO_4 treatment for 6 h, 12 h, 24 h, and 3 days).

Generally, Pi starvation induced the expression of all *EgPHR*s in at least one LP condition (Figure 9A–L). After 6 h of LP condition, Pi starvation significantly ($p < 0.05$) induced the expression of eight *EgPHR*s (*EgPHR2*, *EgPHR4*, *EgPHR5*, *EgPHR6*, *EgPHR9*, *EgPHR10*, *EgPHR11*, *EgPHR12*) (Figure 9B,D,E,F,I–L), and five *EgPHR*s (*EgPHR2*, *EgPHR4*, *EgPHR5*, *EgPHR9*, and *EgPHR11*) dramatically increased their expression ($p < 0.001$) (Figure 9B,D,E,J,L). Similarly, LP induced most *EgPHR*s' (nine *EgPHR*s: *EgPHR1*, *EgPHR2*, *EgPHR6*, *EgPHR7*, *EgPHR8*, *EgPHR9*, *EgPHR10*, *EgPHR11*, and *EgPHR12*) expression under a 12 h LP treatment conditions (Figure 9A,B,F–L). Among them, four *EgPHR*s (*EgPHR2*, *EgPHR6*, *EgPHR11*, and *EgPHR12*) had dramatically ($p < 0.001$) increased gene expression after 12 h of LP treatment (Figure 9B,F,K,L). Interestingly, *EgPHR2* was the only *EgPHR* that increased its expression under a 24 h LP treatment. Similarly, LP induced the expression of only three *EgPHR*s (*EgPHR2*, *EgPHR4* and *EgPHR6*) after 3 days of LP treatment (Figure 9B,D,F). Notably, *EgPHR2* was the only *EgPHR* that was induced by LP under all treatment conditions, and *EgPHR6* was the next *EgPHR* that was regulated by LP under three of the four LP treatment conditions (Figure 9). Overall, these results suggest *EgPHRs* indeed play a vital role in coping with Pi starvation.

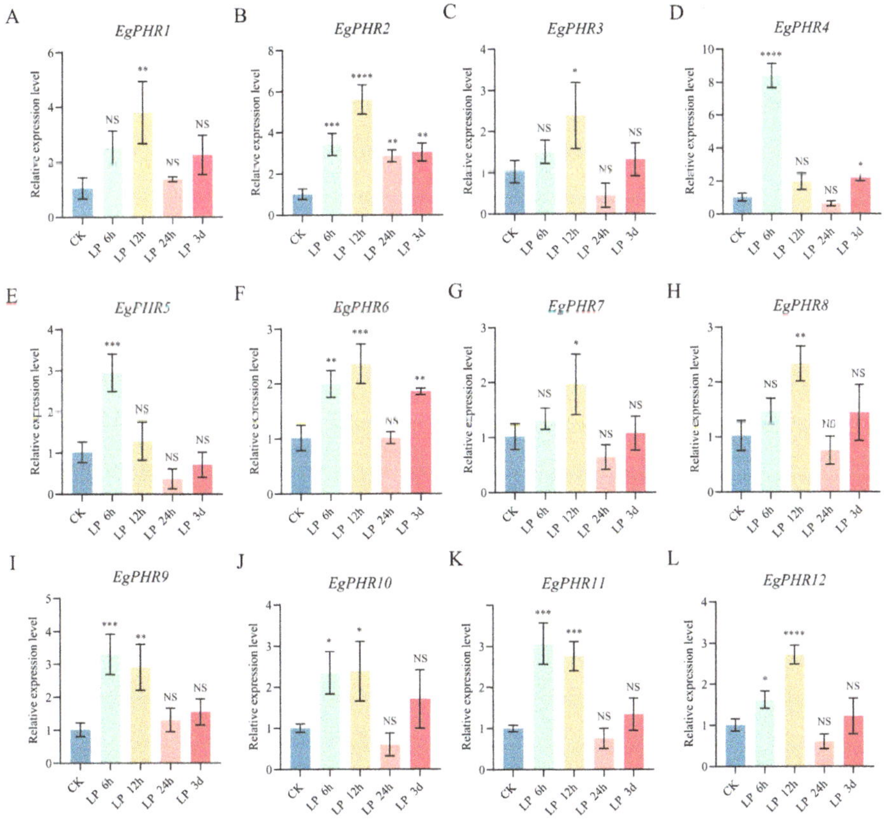

Figure 9. Gene expression analysis of *EgPHR*s in response to phosphate deficiency. (**A–L**) General overview of the expression of the *EgPHRs* in phosphate deficiency treatment in CK, 6 h, 12 h, 24 h and 3 d in roots. Two-way ANOVA test: "NS" represents no significance, "*" represents $p < 0.05$, "**" represents $p < 0.01$, "***" represents $p < 0.001$, "****" $p < 0.0001$.

2.13. Gene Expression Pattern of EgPHRs Under Nitrogen Starvation

Nitrogen and phosphate interact with each other, and the addition of nitrate can partially rescue the Pi starvation phenotype [1,35]. To understand whether nitrogen starvation also affects the Pi starvation response, the gene expression of *EgPHRs* was analyzed by RT-qPCR using the root tissues under nitrogen starvation conditions (control: 10 mM KNO_3, nitrogen starvation, −N: 0 mM KNO_3 for 2 h and 24 h). Generally, nitrogen starvation significantly affected the gene expression of all *EgPHRs* except *EgPHR2* and *EgPHR9* (Figure 10). Nitrogen starvation only significantly ($p < 0.05$) induced the gene expression of *EgPHR5* but not other *EgPHRs* after 2 h of nitrogen starvation (Figure 10E). Conversely, nitrogen starvation significantly ($p < 0.05$) repressed the expression of all *EgPHRs* except *EgPHR2* and *EgPHR9* after 24 h of nitrogen starvation (Figure 10). Overall, nitrogen starvation might negatively regulate the gene expression of *EgPHRs* at the transcriptional level.

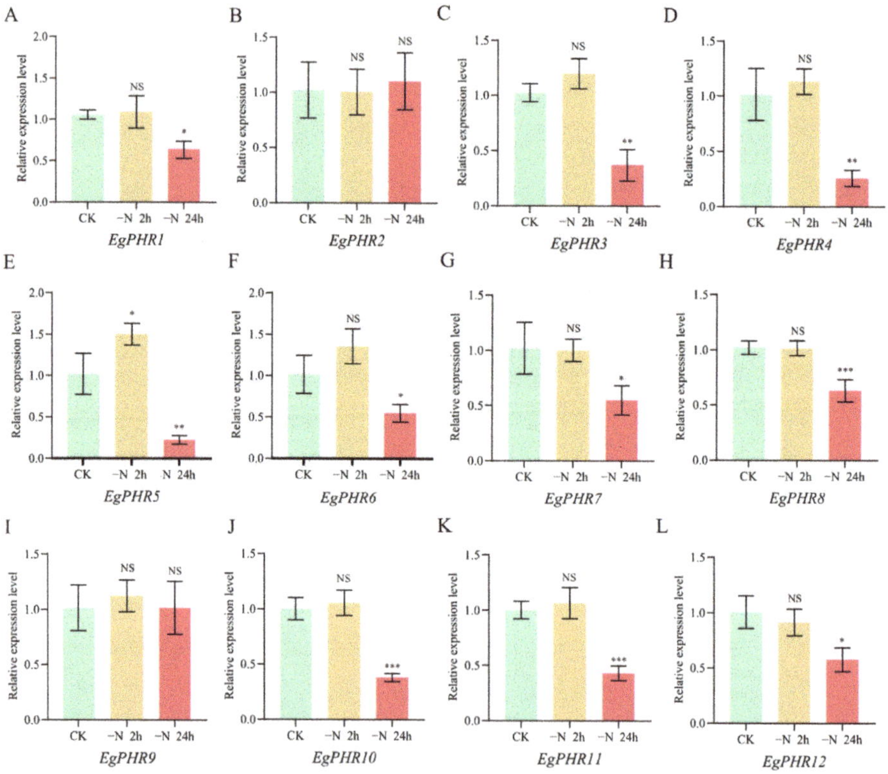

Figure 10. Gene expression analysis of *EgPHRs* in response to nitrogen starvation. (**A–L**) General overview of the expression of the *EgPHRs* in CK, 2 h, and 24 h low nitrogen treatment in roots. CK: 10 mM KNO_3 treatment, −N: 0 mM KNO_3 treatment. One-way ANOVA test: "NS" represents no significance, "*" represents $p < 0.05$, "**" represents $p < 0.01$, "***" represents $p < 0.001$.

2.14. Gene Expression Pattern of EgPHRs Under Boron Deficiency

Boron deficiency greatly reduces the forest productivity by repressing the development of the shoot apex, leading to the top dieback and leaf chlorosis [36]. Boron (B) and P might interact with each other. To test whether B could affect the Pi signaling, the gene expression of *EgPHRs* was assessed via time series experiments in *Eucalyptus* (leaves: mock, 6 h, 24 h,

2 days, 4 days, and 21 days; roots: mock, 6 h, 24 h, 2 days, 4 days, and 21 days) (Figure S4, Table S12).

Generally, *EgPHR*s could be classified into six groups according to their expression pattern under B deficiency. The first group only contained *EgPHR10*, which showed the highest expression in leaves under a 6 h B deficiency treatment (Figure S4). *EgPHR6* was the only member of the second group and showed the lowest expression in roots after 24 h, 2 days, and 4 days) (Figure S4). The third group had seven *EgPHR*s (*EgPHR1*, *EgPHR2*, *EgPHR3*, *EgPHR4*, *EgPHR5*, *EgPHR7*, and *EgPHR9*), and they showed a relative higher expression in leaves after 6 h, 24 h, 2 days, 4 days, or 21 days of B deficiency treatment (Figure S4). *EgPHR12* was the only member in the fourth group and increased its expression in roots under the B deficiency treatment (Figure S4). The fifth group only contained *EgPHR11*, which showed the highest expression in roots under control and decreased expression in roots under B deficiency treatment (Figure S4). *EgPHR8* was the only member of the last group. Its expression was suppressed in leaves after 6 h, 24 h, and 2 days of B deficiency treatment, which increased its expression in roots after 2 days and 21 days of B deficiency treatment (Figure S4). Overall, these results suggest that boron deficiency might also regulate the gene expression of *EgPHR*s at the transcriptional level.

3. Discussion

Sessile plants are subject to many environmental stresses throughout their life cycle, including nutrient deficiencies [37,38]. As a result, plants must coordinate their complex growth with nutrient availability. Phosphorus is an essential macronutrient for all eukaryotes, but how woody plants adapt to phosphorus deficiency remains largely unclear [1]. *Eucalyptus* is one group of the important hardwoods, and is an important source of wood, furniture, and pulp worldwide due to its fast growth rate, short rotation period (*Eucalyptus* plantations are typically managed under short rotation cycles of 5–7 years to maximize biomass yield through rapid growth and frequent harvesting in intensive silvicultural systems), and good wood properties [39]. Therefore, characterization of PHR transcription factors will provide important genetic information for engineered high-phosphate use efficiency (PUE) *Eucalyptus* plants.

A total of twelve EgPHRs were characterized in *Eucalyptus grandis*. The number of EgPHRs were dramatically reduced compared to PHRs from the grass species (14~42 *PHRs* in Arabidopsis, brachypodium, sorghum, *Zea mays*, *G. arboreum*, *G. raimondii*, *G. hirsutum*, and *G. barbadense*) and the woody plant species (21~22 in tea plants and poplar) (Tables 1 and S1) [17–20]. The only exception was rice PHRs (12 PHRs) [17]. It would be very interesting to analyze further in the future why fast-growing species have few PHRs, and the implication of that on high-PUE efficiency.

Phylogenetic analyses from *Eucalyptus*, Arabidopsis, rice, and poplar revealed that PHR proteins could be classified into three groups. The group I PHRs had the most well-studied PHRs that respond to Pi deficiency from model plants, including AtPHR1, AtPHL1, and OsPHR2 (Figure 1) [1,28]. Six *Eucalyptus* PHRs including EgPHR1, EgPHR2, EgPHR4, EgPHR6, EgPHR9, and EgPHR10 belonged to group I PHRs (Figure 1). Taken together with the fact that EgPHR1 and EgPHR6 interacted with SPXs in the protein–protein interaction analysis and the expression of *EgPHR1*, *EgPHR2*, *EgPHR6*, and *EgPHR10* were induced under Pi deficiency (Figure 6A), we then speculated that group I PHRs might also be involved in regulating Pi signaling in *Eucalyptus*. Notably, all group I EgPHRs had a unique motif (LAKYMPDSSE) (Figure 2B, Table S2), and it will be interesting to test in the future whether this motif play an important function on the regulation of EgPHR-mediated Pi signaling.

Group II PHRs comprised the smallest subset, with three EgPHR members (EgPHR3, EgPHR5, and EgPHR7) belonging to this category (Figure 1). Functional characterization has been reported for two genes within that group. The GARP-type transcription factor MYR2 (At3G04030) negatively regulates nitrogen reutilization in Arabidopsis by suppressing asparagine synthetase 1 (ASN1) expression, thereby delaying nutrient starvation- and dark-induced leaf senescence under light–dark cycles in vascular tissues [40]. MYR1 (At5G18240) and MYR2 are associated with light intensity responses, repressing flowering and organ elongation under low-light conditions through the inhibition of GA20ox2 expression, which modulates bioactive gibberellin levels [41]. These findings suggest that group II EgPHRs may also play important roles in light-regulated nitrogen reutilization and organ elongation processes.

The Arabidopsis PHR1 family members form two distinct regulatory modules: PHR1/PHL1 primarily drives core phosphate starvation responses (PSR) by activating phosphate uptake genes and maintaining phosphorus homeostasis under Pi deficiency, while PHL2/PHL3 modulates growth regulation under normal conditions and fine-tunes stress adaptation through synergistic or antagonistic interactions with PHR1/PHL1 [19]. Structural divergence underpins their functional specialization—PHR1/PHL1 dimerizes via N-terminal extensions to activate P1BS-dependent PSR pathways, whereas PHL2/PHL3's unique C-terminal domains restrict interactions to their own cluster, enabling the coordinated regulation of secondary metabolism and chromatin remodeling [19]. Given that *Eucalyptus grandis*'s Group III PHRs (EgPHR8, EgPHR11, and EgPHR12) share phylogenetic clustering with Arabidopsis PHL1/PHL3 (Figure 1), these *EgPHRs* may balance growth–stress trade-offs through structural adaptations in N-/C-terminal domains, potentially mediating species-specific PSR strategies in woody plants.

Phosphate and nitrate are two of the most important mineral nutrients for plant development and productivity [1]. More importantly, a proper N:P ratio is critical for plant growth [42]. Previous studies from Arabidopsis and rice have suggested that plants adopt a NRT1.1-SPX module to simultaneously regulate the nitrate and phosphate deficiency signaling [43]. Briefly, OsNRT1.1B interacts with OsSPX4, and OsSPX4 is the negative regulator for OsPHR2 and OsNLP3 [43]. Similarly, a study in Arabidopsis proved that ~85% of phosphate starvation response genes had their expression dependent on nitrate [35]. Therefore, nitrate regulates the expression of phosphate starvation responses genes. In this study, low nitrogen availability significantly suppressed the expression of nine *EgPHRs* (including *EgPHR6* and *EgPHR10*) in roots (Figure 10), demonstrating that nitrogen promotes phosphorus-related responses in *Eucalyptus*, and the repression of PHR expression under nitrogen deficiency aligns with observations in other species such as Arabidopsis. Notably, *EgPHR6* and *EgPHR10* were also induced by Pi deficiency (Figure 9). Therefore, nitrogen might interact with phosphate starvation responses in *Eucalyptus*. The hypothesis were further supported by an observation that nitrate-associated transcription factor NLP6's binding sites were characterized in the promoter of *EgPHR10* (Figure 4, Table S5). This result further proves that N and P might interact with each other in *Eucalyptus*. A future functional analysis might be performed to further prove whether the EgNLP6-EgPHRs module might regulate the *Eucalyptus* N–P balance.

The interaction between boron (B) and phosphate (P) exhibits both antagonistic and synergistic relationships depending on plant species and environmental conditions [44–47]. Notably, in this study, B deficiency suppressed the expression of *EgPHRs* in roots, potentially due to disrupted P transport systems or competitive absorption dynamics (Figure S4). However, B paradoxically upregulated certain *EgPHRs* homologs in leaves, suggesting a compensatory adaptation to maintain P homeostasis under B deficiency (Figure S4). This dual regulation aligns with prior findings: B and P interactions involve shared transport pathways (e.g., competitive inhibition in tomato and maize), while moderate P availability enhances B uptake by improving rhizosphere conditions and transpiration-driven absorption (e.g., in *Brassica napus*) [45–47]. Extreme B deficiency or toxicity exacerbates P imbalance, as seen in *Vicia faba* and groundnuts, whereas excessive P fertilization reduces B accumulation [44], highlighting their tightly coupled yet antagonistic roles in nutrient dynamics. The observed tissue-specific *EgPHR* expression in *Eucalyptus grandis* under B deficiency may reflect evolutionary strategies to optimize P utilization under fluctuating nutrient stresses, consistent with the conserved PHR-mediated crosstalk between environmental cues and nutrient signaling.

The interplay between jasmonate (JA) and phosphate signaling converges on phosphate starvation response's (PHR) transcription factors as evolutionary conserved regulatory hubs. In this study, prolonged exogenous JA treatment induced the upregulation of 11 *EgPHRs* in *Eucalyptus grandis* (Figure 7B), likely reflecting an adaptive strategy to enhance P acquisition/utilization efficiency under JA-mediated stress responses. This aligns with mechanistic insights from other species: Arabidopsis PHR1 coordinates JA signaling by interacting with JAZ repressors and MYC2 to activate JA-responsive genes (e.g., anthocyanin biosynthesis), while tea plant's CsPHRs integrate P signaling with JA pathways through JAZ degradation-triggered activation of the secondary metabolism [48,49]. However, functional specialization among the 11 *EgPHRs* in *Eucalyptus* requires further investigation to clarify their roles in balancing JA-mediated defense and P homeostasis.

Phosphate represses plant height in grass including Arabidopsis and rice as well as woody plants such as poplar and Chinese fir [1,50–52]. However, the molecular mechanism of how Pi regulates cell cycle and shoot apex remain unclear. In particular, it is still not known whether the functions between woody and grass PHRs are different from each other. In this study, the promoter analysis and protein–protein analysis both showed that *Eucalyptus* PHRs might have different functions compared to grass PHRs (Figures 4 and 5). The promoter analysis revealed that tissue differentiation and cell-cycle-related transcription factors such as WUS, KNOX1, WOX13, and WUS had binding sites in the promoters of *EgPHRs* (Figure 5C). WUS is the critical regulator that modulates the cell fate determination in plants [53,54]. KNOX1 is an important transcription factor in tissue differentiation and cell cycle, and the loss of KNOX1 leads to failure of the Shoot Apical Meristem (SAM) establishment [55]. KNOX1 also functions on the generation of diploid plants [55]. WOX13 negatively regulates the SAM formation of WUS [56]. SIP1 has been reported to be upregulated in spindle cell carcinoma of the head and neck in human [57]. RAX1 is required for the diploid bipolar budding pattern in budding yeast [58]. Therefore, tissue differentiation and cell-cycle-related transcription factors might regulate the gene expression of *EgPHRs*. BSL1 regulates cell fate asymmetry and leads the mother cell to produce daughter cells with different fate during mitosis [59]. KCBPs play important roles in the spindle microtubule convergence in anaphase to ensure that each sister chromosome group has coherence kinetochore-fibers [60]. In this study, the protein–protein interaction analysis unveiled that all EgPHRs except EgPHR7 and EgPHR8 could interact with TRAFAC class myosin-kinesin ATPase superfamily proteins such as BSL1, KCBP1, KLP1, and KLP2 in *Eucalyptus* (Figure 5A). Similar observation was also identified in the protein–protein interaction analysis of poplar PHRs (Figure S3B). In addition, EgPHRs also interacted with DIVARICATA-like and WUSCHEL-like proteins (Figure 5C). Therefore, we speculated that PHRs might also be involved in tissue differentiation and cell-cycle regulation in woody plants. However, these results originated from either cis-elements' screening or interacting based on the protein's amino acid sequences. Functional analyses will be required to further prove the hypothesis that woody plant PHRs might differ from grass PHRs and modulate the cell cycle and tissue differentiation. Overall, our results suggest that *EgPHRs* may play an important role in development, abiotic stress tolerance, and mineral nutrient starvation adaptation.

4. Materials and Methods

4.1. Characterization of EgPHRs in Eucalyptus grandis

The *Eucalyptus grandis* genomic dataset utilized in this investigation was acquired from the NCBI Sequence Read Archive under the whole-genome sequencing (WGS) project JABKBJ01. For the systematic identification of PHR family proteins in *Eucalyptus grandis*, two specialized Hidden Markov Model (HMM) profiles—Myb_CC_LHEQLE (PF14379.9) and Myb_DNA-binding (PF00249.34)—were retrieved from the Pfam protein family database (http://pfam-legacy.xfam.org/) (accessed on 21 August 2024). These conserved domain profiles were subsequently employed in HMMER software (v3.0) to perform comprehensive a genome-wide screening and characterization of potential EgPHR candidates through sequence alignment and domain architecture analysis [61].

The candidate EgPHR proteins were then submitted to three different databases: NCBI-Batch CD-search (https://www.ncbi.nlm.nih.gov/Structure/bwrpsb/bwrpsb.cgi) (accessed on 21 August 2024), Pfam (http://pfam-legacy.xfam.org/) (accessed on 21 August 2024), and SMART (https://smart.embl.de/) (accessed on 21 August 2024) to verify its conserved domain. Proteins with all two conserved domains (Myb_CC_LHEQLE and Myb_DNA binding) were considered as EgPHR proteins. Subsequently, comprehensive physicochemical characterization was performed via ExPASy ProtParam (https://web.expasy.org/protparam/) (accessed on 21 August 2024), quantifying critical biochemical parameters including amino acid composition, molecular mass (kDa), theoretical isoelectric point (pI), grand average of hydropathicity (GRAVY), and instability index to assess structural stability [61]. Subcellular localization was predicted by Plant-mPLoc (http://www.csbio.sjtu.edu.cn/bioinf/plant-multi/#) (accessed on 21 August 2024).

4.2. Phylogenetic Analysis

The PHR amino acid sequences of Arabidopsis were downloaded from TAIR (https://www.arabidopsis.org/) (accessed on 22 August 2024) and the amino acid sequences of *Populus trichocarpa* and *Oryza sativa* were downloaded from the Phytozome database (https://phytozome-next.jgi.doe.gov/) (accessed on 22 August 2024), respectively. Pairwise sequence comparison of these three species with *Eucalyptus grandis* PHR protein was performed using MEGA7.0. After a comparison using Quick Run TrimAL tool from TBtools (v2.154) [62], the sequence was pruned to achieve maximum likelihood evolutionary tree construction (1000 replicates of a bootstrap analysis) and the evolutionary tree was beautified on the ITOL website (https://itol.embl.de/) (accessed on 23 August 2024).

4.3. Conserved Domains, Motifs, and Gene Structure Analysis

A multi-sequence comparison of EgPHR proteins was performed using DNAman8 software [63]. We uploaded the EgPHR protein sequences to the MEME webpage—MEME Suite (https://meme-suite.org/meme/) (accessed on 24 August 2024), set the number of patterns we wanted to discover to 10, and clicked the MAST XML output to download the results. We then extracted the annotation information of *EgPHRs* and visualized the gene structure using the GSDS website (https://gsds.gao-lab.org/index.php) (accessed on 24 August 2024).

4.4. Duplication Event Analysis and Multi-Species Analysis of Covariance

A self-comparison of protein sequences from *Eucalyptus grandis* were conducted using the Blast Compare Two Seqs module from TBtools [62]. We used Quick run MCScanX Wrapper (a built-in tool for TBtools) to obtain a link file of the relationships between genes and visualized them with the Circle Gene View. A multi-species analysis of the covariance between rice, Arabidopsis, *Populus trichoria*, and *Eucalyptus* was conducted and visualized using one-step MCScanX-Ultrafast and Dual System Diagram for McscanX tool from TBtools [62].

4.5. Prediction and Analysis of Potential Upstream Regulatory Transcription Factors for EgPHRs

The promoter sequence 2 kb upstream of the *EgPHR* genes was extracted and submitted to the PlantRegMap database (https://plantregmap.gao-lab.org/regulation_prediction.php) (accessed on 3 December 2024) to predict the upstream regulatory transcription factors. We used Origin2024 to draw chord diagrams and TBtools Heatmap tools for classification and heatmap drawing. We plotted word clouds and bar stacks using the ggplot2 package in R [64].

4.6. Protein–Protein Interaction Analysis

Eucalyptus grandis PHR protein sequences were submitted to the STRING website (https://cn.string-db.org/) (accessed on 13 September 2024), and *Eucalyptus grandis*, *Populus trichocarpa*, and *Arabidopsis thaliana* were selected as the reference species to predict the interaction network between EgPHR, PtrPHR, and AtPHR proteins. The detailed parameters included the full STRING network, evidence in meaning of edges, medium confidence (0.400) in minimum required interaction score, and no more than 30 interactors in the max number of interactors to show. TSV result files in string_interactions_short format were downloaded to organize and beautify protein interaction networks using Cytospace (version 3.6.1). All enriched terms through the Analysis module were downloaded, the bubble diagram was plotted using the ggplot2 package in R, and the sankey diagram was plotted using Origin2024.

4.7. Abiotic Stress Treatments and RT-qPCR Analysis

Eucalyptus grandis seeds were harvested in May 2024 from 15-year-old *Eucalyptus grandis* trees at the Baisha State-owned Forest Farm in Minhou, Fujian Province, and stored in a −20 °C freezer. *Eucalyptus grandis* seeds were sown on vermiculite and germinated and grown in a greenhouse with an ambient temperature of 25 °C, 16 h of light/8 h of darkness, with a light intensity of 50 μmol, and a humidity of 60%. *Eucalyptus* seedlings that were 1.5 months old and growing consistently were selected and grown in 1/2 Hoagland nutrient solution (2.5 mM KNO_3, 0.5 mM KH_2PO_4, 1 mM $MgSO_4$, 2.5 mM $Ca(NO_3)_2{\cdot}4H_2O$, pH 5.0) for three weeks. *Eucalyptus* seedlings were divided into three groups: control group (0.5 mM KH_2PO_4, 10 mM KNO_3, hereinafter referred to as CK), nitrogen starvation group (0 mM KNO_3, hereinafter referred to as -N) and LP group (0.005 mM KH_2PO_4, hereinafter referred to as LP) for hydroponic treatment. Root sampling was performed at 2 h and 24 h for the -N treatment, and root sampling was performed at 6 h, 12 h, 24 h, and 3 days for the LP treatment.

The 1.5-month-old seedlings of *Eucalyptus grandis* (watered once a week with 2 L 1/2 Hoagland nutrient solution, pH 5.0) were grown in a 4 °C low-temperature incubator or at 25 °C with identical light and humidity condition, and the leaves were sampled after 24 h. Total RNA extraction from roots and leaves of *Eucalyptus grandis* were performed using the RNAprep Pure Plant Plus Kit (Polysaccharides & Polyphenolics-rich) (TIANGEN, Beijing,

China). Total RNAs were reverse-transcribed using the Evo M-MLV Reverse Transcription Premixed Kit with gDNA removal (Accurate Biotechnology, Changsha, China). RT-qPCR reactions of 12 *EgPHR* genes were performed on a QuantStudio 1 Plus Real-Time PCR System (Thermo Fisher Scientfic, Waltham, MA, USA) using the SYBR Green Pro Taq HS Premixed qPCR Kit (with Rox) (Accurate Biotechnology, Changsha, China). Primers used in this study are listed in Table S13. The RT-qPCR procedure was as follows: 95 °C for 30 s, followed by 40 cycles of 95 °C for 5 s and 58 °C for 30 s, followed by a final step of 72 °C for 30 s. The *ACTIN1* gene (LOC104425496) was used as an internal control, and for each analysis, 3 technical replicates and 3 biological replicates were performed.

4.8. Data Analysis

Gene expression levels were calculated using the $2^{-\Delta\Delta CT}$ method [38,65]. The ΔCt value was the difference between the Ct value of the target gene and the Ct value of the reference gene. The ΔΔCt value was the ΔCt value of the experimental group minus the ΔCt value of the control group. The relative expression level was calculated using the $2^{-\Delta\Delta Ct}$ formula. The statistical analysis was conducted using *t*-tests along with both one-way ANOVA and two-way ANOVA approaches. "NS" represented no significance, "*" represented $p < 0.05$, "**" represented $p < 0.01$, "***" represented $p < 0.001$, "****" $p < 0.0001$. Bar charts were plotted using the ggplot2 (v3.5.1) package in R (v4.3.1).

The *Eucalyptus grandis*-related transcriptome data used in this study were downloaded from the National Genomics Data Center database of the China National Center for Bioinformation, with accession number PRJCA002468 [39]. Quality control was performed using FastQC (version 0.11.9) and MultiQC (version 1.12) [66]. Following raw data preprocessing with Trim Galore (a Cutadapt wrapper, v0.6.10) and Cutadapt (v4.0), genome alignment was performed against the Eucalyptus grandis reference genome using STAR aligner (v2.7.10a) with pre-built genomic indices [67]. The resulting SAM files were converted to BAM format using samtools (v1.16.1), followed by gene-level quantification through FeatureCounts (v1.6.4) with species-specific annotation files [68,69]. Finally, we normalized all samples using R software DEseq2 (v1.42.1) to obtain the final gene expression matrix [70]. Heatmaps were plotted using TBtools Heatmap tools (v2.154).

5. Conclusions

Phosphorus is an essential macronutrient for plant development and productivity, and PHRs are major regulators of Pi-starvation signaling. The identification and functional analysis of PHRs in *Eucalyptus* will provide important information for the future design of high-PUE woody plants. In this study, twelve PHRs were characterized in *Eucalyptus grandis*. The gene expression analysis revealed that *EgPHRs* might be involved in the regulation of plant development, JA response, SA response, cold stress, salt stress, and nutrient availability. Notably, the protein–protein interactions and promoter-binding transcription factor analysis suggest that *EgPHRs* might be involved in mitotic chromosome segregation and tissue differentiation in *Eucalyptus*. Overall, our results suggest that *EgPHRs* may play an important role in coordinating plant development, abiotic stresses, and nutrient availability.

Supplementary Materials: The supporting information can be downloaded at: https://www.mdpi.com/article/10.3390/ijms26072958/s1.

Author Contributions: Conceptualization, D.Y. and L.M.; methodology, H.X.; software, H.X.; formal analysis, H.X., Y.X., G.L., X.W., X.Z. and D.Y.; data curation, H.X.; writing original draft preparation,

H.X.; writing review and editing, D.Y., L.M. and Z.L.; visualization, H.X.; supervision, D.Y. and L.M. All authors have read and agreed to the published version of the manuscript.

Funding: This work was supported by the Fundamental Research Funds for the Central Non-profit Research Institution of CAF, grant number CAFYBB2022SY017, the National Key Research and Development Program of China during the 14th Five-year Plan Period, grant number 2022YFD2200203, the Forestry Peak Discipline Construction Project of Fujian Agriculture and Forestry University, grant number 72202200205.

Institutional Review Board Statement: Not applicable.

Informed Consent Statement: Not applicable.

Data Availability Statement: The transcriptome data used in this study were downloaded and reanalyzed from the National Genomics Data Center database (PRJCA002468 for tissue expression, JA response, SA response, salt stress, and boron deficiency).

Acknowledgments: The authors would like to thank Yu Lin and Hui Yang of Baisha State-owned Forest Farm in Minhou County, Fujian Province, for their assistance in collecting *Eucalyptus grandis* seeds. The authors would like to thank Ping Zheng of Fujian A&F University for her suggestion on cloud mapping, and thank our laboratory members Yi Han, Lichuan Deng, and Shasha Zhang for their assistance and suggestions in the preparation of plant materials and some experiments.

Conflicts of Interest: The authors declare no conflicts of interest.

Abbreviations

The following abbreviations are used in this manuscript:

EgPHRs	*Eucalyptus grandis* phosphate starvation responses
Eg	*Eucalyptus grandis*
At	*Arabidopsis thaliana*
Potri	*Populus trichocarpa*
Os	*Oryza sativa*
Chr	Chromosome
FPKM	Fragments per kilobase of transcript per million mapped reads
HMM	Hidden Markov Model
RT-qPCR	Quantitative reverse transcription PCR
WGS	Whole-genome sequencing
PUE	Phosphorus use efficiency
PPI	Protein–protein interaction
SPX	SYG1, Pho81, XPR1
PAP	Purple acid phosphatase
WUS	WUSCHEL
KNOX1	Knotted1-like homeobox 1
WOX13	WUSCHEL-related homeobox 13
SIP1	Smad-interacting protein 1
BSL1	BSU1-like family of Ser/Thr protein phosphatases 1
KCBP1	Kinesin-like calmodulin binding protein 1

References

1. Fang, X.; Yang, D.; Deng, L.; Zhang, Y.; Lin, Z.; Zhou, J.; Chen, Z.; Ma, X.; Guo, M.; Lu, Z.; et al. Phosphorus uptake, transport, and signaling in woody and model plants. *For. Res.* **2024**, *4*, e017. [CrossRef]
2. Hong, L.; Wang, Q.; Zhang, J.; Chen, X.; Liu, Y.; Asiegbu, F.O.; Wu, P.; Ma, X.; Wang, K. Advances in the beneficial endophytic fungi for the growth and health of woody plants. *For. Res.* **2024**, *4*, e028. [CrossRef]

3. Paz-Ares, J.; Puga, M.I.; Rojas-Triana, M.; Martinez-Hevia, I.; Diaz, S.; Poza-Carrion, C.; Minambres, M.; Leyva, A. Plant adaptation to low phosphorus availability: Core signaling, crosstalks, and applied implications. *Mol. Plant* **2022**, *15*, 104–124. [CrossRef] [PubMed]
4. Wang, Y.; Chen, Y.F.; Wu, W.H. Potassium and phosphorus transport and signaling in plants. *J. Integr. Plant Biol.* **2021**, *63*, 34–52. [CrossRef]
5. Meharg, A.A. *Marschner's Mineral Nutrition of Higher Plants*, 3rd ed.; Marschner, P., Ed.; Elsevier: Amsterdam, The Netherlands; Academic Press: Amsterdam, The Netherlands, 2011; p. 684, ISBN 978-0-12-384905-2.
6. Malhotra, H.; Vandana; Sharma, S.; Pandey, R. Phosphorus Nutrition: Plant Growth in Response to Deficiency and Excess. In *Plant Nutrients and Abiotic Stress Tolerance*; Springer: Singapore, 2018; pp. 171–190.
7. Bayu, D.; Dejene, A.; Alemayehu, R.; Gezahegn, B. Improving available phosphorus in acidic soil using biochar. *JSSEM* **2017**, *8*, 87–94. [CrossRef]
8. Lei, J.; Yin, J.; Chen, S.; Fenton, O.; Liu, R.; Chen, Q.; Fan, B.; Zhang, S. Understanding phosphorus mobilization mechanisms in acidic soil amended with calcium-silicon-magnesium-potassium fertilizer. *Sci. Total Environ.* **2024**, *916*, 170294. [CrossRef] [PubMed]
9. Liu, D. Root developmental responses to phosphorus nutrition. *J. Integr. Plant Biol.* **2021**, *63*, 1065–1090. [CrossRef]
10. Sega, P.; Pacak, A. Plant PHR Transcription Factors: Put on A Map. *Genes* **2019**, *10*, 1018. [CrossRef]
11. Zhu, J.; Lau, K.; Puschmann, R.; Harmel, R.K.; Zhang, Y.; Pries, V.; Gaugler, P.; Broger, L.; Dutta, A.K.; Jessen, H.J.; et al. Two bifunctional inositol pyrophosphate kinases/phosphatases control plant phosphate homeostasis. *eLife* **2019**, *8*, e43582. [CrossRef]
12. Zhou, J.; Hu, Q.; Xiao, X.; Yao, D.; Ge, S.; Ye, J.; Li, H.; Cai, R.; Liu, R.; Meng, F.; et al. Mechanism of phosphate sensing and signaling revealed by rice SPX1-PHR2 complex structure. *Nat. Commun.* **2021**, *12*, 7040. [CrossRef]
13. Guan, Z.; Zhang, Q.; Zhang, Z.; Zuo, J.; Chen, J.; Liu, R.; Savarin, J.; Broger, L.; Cheng, P.; Wang, Q.; et al. Mechanistic insights into the regulation of plant phosphate homeostasis by the rice SPX2–PHR2 complex. *Nat. Commun.* **2022**, *13*, 1581. [CrossRef]
14. Bustos, R.; Castrillo, G.; Linhares, F.; Puga, M.I.; Rubio, V.; Perez-Perez, J.; Solano, R.; Leyva, A.; Paz-Ares, J. A central regulatory system largely controls transcriptional activation and repression responses to phosphate starvation in Arabidopsis. *PLoS Genet.* **2010**, *6*, e1001102. [CrossRef]
15. Lin, W.Y.; Huang, T.K.; Chiou, T.J. Nitrogen limitation adaptation, a target of microRNA827, mediates degradation of plasma membrane-localized phosphate transporters to maintain phosphate homeostasis in Arabidopsis. *Plant Cell* **2013**, *25*, 4061–4074. [CrossRef]
16. Ayadi, A.; David, P.; Arrighi, J.F.; Chiarenza, S.; Thibaud, M.C.; Nussaume, L.; Marin, E. Reducing the genetic redundancy of Arabidopsis PHOSPHATE TRANSPORTER1 transporters to study phosphate uptake and signaling. *Physiol. Plant* **2015**, *167*, 1511–1526. [CrossRef] [PubMed]
17. Xu, Y.; Liu, F.; Han, G.; Cheng, B. Genome-wide identification and comparative analysis of phosphate starvation-responsive transcription factors in maize and three other gramineous plants. *Plant Cell Rep.* **2018**, *37*, 711–726. [CrossRef]
18. Zhao, Y.; Li, P.; Wang, H.; Feng, J.; Li, Y.; Wang, S.; Li, Y.; Guo, Y.; Li, L.; Su, Y.; et al. Genome-wide investigation and expression pattern of PHR family genes in cotton under low phosphorus stress. *PeerJ* **2022**, *10*, e14584. [CrossRef] [PubMed]
19. Wang, Z.; Zheng, Z.; Zhu, Y.; Kong, S.; Liu, D. PHOSPHATE RESPONSE 1 family members act distinctly to regulate transcriptional responses to phosphate starvation. *Plant Physiol.* **2023**, *191*, 1324–1343. [CrossRef]
20. Yue, C.; Chen, Q.; Hu, J.; Li, C.; Luo, L.; Zeng, L. Genome-Wide Identification and Characterization of GARP Transcription Factor Gene Family Members Reveal Their Diverse Functions in Tea Plant (*Camellia sinensis*). *Front. Plant Sci.* **2022**, *13*, 947072. [CrossRef]
21. Li, R.; An, J.-P.; You, C.-X.; Wang, X.-F.; Hao, Y.-J. Overexpression of MdPHR1 Enhanced Tolerance to Phosphorus Deficiency by Increasing MdPAP10 Transcription in Apple (*Malus* × *Domestica*). *J. Plant Growth Regul.* **2020**, *40*, 1753–1763. [CrossRef]
22. Chen, N.; Tong, S.; Yang, J.; Qin, J.; Wang, W.; Chen, K.; Shi, W.; Li, J.; Liu, J.; Jiang, Y. PtoWRKY40 interacts with PtoPHR1-LIKE3 while regulating the phosphate starvation response in poplar. *Plant Physiol.* **2022**, *190*, 2688–2705. [CrossRef]
23. Liu, T.; Xie, Y. Analysis and Prospects of the Drivers of the Rapid Development of Eucalyptus Plantation Forests in China. *Eucalyptus Technol.* **2020**, *37*, 38–47. [CrossRef]
24. Tang, X.; Lei, P.; You, Q.; Liu, Y.; Jiang, S.; Ding, J.; Chen, J.; You, H. Monitoring Seasonal Growth of Eucalyptus Plantation under Different Forest Age and Slopes Based on Multi-Temporal UAV Stereo Images. *Forests* **2023**, *14*, 2231. [CrossRef]
25. Mohanta, T.K.; Khan, A.; Hashem, A.; Abd_Allah, E.F.; Al-Harrasi, A. The molecular mass and isoelectric point of plant proteomes. *BMC Genom.* **2019**, *20*, 631. [CrossRef]
26. Guruprasad, K.; Reddy, B.V.; Pandit, M.W. Correlation between stability of a protein and its dipeptide composition: A novel approach for predicting in vivo stability of a protein from its primary sequence. *Protein Eng.* **1990**, *4*, 155–161. [CrossRef] [PubMed]

27. Ninkuu, V.; Yan, J.; Fu, Z.; Yang, T.; Zhang, L.; Ren, J.; Li, G.; Zeng, H. Genome-wide identification, phylogenomics, and expression analysis of benzoxazinoids gene family in rice (*Oryza sativa*). *Plant Stress* **2023**, *10*, 100214. [CrossRef]
28. Zhou, J.; Jiao, F.; Wu, Z.; Li, Y.; Wang, X.; He, X.; Zhong, W.; Wu, P. OsPHR2 is involved in phosphate-starvation signaling and excessive phosphate accumulation in shoots of plants. *Plant Physiol.* **2008**, *146*, 1673–1686. [CrossRef]
29. Guo, M.; Ruan, W.; Li, C.; Huang, F.; Zeng, M.; Liu, Y.; Yu, Y.; Ding, X.; Wu, Y.; Wu, Z.; et al. Integrative Comparison of the Role of the PHOSPHATE RESPONSE1 Subfamily in Phosphate Signaling and Homeostasis in Rice. *Plant Physiol.* **2015**, *168*, 1762–1776. [CrossRef]
30. Shi, J.; Zhao, B.; Zheng, S.; Zhang, X.; Wang, X.; Dong, W.; Xie, Q.; Wang, G.; Xiao, Y.; Chen, F.; et al. A phosphate starvation response-centered network regulates mycorrhizal symbiosis. *Cell* **2021**, *184*, 5527–5540.e5518. [CrossRef]
31. Naidoo, R.; Ferreira, L.; Berger, D.K.; Myburg, A.A.; Naidoo, S. The identification and differential expression of Eucalyptus grandis pathogenesis-related genes in response to salicylic acid and methyl jasmonate. *Front. Plant Sci.* **2013**, *4*, 42592.
32. Ye, W.; Wang, T.; Wei, W.; Lou, S.; Lan, F.; Zhu, S.; Li, Q.; Ji, G.; Lin, C.; Wu, X.; et al. The Full-Length Transcriptome of Spartina alterniflora Reveals the Complexity of High Salt Tolerance in Monocotyledonous Halophyte. *Plant Cell Physiol.* **2020**, *61*, 882–896. [CrossRef]
33. Chen, B.; Liu, T.; Yang, Z.; Yang, S.; Chen, J. PacBio Full-Length Transcriptome Sequencing Reveals the Mechanism of Salt Stress Response in Sonneratia apetala. *Plants* **2023**, *12*, 3849. [CrossRef] [PubMed]
34. Oberschelp, G.P.J.; Guarnaschelli, A.B.; Teson, N.; Harrand, L.; Podestá, F.E.; Margarit, E. Cold acclimation and freezing tolerance in three Eucalyptus species: A metabolomic and proteomic approach. *Plant Physiol. Biochem.* **2020**, *154*, 316–327. [CrossRef] [PubMed]
35. Medici, A.; Szponarski, W.; Dangeville, P.; Safi, A.; Dissanayake, I.M.; Saenchai, C.; Emanuel, A.; Rubio, V.; Lacombe, B.; Ruffel, S.; et al. Identification of Molecular Integrators Shows that Nitrogen Actively Controls the Phosphate Starvation Response in Plants. *Plant Cell* **2019**, *31*, 1171–1184. [CrossRef]
36. Luo, J.; Liang, Z.; Wu, M.; Mei, L. Genome-wide identification of BOR genes in poplar and their roles in response to various environmental stimuli. *Environ. Exp. Bot.* **2019**, *164*, 101–113. [CrossRef]
37. Wang, T.; Ye, W.; Zhang, J.; Li, H.; Zeng, W.; Zhu, S.; Ji, G.; Wu, X.; Ma, L. Alternative 3′-untranslated regions regulate high-salt tolerance of Spartina alterniflora. *Plant Physiol.* **2023**, *191*, 2570–2587. [CrossRef]
38. Lin, Z.; Guo, C.; Lou, S.; Jin, S.; Zeng, W.; Guo, Y.; Fang, J.; Xu, Z.; Zuo, Z.; Ma, L. Functional analyses unveil the involvement of moso bamboo (*Phyllostachys edulis*) group I and II NIN-LIKE PROTEINS in nitrate signaling regulation. *Plant Sci.* **2021**, *306*, 110862. [CrossRef]
39. Fan, C.; Lyu, M.; Zeng, B.; He, Q.; Wang, X.; Lu, M.Z.; Liu, B.; Liu, J.; Esteban, E.; Pasha, A.; et al. Profiling of the gene expression and alternative splicing landscapes of Eucalyptus grandis. *Plant Cell Environ.* **2024**, *47*, 1363–1378. [CrossRef] [PubMed]
40. Nakano, Y.; Naito, Y.; Nakano, T.; Ohtsuki, N.; Suzuki, K. NSR1/MYR2 is a negative regulator of ASN1 expression and its possible involvement in regulation of nitrogen reutilization in Arabidopsis. *Plant Sci.* **2017**, *263*, 219–225. [CrossRef]
41. Zhao, C.; Hanada, A.; Yamaguchi, S.; Kamiya, Y.; Beers, E.P. The Arabidopsis Myb genes MYR1 and MYR2 are redundant negative regulators of flowering time under decreased light intensity. *Plant J.* **2011**, *66*, 502–515. [CrossRef]
42. Hu, B.; Chu, C. Nitrogen–phosphorus interplay: Old story with molecular tale. *New Phytol.* **2020**, *225*, 1455–1460. [CrossRef]
43. Hu, B.; Jiang, Z.; Wang, W.; Qiu, Y.; Zhang, Z.; Liu, Y.; Li, A.; Gao, X.; Liu, L.; Qian, Y.; et al. Nitrate–NRT1.1B–SPX4 cascade integrates nitrogen and phosphorus signalling networks in plants. *Nat. Plants* **2019**, *5*, 401–413. [CrossRef]
44. Long, Y.; Peng, J. Interaction between Boron and Other Elements in Plants. *Genes* **2023**, *14*, 130. [CrossRef]
45. Kaya, C.; Tuna, A.L.; Dikilitas, M.; Ashraf, M.; Koskeroglu, S.; Guneri, M. Supplementary phosphorus can alleviate boron toxicity in tomato. *Sci. Hortic.* **2009**, *121*, 284–288. [CrossRef]
46. Gunes, A.; Alpaslan, M. Boron uptake and toxicity in maize genotypes in relation to boron and phosphorus supply. *J. Plant Nutr.* **2000**, *23*, 541–550. [CrossRef]
47. Masood, S.; Zhao, X.Q.; Shen, R.F. Bacillus pumilus increases boron uptake and inhibits rapeseed growth under boron supply irrespective of phosphorus fertilization. *AoB Plants* **2019**, *11*, plz036. [CrossRef]
48. He, K.; Du, J.; Han, X.; Li, H.; Kui, M.; Zhang, J.; Huang, Z.; Fu, Q.; Jiang, Y.; Hu, Y. PHOSPHATE STARVATION RESPONSE1 (PHR1) interacts with JASMONATE ZIM-DOMAIN (JAZ) and MYC2 to modulate phosphate deficiency-induced jasmonate signaling in Arabidopsis. *Plant Cell* **2023**, *35*, 2132–2156. [CrossRef]
49. Li, L.; Zhang, X.; Li, D.; Su, H.; He, Y.; Xu, Z.; Zhao, Y.; Hong, Y.; Li, Q.; Xu, P.; et al. CsPHRs-CsJAZ3 incorporates phosphate signaling and jasmonate pathway to regulate catechin biosynthesis in *Camellia sinensis*. *Hortic. Res.* **2024**, *11*, uhae178. [CrossRef] [PubMed]

50. Xu, H.; Deng, L.; Zhou, X.; Xing, Y.; Li, G.; Chen, Y.; Huang, Y.; Ma, X.; Liu, Z.-J.; Li, M.; et al. Unveiling the PHR-centered regulatory network orchestrating the phosphate starvation signaling in Chinese fir (*Cunninghamia lanceolata*). *bioRxiv* **2024**. [CrossRef]
51. Wang, T.; Jin, Y.; Deng, L.; Li, F.; Wang, Z.; Zhu, Y.; Wu, Y.; Qu, H.; Zhang, S.; Liu, Y.; et al. The transcription factor MYB110 regulates plant height, lodging resistance, and grain yield in rice. *Plant Cell* **2024**, *36*, 298–323. [CrossRef] [PubMed]
52. Kavka, M.; Polle, A. Phosphate uptake kinetics and tissue-specific transporter expression profiles in poplar (*Populus* × *canescens*) at different phosphorus availabilities. *BMC Plant Biol.* **2016**, *16*, 206. [CrossRef]
53. Yadav, R.K.; Perales, M.; Gruel, J.; Girke, T.; Jönsson, H.; Reddy, G.V. WUSCHEL protein movement mediates stem cell homeostasis in the Arabidopsis shoot apex. *Genes Dev.* **2011**, *25*, 2025–2030. [PubMed]
54. Long, X.; Zhang, J.; Wang, D.; Weng, Y.; Liu, S.; Li, M.; Hao, Z.; Cheng, T.; Shi, J.; Chen, J. Expression dynamics of WOX homeodomain transcription factors during somatic embryogenesis in Liriodendron hybrids. *For. Res.* **2023**, *3*, 15. [CrossRef]
55. Furumizu, C.; Alvarez, J.P.; Sakakibara, K.; Bowman, J.L. Antagonistic Roles for KNOX1 and KNOX2 Genes in Patterning the Land Plant Body Plan Following an Ancient Gene Duplication. *PLoS Genet.* **2015**, *11*, e1004980. [CrossRef]
56. Ogura, N.; Sasagawa, Y.; Ito, T.; Tameshige, T.; Kawai, S.; Sano, M.; Doll, Y.; Iwase, A.; Kawamura, A.; Suzuki, T.; et al. WUSCHEL-RELATED HOMEOBOX 13 suppresses de novo shoot regeneration via cell fate control of pluripotent callus. *Sci. Adv.* **2023**, *9*, eadg6983. [CrossRef]
57. Kojc, N.; Zidar, N.; Gale, N.; Poljak, M.; Fujs Komloš, K.; Cardesa, A.; Höfler, H.; Becker, K.-F. Transcription factors Snail, Slug, Twist, and SIP1 in spindle cell carcinoma of the head and neck. *Virchows Arch.* **2009**, *454*, 549–555. [CrossRef]
58. Fujita, A.; Lord, M.; Hiroko, T.; Hiroko, F.; Chen, T.; Oka, C.; Misumi, Y.; Chant, J. Rax1, a protein required for the establishment of the bipolar budding pattern in yeast. *Gene* **2004**, *327*, 161–169. [CrossRef] [PubMed]
59. Guo, X.; Park, C.H.; Wang, Z.Y.; Nickels, B.E.; Dong, J. A spatiotemporal molecular switch governs plant asymmetric cell division. *Nat. Plants* **2021**, *7*, 667–680. [CrossRef]
60. Smirnova, E.A.; Reddy, A.S.; Bowser, J.; Bajer, A.S. Minus end-directed kinesin-like motor protein, Kcbp, localizes to anaphase spindle poles in Haemanthus endosperm. *Cell Motil. Cytoskelet.* **1998**, *41*, 271–280. [CrossRef]
61. Wang, T.; Yang, Y.; Lou, S.; Wei, W.; Zhao, Z.; Ren, Y.; Lin, C.; Ma, L. Genome-Wide Characterization and Gene Expression Analyses of GATA Transcription Factors in Moso Bamboo (*Phyllostachys edulis*). *Int. J. Mol. Sci.* **2019**, *21*, 14. [CrossRef]
62. Chen, C.; Wu, Y.; Li, J.; Wang, X.; Zeng, Z.; Xu, J.; Liu, Y.; Feng, J.; Chen, H.; He, Y.; et al. TBtools-II: A "one for all, all for one" bioinformatics platform for biological big-data mining. *Mol. Plant* **2023**, *16*, 1733–1742. [CrossRef]
63. Wang, T.; Li, Q.; Lou, S.; Yang, Y.; Peng, L.; Lin, Z.; Hu, Q.; Ma, L. GSK3/shaggy-like kinase 1 ubiquitously regulates cell growth from Arabidopsis to Moso bamboo (*Phyllostachys edulis*). *Plant Sci.* **2019**, *283*, 290–300. [CrossRef] [PubMed]
64. Lin, J.; Wu, J.; Zhang, D.; Cai, X.; Du, L.; Lu, L.; Liu, C.; Chen, S.; Yao, Q.; Xie, S.; et al. The GRAS gene family and its roles in pineapple (Ananas comosus L.) developmental regulation and cold tolerance. *BMC Plant Biol.* **2024**, *24*, 1204. [CrossRef] [PubMed]
65. Livak, K.J.; Schmittgen, T.D. Analysis of relative gene expression data using real-time quantitative PCR and the $2^{-\Delta\Delta CT}$ Method. *Methods* **2001**, *25*, 402–408. [CrossRef] [PubMed]
66. Ewels, P.; Magnusson, M.; Lundin, S.; Kaller, M. MultiQC: Summarize analysis results for multiple tools and samples in a single report. *Bioinformatics* **2016**, *32*, 3047–3048. [CrossRef]
67. Dobin, A.; Davis, C.A.; Schlesinger, F.; Drenkow, J.; Zaleski, C.; Jha, S.; Batut, P.; Chaisson, M.; Gingeras, T.R. STAR: Ultrafast universal RNA-seq aligner. *Bioinformatics* **2013**, *29*, 15–21. [CrossRef]
68. Liao, Y.; Smyth, G.K.; Shi, W. featureCounts: An efficient general purpose program for assigning sequence reads to genomic features. *Bioinformatics* **2014**, *30*, 923–930. [CrossRef]
69. Li, H.; Handsaker, B.; Wysoker, A.; Fennell, T.; Ruan, J.; Homer, N.; Marth, G.; Abecasis, G.; Durbin, R.; Genome Project Data Processing, S. The Sequence Alignment/Map format and SAMtools. *Bioinformatics* **2009**, *25*, 2078–2079. [CrossRef]
70. Love, M.I.; Huber, W.; Anders, S. Moderated estimation of fold change and dispersion for RNA-seq data with DESeq2. *Genome Biol.* **2014**, *15*, 550. [CrossRef]

International Journal of *Molecular Sciences*

MDPI

Article

Chlorine Modulates Photosynthetic Efficiency, Chlorophyll Fluorescence in Tomato Leaves, and Carbohydrate Allocation in Developing Fruits

Longpu Su †, Tao Lu †, Qiang Li, Yang Li, Xiaoyang Wan, Weijie Jiang * and Hongjun Yu *

State Key Laboratory of Vegetable Biobreeding, Institute of Vegetables and Flowers, Chinese Academy of Agricultural Sciences, Beijing 100081, China; sulongpu@163.com (L.S.); lutao@caas.cn (T.L.)

* Correspondence: jiangweijie@caas.cn (W.J.); yuhongjun@caas.cn (H.Y.); Tel.: +86-10-82108797 (W.J. & H.Y.)

† These authors contributed equally to this work.

Abstract: Chlorine (Cl) is an essential nutrient for higher plants, playing a critical role in their growth and development. However, excessive Cl application can be detrimental, leading to its limited use in controlled-environment agriculture. Recently, Cl has been recognized as a beneficial macronutrient, yet studies investigating its impact on plant growth and fruit quality remain scarce. In this study, we determined the optimal Cl concentration in nutrient solutions through a series of cultivation experiments. A comparative analysis of Cl treatments at 1, 2, and 3 mM revealed that 3 mM Cl^- significantly enhanced chlorophyll content, biomass accumulation, and yield. Furthermore, we examined the effects of 3 mM Cl^- (supplied as 1.5 mM $CaCl_2$ and 3 mM KCl) on leaf photosynthesis, chlorophyll fluorescence, and fruit sugar metabolism. The results demonstrated that Cl^- treatments enhanced the activity of Photosystem I (PS I) and Photosystem II (PS II), leading to a 25.53% and 28.37% increase in the net photosynthetic rate, respectively. Additionally, Cl^- application resulted in a 12.3% to 16.5% increase in soluble sugar content in mature tomato fruits. During fruit development, Cl^- treatments promoted the accumulation of glucose, fructose, and sucrose, thereby enhancing fruit sweetness and overall quality. The observed increase in glucose and fructose levels was attributed to the stimulation of invertase activity. Specifically, acidic invertase (AI) activity increased by 61.6% and 62.6% at the green ripening stage, while neutral invertase (NI) activity was elevated by 56.2% and 32.8% in the $CaCl_2$ and KCl treatments, respectively, at fruit maturity. Furthermore, sucrose synthase (SS-I) activity was significantly upregulated by 1.5- and 1.4-fold at fruit maturity, while sucrose phosphate synthase (SPS) activity increased by 76.4% to 77.8% during the green ripening stage. These findings provide novel insights into the role of Cl^- in tomato growth and metabolism, offering potential strategies for optimizing fertilization practices in protected horticulture.

Keywords: chlorine; photosynthesis; sucrose metabolism; yield; quality

1. Introduction

Chloride (Cl^-) serves not only as a critical micronutrient but also as an essential macronutrient for plants [1–5]. As the anionic form of Cl^-, Cl^- represents the most abundant soluble anion in plant tissues, typically ranging from 2 to 20 mg/g dry weight [6]. Recognized as an essential plant micronutrient in 1954, its fundamental role in plant physiology was initially established through studies demonstrating growth impairment in plants cultivated under chloride-deficient conditions [7]. Despite its widespread availability in nature, contemporary research suggests that its classification as merely a micronutrient underestimates its physiological significance, as optimal plant growth frequently necessitates Cl^- concentrations surpassing the conventional micronutrient threshold of 100 mg/kg dry weight [3]. Plants predominantly acquire Cl^- from soil solutions via passive uptake through anion channels and active transport systems in root cells, with additional mechanisms facilitating selective compartmentalization within tissues and organelles [8]. In agricultural systems, Cl^- is commonly supplied through irrigation water, organic amendments, and various fertilizer formulations, including potassium chloride, calcium chloride, and ammonium chloride, with application rates typically ranging from 5 to 50 kg Cl^-/ha, depending on the crop species, soil characteristics, and environmental conditions [1]. The antagonistic interaction between Cl^- and NO_3^- during root uptake necessitates careful nutrient management, particularly in Cl^--sensitive crops such as tobacco, legumes, and certain fruit trees [5]. Modern precision agriculture approaches integrate soil and tissue testing for Cl^- status to optimize fertilization strategies, maximizing its beneficial effects while mitigating potential risks associated with salt stress in susceptible crops [9].

Cl^- plays a pivotal role in plant development, functioning as a key regulator of enzymatic activity and physiological processes [10,11]. For instance, Cl^- has been shown to modulate the activity of enzymes such as asparagine synthetase, amylases, and tonoplast H^+-ATPase [12]. Additionally, Cl^- contributes significantly to maintaining charge balance with essential cations, including K^+ and H^+, thereby stabilizing membrane electrical potential and modulating pH gradients [1–3,13]. Beyond its roles in enzyme regulation and ion homeostasis, Cl^- is crucial for osmotic regulation and water relations within plant cells [8]. As a highly mobile anion, it facilitates rapid osmotic adjustments, particularly in guard cells during stomatal movements [14]. Studies have demonstrated that Cl^- influx into guard cells is integral to stomatal opening, acting synergistically with K^+ to regulate turgor pressure [15]. This stomatal regulation directly influences water-use efficiency and carbon assimilation rates [3]. Cl^- has also been implicated in nitrogen utilization efficiency, influencing nitrate uptake and assimilation pathways [9]. Furthermore, under conditions of water scarcity, the application of Cl^- reduces stress symptoms, promotes favorable plant growth conditions, and enhances resilience to abiotic stresses [3,16–18].

Photosynthesis, one of the most fundamental biochemical processes on Earth, underpins all stages of plant development and productivity [19]. Previous studies on Cl^- have predominantly focused on its role in PS II, where only trace amounts are required to sustain effective leaf photosynthesis [10,11,20]. However, the influence of Cl^- on photosynthetic efficiency extends beyond its structural role in PS II [4]. Experiments utilizing Cl^--deficient plants have revealed substantial reductions in both the light-dependent and light-independent reactions of photosynthesis [21]. Specifically, Cl^- deficiency has been associated with up to a 30% decline in PS II quantum yield in certain species [22]. Moreover, Cl^- influences electron transport rates and carbon fixation efficiency, underscoring their broader significance in photosynthetic metabolism [21]. Adequate Cl^- supply has been demonstrated to enhance chlorophyll content and maintain thylakoid membrane integrity, both of which are critical for efficient light harvesting [22]. Furthermore, Cl^- indirectly modulates photosynthesis by influencing stomatal conductance and mesophyll resistance to CO_2 diffusion [3]. In addition to its role as a bridging ligand between manganese atoms in PS II, Cl^- serves as a crucial cofactor in the oxygen-evolving complex, facilitating the hydrolysis required for oxygen evolution during photosynthesis [11,13]. Consequently, Cl^- deficiency can lead to compromised photosynthetic efficiency, ultimately impacting crop productivity [10]. Photosynthesis is a primary determinant of crop yield, serving as the driving force behind biomass accumulation and fruit development in tomatoes [19,23,24]. Sugars in tomato fruits are predominantly derived from photosynthetic carbon assimilation, highlighting the importance of optimizing photosynthetic efficiency to improve fruit quality [25]. Recent studies indicate that the exogenous application of 5 mM Cl^- can enhance net photosynthesis and water-use efficiency in tobacco plants, with similar effects reported in other crops [9,26]. Moreover, an increase in shoot carbon, sulfur, zinc, and copper content has been observed following the application of 5 mM NaCl, suggesting that Cl^-, in conjunction with these elements, contributes to enhanced shoot growth and carbon metabolism [18]. Thus, maintaining optimal Cl^- concentrations is crucial for sustaining efficient photosynthesis and improving overall plant performance [20,27].

Tomato (*Solanum lycopersicum*) is a nutritionally rich berry crop characterized by high adaptability and yield potential [28–30]. Consumed both fresh and in processed forms—such as sauces, pastes, and juices—tomatoes hold significant economic and dietary importance worldwide [31]. Fruit quality in tomatoes is influenced by multiple factors, including sugar content, acidity, texture, aroma, and phytochemical composition, particularly lycopene and other carotenoids [32]. The balance between sugars (primarily glucose and fructose) and organic acids (mainly citric and malic acids) is a key determinant of tomato flavor [32]. However, modern breeding programs have often prioritized yield, disease resistance, and shelf life over flavor attributes, leading to consumer dissatisfaction with commercial cultivars [33]. As greenhouse tomato cultivation expands and consumer preferences evolve, there is increasing demand for tomatoes with superior organoleptic properties, including enhanced sweetness and aroma [34]. However, modern cultivation practices, particularly those emphasizing high-yield production, have been criticized for their negative impact on fruit flavor [33]. Sugar content, a primary determinant of tomato sweetness, is of considerable commercial significance [35].

Previous research on the effects of Cl^- on crop quality has primarily focused on comparisons between chloride- and non-chloride-based fertilization regimes. While studies investigating the direct influence of Cl^- on tomato sugar metabolism remain limited, most have examined the combined effects of cations and Cl^-, often attributing observed changes to cation availability rather than Cl^- itself. Understanding the specific role of Cl^- in tomato metabolism has significant implications for optimizing fertilization strategies [9]. Current agricultural practices tend to emphasize nitrogen, phosphorus, and potassium requirements while largely overlooking the contributions of anions, including Cl^- [14]. Deciphering the precise mechanisms by which Cl^- influences photosynthesis and sugar accumulation could inform the development of fertilization protocols aimed at simultaneously enhancing yield and fruit quality [9]. Moreover, in light of increasing soil salinization and water scarcity, elucidating how plants efficiently utilize Cl^- could contribute to the development of more resilient and resource-efficient cropping systems [4]. By systematically isolating the effects of Cl^- from those of accompanying cations, this study aims to address a fundamental knowledge gap in plant nutrition research, providing novel insights into the role of Cl^- in photosynthesis and sugar metabolism in tomatoes [14].

2. Results

2.1. Effect of Exogenous Chlorine on Tomato Growth and Yield

During the fruit-setting stage, the application of $CaCl_2$ and KCl resulted in a significant increase in plant height, with increments of 13.80% and 10.60% ($p < 0.05$) compared to the control, respectively. However, no significant effect of exogenous Cl^- on stem diameter was observed (Figure 1). In the pre-screening concentration experiments, the biomass of tomato plants in the T3 treatment showed a significant increase of 11.48% compared to the control

(Supplementary Figure S1), suggesting that Cl^- enhances plant growth. Furthermore, the T3 treatment led to a 14.82% increase in total fruit yield compared to the control (CK) (Supplementary Figure S2). Additionally, variations in chlorophyll content were observed at different developmental stages, with the T3 treatment resulting in a 13.77% increase in chlorophyll content relative to the control on day 49 (Supplementary Table S2). These findings indicate that Cl^- application enhances chlorophyll accumulation in tomato leaves, potentially contributing to improved photosynthetic efficiency.

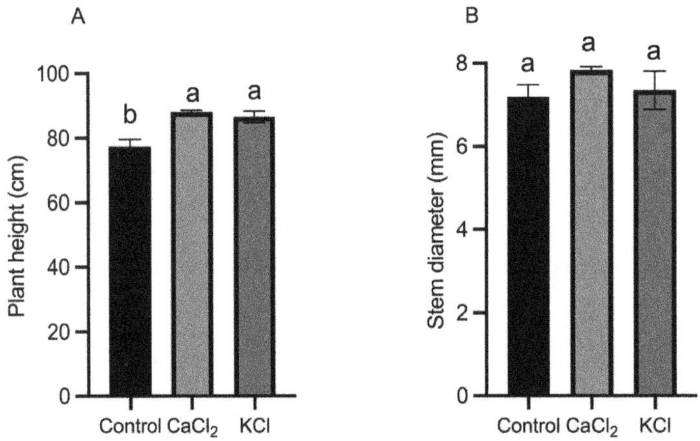

Figure 1. Effects of Cl^- on plant height and stem diameter of tomato plants. (**A**) Plant height and (**B**) stem diameter. Plants were subjected to three treatments: control (Hoagland solution), 1.5 mM $CaCl_2$ (3 $mmol \cdot L^{-1}$ Cl^-), and 3 mM KCl (3 $mmol \cdot L^{-1}$ Cl^-). Data are expressed as means ± SD. Different letters above bars indicate significant differences at $p < 0.05$.

2.2. *Effect of Exogenous Chlorine on Photosynthetic Performance*

To assess the impact of Cl^- on photosynthetic activity, gas exchange parameters were measured. As shown in Table 1, while the intercellular CO_2 concentration (Ci) remained unchanged, the net photosynthetic rate (Pn), transpiration rate (Tr), and stomatal conductance (Gs) increased significantly following Cl^- treatment. Specifically, $CaCl_2$ and KCl application led to Pn enhancements of 25.53% and 28.37%, respectively. In addition, the leaf Tr increased by 32.97% and 44.23%, while the Gs exhibited considerable increases of 92.31% and 101.44% under $CaCl_2$ and KCl treatments, respectively.

Table 1. Effects of Cl^- on gas exchange parameters of tomato plants. Control (0 $mmol \cdot L^{-1}$ Cl^-), $CaCl_2$ (3 $mmol \cdot L^{-1}$ Cl^-), and KCl (3 $mmol \cdot L^{-1}$ Cl^-). Data represent the average of three replicates ± SD. Different letters indicate significant differences at $p < 0.05$.

Treatments	Pn ($\mu mol \cdot m^{-2} \cdot s^{-1}$)	Ci ($\mu mol \cdot mol^{-1}$)	Tr ($mmol \cdot m^{-2} \cdot s^{-1}$)	Gs ($mmol \cdot m^{-2} \cdot s^{-1}$)
Control	14.1 ± 1.31 b	264.83 ± 25.38 a	7.28 ± 0.38 b	336.00 ± 28.46 b
$CaCl_2$	17.7 ± 0.71 a	269.50 ± 9.54 a	9.68 ± 1.03 a	646.17 ± 191.57 a
KCl	18.1 ± 1.30 a	284.67 ± 0.87 a	10.05 ± 0.32 a	676.83 ± 38.68 a

Furthermore, Figure 2 illustrates the Pn–light response curve of tomato plants under Cl^- treatment. The net photosynthetic rate initially increased and then plateaued with rising light intensity. Compared to the control, the maximum net photosynthetic rate (Pmax) was elevated by 38.8% and 29.0% under the $CaCl_2$ and KCl treatments, respectively.

These results suggest that exogenous Cl^- improves photosynthetic capacity by enhancing gas exchange efficiency.

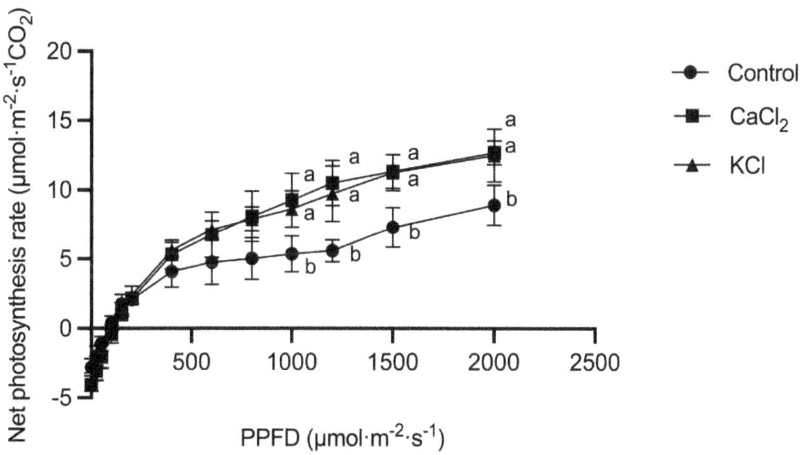

Figure 2. Effects of Cl^- on Pn–light response curve of tomato plants. Plants were subjected to three treatments: control (0 mmol·L^{-1} Cl^-), $CaCl_2$ (3 mmol·L^{-1} Cl^-), and KCl (3 mmol·L^{-1} Cl^-). The values are the means ± SDs (n = 3). Different letters above bars indicate significant differences at $p < 0.05$.

2.3. *Effect of Exogenous Chlorine on Activity of Photosystems*

The maximum quantum efficiency of PS II (Fv/Fm) increased by 1.2% and 1.9% under $CaCl_2$ and KCl treatments, respectively, compared to the control (Figure 3). Additionally, as tomato plants matured, the maximal oxidation state of PS I (Pm) exhibited significant increases of 17.84% and 25.78% under $CaCl_2$ and KCl treatments, respectively, indicating that Cl^- application enhances light energy utilization efficiency in plant leaves.

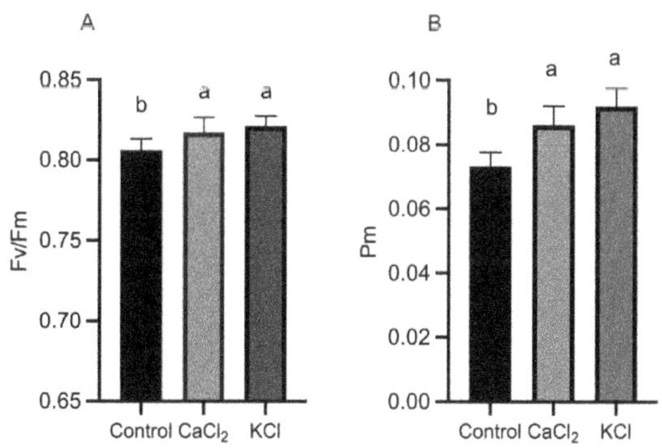

Figure 3. Effects of Cl^- on Fv/Fm and Pm of tomato plants. (**A**) Maximum photochemical efficiency, Fv/Fm, and (**B**) PS I maximum oxidation state, Pm. Plants were subjected to three treatments: control (Hoagland solution), 1.5 mM $CaCl_2$ (3 mmol·L^{-1} Cl^-), and 3 mM KCl (3 mmol·L^{-1} Cl^-). Data are expressed as means ± SD. Different letters above bars indicate significant differences at $p < 0.05$.

As shown in Figure 4, exogenous Cl^- treatments significantly increased the quantum efficiency of PS II photochemistry (Y(II)) and the quantum efficiency of regulated energy dissipation in PS II (Y(NPQ)) while reducing the quantum efficiency of non-regulated energy dissipation in PS II (Y(NO)) compared to the control. Specifically, the Y(II) increased by 29.51% and 32.65% under $CaCl_2$ and KCl treatments, respectively, while the Y(NPQ) increased by 9.62% and 4.44%. Conversely, the Y(NO) decreased by 15.93% and 13.71%, suggesting that Cl^- enhances the actual photochemical efficiency of PS II, thereby improving the plant's ability to sustain photosynthetic activity. Additionally, a significant increase in the PS I photochemical yield (Y(I)) was observed, along with a notable decrease in PS I donor-side limitation (Y(ND)) and PS I acceptor-side limitation (Y(NA)). Specifically, the Y(I) increased by 54.6% and 51.2% under $CaCl_2$ and KCl treatments, respectively, while Y(ND) decreased by 16.5% and 13.5% (Figure 5). These results suggest that Cl^- enhances PS I photochemical efficiency while reducing PS I-driven electron transport limitations.

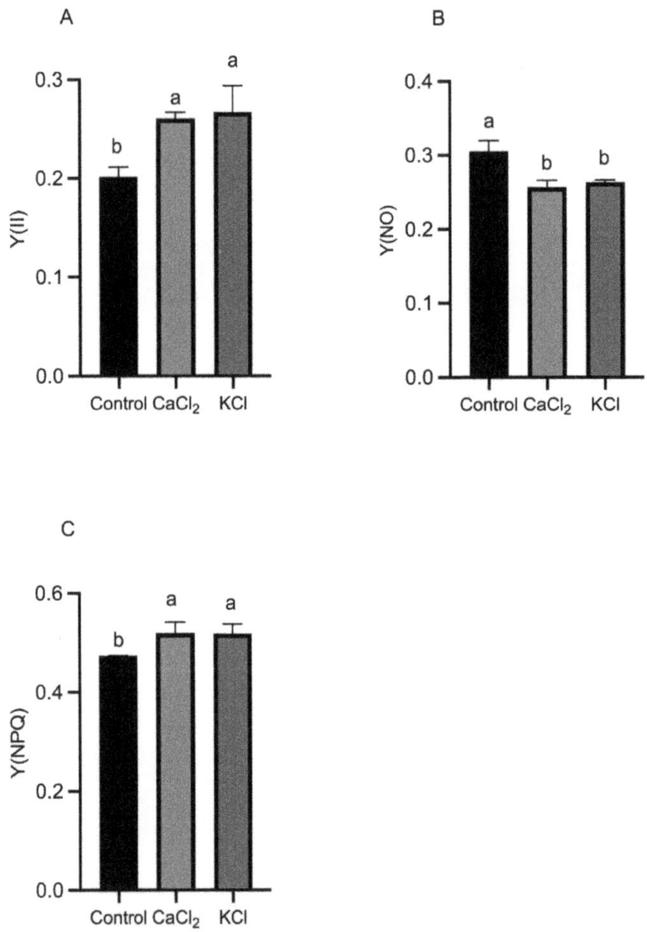

Figure 4. Effects of Cl^- on PS II reaction center of tomato plants. Plants were subjected to three treatments: control (Hoagland solution), 1.5 mM $CaCl_2$ (3 $mmol \cdot L^{-1}$ Cl^-), and 3 mM KCl (3 $mmol \cdot L^{-1}$ Cl^-). (**A**) Quantum efficiency of PS II photochemistry (Y(II)). (**B**) Quantum efficiency of non-regulated energy dissipation in PS II (Y(NO)). (**C**) Quantum efficiency of regulated energy dissipation in PS II (Y(NPQ)). Data are expressed as means ± SD. Different letters above bars indicate significant differences at $p < 0.05$.

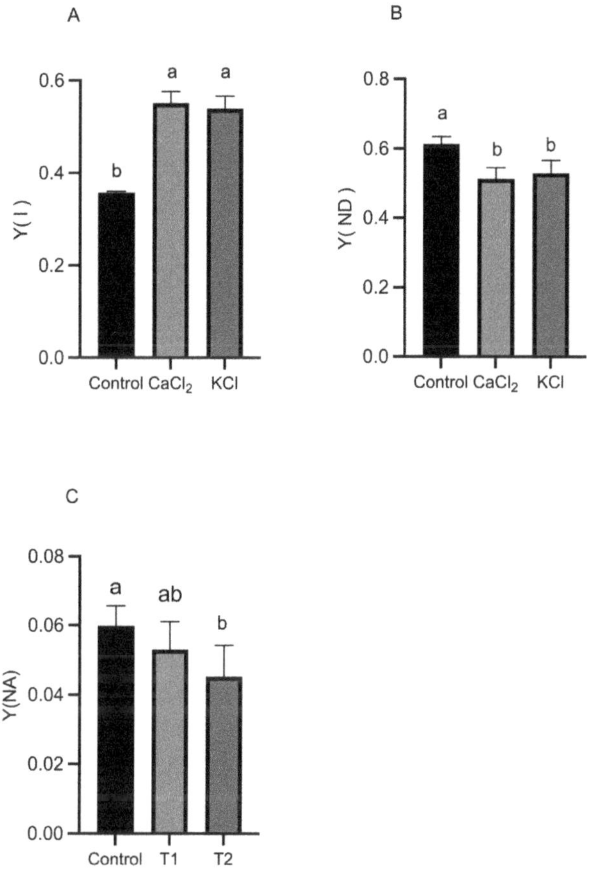

Figure 5. Effects of Cl^- on PS I reaction center of tomato plants. Plants were subjected to three treatments: control (Hoagland solution), 1.5 mM $CaCl_2$ (3 mmol·L^{-1} Cl^-), and 3 mM KCl (3 mmol·L^{-1} Cl^-). (**A**) Photochemical quantum yield of PS I in the light Y(I). (**B**) Quantum yield of non-photochemical energy dissipation at PS I due to donor-side confinement (Y(ND)). (**C**) Quantum yield of non-photochemical energy dissipation at PS-I due to the acceptor side (Y(NA)). Data are expressed as means ± SD. Different letters above bars indicate significant differences at $p < 0.05$.

2.4. Effect of Exogenous Chlorine on Photosynthetic Electron Transport

To further elucidate the role of exogenous Cl^- in photosynthetic electron transport, the electron transport rate in PS II (ETR(II)) was evaluated (Figure 6). Across various light intensities, the ETR(II) in Cl^--treated plants consistently exceeded that of the control. The fast optical response curve of the ETR(II) revealed that the curve slope α increased by 15.2% and 21.2% under $CaCl_2$ and KCl treatments, respectively, compared to the control. Additionally, the light saturation point (Ek) increased by 14.1% and 42.8% in $CaCl_2$ and KCl treatments, respectively. Notably, the maximum ETR(II) value in the KCl treatment reached 216.2 μmol m^{-2} s^{-1}, representing a 62.9% increase compared to the control. Furthermore, as the light intensity increased from low to high, the ETR(II) rapidly rose before stabilizing. At a light intensity of 300 μmol photons m^{-2}·s^{-1}, the ETR(II) values in Cl^--treated plants remained higher than those in the control, indicating that Cl^- enhances PS II electron transport efficiency and overall photosynthetic performance.

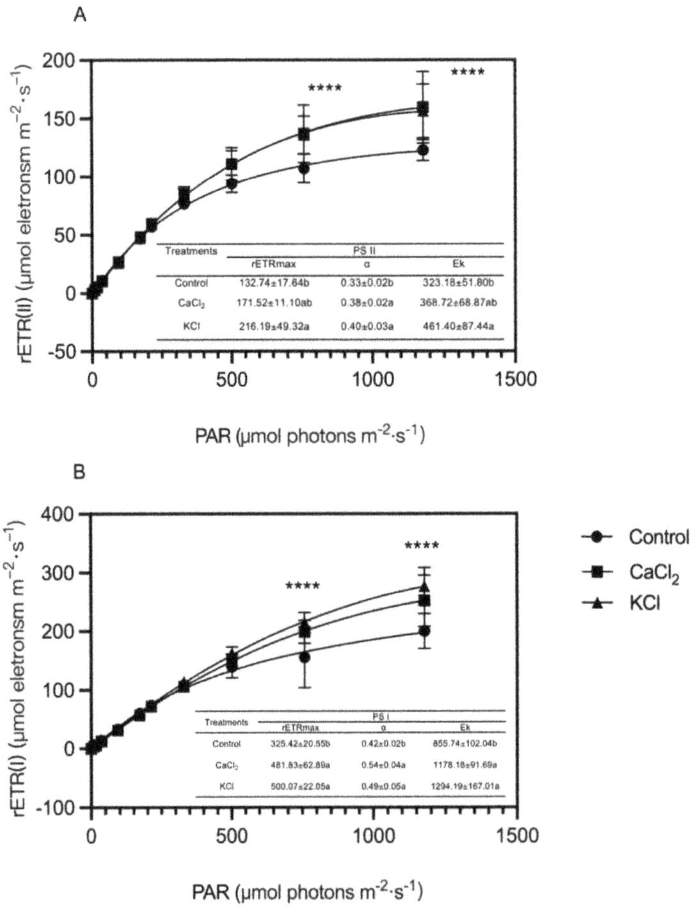

Figure 6. Effects of Cl^- on photosynthetic rETR-PAR response curves of tomato plants. Plants were subjected to three treatments: control (Hoagland solution), 1.5 mM $CaCl_2$ (3 mmol·L^{-1} Cl^-), and 3 mM KCl (3 mmol·L^{-1} Cl^-). (**A**) Photosynthetic rETR(I)–PAR response curve and fitting parameters of tomato plants. (**B**) Photosynthetic rETR(II)–PAR response curve and fitting parameters of tomato plants. Data are expressed as means ± SD. **** above the curve indicates significant differences at $p < 0.01$. Within each row, means followed by different letters are significantly different ($p < 0.05$).

2.5. Effect of Exogenous Chlorine on Fruit Sugar Content and Sugar Ratio

As the tomato fruits matured, the soluble sugar content steadily increased. As shown in Figure 7, fruits from Cl^--treated plants exhibited a significantly higher accumulation of soluble sugars compared to the control. Specifically, soluble sugar content in mature fruits increased by 12.3% and 16.5% under $CaCl_2$ and KCl treatments, respectively, indicating that exogenous Cl^- improves fruit quality.

To assess sugar metabolism dynamics, the sucrose, glucose, and fructose levels were measured at three key developmental stages: fruit expansion, green ripening, and full ripening (Figure 8). Our results demonstrated that Cl^--treated plants exhibited significantly higher sugar content during the early stages of fruit expansion. As fruits ripened, the sucrose, glucose, and fructose content progressively increased. At full ripeness, the sucrose content in fruits treated with $CaCl_2$ and KCl increased by 16.6% and 14.1%, respectively, relative to the control. Additionally, the glucose content in fruits treated with $CaCl_2$ and KCl was elevated by 14.4% and 15.6%, respectively. Although no significant differences in the fructose content were observed between treatments at full ripeness, the total fructose and glucose content significantly increased by 8.7% and 10.5%, respectively, compared to the control. These findings suggest that Cl^- supplementation enhances sucrose accumulation and its subsequent conversion into glucose and fructose, ultimately improving the overall fruit sugar content.

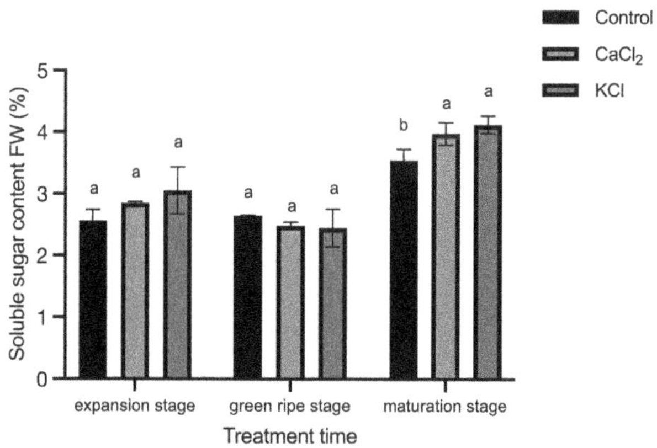

Figure 7. Effects of Cl^- on sugar content of fruits. Plants were subjected to three treatments: control (Hoagland solution), 1.5 mM $CaCl_2$ (3 mmol·L^{-1} Cl^-), and 3 mM KCl (3 mmol·L^{-1} Cl^-). Data are expressed as means ± SD. Different letters above bars indicate significant differences at $p < 0.05$.

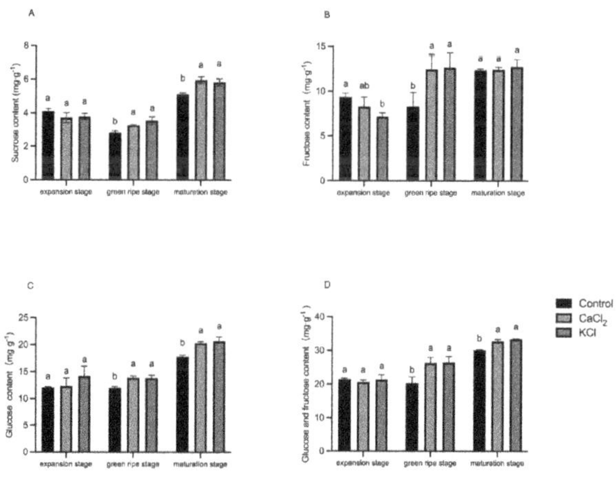

Figure 8. Effects of Cl^- on sucrose, glucose, fructose content of fruits in different growth stages. Plants were subjected to three treatments: control (Hoagland solution), 1.5 mM $CaCl_2$ (3 mmol·L^{-1} Cl^-), and 3 mM KCl (3 mmol·L^{-1} Cl^-). (**A**) Effect on sucrose content of fruits. (**B**) Effect on glucose content of fruits. (**C**) Effect on fructose content of fruits. (**D**) Effect on glucose and fructose content of fruits. Data are expressed as means ± SD. Different letters above bars indicate significant differences at $p < 0.05$.

2.6. Effect of Exogenous Chlorine on Enzyme Activity Related to Sugar Metabolism

To further investigate the mechanism by which Cl^- influences sugar accumulation, we analyzed key enzymes involved in sugar metabolism at different fruit developmental stages.

Specifically, we evaluated the activities of invertase (Ivr), including NI and AI, as well as SS-I. As shown in Figure 9, the SS-I, AI, and NI activities exhibited a continuous upward trend throughout fruit development. Although the SS-I activity did not significantly differ between treatments during the expansion stage, it increased by 24.5% and 46.4% under $CaCl_2$ and KCl treatments, respectively, during the green ripening stage. At full maturity, the SS-I activity increased by 1.5-fold and 1.4-fold, respectively, compared to the control. The AI activity did not differ significantly among treatments during fruit growth and maturation. However, during the green ripening stage, the AI activity increased by 61.6% and 62.6% under $CaCl_2$ and KCl treatments, respectively. The NI activity also increased significantly at fruit maturity, with increments of 56.2% and 32.8%. These results indicate that Cl^- enhances sucrose hydrolysis and hexose accumulation by modulating the activities of sucrose-related enzymes.

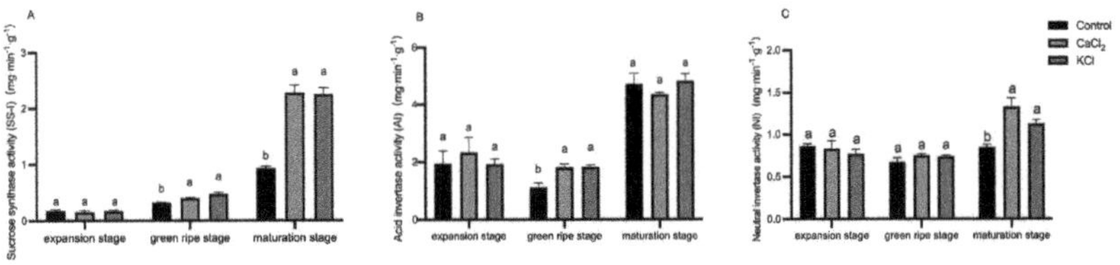

Figure 9. Effect of Cl^- on the activities of key enzymes of sucrose metabolism. Plants were subjected to three treatments: control (Hoagland solution), 1.5 mM $CaCl_2$ (3 mmol·L^{-1} Cl^-), and 3 mM KCl (3 mmol·L^{-1} Cl^-). (**A**) Effect on sucrose synthetase (SS-I) activity. (**B**) Effect on acidic invertase (AI) activity. (**C**) Effect on neutral invertase (NI) activity. Data are expressed as means ± SD. Different letters above bars indicate significant differences at $p < 0.05$.

Throughout fruit development, the SPS activity in leaves exhibited a declining trend. The SPS activity was highest during fruit expansion but gradually decreased as the fruit matured, becoming nearly absent by the ripening stage. Cl^- treatment effectively increased the SPS activity during both fruit expansion and green ripening. The SPS activity increased by 14.5% and 30.0% under $CaCl_2$ and KCl treatments, respectively, and at the green ripening stage, it was significantly higher than in the control, with increases of 76.4% and 77.8% (Figure 10).

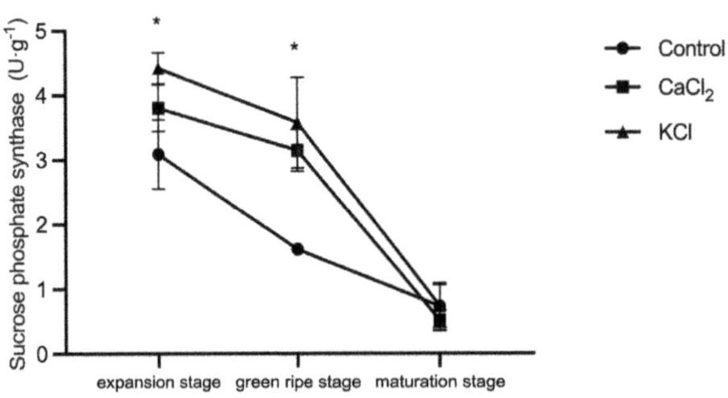

Figure 10. Effect of Cl^- on sucrose phosphate synthase activity. Plants were subjected to three treatments: control (Hoagland solution), 1.5 mM $CaCl_2$ (3 mmol·L^{-1} Cl^-), and 3 mM KCl (3 mmol·L^{-1} Cl^-). Data are expressed as means ± SD. * above the line in the line graph indicate significant differences at $p < 0.05$.

3. Discussion

3.1. Cl^- Has a Positive Effect on the Growth of Tomato Plants

Our study demonstrates that short-term exposure to low Cl^- concentrations (3 mM) significantly enhances dry matter accumulation and fruit yield in tomato plants (Supplementary Figures S1 and S2). These findings align with previous observations in *Arabidopsis* thaliana, where 5 mM NaCl increased shoot and root biomass, whereas concentrations exceeding 10 mM suppressed growth [18]. Notably, plants treated with Cl^- exhibited consistently higher biomass compared to those receiving only basic nutrient solutions or phosphate (PO_4^{3-}) and sulfate (SO_4^{2-}) treatments, even under fluctuating environmental conditions [3]. This contrast highlights the specific role of Cl^-—rather than generic anion effects or cation cofactors (e.g., K^+)—in driving growth promotion. The growth-enhancing effects of Cl^- appear conserved across species: Franco-Navarro et al. (2016, 2021) reported that 1–5 mM Cl^- improved water-use efficiency in tobacco and enhanced drought resilience in tomatoes [3,20].

The growth-enhancing effects of Cl^- likely operate through two interconnected mechanisms. One possible mechanism underlying Cl^--induced growth enhancement is its interaction with nitrate (NO_3^-). When Cl^- is present at macronutrient levels, it may regulate NO_3^- function, preventing its role as a mere osmotic or charge-balancing molecule. This regulation facilitates improved nitrogen absorption and utilization, which is critical for plant productivity [36,37]. Furthermore, Cl^- accumulation has been associated with reduced nitrate compartmentalization in vacuoles, thereby enhancing nitrogen assimilation. Studies have reported that Cl^- supplementation increases nitrogen uptake by 60–80% in tomato and tobacco plants and by up to 22% in citrus and olive crops [26]. Another potential explanation for the growth-promoting effects of Cl^- is its influence on leaf morphology. Cl^--treated plants exhibit juicier, thicker leaves with larger leaf cells and a higher density of small chloroplasts in mesophyll cells [3,38]. Leaf growth promotes an increase in biomass with the expansion of leaf cells.

Tomato biomass and yield are closely linked to photosynthetic efficiency, root nutrient uptake, and the allocation of photosynthetic products [39]. Fruit yield, in particular, correlates strongly with plant size, including stem diameter [40]. Our findings suggest that the application of 3 mM Cl^- enhances plant growth and chlorophyll content (Supplementary Figures S1 and S2 and Tables S1 and S2), although the stem diameter remains unaffected. Chlorophyll, the key pigment in photosynthesis, plays a crucial role in energy capture and conversion [41]. Previous research has also shown that Cl^- treatment increases chloroplast distribution in tobacco, primarily by increasing the number of small chloroplasts within cells [38]. This suggests that Cl^- promotes biomass accumulation in tomato plants primarily by enhancing photosynthetic efficiency [42], ultimately leading to higher fruit yields (Supplementary Figure S2).

Research has demonstrated that chloride's effects on plants follow a biphasic pattern, transitioning from beneficial to toxic at species-specific thresholds. Most plants require Cl^- at concentrations of 0.2–0.4 mg/g dry weight for normal growth [2], and moderate supplementation (1–5 mM) has been shown to enhance physiological performance. Franco-Navarro et al. (2016) reported improved water-use efficiency in tobacco at 1–5 mM Cl^- [3], and subsequent research by Franco-Navarro et al. (2021) confirmed that similar benefits extended to tomatoes at 5 mM Cl^- under drought conditions [20]. However, the optimal range for chloride nutrition is relatively narrow, as concentrations exceeding 4–7 mg/g dry weight often induce toxicity symptoms in sensitive species [1]. This toxicity threshold varies significantly among species. For instance, *Arabidopsis* exhibited positive growth responses at 5 mM NaCl but showed biomass reduction at concentrations above 10 mM [18].

In contrast, studies on persimmon (*Diospyros kaki*) identified a precise toxicity threshold at 17.6 mg/g dry weight in leaf tissue [43]. In our preliminary experiments on cherry tomatoes, we supplemented the nutrient solution with Cl^- at concentrations ranging from 0 to 3 mM to evaluate potential improvements in fruit quality. The 3 mM Cl^- treatment significantly enhanced fruit quality but concurrently reduced overall plant growth under field conditions. Subsequent trials continued to explore the 0–3 mM Cl^- concentration range to optimize the balance between growth and fruit quality.

3.2. Cl^- Promotes Photosynthesis in the Leaves of Tomato Plants and Improves the Photosynthetic Performance of the Leaves

The application of 3 $mmol \cdot L^{-1}$ Cl^- significantly increased the Pn, Gs, and Tr in tomato leaves (Table 1). These findings highlight the critical role of water and ion homeostasis in regulating plant growth and photosynthetic efficiency [24]. Cl^- influences anion transport in epidermal cells, leading to elevated anion concentrations in the apoplast, which in turn affects stomatal regulation. This mechanism enhances leaf cell expansion, optimizes water-use efficiency, and improves overall photosynthetic performance [38]. Gas exchange parameters, including the Gs, Tr, and Pn, are fundamental indicators of plant photosynthetic capacity, as they directly impact carbon assimilation [44]. The strong correlation observed between these parameters in Cl^--treated plants suggests that the increase in the Pn is primarily driven by improved stomatal conductance and transpiration. The observed rise in the Gs indicates that Cl^- may regulate ion transport in guard cells, potentially through K^+-Cl^- co-transport mechanisms, which facilitate water influx and stomatal opening. This process enhances CO_2 assimilation efficiency, thereby promoting photosynthetic carbon fixation and overall plant growth [45,46].

The observed increase in Fv/Fm (>0.8) suggests that Cl^- alleviates photoinhibition and stabilizes the PS II reaction center, particularly the D1 protein [47]. The Pm reflects its electron transfer capacity, indicating potential improvements in the efficiency of the electron transport chain (ETC). Cl^- may facilitate PS I acceptor-side reoxidation by optimizing the activity of the oxygen-evolving complex (OEC) and cytochrome b6/f complex, which are critical components regulating linear and cyclic electron flow [48]. Under optimal light conditions, PS I efficiently drives electron transport to generate NADPH and ATP, essential substrates for carbon fixation and energy metabolism [49]. During this energy conversion process, electrons traverse the thylakoid membrane through PS II and PS I before ultimately reducing $NADP^+$ to NADPH in the stroma [50]. Cl^- supplementation led to a significant increase in the Y(II), indicating enhanced photochemical conversion efficiency in PS II reaction centers (Figure 4A). This suggests that Cl^- treatment redirects more excitation energy toward photochemical reactions rather than dissipative pathways. The stabilization of the PS II OEC by Cl^- may facilitate water splitting and electron transport initiation via interactions with the Mn_4CaO_5 cluster [51]. This mechanism underscores the essential regulatory role of Cl^- in photosynthesis and provides insights for the development of artificial photosynthetic catalysts [51].

Light quanta are essential in PS II, where they convert light energy into stored chemical energy [52]. Cl^- application significantly increased the Y(NPQ) while reducing the Y(NO), both of which remained within physiologically normal ranges (Figure 4B,C). The increase in the Y(NPQ) suggests that Cl^- enhances non-photochemical quenching (NPQ), thereby improving photoprotection and mitigating excess light-induced damage [53]. Meanwhile, the Y(NO), an indicator of passive energy dissipation and photoinhibition, was markedly reduced [54]. The Y(NPQ) and Y(NO) serve as critical indicators of photoprotective capacity in plants. According to Klughammer et al. (2008), the Y(NPQ) represents the capacity to dissipate excess light energy through regulated mechanisms, such as heat dissipation, while the Y(NO) quantifies uncontrolled energy losses that can lead to photodamage [55]. An increase in the Y(NPQ) coupled with a decrease in the Y(NO) suggests that Cl^- enhances regulated dissipation pathways, thereby improving photoprotection. This effect may be attributed to Cl^--induced conformational changes in the light-harvesting complex II (LHCII), leading to increased aggregation states that strengthen NPQ machinery [56]. Cl^- supplementation significantly increased the Y(II) and ETR(II) (Figures 4A and 6A) while reducing the Y(NO) (Figure 4B). These findings indicate that Cl^- enhances photosynthetic electron transfer and stabilizes the PS II complex, ultimately improving tomato plant photosynthetic metabolism [57].

Damage to PS I has severe consequences for plant photosynthetic performance, as PS I recovery is slow and its impairment can disrupt the entire light-harvesting system [58]. Our results demonstrated that Cl^--treated plants exhibited higher Y(I) levels and reduced Y(ND) (Figure 5A,B), indicating that Cl^- exerts a protective effect on PS I, potentially shielding it from photoinhibition stress. Cl^- may facilitate electron transport chain function by enhancing NADPH generation and accelerating Fd reoxidation [59,60]. Furthermore, elevated Y(I) and Pm values suggest an increase in cytochrome b6/f complex-mediated electron transport efficiency. One proposed mechanism involves Cl^- stabilizing PQH_2 binding to the cytochrome b6/f complex, thereby optimizing the Q-cycle and maintaining the transmembrane proton gradient (ΔpH) necessary for ATP synthesis [61]. An alternative hypothesis suggests that Cl^- regulates the balance between cyclic electron transport (CET) and linear electron transport (LET), preventing PS I over-reduction (e.g., excessive Y(NA)) and ensuring a balanced ATP/NADPH supply [62].

Our findings indicate that Cl^- enhances PS II and PS I activity, improving photoprotection and electron transport efficiency in tomato plants. This enhancement may be associated with specific Cl^- transmembrane transport proteins. For instance, the carotenoid protein VCCN1 in *Arabidopsis* thaliana functions as a voltage-dependent Cl^- channel, facilitating Cl^- transport into the thylakoid lumen and regulating photoprotective and electron transport processes [63]. Additionally, the Cl^- transporter CLCe may help maintain Cl^- homeostasis, accelerating NPQ activation and optimizing electron transport [63]. During transitions from darkness to low light, VCCN1 enhances NPQ activation, which is further stimulated by CLCe, thereby promoting electron transport once NPQ reaches homeostasis. Conversely, under high-light conditions, VCCN1 predominantly accelerates NPQ activation with minimal impact on electron transport. These findings suggest that VCCN1 and CLCe regulate NPQ and electron transport in response to fluctuating light conditions [64]. Recent studies on Cl^- channels suggest that Cl^- fluxes are influenced by light and correlate with the partitioning of the electrical potential (Δψ) and chemical (ΔpH) components of the proton motive force (PMF). These fluxes contribute to the regulation of photosynthetic electron transport and photoprotection [65,66]. Understanding the role of Cl^- transport proteins in these processes represents a key direction for future research.

3.3. Cl^- Promotes Fruit Sugar Metabolism

The ripening of tomato fruits is closely associated with sugar accumulation, which plays a crucial role in determining fruit quality [67]. Sugar content is a key parameter influencing fruit sweetness, an important trait in commercial markets. In this study, we utilized a PAL-1 digital refractometer to measure the soluble solids content of tomato fruits at various developmental stages. Our results indicated a steady increase in soluble sugar content as the fruits matured, with a significant elevation observed in fruits subjected to Cl^- treatment (Figure 7). These findings suggest that the external application of Cl^-

effectively enhances sugar accumulation, thereby improving fruit quality. Previous studies have demonstrated a close genetic linkage between loci regulating soluble solids content and fruit weight in tomatoes, indicating that soluble solids content is generally inversely correlated with fruit weight. In most cases, an increase in fruit yield is associated with a reduction in soluble solids content [28,34]. However, our experimental results revealed that Cl^- application not only increased soluble sugar content but also enhanced overall fruit yield, suggesting that Cl^- may mitigate the typical trade-off between these two traits.

Mature tomato fruits consist of approximately 90–95% water, with the remaining 5–10% primarily composed of carbohydrates [68]. The accumulation of sugars, particularly glucose, fructose, and sucrose, is essential for fruit quality development [69]. Sucrose serves as the primary assimilate transported from source organs and plays a central role in carbohydrate translocation in tomato plants [34,70]. However, ripe tomato fruits generally contain relatively low sucrose levels, with higher proportions of glucose and fructose (hexose sugars), which influence fruit structure, flavor synthesis, ripening, and senescence. Our results showed that Cl^- treatment significantly increased glucose and fructose content during fruit ripening (Figure 8), further supporting the role of Cl^- in enhancing fruit sugar metabolism.

Numerous studies have indicated that the majority of photosynthetic carbon assimilates in tomato fruits originate from source leaves rather than being synthesized de novo within the fruit tissues [71]. Photosynthesis in source organs, such as leaves, contributes to nearly 80% of the total carbon content in fruits [30,72]. The metabolism of sucrose plays a pivotal role in sugar accumulation, as its hydrolysis into hexoses is a key step in fruit development. This process is mediated by various enzymes and regulatory genes [73]. Additionally, sucrase activity is considered an indicator of sink strength [74]. Sucrose-cleaving enzymes, particularly Ivr and Sus, are critical determinants of sugar composition in tomato fruits. Invertase catalyzes the irreversible hydrolysis of sucrose into glucose and fructose, while sucrose synthase facilitates the conversion of sucrose into fructose and UDP-glucose [75,76]. Our study revealed that sucrose synthase activity in fruits during the green and mature ripening stages was significantly enhanced under Cl^- treatment (Figure 9), suggesting that Cl^- effectively alters sugar metabolism by modulating sucrase activity. These enzymes play crucial roles in fructose metabolism, sugar regulation, and overall fruit growth and development. Previous studies have shown that an increase in acid invertase activity during fruit ripening leads to enhanced hexose accumulation in cultivated tomatoes [77]. Wang et al. [78] demonstrated that knocking down INVINH1, an inhibitor of the acidic invertase gene Lin5, resulted in a significant increase in hexose content, further corroborating the role of invertase in sugar accumulation. Invertase activity is a key determinant of sucrose metabolism and sugar partitioning in developing fruits. The increased AI activity during fruit growth promotes sucrose conversion into hexoses, which are essential for maintaining the sink strength of the fruit [76]. Our experiments demonstrated that Cl^- supplementation significantly enhanced AI activity during the green ripening stage, thereby facilitating the accumulation of hexose sugars in tomato fruits. Interestingly, Cl^- also increased NI activity, particularly during the early and late stages of fruit development; however, the overall activity of NI remained lower than that of AI.

The flux of NI is primarily regulated by its intrinsic enzymatic activity, particularly during the initial and final phases of fruit development. In contrast, AI activity during the fruit expansion phase is predominantly influenced by substrate availability and the activity of plasma membrane sucrose transporters [79]. These findings indicate that Cl^- enhances sugar metabolism in tomato fruits by modulating sucrose cleavage enzyme activity, thereby promoting the conversion of sucrose into glucose and fructose, which are the primary determinants of fruit sweetness and quality.

The net synthesis of sucrose is primarily mediated by SPS, which serves as a key regulatory enzyme in both sucrose biosynthesis and photosynthesis. SPS activity is typically elevated in source tissues while being lower in sink organs [80,81]. Exogenous Cl^- application may enhance sucrose accumulation in leaves by modulating SPS activity, thereby facilitating sucrose transport from leaves to fruits and ultimately increasing fruit sugar content (Figure 8). These findings are consistent with previous studies [76,82]. Worrell et al. (1991) genetically modified tomato plants to overexpress maize SPS cDNA, driven by the tobacco RuBisCO small subunit promoter (rbcs) [83]. This genetic modification resulted in increased SPS activity in leaves compared to control plants. Furthermore, the photosynthetic efficiency of the transgenic plants was enhanced by approximately 20% under both light-saturated and CO_2-saturated conditions. The overexpression of SPS also led to a higher sucrose-to-starch ratio in leaves, with transformed plants producing 1.5-fold more fruit than the control group [84–87]. These findings provide a promising direction for future research on the molecular regulatory mechanisms by which Cl^- influences sugar accumulation, transformation, and transport.

The activity of certain enzymes (e.g., α-amylase) is Cl^--dependent [88]. Previous research has primarily focused on the indirect effects of Cl^- on plant enzymes, particularly through mechanisms such as osmotic regulation, ion homeostasis, and signal transduction, rather than direct enzymatic interactions. Elevated Cl^- concentrations promote cell elongation by maintaining vacuolar osmotic pressure and may alter the cytoplasmic microenvironment (e.g., hydration status), thereby influencing enzymatic reaction rates [14]. Under drought stress, plants accumulate soluble sugars such as sucrose and fructose to maintain osmotic balance, while Cl^- indirectly regulates SPS and SS activity by modulating stomatal aperture and water status [76,89]. Additionally, Cl^- participates in vacuolar membrane ion transport through Cl^-/H^+ exchange mechanisms, potentially involving CLC antiporters, thereby regulating vacuolar pH [90,91]. Alterations in local pH homeostasis may indirectly affect enzyme activity. Furthermore, Cl^- modulates Ca^{2+} channel activity by influencing the membrane potential, which in turn activates calcium-dependent protein kinases (CDPKs), regulating the gene expression of SS or Ivr [92]. Experimental studies demonstrate that Cl^- enhances sucrose synthesis by inhibiting NO_3^- uptake [93], thereby alleviating the competition between nitrogen assimilation and carbon metabolism. Given that SS is tightly regulated by the carbon–nitrogen balance, Cl^- may indirectly influence sucrose accumulation by modulating the activity of nitrogen metabolism-associated enzymes [94,95]. Furthermore, Cl^- regulates the activity of nitrogen metabolism enzymes (e.g., glutamine synthetase) by modulating the uptake equilibrium of intracellular cations (e.g., K^+, NH_4^+). This mechanism impacts nitrogen metabolism while also indirectly influencing sucrose synthesis through carbon–nitrogen balance adjustments. These findings reveal that Cl^- acts as a pivotal bridge in plant carbon–nitrogen metabolism, engaging multiple metabolic pathways and regulating key enzymatic activities [96,97].

4. Materials and Methods

4.1. Growth Conditions and Experimental Materials

This experiment was carried out in a glass greenhouse at the Chinese Academy of Agricultural Sciences (Haidian 39.97° N, 116.33° E, Beijing). Two varieties of tomato plants, 'Yuanwei No. 1' and 'Money Maker', were used in this study. The plants were cultivated under natural light and photoperiod conditions. The nutrient solution formula of the substrate was modified Hoagland nutrient solution, as follows: 614 mg/L $Ca(NO_3)_2{\cdot}4H_2O$,

430 mg/L KNO_3, 267 mg/L KH_2PO_4, 33 mg/L$(NH_4)_2SO_4$, 430 mg/L $Mg(SO_4)_2$ $7H_2O$, 379 mg/L K_2SO_4, 6.3 mg/L FeNa-ethylenediaminetetraacetic acid (EDTA), 1.7 mg/L $MnSO_4{\cdot}H_2O$, 1.4 mg/L $ZnSO_4{\cdot}7H_2O$, 2.3 mg/L $Na_2B_4O_7{\cdot}10H_2O$, 0.2 mg/L $CuSO_4{\cdot}5H_2O$, and 0.2 mg/L $Na_2MoO_4{\cdot}2H_2O$.

The tomato variety 'Yuanwei No. 1' was selected for this study, and various levels of Cl^- were incorporated into the basal nutrient solution. The experiment included four levels of Cl^-: CK (0 $mmol{\cdot}L^{-1}$ Cl^-), T1 (1 $mmol{\cdot}L^{-1}$ Cl^-), T2 (2 $mmol{\cdot}L^{-1}$ Cl^-), and T3 (3 $mmol{\cdot}L^{-1}$ Cl^-). Building on the initial observed effects on tomato plant growth and fruit yield (Supplementary Figures S1 and S2), follow-up experiments were conducted to investigate the impact of 3 mM Cl^- on photosynthesis, fruit quality, and metabolism. Three salt solutions that preserved the same cationic balance were added to a basic nutrient solution used to water the plants: 0 mM Cl^- (control), 3 mM Cl^- with 1.5 mM $CaCl_2$, and 3 mM KCl. In a complete block design, the treatments were dispersed at random and duplicated three times, with five plants in each replicate.

Tomato seeds were sown in trays filled with peat and allowed to germinate until they developed five true leaves. Three days after flowering, the Cl^- treatment commenced, and the seedlings were subsequently transferred to plastic pots containing seven to eight liters of substrate.

4.2. Plant Growth Indicators

A ruler and an electronic caliper were employed to measure plant height and stem diameter, respectively, to evaluate growth indicators. The height and biomass were measured by a tape measure and scales. The soluble solids content of the fruit was determined using a PAL-1 digital refractometer (ATAGO Co., Ltd., Tokyo, Japan).

4.3. Gas Exchange and Chlorophyll Fluorescence Parameters

Following the methodology of Lu et al. [53], gas exchange parameters were measured using a Li-6400XT instrument (Li-Cor Inc., St. Lincoln, NE, USA) at a light intensity of 800 $\mu mol{\cdot}m^{-2}{\cdot}s^{-1}$. The Dual PAM-100F (Heinz Walz, Efffeltrich, Germany) was utilized to assess the P700 redox status and chlorophyll fluorescence in vivo, following the procedures outlined in previous studies [98]. After exposure to various photosynthetic photon flux densities (PPFDs) for two minutes, light-adapted curves were recorded [53]. The plants were dark-adapted for 20 min prior to measurement. To determine the minimal fluorescence (Fo) in the dark-adapted state, a detecting light with a PPFD of less than 0.1 $\mu mol{\cdot}m^{-2}{\cdot}s^{-1}$ and a frequency of 0.6 kHz was activated. To obtain the maximum fluorescence (Fm) in the dark-adapted condition, a saturated pulse of light (PPFD = 600 $\mu mol{\cdot}m^{-2}{\cdot}s^{-1}$, white light, 20 kHz, 1 pulse) was then activated. Following this, the actinic light (PPFD = 535 $\mu mol{\cdot}m^{-2}{\cdot}s^{-1}$) was illuminated, and after the light-stable fluorescence (Fs) stabilized, the saturated pulsed light (PPFD = 10,000 $\mu mol{\cdot}m^{-2}{\cdot}s^{-1}$, 300 ms) was activated to determine the maximum fluorescence (Fm′) in the light. Finally, the actinic light was turned off, and far-red light (PPFD = 5 $\mu mol{\cdot}m^{-2}{\cdot}s^{-1}$) was applied to measure the initial fluorescence (Fo′) under light after a duration of 3 s [34,70]. The formula Fv/Fm = (Fm − Fo)/Fm represents the maximum photochemical efficiency, indicating that the PS II reaction center is fully open after sufficient dark adaptation. The P700 signal was assessed using the light absorption technique at wavelengths between 830 and 875 nm. After pre-irradiation with far-red light, the Pm represents the maximum oxidation value of P700 under saturated pulsed light. This value indicates the quantity of effective PS I complexes and the extent of total oxidation of P700.

Chlorophyll fluorescence parameters and the P700 redox state were measured using a Dual PAM-100F fluorometer (Heinz Walz, Effenberg, Germany) following established protocols [98]. Plants were dark-adapted for 30 min to ensure the complete recovery of PS II from light stress. Afterward, leaves were exposed to a series of light treatments (0–1200 $\mu mol \cdot m^{-2} \cdot s^{-1}$ photosynthetically active radiation) for 2 min to record light-adapted fluorescence curves [53].

Calculation of fluorescence parameters:

Maximum photochemical efficiency of PS II (Y(II)):

$$Y(II) = (Fm' - Fs)/Fm'$$

where Fm′ is the maximum fluorescence in the light-adapted state, and Fs is the light-stable fluorescence.

Non-photochemical quenching (NPQ):

$$NPQ = (Fm - Fm')/Fm'$$

This parameter quantifies the energy dissipated as heat via regulated (NPQ) and non-regulated (Y(NO)) pathways.

Quantum yield of non-regulated energy dissipation in PS II (Y(NO)):

$$Y(NO) = Fs/Fm$$

PS I energy partitioning parameters: Acceptor-side non-photochemical energy dissipation (Y(NA)):

$$Y(NA) = (Pm - Pm')/Pm$$

where Pm and Pm′ represent the maximum and minimum P700 oxidation states, respectively.

Donor-side confinement energy dissipation (Y(ND)):

$$Y(ND) = 1 - P700red$$

P700red is the P700 redox ratio under far-red light illumination.

Photochemical quantum yield of PS I (Y(I)):

$$Y(I) = 1 - Y(ND) - Y(NA)$$

Technical notes: chlorophyll fluorescence signals were collected from the upper leaf surface, while P700 measurements reflect the entire leaf tissue due to the deeper tissue penetration of near-infrared light [53].

After a two-minute exposure to various photosynthetic photon flux densities (PPFDs), the light-adapted curves were recorded [34]. The light intensity gradient in this experiment was set to 2000, 1500, 1200, 1000, 800, 600, 400, 200, 150, 100, 50, 25, and 0 $\mu mol \cdot m^{-2} \cdot s^{-1}$, with each light intensity lasting 20 s. The saturated pulsed light was set at 10,000 $\mu mol \cdot m^{-2} \cdot s^{-1}$ and was applied for 300 milliseconds. The software's built-in model was employed to fit the ETR(II) and ETR(I) curves. The results yielded the maximum electron transfer rate ETR_{max}, the half-saturated light intensity Ik, and the initial slope α of the rapid light response curve for the electron transfer rate.

Samples with uniform leaf age and growth conditions were selected, and chlorophyll content (measured as SPAD) was assessed between 9:00 and 11:30 a.m. on sunny days. The relative chlorophyll content was assessed using a Soil Plant Analysis Development (SPAD) meter (502 Plus Chlorophyll Meter, Spectrum Technologies, Bridgend, UK). This SPAD analyzer employs two wavelengths—650 nm and 940 nm—to evaluate leaf transmission [99].

4.4. Sucrose Metabolism Enzyme Activity

Soluble sugar content was assessed using the anthrone colorimetric method, with the percentage of soluble sugars calculated according to the formula (C × V × N/W × V1 × 106) × 100. In this equation, C denotes the sugar concentration (μg) derived from the standard curve, V represents the total volume of the extract (mL), V1 indicates the volume of liquid absorbed during the determination (ml), N signifies the dilution ratio, and W refers to the weight of the sample (g). Using a spectrophotometer in conjunction with a specialized detection kit from Beijing Solaibao Technology Co., Ltd., Beijing, China, we assessed the contents of sucrose, fructose, and glucose in the fruit at various developmental stages: expanding, green ripening, and full ripening. The second ear fruit of each plant was selected for analysis. A comprehensive overview of the procedure is outlined in the Solaibao test kit manual. During this period, the activities of SS-I, AI, NI, and SPS were measured using the Solaibao kit (Beijing Solaibao Technology Co., Ltd., Beijng, China).

SS-I activity was quantified using a Solarbio kit (BC4315). Fresh tissue (0.1 g) was homogenized in ice-cold extraction buffer (1:5–10 *w*/*v*; 50 mM Tris-HCl, pH 7.5, 5 mM $MgCl_2$, 0.1% Triton X-100) and centrifuged (8000× *g*, 10 min, 4 °C), and the supernatant was used as the enzyme extract. Reactions (90 μL total) contained 10 μL extract, 40 μL reaction buffer, and 40 μL UDP substrate solution. Controls substituted the substrate with distilled water. After 30 min at 30 °C, reactions were terminated at 95 °C (10 min), mixed with 3,5-dinitrosalicylic acid (DNSA) chromogenic agent, and incubated at 95 °C for 5 min. Absorbance at 540 nm was measured after cooling and dilution. Activity was calculated as 33.33 × (ΔA540/tissue weight), where one unit (U) equals 1 μg fructose produced per gram of tissue per minute. AI and NI activities were determined using kits BC0560 and BC0575 (Solarbio), respectively. Tissues were homogenized similarly but centrifuged at 12,000× *g* to isolate subcellular fractions. AI reactions (1.8 mL) used pH 4.8 buffer with sucrose substrate, while NI used pH 7.0 buffer. Both assays involved 30 min incubation at 37 °C, followed by reaction termination (95 °C, 10 min), DNSA color development (10 min at 95 °C), and 540 nm detection. Activities were expressed as 33.3 × (ΔA540/tissue weight), with one U defined as 1 μg reducing sugar generated per gram of tissue per minute. Quality controls included triplicate measurements, heat-inactivated enzyme blanks, and strict standardization of heating/cooling times. All boiling steps used sealed tubes to prevent evaporation, with <5% coefficient of variation across replicates.

The activity of SPS was determined using a commercial kit (Cat. No. BC0600, Beijing Solarbio Science & Technology Co., Ltd., Beijing, China). Briefly, 0.1 g of plant tissue was homogenized in pre-cooled extraction buffer at a ratio of 1:5–10 (*w*/*v*) and centrifuged at 8000× *g* for 10 min at 4 °C. The supernatant was collected as the enzyme extract. The reaction mixture contained 30 μL enzyme extract, 150 μL Reagent I (buffer), and 30 μL Reagent II (500 μg/mL sucrose solution, freshly prepared). For the control group, 150 μL distilled water replaced Reagent I. After incubation at 25 °C for 10 min, the reaction was terminated by adding 50 μL Reagent III, followed by boiling for 10 min. Subsequently, 700 μL Reagent IV and 200 μL Reagent V were added, and the mixture was incubated at 80 °C for 20 min for color development. After centrifugation at 12,000× *g* for 10 min, the supernatant absorbance was measured at 480 nm. A standard curve was established using 500 μg/mL sucrose solution (ΔA = Astandard − Ablank). SPS activity was calculated as 50 × (ΔAsample/ΔAstandard) ÷ protein concentration (mg/mL) or

50 × (ΔAsample/ΔAstandard) ÷ sample weight (g), defining one unit as the amount of enzyme producing 1 μg sucrose per mg of protein or per gram of tissue per minute. All tubes were tightly sealed to prevent cap rupture during boiling, and temperature was strictly controlled during color development.

4.5. Statistical Analysis

Data are presented as the mean ± standard deviation (SD) of three biological replicates. Prior to analysis, datasets underwent quality control for missing values and outliers. Missing data were excluded or interpolated for time-series continuity. Outliers were identified via statistical thresholds (IQR or Grubbs' test). Statistical analyses were conducted using SPSS version 27.0 for normality (Shapiro–Wilk test for small samples or D'Agostino–Pearson omnibus test for larger datasets) and variance homogeneity (Levene's test) assessments and GraphPad Prism version 6.01 for one-way ANOVA (if data met normality and homogeneity assumptions) or Welch's ANOVA (for heterogeneous variances), followed by Duncan's multiple range test (α = 0.05) for identifying significant differences among treatments.

5. Conclusions

This study demonstrates that exogenous Cl^- supplementation is beneficial for crop growth. Among the tested concentrations, 3 mM Cl^- proved to be the most effective. The results indicate that exogenous Cl^- enhances leaf photochemical efficiency and photosystem activity, thereby promoting photosynthetic product synthesis and ultimately increasing biomass accumulation. Furthermore, Cl^- primarily facilitates sucrose accumulation in developing fruits and its subsequent conversion into glucose and fructose, leading to an increase in total sugar content and improved fruit sweetness. While this study provides preliminary insights into the effects of Cl^- on photosynthesis and sucrose metabolism in tomato plants, further research is required to elucidate the underlying molecular mechanisms.

Supplementary Materials: The following supporting information can be downloaded at https://www.mdpi.com/article/10.3390/ijms26072922/s1.

Author Contributions: Conceptualization, L.S. and T.L.; methodology, L.S., T.L., X.W., Q.L. and Y.L.; validation, H.Y. and W.J.; formal analysis, L.S. and T.L.; investigation, L.S.; resources, H.Y. and W.J.; data curation, L.S.; writing—original draft preparation, L.S.; writing—review and editing, T.L.; visualization, H.Y. and W.J.; project administration, Q.L. and Y.L. All authors have read and agreed to the published version of the manuscript.

Funding: This research was funded by the National Key Research and Development Program of China, grant numbers 2024YFD2300702, 2023YFD2300700, 2021YFD1600303 and 2022YFD1602400. This research was funded by the National Natural Science Foundation of China, grant number 32002115.

Institutional Review Board Statement: Not applicable.

Informed Consent Statement: Not applicable.

Data Availability Statement: Data is contained within the article and Supplementary Materials.

Conflicts of Interest: The authors declare no conflicts of interest.

Abbreviations

The following abbreviations are used in this manuscript:

AI	Acid invertase
AQE	Apparent quantum efficiency
Ci	Intercellular CO_2 concentration
Cl	Chlorine
ETR(I)	PS I rate of electronic transmission
ETR(II)	PS II rate of electronic transmission
Fv/Fm	Maximal photochemical efficiency of PS II
Gs	Stomatal conductance
Ivr	Invertase
NI	Neutral invertase
Pm	PS I maximum oxidation state
Pn	Net photosynthesis rate
PS I	Photosystem I
PS II	Photosystem II
SPS	Sucrose phosphate synthase
SS-I	Sucrose synthase
Tr	Transpiration rate
Y(I)	Quantum efficiency of PS I photochemistry
Y(II)	Quantum efficiency of PS II photochemistry
Y(NA)	Acceptor-side limitation of PS I
Y(ND)	Donor-side limitation of PS I
Y(NO)	Yield of non-regulated energy dissipation
Y(NPQ)	Yield of regulated energy dissipation

References

1. Xu, G.; Magen, H.; Tarhitzkya, J. Advances in chloride nutrition of plants. *Adv. Agron.* **2000**, *68*, 97–110. [CrossRef]
2. Broadley, M.; Brown, P.; Cakmak, I.; Rengel, Z.; Zhao, F. Function of Nutrients. In *Marschner's Mineral Nutrition of Higher Plants*; Elsevier: Amsterdam, The Netherlands, 2012; pp. 191–248, ISBN 978-0-12-384905-2.
3. Franco-Navarro, J.D.; Brumós, J.; Rosales, M.A.; Cubero-Font, P.; Talón, M.; Colmenero-Flores, J.M. Chloride Regulates Leaf Cell Size and Water Relations in Tobacco Plants. *J. Exp. Bot.* **2016**, *67*, 873–891. [CrossRef]
4. Raven, J.A. Chloride: Essential Micronutrient and Multifunctional Beneficial Ion. *J. Exp. Bot.* **2016**, *68*, 359–367. [CrossRef] [PubMed]
5. Geilfus, C.-M. Review on the Significance of Chlorine for Crop Yield and Quality. *Plant Sci.* **2018**, *270*, 114–122. [CrossRef]
6. White, P. Chloride in Soils and Its Uptake and Movement Within the Plant: A Review. *Ann. Bot.* **2001**, *88*, 967–988. [CrossRef]
7. Broyer, T.C.; Carlton, A.B.; Johnson, C.M.; Stout, P.R. Chlorine—A Micronutrient Element for Higher Plants. *Plant Physiol.* **1954**, *29*, 526–532. [CrossRef] [PubMed]
8. Teakle, N.L.; Tyerman, S.D. Mechanisms of Cl^- Transport Contributing to Salt Tolerance. *Plant Cell Environ.* **2010**, *33*, 566–589.
9. Colmenero-Flores, J.M.; Franco-Navarro, J.D.; Cubero-Font, P.; Peinado-Torrubia, P.; Rosales, M.A. Chloride as a Beneficial Macronutrient in Higher Plants: New Roles and Regulation. *Int. J. Mol. Sci.* **2019**, *20*, 4686. [CrossRef]
10. Terry, N. Photosynthesis, Growth, and the Role of Chloride 12. *Plant Physiol.* **1977**, *60*, 69–75.
11. Kawakami, K.; Umena, Y.; Kamiya, N.; Shen, J.-R. Location of Chloride and Its Possible Functions in Oxygen-Evolving Photosystem II Revealed by X-Ray Crystallography. *Proc. Natl. Acad. Sci. USA* **2009**, *106*, 8567–8572. [CrossRef]
12. Pokhrel, R.; McConnell, I.L.; Brudvig, G.W. Chloride Regulation of Enzyme Turnover: Application to the Role of Chloride in Photosystem II. *Biochemistry* **2011**, *50*, 2725–2734. [CrossRef] [PubMed]
13. Hänsch, R.; Mendel, R.R. Physiological Functions of Mineral Micronutrients (Cu, Zn, Mn, Fe, Ni, Mo, B, Cl). *Curr. Opin. Plant Biol.* **2009**, *12*, 259–266. [CrossRef] [PubMed]
14. Wege, S.; Gilliham, M.; Henderson, S.W. Chloride: Not Simply a 'Cheap Osmoticum', but a Beneficial Plant Macronutrient. *J. Exp. Bot.* **2017**, *68*, 3057–3069. [CrossRef]
15. Chen, Z.-H.; Hills, A.; Bätz, U.; Amtmann, A.; Lew, V.L.; Blatt, M.R. Systems Dynamic Modeling of the Stomatal Guard Cell Predicts Emergent Behaviors in Transport, Signaling, and Volume Control. *Plant Physiol.* **2012**, *159*, 1235–1251. [CrossRef]
16. Schwenke, G.D.; Simpfendorfer, S.R.; Collard, B.C.Y. Confirmation of Chloride Deficiency as the Cause of Leaf Spotting in Durum Wheat Grown in the Australian Northern Grains Region. *Crop Pasture Sci.* **2015**, *66*, 122. [CrossRef]

17. Saito, S.; Uozumi, N. Guard Cell Membrane Anion Transport Systems and Their Regulatory Components: An Elaborate Mechanism Controlling Stress-Induced Stomatal Closure. *Plants* **2019**, *8*, 9. [CrossRef]
18. Hongqiao, L.; Suyama, A.; Mitani-Ueno, N.; Hell, R.; Maruyama-Nakashita, A. A Low Level of NaCl Stimulates Plant Growth by Improving Carbon and Sulfur Assimilation in *Arabidopsis thaliana*. *Plants* **2021**, *10*, 2138. [CrossRef] [PubMed]
19. Wang, C.; Liu, X.; Li, J.; Yue, L.; Yang, H.; Zou, H.; Wang, Z.; Xing, B. Copper Nanoclusters Promote Tomato (*Solanum lycopersicum* L.) Yield and Quality Through Improving Photosynthesis and Roots Growth. *Environ. Pollut.* **2021**, *289*, 117912. [CrossRef]
20. Franco-Navarro, J.D.; Díaz-Rueda, P.; Rivero-Núñez, C.M.; Brumós, J.; Rubio-Casal, A.E.; De Cires, A.; Colmenero-Flores, J.M.; Rosales, M.A. Chloride Nutrition Improves Drought Resistance by Enhancing Water Deficit Avoidance and Tolerance Mechanisms. *J. Exp. Bot.* **2021**, *72*, 5246–5261. [CrossRef]
21. Robinson, S.P.; Downton, W.J.S.; Millhouse, J.A. Photosynthesis and Ion Content of Leaves and Isolated Chloroplasts of Salt-Stressed Spinach. *Plant Physiol.* **1983**, *73*, 238–242. [CrossRef]
22. Winter, K. Sodium, Potassium, Chloride and Proline Concentrations of Chloroplasts Isolated from a Halophyte, *Mesembryanthemum crystmlinum* L. *Planta* **1986**, *168*, 421–426.
23. Petridis, A.; Van Der Kaay, J.; Chrysanthou, E.; McCallum, S.; Graham, J.; Hancock, R.D. Photosynthetic Limitation as a Factor Influencing Yield in Highbush Blueberries (*Vaccinium corymbosum*) Grown in a Northern European Environment. *J. Exp. Bot.* **2018**, *69*, 3069–3080. [CrossRef]
24. Wang, M.; Dong, C.; Gao, W. Evaluation of the Growth, Photosynthetic Characteristics, Antioxidant Capacity, Biomass Yield and Quality of Tomato Using Aeroponics, Hydroponics and Porous Tube-Vermiculite Systems in Bio-Regenerative Life Support Systems. *Life Sci. Space Res.* **2019**, *22*, 68–75. [CrossRef]
25. Klee, H. Genetics and Control of Tomato Fruit Ripening and Quality Attributes. *Annu. Rev. Genet.* **2011**, *45*, 41–59. [CrossRef]
26. Rosales, M.A.; Franco-Navarro, J.D.; Peinado-Torrubia, P.; Díaz-Rueda, P.; Álvarez, R.; Colmenero-Flores, J.M. Chloride Improves Nitrate Utilization and NUE in Plants. *Front. Plant Sci.* **2020**, *11*, 442. [CrossRef]
27. Yuan, G.-F.; Jia, C.-G.; Li, Z.; Sun, B.; Zhang, L.-P.; Liu, N.; Wang, Q.-M. Effect of Brassinosteroids on Drought Resistance and Abscisic Acid Concentration in Tomato under Water Stress. *Sci. Hortic.* **2010**, *126*, 103–108. [CrossRef]
28. Tanksley, S.D. The Genetic, Developmental, and Molecular Bases of Fruit Size and Shape Variation in Tomato. *Plant Cell Online* **2004**, *16*, S181–S189. [CrossRef]
29. Bertin, N.; Génard, M. Tomato Quality as Influenced by Preharvest Factors. *Sci. Hortic.* **2018**, *233*, 264–276. [CrossRef]
30. Yuan, Y.; Mei, L.; Wu, M.; Wei, W.; Shan, W.; Gong, Z.; Zhang, Q.; Yang, F.; Yan, F.; Zhang, Q.; et al. SlARF10, an Auxin Response Factor, Is Involved in Chlorophyll and Sugar Accumulation during Tomato Fruit Development. *J. Exp. Bot.* **2018**, *69*, 5507–5518. [CrossRef]
31. Kimura, S.; Sinha, N. Tomato (*Solanum lycopersicum*): A Model Fruit-Bearing Crop. *Cold Spring Harb. Protoc.* **2008**, *2008*, pdb-emo105.
32. Causse, M.M.; Friguet, C.; Coiret, C.; Lépicier, M.; Navez, B.; Lee, M.; Holthuysen, N.; Sinesio, F.; Moneta, E.; Grandillo, S. Consumer Preferences for Fresh Tomato at the European Scale: A Common Segmentation on Taste and Firmness. *J. FoodSci.* **2010**, *75*, S531–S541.
33. Tieman, D.; Zhu, G.; Resende, M.F.R.; Lin, T.; Nguyen, C.; Bies, D.; Rambla, J.L.; Beltran, K.S.O.; Taylor, M.; Zhang, B.; et al. A Chemical Genetic Roadmap to Improved Tomato Flavor. *Science* **2017**, *355*, 391–394. [CrossRef]
34. Lu, T.; Yu, H.; Li, Q.; Chai, L.; Jiang, W. Improving Plant Growth and Alleviating Photosynthetic Inhibition and Oxidative Stress From Low-Light Stress With Exogenous GR24 in Tomato (*Solanum lycopersicum* L.) Seedlings. *Front. Plant Sci.* **2019**, *10*, 490. [CrossRef]
35. Matsukura, C. Sugar Accumulation in Tomato Fruit and Its Modification Using Molecular Breeding Techniques. In *Functional Genomics and Biotechnology in Solanaceae and Cucurbitaceae Crops*; Ezura, H., Ariizumi, T., Garcia-Mas, J., Rose, J., Eds.; Biotechnology in Agriculture and Forestry; Springer: Berlin/Heidelberg, Germany, 2016; Volume 70, pp. 141–154, ISBN 978-3-662-48533-0.
36. Krapp, A. Plant Nitrogen Assimilation and Its Regulation: A Complex Puzzle with Missing Pieces. *Curr. Opin. Plant Biol.* **2015**, *25*, 115–122. [CrossRef]
37. Guan, F.; Xia, M.; Tang, X.; Fan, S. Spatial Variability of Soil Nitrogen, Phosphorus and Potassium Contents in Moso Bamboo Forests in Yong'an City, China. *Catena* **2017**, *150*, 161–172. [CrossRef]
38. Franco-Navarro, J.D.; Rosales, M.A.; Cubero-Font, P.; Calvo, P.; Álvarez, R.; Diaz-Espejo, A.; Colmenero-Flores, J.M. Chloride as a Macronutrient Increases Water-Use Efficiency by Anatomically Driven Reduced Stomatal Conductance and Increased Mesophyll Diffusion to CO_2. *Plant J.* **2019**, *99*, 815–831. [CrossRef]
39. Parvizi, H.; Sepaskhah, A.R. Effect of Drip Irrigation and Fertilizer Regimes on Fruit Quality of a Pomegranate (*Punica granatum* (L.) Cv. Rabab) Orchard. *Agric. Water Manag.* **2015**, *156*, 70–78. [CrossRef]

40. Koyama, K.; Ikeda, H.; Poudel, P.R.; Goto-Yamamoto, N. Light Quality Affects Flavonoid Biosynthesis in Young Berries of Cabernet Sauvignon Grape. *Phytochemistry* **2012**, *78*, 54–64. [CrossRef] [PubMed]
41. Chen, B.; Huang, G.; Lu, X.; Gu, S.; Wen, W.; Wang, G.; Chang, W.; Guo, X.; Zhao, C. Prediction of Vertical Distribution of SPAD Values Within Maize Canopy Based on Unmanned Aerial Vehicles Multispectral Imagery. *Front Plant Sci* **2023**, *14*, 1253536. [CrossRef]
42. D'Esposito, D.; Cappetta, E.; Andolfo, G.; Ferriello, F.; Borgonuovo, C.; Caruso, G.; De Natale, A.; Frusciante, L.; Ercolano, M.R. Deciphering the Biological Processes Underlying Tomato Biomass Production and Composition. *Plant Physiol. Biochem.* **2019**, *143*, 50–60. [CrossRef]
43. Besada, C.; Gil, R.; Bonet, L.; Quiñones, A.; Intrigliolo, D.; Salvador, A. Chloride Stress Triggers Maturation and Negatively Affects the Postharvest Quality of Persimmon Fruit. Involvement of Calyx Ethylene Production. *Plant Physiol. Biochem.* **2016**, *100*, 105–112. [CrossRef] [PubMed]
44. Shen, Z.; Cheng, X.; Li, X.; Deng, X.; Dong, X.; Wang, S.; Pu, X. Effects of Silicon Application on Leaf Structure and Physiological Characteristics of *Glycyrrhiza uralensis* Fisch. and *Glycyrrhiza inflata* Bat. under Salt Treatment. *BMC Plant Biol* **2022**, *22*, 390. [CrossRef]
45. Hanstein, S.M.; Felle, H.H. CO_2-Triggered Chloride Release from Guard Cells in Intact Fava Bean Leaves. Kinetics of the Onset of Stomatal Closure. *Plant Physiol.* **2002**, *130*, 940–950. [PubMed]
46. Abou Seeda, M.A.; Abou El-Nour EA, A.; Hammad, S.A.; Yassen, A.A. Chloride Ions as a Beneficial and Essential Micronutrient Multifunctional, Role and Regulation in Plant Physiology: A Review. *Middle East J. Appl. Sci.* **2021**, *11*, 76–125. [CrossRef]
47. Goutam Kumar Dash, S.P.; Soumya Kumar Sahoo, M.B.; Mirza Jaynul Baig, R.K.P.; Swain, P. Photosystem II Photochemistry and Chlorophyll Intensity in Rice (*Oryza sativa* L.) under Drought at Flowering Stage. *Int. J. Curr. Microbiol. App. Sci.* **2021**, *10*, 746–756. [CrossRef]
48. Anderson, J.M. Cytochrome B6f Complex: Dynamic Molecular Organization, Function and Acclimation. *Photosynth. Res.* **1992**, *34*, 341–357. [CrossRef]
49. Klughammer, C.; Schreiber, U. Saturation Pulse Method for Assessment of Energy Conversion in PS I. *Planta* **1994**, *192*, 261–268.
50. Yamori, W.; Sakata, N.; Suzuki, Y.; Shikanai, T.; Makino, A. Cyclic Electron Flow Around Photosystem I via Chloroplast NAD(P)H Dehydrogenase (NDH) Complex Performs a Significant Physiological Role During Photosynthesis and Plant Growth at Low Temperature in Rice. *Plant J.* **2011**, *68*, 966–976. [CrossRef]
51. Haddy, A.; Beravolu, S.; Johnston, J.; Kern, H.; McDaniel, M.; Ore, B.; Reed, R.; Tai, H. Exploring the Interdependence of Calcium and Chloride Activation of O_2 Evolution in Photosystem II. *Photosynth. Res.* **2024**, *162*, 385–400.
52. Gorbe, E.; Calatayud, A. Applications of Chlorophyll Fluorescence Imaging Technique in Horticultural Research: A Review. *Sci. Hortic.* **2012**, *138*, 24–35. [CrossRef]
53. Lu, T.; Song, Y.; Yu, H.; Li, Q.; Xu, J.; Qin, Y.; Zhang, G.; Liu, Y.; Jiang, W. Cold Stress Resistance of Tomato (*Solanum lycopersicum*) Seedlings Is Enhanced by Light Supplementation from Underneath the Canopy. *Front. Plant Sci.* **2022**, *13*, 831314. [CrossRef] [PubMed]
54. Zhou, Y.; Schideman, L.C.; Park, D.S.; Stirbet, A.; Govindjee; Rupassara, S.I.; Krehbiel, J.D.; Seufferheld, M.J. Characterization of a Chlamydomonas Reinhardtii Mutant Strain with Improved Biomass Production Under Low Light and Mixotrophic Conditions. *Algal Res.* **2015**, *11*, 134–147. [CrossRef]
55. Klughammer, C.; Schreiber, U. Complementary PS II Quantum Yields Calculated from Simple Fluorescence Parameters Measured by PAM Fluorometry and the Saturation Pulse Method. *PAM Appl. Notes* **2008**, *1*, 201–247.
56. Tang, Y.; Wen, X.; Lu, Q.; Yang, Z.; Cheng, Z.; Lu, C. Heat Stress Induces an Aggregation of the Light-Harvesting Complex of Photosystem II in Spinach Plants. *Plant Physiol.* **2007**, *143*, 629–638. [CrossRef] [PubMed]
57. Turan, S.; Kumar, S.; Cornish, K. Photosynthetic Response of In Vitro Guayule Plants in Low and High Lights and the Role of Non-Photochemical Quenching in Plant Acclimation. *Ind. Crops Prod.* **2014**, *54*, 266–271. [CrossRef]
58. Nishiyama, Y.; Allakhverdiev, S.I.; Murata, N. Protein Synthesis Is the Primary Target of Reactive Oxygen Species in the Photoinhibition of Photosystem II. *Physiol. Plant.* **2011**, *142*, 35–46. [CrossRef]
59. Feng, Y.; Wu, H.; Liu, H.; He, Y.; Yin, Z. Effects of OsRCA Overexpression on Rubisco Activation State and Photosynthesis in Maize. *Plants* **2023**, *12*, 1614. [CrossRef]
60. Storti, M.; Puggioni, M.P.; Segalla, A.; Morosinotto, T.; Alboresi, A. The Activity of Chloroplast NADH Dehydrogenase-like Complex Influences the Photosynthetic Activity of the Moss Physcomitrella Patens. *bioRxiv* **2020**. [CrossRef]
61. Schumaker, M.F.; Kramer, D.M. Comparison of Monte Carlo Simulations of Cytochrome B6f with Experiment Using Latin Hypercube Sampling. *Bull. Math. Biol.* **2011**, *73*, 2152–2174. [CrossRef]
62. Johnson, G.N. Physiology of PSI Cyclic Electron Transport in Higher Plants. *Biochim. Biophys. Acta (BBA)-Bioenerg.* **2011**, *1807*, 384–389.

63. Herdean, A.; Teardo, E.; Nilsson, A.K.; Pfeil, B.E.; Johansson, O.N.; Ünnep, R.; Nagy, G.; Zsiros, O.; Dana, S.; Solymosi, K.; et al. A Voltage-Dependent Chloride Channel Fine-Tunes Photosynthesis in Plants. *Nat. Commun.* **2016**, *7*, 11654. [CrossRef] [PubMed]
64. Duan, Z.; Kong, F.; Zhang, L.; Li, W.; Zhang, J.; Peng, L. A Bestrophin-like Protein Modulates the Proton Motive Force across the Thylakoid Membrane in *Arabidopsis*. *J. Integr. Plant Biol.* **2016**, *58*, 848–858. [CrossRef]
65. Hagino, T.; Kato, T.; Kasuya, G.; Kobayashi, K.; Kusakizako, T.; Hamamoto, S.; Sobajima, T.; Fujiwara, Y.; Yamashita, K.; Kawasaki, H.; et al. Cryo-EM Structures of Thylakoid-Located Voltage-Dependent Chloride Channel VCCN1. *Nat. Commun.* **2022**, *13*, 2505. [CrossRef]
66. Imaizumi, K.; Nishimura, T.; Nagao, R.; Saito, K.; Nakano, T.; Ishikita, H.; Noguchi, T.; Ifuku, K. D139N Mutation of PsbP Enhances the Oxygen-Evolving Activity of Photosystem II through Stabilized Binding of a Chloride Ion. *PNAS Nexus* **2022**, *1*, pgac136. [CrossRef] [PubMed]
67. Centeno, D.C.; Osorio, S.; Nunes-Nesi, A.; Bertolo, A.L.F.; Carneiro, R.T.; Araújo, W.L.; Steinhauser, M.-C.; Michalska, J.; Rohrmann, J.; Geigenberger, P.; et al. Malate Plays a Crucial Role in Starch Metabolism, Ripening, and Soluble Solid Content of Tomato Fruit and Affects Postharvest Softening. *Plant Cell* **2011**, *23*, 162–184. [CrossRef] [PubMed]
68. Chen, J.; Kang, S.; Du, T.; Guo, P.; Qiu, R.; Chen, R.; Gu, F. Modeling Relations of Tomato Yield and Fruit Quality with Water Deficit at Different Growth Stages Under Greenhouse Condition. *Agric. Water Manag.* **2014**, *146*, 131–148. [CrossRef]
69. Dou, J.; Wang, J.; Tang, Z.; Yu, J.; Wu, Y.; Liu, Z.; Wang, J.; Wang, G.; Tian, Q. Application of Exogenous Melatonin Improves Tomato Fruit Quality by Promoting the Accumulation of Primary and Secondary Metabolites. *Foods* **2022**, *11*, 4097. [CrossRef]
70. Grieco, M.; Tikkanen, M.; Paakkarinen, V.; Kangasjärvi, S.; Aro, E.-M. Steady-State Phosphorylation of Light-Harvesting Complex II Proteins Preserves Photosystem I Under Fluctuating White Light. *Plant Physiol.* **2012**, *160*, 1896–1910. [CrossRef]
71. Ruan, Y.-L.; Patrick, J.W.; Bouzayen, M.; Osorio, S.; Fernie, A.R. Molecular Regulation of Seed and Fruit Set. *Trends Plant Sci.* **2012**, *17*, 656–665. [CrossRef]
72. Powell, A.L.T.; Nguyen, C.V.; Hill, T.; Cheng, K.L.; Figueroa-Balderas, R.; Aktas, H.; Ashrafi, H.; Pons, C.; Fernández-Muñoz, R.; Vicente, A.; et al. *Uniform Ripening* Encodes a *Golden 2-like* Transcription Factor Regulating Tomato Fruit Chloroplast Development. *Science* **2012**, *336*, 1711–1715. [CrossRef]
73. Li, R.; Wang, J.; Yuan, H.; Niu, Y.; Sun, J.; Tian, Q.; Wu, Y.; Yu, J.; Tang, Z.; Xiao, X.; et al. Exogenous Application of ALA Enhanced Sugar, Acid and Aroma Qualities in Tomato Fruit. *Front. Plant Sci.* **2023**, *14*, 1323048. [CrossRef]
74. Yamaki, S. Metabolism and Accumulation of Sugars Translocated to Fruit and Their Regulation. *J. Japan. Soc. Hortic. Sci.* **2010**, *79*, 1–15. [CrossRef]
75. Saito, T.; Matsukura, C. Effect of Salt Stress on the Growth and Fruit Quality of Tomato Plants. In *Abiotic Stress Biology in Horticultural Plants*; Kanayama, Y., Kochetov, A., Eds.; Springer: Tokyo, Japan, 2015; pp. 3–16, ISBN 978-4-431-55250-5.
76. Luo, A.; Zhou, C.; Chen, J. The Associated with Carbon Conversion Rate and Source–Sink Enzyme Activity in Tomato Fruit Subjected to Water Stress and Potassium Application. *Front. Plant Sci.* **2021**, *12*, 681145. [CrossRef]
77. Biais, B.; Bénard, C.; Beauvoit, B.; Colombié, S.; Prodhomme, D.; Ménard, G.; Bernillon, S.; Gehl, B.; Gautier, H.; Ballias, P.; et al. Remarkable Reproducibility of Enzyme Activity Profiles in Tomato Fruits Grown Under Contrasting Environments Provides a Roadmap for Studies of Fruit Metabolism. *Plant Physiol.* **2014**, *164*, 1204–1221. [CrossRef]
78. Wang, B.; Li, N.; Huang, S.; Hu, J.; Wang, Q.; Tang, Y.; Yang, T.; Asmutola, P.; Wang, J.; Yu, Q. Enhanced Soluble Sugar Content in Tomato Fruit Using CRISPR/Cas9-Mediated SlINVINH1 and SlVPE5 Gene Editing. *PeerJ* **2021**, *9*, e12478. [CrossRef]
79. Beauvoit, B.P.; Colombié, S.; Monier, A.; Andrieu, M.-H.; Biais, B.; Bénard, C.; Chéniclet, C.; Dieuaide-Noubhani, M.; Nazaret, C.; Mazat, J.-P.; et al. Model-Assisted Analysis of Sugar Metabolism throughout Tomato Fruit Development Reveals Enzyme and Carrier Properties in Relation to Vacuole Expansion. *Plant Cell* **2014**, *26*, 3224–3242. [CrossRef] [PubMed]
80. Nguyen-Quoc, B.; N'Tchobo, H.; Foyer, C.H.; Yelle, S. Overexpression of Sucrose Phosphate Synthase Increases Sucrose Unloading in Transformed Tomato Fruit. *J. Exp. Bot.* **1999**, *50*, 785–791. [CrossRef]
81. Rojas-González, J.A.; Soto-Súarez, M.; García-Díaz, Á.; Romero-Puertas, M.C.; Sandalio, L.M.; Mérida, Á.; Thormählen, I.; Geigenberger, P.; Serrato, A.J.; Sahrawy, M. Disruption of Both Chloroplastic and Cytosolic FBPase Genes Results in a Dwarf Phenotype and Important Starch and Metabolite Changes in *Arabidopsis thaliana*. *J. Exp. Bot.* **2015**, *66*, 2673–2689. [CrossRef]
82. Gerardeaux, E.; Jordan-Meille, L.; Constantin, J.; Pellerin, S.; Dingkuhn, M. Changes in Plant Morphology and Dry Matter Partitioning Caused by Potassium Deficiency in *Gossypium hirsutum* (L.). *Environ. Exp. Bot.* **2010**, *67*, 451–459. [CrossRef]
83. Worrell, A.C.; Bruneau, J.-M.; Summerfelt, K.; Boersig, M.; Voelker, T.A. Expression of a Maize Sucrose Phosphate Synthase in Tomato Alters Leaf Carbohydrate Partitioning. *Plant Cell* **1991**, *3*, 1121–1130.
84. Galtier, N.; Foyer, C.H.; Huber, J.; Voelker, T.A.; Huber, S.C. Effects of Elevated Sucrose-Phosphate Synthase Activity on Photosynthesis, Assimilate Partitioning, and Growth in Tomato (*Lycopersicon esculentum* Var UC82B). *Plant Physiol.* **1993**, *101*, 535–543. [CrossRef]

85. Galtier, N.; Foyer, C.H.; Murchie, E.; Aired, R.; Quick, P.; Voelker, T.A.; Thépenier, C.; Lascève, G.; Betsche, T. Effects of Light and Atmospheric Carbon Dioxide Enrichment on Photosynthesis and Carbon Partitioning in the Leaves of Tomato (*Lycopersicon esculentum* L.) Plants over-Expressing Sucrose Phosphate Synthase. *J. Exp. Bot.* **1995**, *46*, 1335–1344. [CrossRef]
86. Micallef, B.J.; Haskins, K.A.; Vanderveer, P.J.; Roh, K.-S.; Shewmaker, C.K.; Sharkey, T.D. Altered Photosynthesis, Flowering, and Fruiting in Transgenic Tomato Plants That Have an Increased Capacity for Sucrose Synthesis. *Planta* **1995**, *196*, 327–334. [CrossRef]
87. Signora, L.; Galtier, N.; Skot, L.; Lucas, H.; Foyer, C.H. Over-Expression of Sucrose Phosphate Synthase in *Arabidopsis thaliana* Results in Increased Foliar Sucrose/Starch Ratios and Favours Decreased Foliar Carbohydrate Accumulation in Plants after Prolonged Growth with CO_2 Enrichment. *J. Exp. Bot.* **1998**, *49*, 669–680. [CrossRef]
88. Markússon, S.; Hjörleifsson, J.G.; Kursula, P.; Ásgeirsson, B. Structural Characterization of Functionally Important Chloride Binding Sites in the Marine Vibrio Alkaline Phosphatase. *Biochemistry* **2022**, *61*, 2248–2260.
89. Amoah, J.N. Effect of Drought Acclimation on Sugar Metabolism in Millet. *Protoplasma* **2025**, *262*, 35–49. [CrossRef]
90. Pottosin, I.I.; Muñiz, J. Higher plant vacuolar ionic transport in the cellular context. *Acta Bot. Mex.* **2002**, *60*, 37–77. [CrossRef]
91. Mansour, M.M.F. Role of Vacuolar Membrane Transport Systems in Plant Salinity Tolerance. *J. Plant Growth Regul.* **2023**, *42*, 1364–1401. [CrossRef]
92. Valdivieso, Á.G. The Chloride Anion as a Signalling Effector. *Biol. Rev.* **2019**, *94*, 1839–1856. [CrossRef]
93. Zhang, X.; Franzisky, B.L.; Eigner, L.; Geilfus, C.; Zörb, C. Antagonism of Chloride and Nitrate Inhibits Nitrate Reductase Activity in Chloride-Stressed Maize. *Plant Growth Regul.* **2021**, *93*, 279–289. [CrossRef]
94. Xu, Y.; Fu, X.; Sharkey, T.D.; Shachar-Hill, Y.; Walker, B.J. The Metabolic Origins of Non-Photorespiratory CO_2 Release during Photosynthesis: A Metabolic Flux Analysis. *Plant Physiol.* **2021**, *186*, 297–314. [PubMed]
95. White, A.C.; Rogers, A.; Rees, M.; Osborne, C.P. How Can We Make Plants Grow Faster? A Source–Sink Perspective on Growth Rate. *J. Exp. Bot.* **2016**, *67*, 31–45. [PubMed]
96. Duran, C. Chloride Channels: Often Enigmatic, Rarely Predictable. *Annu. Rev. Physiol.* **2010**, *72*, 95–121. [CrossRef]
97. Xin, W.; Zhang, L.; Zhang, W.; Gao, J.; Yi, J.; Zhen, X.; Li, Z.; Zhao, Y.; Peng, C.; Zhao, C. An Integrated Analysis of the Rice Transcriptome and Metabolome Reveals Differential Regulation of Carbon and Nitrogen Metabolism in Response to Nitrogen Availability. *Int. J. Mol.Sci.* **2019**, *20*, 2349. [PubMed]
98. Pietrzykowska, M.; Suorsa, M.; Semchonok, D.A.; Tikkanen, M.; Boekema, E.J.; Aro, E.-M.; Jansson, S. The Light-Harvesting Chlorophyll a/b Binding Proteins Lhcb1 and Lhcb2 Play Complementary Roles during State Transitions in *Arabidopsis*. *Plant Cell* **2014**, *26*, 3646–3660. [CrossRef]
99. Pružinská, A.; Shindo, T.; Niessen, S.; Kaschani, F.; Tóth, R.; Millar, A.H.; van der Hoorn, R.A.L. Major Cys Protease Activities Are Not Essential for Senescence in Individually Darkened *Arabidopsis* Leaves. *BMC Plant Biol.* **2017**, *17*, 4. [CrossRef]

International Journal of
Molecular Sciences

MDPI

Article

Strain-Specific Features of Primary Metabolome Characteristic for Extremotolerant/Extremophilic Cyanobacteria Under Long-Term Storage

Tatiana Bilova [1,2,*], Nikita Golushko [1,2], Nadezhda Frolova [1], Alena Soboleva [1], Svetlana Silinskaia [1], Anna Khakulova [3], Anastasia Orlova [1], Maria Sinetova [4], Dmitry Los [4] and Andrej Frolov [1,*]

1 Laboratory of Analytical Biochemistry and Biotechnology, K.A. Timiryazev Institute of Plant Physiology Russian Academy of Sciences, 127276 Moscow, Russia; st078415@student.spbu.ru (N.G.); frolovanadja@yandex.ru (N.F.); soboleva@ifr.moscow (A.S.); svetlanasilsv@mail.ru (S.S.); orlova@ifr.moscow (A.O.)

2 Department of Plant Physiology and Biochemistry, St. Petersburg State University, 199034 St. Petersburg, Russia

3 Chemical Analysis and Materials Research Core Facility Center, Research Park, St. Petersburg State University, 199034 St. Petersburg, Russia; a.khakulova@spbu.ru

4 Laboratory of Intracellular Regulation, K.A. Timiryazev Institute of Plant Physiology Russian Academy of Sciences, 127276 Moscow, Russia; maria.sinetova@mail.ru (M.S.); losda@ippras.ru (D.L.)

* Correspondence: t.bilova@spbu.ru (T.B.); frolov@ifr.moscow (A.F.)

Abstract: Cyanobacteria isolated from extreme habitats are promising in biotechnology due to their high adaptability to unfavorable environments and their specific natural products. Therefore, these organisms are stored under a reduced light supply in multiple collections worldwide. However, it remains unclear whether these strains maintain constitutively expressed primary metabolome features associated with their unique adaptations. To address this question, a comparative analysis of primary metabolomes of twelve cyanobacterial strains from diverse extreme habitats was performed by a combined GC-MS/LC-MS approach. The results revealed that all these cyanobacterial strains exhibited clear differences in their patterns of primary metabolites. These metabolic differences were more pronounced for the strains originating from ecologically different extreme environments. Extremotolerant terrestrial and freshwater strains contained lower strain-specifically accumulated primary metabolites than extremophilic species from habitats with high salinity and alkalinity. The latter group of strains was highly diverse in amounts of specific primary metabolites. This might indicate essentially different molecular mechanisms and metabolic pathways behind the survival of the microorganisms in saline and alkaline environments. The identified strain-specific metabolites are discussed with respect to the metabolic processes that might impact maintaining the viability of cyanobacteria during their storage and indicate unique adaptations formed in their original extreme habitats.

Keywords: cyanobacterial strains; extremophiles; haloalkaliphiles and natronophiles; adaptation to extreme conditions; metabolomics; GC-MS; LC-MS; primary metabolites; metabolic pathways

1. Introduction

Cyanobacteria are an ancient group of photosynthetic microorganisms found in diverse terrestrial and water environments, significantly impacting water quality and the functioning of aquatic ecosystems [1]. In addition to their ecological roles, they are essential components of food webs, serving as the only prokaryotic organisms, primary producers performing oxygenic photosynthesis [2]. Another noteworthy attribute of cyanobacteria is their remarkable adaptability to a diverse range of extreme environments such as drought, high or low temperatures, ultraviolet radiation, salinity and alkalinity [3]. Some cyanobacteria, considered extremophilic, have developed unique adaptation strategies during their extensive evolutionary history that allow them to fine-tune their metabolism to 'extreme' conditions. It is important to note that 'extreme' conditions represent the norm at which adapted organisms are able to function metabolically and biochemically without experiencing a state of stress. Other cyanobacteria considered as extremotolerant have adapted to tolerate a wide range of variation in one or more life-limiting parameters (e.g., temperature variation, water availability) [4].

Different molecular mechanisms mediating cyanobacteria adaptations are known. As they are exposed to UV radiation in their natural habitats, the majority of cyanobacteria developed a few UV-stress mitigation mechanisms: enhanced biosynthesis of carotenoids [5], mycosporine-like amino acids [6], polyphenols [7] and polyamines to protect the photosynthetic apparatus; up-regulation of scavengers for reactive oxygen species (ROS) and antioxidant machinery [8,9]; induction of repairing systems for UV-damaged biopolymers [10]. Additional strategies used by desiccation-tolerant cyanobacteria include the production of extracellular polymeric substances, recruitment of chaperones to maintain protein integrity, activation of ion channels, synthesis of compatible solutes to overcome low water availability [11] and synthesis of the non-reducing sugar trehalose to protect membranes during desiccation [12]. To withstand low temperatures, cyanobacteria are known to produce cold shock proteins, antifreeze proteins and cryoprotectants (e.g., dimethylsulfoniopropionate [13,14]), protecting against freezing damage [15]. The adaptations that cyanobacteria have evolved to survive in saline environments (seawater, hypersaline lakes) include a profound reorganization of regulatory systems. These ensure the functioning of fundamental processes such as the maintenance of osmotic pressure [16–18], cellular homeostasis, energy production and the excretion of sodium ions [19]. Indeed, the genomes of the halotolerant cyanobacteria carry clusters of genes encoding proteins involved in the adaptation to high salinity. Interestingly, in some cases, the metabolic adjustment to extreme conditions enables cyanobacteria to produce highly valuable substances with functional importance and potential human health benefits [18]. This makes cyanobacteria a source of natural compounds with characteristic biological properties and a potential candidate for pharmaceuticals.

The current work was focused on comparative profiling primary metabolites of cultivated cyanobacterial strains previously extracted from various extreme environments (high or low temperatures, drought, increased salinity and alkalinity). Primary metabolites, the small molecules involved in major metabolic pathways crucial for energy production, carbon fixation, the synthesis of essential biopolymers and secondary compounds [20], are the major players mediating the diversity of adaptive metabolic responses in cyanobacteria. They appeal to scientific interest for at least two reasons. First, these adaptations hold potential for biotechnological applications due to the production of unique biologically active compounds [21]. Second, the cyanobacterial strains can be used as models to address the molecular basis of their exceptional adaptivity and transfer the knowledge gained for improving the tolerance of crops to adverse conditions [19]. However, despite progress in understanding cyanobacterial adaptations to extreme environments, many aspects remain uncertain. Specifically, the intricate details of the molecular mechanisms and regulatory networks involved in these adaptive processes in changeable environments are not fully elucidated. Additionally, the comprehensive profiles of primary metabolites and their specific roles in enabling cyanobacterial survival under such conditions are still largely unknown [22,23].

To effectively study the primary metabolism of these cyanobacteria, their primary metabolomes need to be addressed. This can be done using modern high-throughput analytical techniques such as gas chromatography–mass spectrometry and liquid chromatography–mass spectrometry (GC-MS and LC-MS). GC-MS analyzes small molecules (e.g., amino acids, sugars) by converting them into volatile thermally stable derivatives through a derivatization process [24]. LC-MS, on the other hand, analyzes small molecules not suitable for GC-MS due to their thermal instability [25]. This combination of methods acquires a comprehensive metabolite profile, addressing the limitations of each technique [26–28]. Here, we employed the advanced GC-MS and LC-MS techniques to perform a comparative analysis of the primary metabolomes of twelve cyanobacterial strains. The strains, isolated from different extreme habitats, were cultivated in strain-specific media under conditions suboptimal for growth, which were provided by a reduction in light and the absence of an additional CO_2 supply. These cultivation conditions guaranteed the long-term maintenance (storage) of the strains in a metabolically active state [29]. The long-term storage of metabolically active cyanobacterial cultures is of great importance in order to guarantee the sustainable availability of their diversity for fundamental and applied research purposes, as well as for the biotechnological production of unique cyanobacterial bioactive compounds [30]. This exploratory study is an attempt to investigate whether strains cultivated under long-term storage conditions retain a set of primary metabolites that may be related to the unique adaptations formed in the specific extreme conditions in which the cyanobacteria originally evolved.

2. Results

2.1. Characterization of Studied Extremophilic and Extremotolerant Cyanobacteria Strains

The studied strains were isolated from different extreme habitats (Table 1) and were maintained in the IPPAS collection by periodic transfer as unialgal non-axenic cultures for 4–50 years before the present study. The strains of genus *Sodalinema*, IPPAS B-2050, IPPAS B-2037 and IPPAS B-353, are the reference strains of *S. stalii*, *S. orleanskyi* and *S. gerasimenkoae*, respectively, which were described based on a polyphasic approach by Samylina et al. [31]. Other strains were tentatively identified based on their partial 16S rRNA sequences and morphology (Supplementary S1, Figures S1-1 and S1-2).

Based on their habitat, the studied strains can be divided into desiccation-tolerant, high- and low-temperature-tolerant, and halophilic, haloalkaliphilic and natronophilic groups. Desiccation-tolerant strains of *Nostoc commune* B-1519 and B-1520 were isolated from the same macrocolony collected from the soil surface. Such macrocolonies are subject to periodic desiccation during the summer period. The thermotolerant freshwater strain *Dolichospermum* sp. IPPAS B-1213 was isolated from a hot spring and has an optimal growth temperature of 38 °C [32]. The cold-tolerant strains *Anabaena* sp. IPPAS B-1535 and *A.* cf. pirinica IPPAS B-1533 were isolated from the Siberian river Yenesei, where the water temperature ranges from 5 to 14 °C in spring–summer period and the flow velocity is up to 2 m s^{-1} [33]. Haloalkaliphilic and natronophilic strains of *Sodalinema orleanskyi* B-2037, *S. gerasimenkoae* B-353, *Limnospira* sp. B-256, B-287, and B-1526, and *Nodularia* sp. B-1529 were isolated from saline–alkaline and soda lakes. Such mineral lakes are characterized by alkaline pH (9–11.5) and high concentrations of Na^+ (K^+) and Cl^-; soda lakes additionally contain bicarbonates and carbonates [34,35]. It is important to note that salinity in some mineral lakes (e.g., soda lakes in the Altai Region, Russia) can exhibit dramatic fluctuations depending on the precipitation–evaporation balance [36]. Halophilic strain *S. stalii* B-2050 was isolated from the North Sea.

Table 1. List of investigated extremotolerant and extremophilic cyanobacteria.

Tolerance Group	IPPAS ID	Species Name	Extreme Environment	Storage Conditions for Cyanobacteria Cultivation
Desiccation-tolerant	B-1520	*Nostoc commune*	Macrocolony collected from soil surface, Gorodets village, Kaluga obl., Russia. Heterocystous diazotroph. Resistant to desiccation.	BG-11 medium without nitrogen (pH 7.5) [37], light intensity 50 µmol photons m^{-2} s^{-1}, in an orbital shaker at 22 °C for 3 months.
	B-1519	*Nostoc commune*	Macrocolony collected from soil surface, Gorodets village, Kaluga obl., Russia. Heterocystous diazotroph. Resistant to desiccation.	BG-11 medium without nitrogen (pH 7.5) [37], light intensity 50 µmol photons m^{-2} s^{-1}, in an orbital shaker at 22 °C for 3 months.
High- and low-temperature-tolerant	B-1213	*Dolichospermum* sp.	Hot springs, Karlovy Vary, Czech Republic. Heterocystous diazotroph. Thermophile.	BG-11 medium without nitrogen (pH 7.5) [37], light intensity 50 µmol photons m^{-2} s^{-1}, in an orbital shaker at 22 °C for 8 months.
	B-1533	*Anabaena* cf. pirinica	Yenisei river, Krasnoyarsk, Russia. Cold-tolerant: able to survive at low temperatures (up to 10–11 °C) and high water flow rate. Heterocystous diazotroph.	No. 6 medium without nitrogen (pH 7.2) [38], light intensity 50 µmol photons m^{-2} s^{-1}, in an orbital shaker at 22 °C for 3 months.
	B-1535	*Anabaena* sp.	Yenisei river, Krasnoyarsk, Russia. Cold-tolerant: able to survive at low temperatures (up to 10–11 °C) and high water flow rate. Heterocystous diazotroph.	No. 6 medium without nitrogen (pH 7.2) [38], light intensity 50 µmol photons m^{-2} s^{-1}, in an orbital shaker at 22 °C for 3 months.
Halophilic, haloalkaliphilic and natronophilic	B-2050	*Sodalinema stalii*	Coastal shoals, Mellum Island, North Sea, Germany. Salinity about 30 g/L. Halophilic.	ASNIII medium (pH 7.5) [39], light intensity 50 µmol photons m^{-2} s^{-1}, 27 °C in a growth chamber MLR-352-PE (Panasonic, Osaka, Japan) for 3 weeks, then 22 °C for 2 months.
	B-2037	*Sodalinema orleanskyi*	Salt alkaline lake Eyasi, Tanzania. Haloalkaliphile, natronophile (pH_{opt} 9–10, growth requires 0.2 M $NaHCO_3$ in the medium).	S medium (pH 9.0–9.5) [31], light intensity 50 µmol photons m^{-2} s^{-1}, 32 °C in a growth chamber MLR-351 (SANYO, Osaka, Japan) for 3 weeks, then 22 °C for 2 months.
	B-353	*Sodalinema gerasimenkoae*	Salt alkaline lake Khilganta, Transbaikal Territory, Russia. Haloalkaliphile, natronophile (pH_{opt} 9–10, growth requires 0.2 M $NaHCO_3$ in the medium).	S medium (pH 9.0–9.5) [31], light intensity 50 µmol photons m^{-2} s^{-1}, 27 °C in a growth chamber MLR-352-PE (Panasonic, Osaka, Japan) for 3 weeks, then 22 °C for 2 months.
	B-1526	*Limnospira* sp.	Soda Lake Gorchina I, Altai Region, Russia. Haloalkaliphile, natronophile (pH_{opt} 9–10, growth requires 0.2 M $NaHCO_3$ in the medium).	Zarrouk medium (pH 9.5) [40], light intensity 50 µmol photons m^{-2} s^{-1}, 32 °C in a growth chamber MLR-351 (SANYO, Osaka, Japan) for 6 weeks.
	B-287	*Limnospira* sp.	The origin of the strain is not precisely known. Haloalkaliphile, natronophile (pH_{opt} 9–10, growth requires 2 M $NaHCO_3$ in the medium).	Zarrouk medium (pH 9.5) [40], light intensity 50 µmol photons m^{-2} s^{-1}, 32 °C in a growth chamber MLR-351 (SANYO, Osaka, Japan) for 6 weeks.
	B-256	*Limnospira* sp.	Bodou Soda Lake, Chad. Haloalkaliphile, natronophile (optimum pH 9–10, growth requires 0.2 M $NaHCO_3$ in the medium).	Zarrouk medium (pH 9.5) [40], light intensity 50 µmol photons m^{-2} s^{-1}, in a growth chamber MLR-351 (SANYO, Osaka, Japan) 32 °C for 6 weeks.
	B-1529	*Nodularia* sp.	Soda Lake Gorchina I, Altai Region, Russia. Haloalkaliphile, natronophile (pH_{opt} 9–10, growth requires 0.2 M $NaHCO_3$ in the medium). Heterocystous diazotroph.	Zarrouk medium without nitrogen (pH 9.5) [40], light intensity 50 µmol photons m^{-2} s^{-1}, 27 °C in a growth chamber MLR-352-PE (Panasonic, Japan) for 3 weeks, then 22 °C for 3 weeks.

2.2. Characterization of Extremophilic and Extremotolerant Cyanobacteria Polar Metabolite Patterns

The GC-MS-based untargeted analysis of derivatized aq. methanolic extracts of all extremophilic and extremotolerant cyanobacterial strains investigated here allowed the detection of a total of 358 total ion current (TIC) peaks of thermally stable polar metabolite trimethylsilyl (TMS) and methoxy-(MEOX)-TMS derivatives (features) (Supplementary S2, Table S2-1). Out of this number, 75 features were identified by co-elution with authentic standards from an in-house authentic standard library (ihASL). Structural annotation of the other 74 features relied on close similarity in experimentally obtained retention index (RI) and electron ionization mass spectrum (EI-MS) with those GC-MS data available for standard compounds from known spectral libraries, GMD and NIST. The identified metabolites were represented by following classes: 15 organic acids (lactic, oxalic, glycolic, succinic, glyceric, itaconic, fumaric, salicylic, *D*-erythronic, 2-ketoglutaric, citric, isocitric, propanoic, gluconic, mucic), 15 amino acids (Val, Ala, Gly, Tyr, Leu, Ile, Ser, Asp, Glu, Phe, Lys, Trp, Orn, homoserine, γ-aminobutyric acid), 11 sugars and derivatives (ribose, xylose, fructose, mannose, galactose, glucose, *N*-acetyl-glucosamine, sucrose, maltose, isomaltose, α,α-trehalose), 8 polyols (glycerol, erythritol, 1-deoxypentitol, arabitol, ribitol, mannitol, sorbitol, *myo*-inositol), 4 sugar phosphates and other organic phosphates (*myo*-inositol phosphate, fructose 6-phosphate, glucose 6-phosphate, methyl-phosphate), 20 fatty acids and derivatives (nonanoic acid, dodecanoic acid, 1-dodecanol, tetradecanoic acid, *n*-pentadecanoic acid, palmitic acid, methyl palmitate, *trans*-9-hexadecenoic acid, *cis*-9-hexadecenoic acid, heptadecanoic acid, *cis*-10-heptadecenoic acid, stearic acid and its methyl ester, γ-linolenic acid and its methyl ester, linoleic acid and its methyl, oleic acid, *trans*-9-octadecenoic acid, nonadecan-1-ol) and 29 representatives of other classes. The third group of 100 features with structures that could not be established by similarity with library data was annotated by a presence in their EI-MS fragment ions that might be considered according to Harvey and Vouros [41] as diagnostic for specific chemical classes as follows: monosaccharides and their derivatives (m/z 204, 217, 319), di-, tri- or oligosaccharides (m/z 204, 217, 361), amino acids and their derivatives (m/z 100, 174), phosphate-containing substances (m/z 299, 315, 357), fatty acids (m/z 117, 129). Furthermore, 109 features could not be annotated to any chemical class and were categorized as unknowns. Annotations of the unknowns and features related to a specific chemical class as well were assigned with their unique RI.

LC-MS analysis based on the target MRM mode allowed the detection and identification of a total of 183 thermally labile analytes in polar extracts of all investigated extremophilic and extremotolerant cyanobacterial strains. The identified metabolites were representatives of chemical classes such as amino acids and their derivatives (Arg, Gln, Gly, Met, Ser, Thr, Tyr, Phe, Asp, Glu, Trp, citrulline, S-adenosyl-*L*-homocysteine), sugars and their derivatives (sucrose, glucopyranuronic acid, galactopyranuronic acid, galactonic acid/gluconic acid), sugar phosphates and other organic phosphates (21 identified metabolites), nucleotides (32 identified metabolites), coenzyme A thioesters (6 identified metabolites) and other compounds involved in energy metabolic pathways (Supplementary S2, Table S2-2). Importantly, patterns of metabolites detected by LC-MS and GC-MS were partially overlapped for some amino acids, organic acids, carbohydrates and representatives of other chemical classes. Although all these metabolites have been isolated from cyanobacteria, it cannot be excluded, particularly in the case of *trans*-fatty acids, that they may also originate from bacteria that accompany cyanobacterial cultures.

2.3. Identification of Strain-Dependent Variability in the Primary Metabolome

The combination of the GC-MS- and LC-MS-originated datasets resulted in a merged matrix consisting of 530 analytes in total. This matrix was subjected to statistical processing to estimate the magnitude of metabolome variance among the studied extremophilic and extremotolerant cyanobacteria strains. PCA built for the first two PCs, explaining a total of 49.7% of the variance, revealed the following regularities in sample clustering (Figure 1):

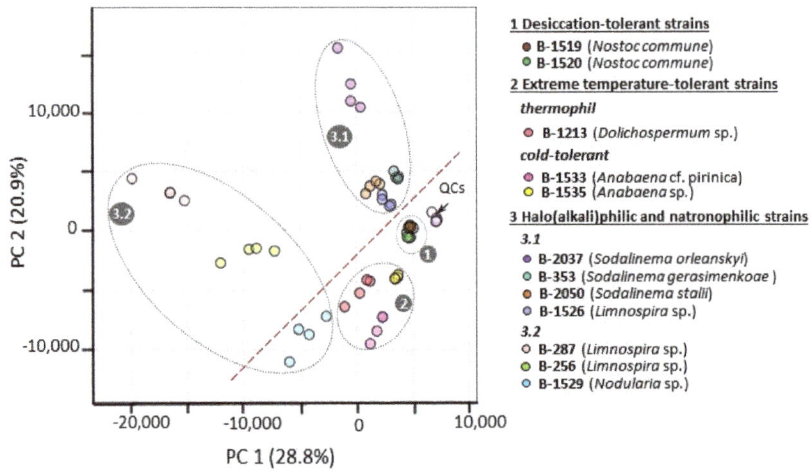

Figure 1. PCA presenting score plot built in coordinates of first two PCs for primary metabolomes of extremophilic and extremotolerant cyanobacteria strains. Input data for the PCA included 530 thermally stable and thermally labile polar metabolites detected by GC-MS and LC-MS in cyanobacterial extracts. Numbers in grey circles and dotted gray line mark clusters formed by metabolomes of cyanobacteria strains adapted to survive in similar extreme environments as follows: desiccation-tolerant strains (1); high-and low-temperature-tolerant strains (2); halo(alkali)philic and natronophilic strains (3) are indicated as two subgroups 3.1 and 3.2 for strains without or with considerable intergroup variability, respectively. The dotted red line separates the groups of strains possessing nitrogen-fixation ability (heterocyst-forming diazotrophs, left) and those lacking this ability (right). QCs—quality control samples. The corresponding loading plot is presented in Supplementary S2, Figure S2-1.

(1) Quality control samples (QCs) whose metabolite compositions were analyzed by both GC-MS- and LC-MS-based methods displayed a clear cluster separated from all other studied sample groups. The QC cluster was characterized by low in-group variance that serves as validation of the high quality of the data acquisitions performed without any serious technical problems during GC-MS and LC-MS analyses.

(2) The sample metabolomes formed distinct clusters attributed to the studied cyanobacterial strains. Furthermore, the metabolomes of the cyanobacterial strains exhibited additional clustering according to the strain extrema ecological groups. Indeed, in the PCA model, two distinct clusters were formed by desiccation-tolerant strains (B-1520 and B-1519, Figure 1, cluster 1) and temperature-tolerant strains (low-temperature B-1533, B-1535 and high-temperature B-1213, Figure 1, cluster 2). The clustering of haloalkaliphilic and natronophilic cyanobacteria strains (B-2037, B-1529, B-353, B-1526, B-256, B-287, B-2050, Figure 1, cluster 3) was more complicated and could be described by dividing the strains into two subgroups. The first subgroup denoted strains B-2050, B-353, B-1526 and B-2037 (Figure 1, cluster 3.1) without considerable intergroup metabolome variability. Conversely,

strains, B-287, B-256 and B-1529 (Figure 1, cluster 3.2) within the second subgroup were characterized by essential intergroup variability. Additionally, the observed difference in distances between the strain clusters in the first and second subgroups indicates distinctions and similarities, respectively, in metabolite patterns in strains in this haloalkaliphilic and natronophilic ecological group.

(3) PCA also showed that cyanobacterial strains with heterocyst-forming ability (heterocystous diazotrophs B-1520, B-1519, B-1213, B-1533, B-1535, B-1529) formed a cluster separated from the other strains without this ability.

(4) An analysis of the PCA loadings plot revealed metabolites such as succinate, sugar (RI2304), trehalose, malate and glucosylglycerol contributing to the observed clustering of studied strain metabolomes (Supplementary S2, Figure S2-1).

In the next step of data analysis, a heatmap was created. This method is very effective in drawing attention to strains characterized by rich patterns of accumulated metabolites and assessing potential relationships between the strains (Figure 2). Indeed, the heatmap has clustered a group of three strains (B-1529, B-256, B-287) out of the majority of other cyanobacteria. The strains were assigned to the haloalkaliphilic and natronophilic ecological group and were highlighted by the highest number of accumulated compounds. It is worth remembering that these strains were also selected by PCA as the most distinguished ones (i.e., characterized with the highest intergroup variability) from other strains. Further clustering of the remaining majority of strains resulted in the isolation of the two cold-tolerant strains (B-1533 and B-1535) from a cluster of the remaining seven strains representing groups resistant to desiccation, high temperature, and haloalkaliphilic and natronophilic.

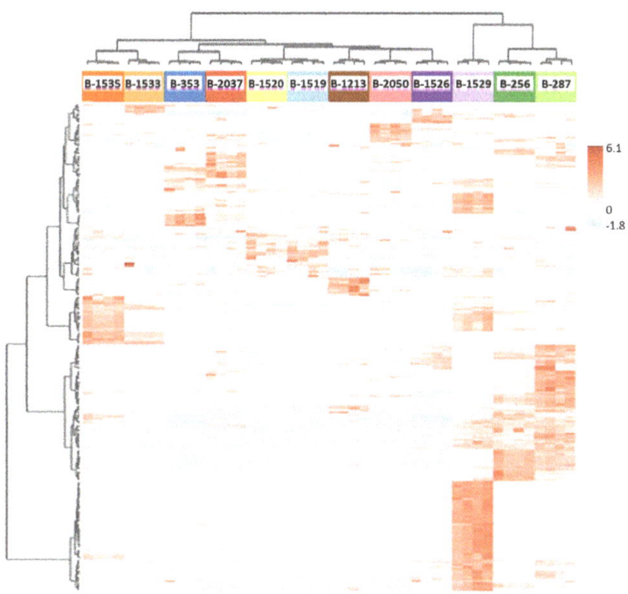

Figure 2. Heatmap representing the overall view of relative abundances of 530 metabolite features (both annotated and nonannotated) detected by GC-MS and LC-MS in polar extracts of twelve studied strains of extremophilic cyanobacteria. Each color marks four samples attributed to the indicated specified cyanobacteria strain. The strain descriptions are given in Table 1 and Figure 1.

Overall, the obtained results indicated similarities and distinctions in metabolite profiles among the studied cyanobacterial strains. Despite the lengthy period during which cyanobacterial strains have been cultivated under storage conditions, it was observed that metabolomic similarities were more characteristic of strains originating from the same ecological extreme habitats. In contrast, differences in metabolite profiles were more characteristic of strains from ecologically different extreme habitats. Nevertheless, the exception to this regularity was a few strains of the haloalkaliphilic and natronophilic ecological group which demonstrated distinct clustering between themselves and strains of other extreme habitats. Among them, three strains, B-1529, B-256 and B-287, contained the highest number of accumulated metabolites.

2.4. Identifying Patterns of Strain Differences in Primary Metabolome Associated with Cyanobacteria Inhabiting Extreme Environments

To elucidate the key metabolites that could potentially be involved in the adaptations formed by cyanobacteria under extreme conditions, a comparative pairwise analysis of the primary metabolome of each of the strains studied was performed with the metabolomes of all other strains. To facilitate the identification of regularities that may be observed in metabolite patterns and be associated with a strain's extreme habitat, the comparative analysis results are described for strains grouped according to their extreme habitats.

2.4.1. Desiccation-Tolerant Cyanobacteria Strains B-1520 and B-1519

PCA performed for samples of two groups, where the first one was attributed to strain B-1520 and the second to the samples of all other strains studied, did not show their clear clustering (Supplementary S2, Figure S2-2A). In contrast, PLS-DA, a statistical method that enhances intergroup variance and decreases the impact of in-group variance, demonstrated the separation of the B-1520 group from the sample population of other strains (Figure S2-2B). This analysis also identified some metabolites in B-1520 samples that contribute to the differentiation of this strain from other cyanobacteria (Figure S2-2C). To select metabolites of B-1520 exerting a significant difference in abundance in comparison with other strains, Volcano plot analysis was applied. This analysis indicated six metabolite features. Further, taking into consideration the different variances of compared groups and expected non-normally distributed data, the significance of the difference between groups in the abundance of metabolites indicated by Volcano analysis was additionally estimated by Welch's *t*-test and the Mann–Whitney U-test. This approach resulted in four metabolite features (salicylic acid, erythritol, ethanolamine and an unidentified compound with RI1501) showing a marked increase in relative content in B-1520 compared to their average levels in other strains (Table 2, Supplementary S2, Table S2-3). Among these compounds, salicylic acid demonstrated the most substantial fold difference (14-fold, $p < 0.001$), underscoring its potential role in the strain's metabolic uniqueness.

PCA performed with respect to strain B-1519 did not reveal clustering with other strains. As in the previous case of B-1520, only the PLS-DA was able to detect differences important for separating strain B-1519 from other samples (Supplementary S2, Figure S2-3A–C). The Volcano plot with subsequent significance validation of its results with Welch's and Mann–Whitney tests revealed four compounds present at higher levels in this strain than in others (Table 2, Supplementary S2, Figure S2-3D, Table S2-3). Among these metabolites, the highest difference in abundance was found for erythritol. The level of the compound was 6.5-fold higher compared to its average content in other strains.

Interestingly, the revealed metabolite patterns for the desiccation-tolerant strains B-1519 and B-1520 shared a common propensity to accumulate two specific substances, namely erythritol and the unidentified compound RI1501.

Table 2. Patterns of major metabolites * associated with desiccation-tolerant cyanobacterial strains *N. commune* B-1520 and B-1519.

Metabolite [a]	Chemical Class [b]	TMS Derivative (Feature) [c]	Strains Tolerant to Desiccation				Method
			B-1520		B-1519		
			FC	p [d]	FC	p	
Salicylic acid	Ph	2TMS	14	<0.001			GC-MS
Erythritol	P	4TMS	4.4	0.02	6.5	<0.001	GC-MS
Ethanolamine	A	3TMS	4.2	<0.001			GC-MS

* The metabolites were detected by GC-MS and LC-MS in polar extracts of cyanobacteria cultivated under storage conditions and showed significant differences (FC $\geq$ 2, $p \leq 0.05$) in relative content in specified cyanobacterial strains compared to the average content of these metabolites in all extremophilic strains studied. [a] Only structurally annotated metabolites are presented. The whole pattern including nonannotated compounds (unknowns) is presented in Supplementary S2, Table S2-3. The metabolites are arranged according to their chemical structure: [b] phenolic compound (Ph), polyol (P), amine (A). [c] Specific metabolite features representing each metabolite. TMS—trimethylsilyl group. [d] The Student's *t*-test *p*-value was calculated with the false discovery rate (FDR) obtained according to the Benjamini–Hochberg approach and did not exceed the *p*-value threshold 0.05 after applying the FDR correction. In addition, Welch's *t*-test ($p \leq 0.05$) and the Mann–Whitney U-test ($p \leq 0.01$) were used to estimate the significance of the difference between the compared groups for each metabolite reported here and are presented in Supplementary S2, Table S2-3.

2.4.2. Strains Tolerant to High or Cold Temperatures

When comparing metabolomes of thermotolerant B-1213 against other extremophilic and extremotolerant strains, only PLS-DA, but not PCA, was effective in clustering, and therefore in finding meaningful differences in its metabolite profiles from other strains (Supplementary S2, Figure S2-4A–C). Further application of a Volcano plot supplemented with additional abovementioned statistical tests allowed the identification of nine metabolites in this strain characterized by enhanced abundances compared to the average content of these metabolites in other strains (Figure S2-4D, Table S2-4). In this pattern, the following metabolites displayed the highest relative levels: ADP-ribose (27-fold), RI3462 trisaccharide (25-fold), unknown compound RI1022 (56-fold) (Table 3, Supplementary S2, Table S2-4).

In similar comparative analyses performed for cold-tolerant strains B-1533 and B-1535, again, PLS-DA, but not PCA, was able to efficiently identify metabolite profiles that distinguished those strains from other groups. Univariate statistical analysis revealed 15 metabolites (represented by 19 features) in B-1533 and 9 metabolites in B-1535 with significantly elevated abundances compared to the average levels of those compounds observed in other strains (Table 3, Supplementary S2, Figures S2-5 and S2-6, Table S2-4). This pattern for B-1533 was dominated by monosaccharides and their phosphate derivatives, whereas the pattern for B-1535 included an equal number of representatives of different chemical classes. The highest (at least 5-fold higher than in the group of samples from other strains) relative abundance levels were characteristic as follows: for B-1533, 2C-methyl-*D*-erythritol (17-fold), glucose (12-fold), nonadecan-1-ol (9-fold), ureidosuccinic acid (9-fold), aconitic acid (9-fold) and digalacturonic acid (5-fold); for B-1535, mannitol (10-fold), ureidosuccinic acid (10-fold), cGMP (6-fold) and MEP (5-fold). The cold-tolerant strains accumulated four common metabolites, namely 3-dehydroshikimic acid, digalacturonic acid, ureidosuccinic acid and MEP (Table 3). However, no shared metabolites were found between high- and low-temperature-tolerant strains.

Table 3. Patterns of major metabolites * associated with each specified cyanobacterial strain B-1213, B-1533 and B-1535 adapted to extreme temperature environment.

Metabolite [a]	Chem. Class [b]	Feature [c]	Strains Tolerant to Extreme High or Low Temperatures						Method
			Dolichospermum sp., B-1213 (Thermotolerant)		*Anabaena* cf. *pirinica*, B-1533 (Cold-Tolerant)		*Anabaena* sp., B-1535 (Cold-Tolerant)		
			FC	*t*-Test *p* [d]	FC	*t*-Test *p*	FC	*t*-Test *p*	
Glycolic acid	CA						2.4	<0.001	LC-MS
Aconitic acid	CA				9	<0.001			LC-MS
3-Dehydroshikimic acid	CA				2.3	0.001	2.3	0.03	LC-MS
Glucose	S	1MEOX, 5TMS (1)			12	<0.001			GC-MS
		1MEOX, 5TMS (2)			8.4	<0.001			GC-MS
Mannitol	P	6TMS					10	<0.001	GC-MS
2C-methyl-*D*-erythritol	P				17	<0.001			LC-MS
Digalacturonic acid	SA				5.1	<0.001	3.8	0.002	LC-MS
Glucose 1-phosphate	SP		4	<0.001					LC-MS
					5.9	<0.001			LC-MS
Glucose 6-phosphate		1MEOX, 6TMS (1)			5.9	<0.001			GC-MS
		1MEOX, 6TMS (2)			6.1	<0.001			GC-MS
Fructose 6-phosphate	SP				6.1	<0.001			LC-MS
		1MEOX, 6TMS			5.1	<0.001			GC-MS
2-Keto-3-deoxy-6-phosphogluconate	SP				6	<0.001			LC-MS
2C-methyl-*D*-erythritol-4-phosphate (MEP)	SP				4.2	0.03	5.4	<0.001	LC-MS
Ribulose 5-phosphate /Xylulose 5-phosphate	SP				3.6	<0.001			LC-MS
Phosphate	P_n		4	<0.001					LC-MS
Glutamic acid	AA		5.3	<0.001					LC-MS
Ureidosuccinic acid	AAd				9.3	<0.001	9.7	<0.001	LC-MS
NADH	Nuc		4.2	0.008					LC-MS
cGMP	Nuc						5.8	<0.001	LC-MS
ADP-ribose	NucS		27	<0.001					LC-MS
2-Hydroxypyridine	ON	1TMS	5.6	<0.001					GC-MS
Dihydroorotic acid	ON						4.4	<0.001	LC-MS
Nonadecan-1-ol	FAl	1TMS			9.4	<0.001			GC-MS

* The metabolites were detected by GC-MS and LC-MS in polar extracts of cyanobacteria cultivated under storage conditions and showed significant differences (FC $\geq$ 2, $p \leq 0.05$) in relative content in specified cyanobacterial strains compared to the average content of these metabolites in all extremophilic strains studied. [a] Only structurally annotated metabolites are presented and arranged according to their chemical structure: [b] carboxylic acids (CAs), sugars (Ss), polyols (Ps), sugar-derived acids (SAs), sugar phosphates and derivatives (SPs), phosphate (P_n), amino acids (AAs) and derivatives (AAds), nucleotides (Nucs), sugar nucleotides (NucSs), other N-containing compounds (ONs), fatty alcohol (FAl). [c] The features representing each metabolite are specified with characteristic methyloxime (MEOX) and trimethylsilyl (TMS) derivatization groups. The whole pattern including compounds annotated to chemical class and nonannotated compounds (unknowns) is presented in Supplementary S2, Table S2-4. [d] The Student's *t*-test *p*-value was calculated with the FDR obtained according to the Benjamini–Hochberg approach [42]. Welch's *t*-test ($p \leq 0.05$) and the Mann–Whitney U-test ($p \leq 0.01$) additionally applied to estimate the significance of the difference between the compared groups for each metabolite reported here are presented in Supplementary S2, Table S2-4.

2.4.3. Halo(alkali)philic and Natronophilic Cyanobacteria Strains

The group of halo(alkali)philic and natronophilic strains of cyanobacteria was separated into two subgroups according to the PCA results (Figure 1). The first included strains characterized by metabolome similarities, while the second one included strains displaying significant metabolome differences. Therefore, to better reflect metabolites contributing to the observed metabolome similarities and differences in the respective subgroups, the results of the comparative analysis for the strains of these two subgroups were also considered separately.

Strains Demonstrating Similarity in Metabolomes

For the four strains (B-2037, B-353, B-2050 and B-1526) of the first subgroup of halo(alkali)- and natronophilic cyanobacteria, PCA performed to distinguish each specified strain did not reveal (with an exception for B-2037) essential differences in their metabolite profiles compared to other cyanobacteria strains (Supplementary S2, Figures S2-7A–S2-10A). However, employing the PLS-DA method confirmed the uniqueness of each of these strains by clustering metabolomes of their samples from the sample population of other strains (Figures S2-7B,C–S2-10B,C). For each of these strains, further univariate analysis revealed a pattern of specific metabolites, which mainly exhibited significantly (*t*-test $p \leq 0.05$, U-test $p \leq 0.01$) more than 2-fold increased abundances in the specified strains compared to the average level of the compounds in other investigated cyanobacteria (Figures S2-7D,E–S2-10D,E). Thus, the pattern with the highest number of strain-distinguished metabolites within this subgroup, namely eleven, was found in strain B-2037 (Figure S2-7D–F, Table S2-5). In this pattern, ten metabolites displayed 2–10-fold accumulation in the strain. The abundance of another item (3-hydroxypiruvate) was significantly lower than the average level of the compounds in other strains. Among the accumulated metabolites, nucleotides and sugars dominated (Table 4, Supplementary S2, Figure S2-7E,F). The highest abundance levels were found for glucosylglycerol (10-fold) and monounsaturated *trans*-9-hexadecenoic acid (9-fold) (Table 4, Supplementary S2, Table S2-5).

The metabolite patterns associated with other strains of this halo(alkali)philic and natronophilic cyanobacteria subgroup such as B-353, B-2050 and B-1526 included eight, nine and six compounds, respectively. For strain B-353, the metabolite displaying the highest abundance was adenosine (Figure S2-8E). The relative level of this compound was 41-fold higher its average content in other cyanobacteria (Table 4). Other metabolites accumulated in B-353 more than 5-fold were three nonannotated compounds, RI1349, RI1352 and RI1219 (Table S2-5).

Table 4. Patterns of major metabolites * associated with halo(alkali)philic and natronophilic cyanobacteria strains from subgroup without appearing considerable intergroup variability (B-2037, B-353, B-2050 and B-1526).

Metabolite [a]	Chem. Class [b]	Feature [c]	*Sodalinema orleanskyi*, B-2037		*S. gerasimenkoae*, B-353		*S. stalii*, B-2050		*Limnospira* sp., B-1526		Method
			FC	*t*-Test [d]	FC	*t*-Test *p*	FC	*t*-Test *p*	FC	*t*-Test *p*	
3-Hydroxypyruvate	CA		0.34	0.04							LC-MS
Lactic acid	CA	2TMS							3	0.01	GC-MS
3-Hydroxybutyric acid	CA	2TMS							20	<0.001	GC-MS
Isocitric acid	CA		4.9	<0.001			4.9	<0.001			LC-MS
Shikimic acid	CA				4.4	<0.001					LC-MS
2-Phosphoglycolate	CAP		6.1	<0.001							LC-MS
Ribose	S	1MEOX, 4TMS					4.1	<0.001			GC-MS
RI2255 Glucosylglycerol 1	S	6TMS	10	<0.001							GC-MS
RI2310 Glucosylglycerol 2	S	6TMS							8.5	<0.001	GC-MS
Ribonic acid	SA						6.5	<0.001			LC-MS
D-Erythronic acid	SA	4TMS					2.8	0.001			GC-MS
Galacturonic acid	SA						13	<0.001			LC-MS
Glycerol	P	3TMS					3.5	<0.001			GC-MS
Sedoheptulose-1,7-biphosphate	SP		5.6	<0.001							LC-MS

Table 4. *Cont.*

Metabolite [a]	Chem. Class [b]	Feature [c]	*Sodalinema orleanskyi*, B-2037		*S. gerasimenkoae*, B-353		*S. stalii*, B-2050		*Limnospira* sp., B-1526		Method
			FC	*t*-Test [d]	FC	*t*-Test *p*	FC	*t*-Test *p*	FC	*t*-Test *p*	
Pyroglutamic acid	AA	2TMS							2.7	0.01	GC-MS
NADPH	Nuc						8.6	<0.001			LC-MS
ADP	Nuc		5.7	<0.001							LC-MS
CDP	Nuc		5.8	<0.001							LC-MS
GDP	Nuc		4.9	<0.001							LC-MS
Adenosine	Nuc				41	<0.001					LC-MS
Orotic acid	ON						5.0	<0.001			LC-MS
trans-9-Hexadecenoic acid	FA	1TMS	9.1	<0.001							GC-MS
Phosphoric acid	IO	3TMS	2.2	0.03							GC-MS
Chloride	IO		2.6	<0.001	2.5	<0.001					LC-MS

* The metabolites were detected by GC-MS and LC-MS in polar extracts of cyanobacteria cultivated under storage conditions and showed significant differences (FC ≥ 2, $p \leq 0.05$) in relative content in specified cyanobacterial strains compared to the average content of these metabolites in all extremophilic strains studied. [a] Only structurally annotated metabolites presented and arranged according to their chemical structure: [b] carboxylic acids (CAs) and their phosphate derivatives (CAPs), sugars (Ss), sugar-derived acids (SAs), polyols (Ps), sugar phosphates (SPs), amino acids (AAs), nucleotides and nucleosides (Nucs), other N-containing compounds (ONs), fatty acids (FAs), inorganic ions (IOs). [c] The features representing each metabolite are specified with characteristic methyloxime (MEOX) and trimethylsilyl (TMS) derivatization groups. The whole pattern including compounds annotated to chemical class and unknowns is presented in Supplementary S2, Table S2-5. [d] The Student's *t*-test *p*-value was calculated with the FDR obtained according to the Benjamini–Hochberg approach [42]. Welch's *t*-test ($p \leq 0.05$) and the Mann–Whitney U-test ($p \leq 0.01$) additionally applied to estimate the significance of the difference between the compared groups are presented in Supplementary S2, Table S2-5.

In the pattern of strain B-2050, the nine metabolite features that exhibited significant enrichment were isocitric acid, glycerol and representatives of sugars, nucleobases, nucleotides and their derivatives, and the unknown compound RI3201. Among these compounds, more than 5-fold accumulation was observed for digalacturonic acid (13-fold), NADPH (8.8-fold), ribonic acid (6.5-fold) and orotic acid (5-fold). The metabolic profile of strain B-1526 comprised carboxylic acids (lactic acid, 3-hydroxybutyric acid), sugars and derivatives (glucosylglycerol, RI3124 disaccharide-derivate) and pyroglutamic acid. Among them, the highest abundance in comparison with other cyanobacteria was observed for 3-hydroxybutyric acid (20-fold). Another metabolite with an 8.5-fold increase above its average level was glucosylglycerol (Table 4). Interestingly, among these four strains considered, the observed specific metabolite patterns had only a couple of common metabolites: glucosylglycerol, with elevated contents in B-2037 and B-1526, and chloride, whose relative contents were found to be the highest in B-2037 and B-353 (Table 4, Supplementary S2, Table S2-5).

Cyanobacterial Strains Demonstrating Considerable Differences in Metabolomes

PCA and particular PLS-DA performed in regard to strain B-1529 showed clustering metabolomes of its samples from the majority of other cyanobacterial strains (Supplementary S2, Figure S2-11A–C). This indicates substantial differences in metabolite profiles between these compared groups. Indeed, further investigation using Volcano analysis and statistical Welch's and Mann–Whitney tests revealed 67 metabolites (presented by 76 features) significantly enriched in strain B-1529 compared to their average levels found in other cyanobacterial strains. The compounds were mainly representatives of amino acids, sugars, sugar phosphates, fatty acids, nucleosides and nucleosides (Table 5, Supplementary S2, Figure S2-11D–F, Table S2-6). Interestingly, about half of the compounds displayed more than 10-fold accumulation in this strain.

Among them, there were 14 amino acids. The amino acids exhibiting the highest abundances (30–80-fold) were Try, Gln, The, Phe, Met.

Table 5. Patterns of major metabolites * associated with specified haloalkaliphilic and natronophilic cyanobacteria strains B-1529, B-287 and B-256 demonstrating considerable intergroup metabolome variability.

Metabolite [a]	Chem. Class [b]	Feature [c]	*Nodularia* sp., B-1529		*Limnospira* sp., B-287		*Limnospira* sp., B-256		Method
			FC	*t*-Test *p* [d]	FC	*t*-Test *p*	FC	*t*-Test *p*	
3-Phosphoglyceric acid	CAP				14	<0.001			LC-MS
2-Phosphoglyceric acid	CAP				18	<0.001			LC-MS
Phosphoenolpyruvic acid	CAP				27	<0.001			LC-MS
Fructose	S	1MEOX, 5TMS (1)	32	<0.001					GC-MS
		1MEOX, 5TMS (2)	35	<0.001					GC-MS
Mannose	S	1MEOX, 5TMS	17	<0.001					GC-MS
Galactose	S	1MEOX, 5TMS	17	<0.001					GC-MS
Glucose	S	1MEOX, 5TMS	14	<0.001					GC-MS
α,α-Trehalose	S	8TMS			26	<0.001			GC-MS
D-galactonic acid	SA		13	<0.001					LC-MS
1-Deoxyxylulose 5-phosphate	SP						13	<0.001	LC-MS
2-Deoxyribose 5-phosphate	SP						12	<0.001	LC-MS
Sedoheptulose 7-phosphate	SP		11	<0.001					LC-MS
Glucosamine 6-phosphate	SP						14	<0.001	LC-MS
Alanine	AA	2TMS	10	<0.001					GC-MS
Arginine	AA		17	<0.001					LC-MS
Argininosuccinate	AA		11	<0.001					LC-MS
Aspartic acid	AA		11	<0.001					LC-MS
		3TMS	25	<0.001					GC-MS
Serine	AA		25	<0.001					LC-MS
Threonine	AA		46	<0.001					LC-MS
Valine	AA	2TMS	16	<0.001					GC-MS
Isoleucine	AA	2TMS	25	<0.001					GC-MS
Glycine	AA		15	<0.001					LC-MS
		2TMS	24	<0.001					GC-MS
Phenylalanine	AA		43	<0.001					LC-MS
Tryptophan	AA		33	<0.001					LC-MS
Tyrosine	AA	2TMS	30	<0.001					GC-MS
		3TMS	12	<0.001					GC-MS
Methionine	AA		80	<0.001					LC-MS
Glutamine	AA		34	<0.001					LC-MS
Cytidine	Nuc		14	<0.001					LC-MS
Uridine	Nuc		21	<0.001					LC-MS
Guanosine	Nuc		22	<0.001					LC-MS
2′-Deoxyguanosine	Nuc		19	<0.001					LC-MS
ATP	Nuc				10	<0.001			LC-MS
GTP	Nuc				10	<0.001			LC-MS
3-Ureidopropionic acid	ON		74	<0.001					LC-MS
Uric acid	ON		42	<0.001					LC-MS
Isovaleryl-CoA	CoAt		13	<0.001					LC-MS
Acetoacetyl-CoA	CoAt		13	<0.001					LC-MS
S-acetyl-CoA	CoAt				11	<0.001	7.0	<0.001	LC-MS

Table 5. *Cont.*

Metabolite [a]	Chem. Class [b]	Feature [c]	*Nodularia* sp., B-1529		*Limnospira* sp., B-287		*Limnospira* sp., B-256		Method
			FC	*t*-Test *p* [d]	FC	*t*-Test *p*	FC	*t*-Test *p*	
trans-9-Octadecenoic acid	FA	1TMS	18	<0.001					GC-MS
Linoleic acid	FA	1TMS	18	<0.001					GC-MS
Oleic acid	FA	1TMS	30	<0.001					GC-MS
cis-9-Hexadecenoic acid	FA	1TMS	14	<0.001					GC-MS
delta3-isopentenyl pyrophosphate	PP				10	<0.001			LC-MS
Dimethylallylpyrophosphate	PP				10	<0.001			LC-MS

* Metabolites were detected by GC-MS and LC-MS in polar extracts of cyanobacteria cultivated under storage conditions and showed significant differences ($FC \geq 10$, $p \leq 0.001$) in relative content in specified cyanobacterial strains compared to the average content of these metabolites in all extremophilic strains studied. [a] Only structurally annotated metabolites exhibiting a considerable increase (≥10-fold) in abundance are presented and arranged according to their chemical structure: [b] carboxylic acid phosphate derivatives (CAPs), sugars (Ss), sugar-derived acids (SAs), sugar phosphates (SPs), amino acids (AAs), nucleotides and nucleosides (Nucs), other N-containing compounds (ONs), CoA thioesters (CoAts), fatty acids (FAs), prenyl phosphates (PPs). [c] Metabolite features representing each metabolite are specified with characteristic MEOX and TMS derivatization groups. The whole pattern including all compounds (annotated and nonannotated) demonstrating ≥ 2-fold difference (Student's *t*-test *p* (FDR-adjusted) ≤ 0.05, Welch's *t*-test $p \leq 0.05$, Mann–Whitney U-test $p \leq 0.01$) in abundance compared to their average levels is presented in Supplementary S2, Table S2-6. [d] The Student's *t*-test *p*-value was calculated with the FDR obtained according to the Benjamini–Hochberg approach [42].

Similar to B-1529, other two strains of this haloalkaliphilic and natronophilic subgroup, namely B-287 and B-256, also demonstrated apparent clustering from the major population of compared cyanobacterial metabolomes in corresponding PCA and PLS-DA models (Supplementary S2, Figures S2-12 and S2-13A–C). However, metabolite patterns revealed by univariate analysis for each of the cyanobacterial strains contained a fewer strain-distinguished compounds than B-1529. Thus, for the strains B-287 and B-256, the patterns included 53 and 25 metabolites, respectively (Table S2-6).

In the metabolite pattern associated with B-287, nucleotides, carboxylic acids and C2-, C3-phosphates dominated (Figure S2-12E, Table S2-6). In this strain, the levels of ten metabolites were over 10-fold higher compared to their average levels of other investigated strains. Among them, the most accumulated compounds were phosphoenolpyruvate (27-fold), trehalose (26-fold), a few participants of glycolysis such as 3- and 2-phosphoglycerates (14- and 18-fold) and an unidentified sugar (RI2304, 12-fold) (Table 5, Supplementary S2, Table S2-6).

The metabolite pattern associated with B-256 was dominated by nucleotides and sugar phosphates (Supplementary S2, Figure S2-13E, Table S2-6). In this pattern, there were only three metabolites, representatives of sugar phosphate group, namely 2-deoxyribose 5-phosphate, 1-deoxyxylulose 5-phosphate and glucosamine 6-phosphate, whose abundance levels exceeded the average content of the compounds in other strains by more than 10-fold (Table 5, Supplementary S2, Table S2-6).

Thus, the indicated metabolite patterns for each strain from this subgroup of haloalkaliphilic and natronophilic cyanobacteria showed substantial differences in the composition and abundance levels of individual metabolites. This may point to specific metabolic strategies that these strains specifically adapted to survive under extreme conditions of high salinity and alkalinity. Nevertheless, these strains had also some degree of overlap in their metabolite patterns. For example, patterns of strains B-287 and B-256 had twelve shared metabolites, mainly represented by sugar phosphates, nucleotides and carboxylic acids (Supplementary S2, Table S2-6). Other pairs of strains such as B-287 and B-1529 had five common metabolites among which nucleotides dominated, while strains B-256 and B-1529 shared just one metabolite, dTMP. Importantly, no metabolites were found to be common to the patterns of the all three strains (B-1529, B-256, B-287), indicating a substantial difference in the metabolite patterns of the studied strains of this subgroup.

Overall, among all twelve strains considered here, B-1529 was characterized by the richest pattern of metabolites, the majority of which were representatives of amino acids. This abundance of amino acids is probably accounted for by its nitrogen-fixing ability. Notably, B-1529 is the sole diazotrophic heterocystous strain that represents the order *Nostocales* among the haloalkaliphiles and natronophiles. Here, it is important to remember that other nitrogen-fixators among the strains considered here, belonging to the desiccation- and extreme-temperature-tolerant ecological groups, were also heterocystous representatives of the order *Nostocales*. Furthermore, out of the all heterocystous diazotrophs, strain B-1529 demonstrated the most intensive accumulation (>10-fold) of rich patterns of amino acids and other many metabolites. Thus, strain B-1529 exhibits uniqueness in the diversity of enriched metabolites and in the magnitude of their enrichment.

In contrast, among the strains considered here, B-1520 from the desiccation-tolerant group (exhibiting the same heterocyst-forming capacity as B-1529) was characterized by the scarcest strain-specific pattern, containing only four metabolites.

2.5. *Identifying Primary Metabolomic Characteristics Associated with Heterocyst-Forming Cyanobacterial Strains*

As the PCA separated the six diazotrophic heterocystous strains (B-1520, B-1519, B-1213, B-1533, B-1535, B-1529) from the other strains examined in this study (Figure 1), and these nitrogen-fixing cyanobacteria appeared to be very distinct in terms of the numbers of strain-specific metabolites, this prompted us to ascertain the metabolomic signatures that differentiate the heterocystous strains and strains unable to form heterocysts. Towards this end, the primary metabolomes of these two groups of strains were compared using Volcano analysis with predefined thresholds of FC ≥ 2 and Student's *t*-test *p*-value (FDR-adjusted) ≤ 0.05. This comparison revealed a rich pattern of metabolites, among which 23 and 41 compounds displayed higher and lower abundances, respectively, in the heterocyst-forming strains (Table 6, Supplementary S2, Table S2-7). It is noteworthy that sugars, polyols and sugar phosphates, but not nitrogen-containing compounds, were predominant among metabolites displaying enhanced abundance in the heterocystous diazotrophs. The nitrogen-containing group of metabolites was represented by only four compounds, including ethanolamine, tyrosine, argininosuccinate and ureidosuccinate. Notably, the levels of the last two of these metabolites increased considerably (28- and 37-fold) in the heterocyst-forming strains (Table 6). In addition, the presence of a variety of nucleotides among compounds displaying decreased abundances was unexpected. The relative contents of the nucleotides were 3–17-fold lower in heterocystous than in non-heterocystous strains.

Other metabolites whose levels were found to be substantially elevated in the nitrogen-fixing heterocyst-forming cyanobacteria included salicylic acid (18-fold) and metabolites of isoprenoid backbone (MEP) biosynthesis such as 2C-methyl-*D*-erythritol (43-fold) and 2C-methylerythritol 4-phosphate (31-fold). Conversely, the levels of two other MEP biosynthesis downstream metabolites, namely HMBDP and isopentenyl pyrophosphate, were 8.9- and 10-fold lower, respectively, in these strains. Furthermore, the levels of C2-, C3-carboxylic acid phosphates, sugar biphosphates, trehalose, and glucosylglycerol were found to be more than fivefold lower in heterocystous than in non-heterocystous strains (Table 6).

Table 6. Patterns of major differentially expressed metabolites * in heterocyst-forming and non-heterocystous strains.

Metabolite [a]	Chem. Class [b]	Feature [c]	FC	*t*-Test *p* [d]	Method
Metabolites with higher relative content in heterocyst-forming compared to non-heterocystous strains					
Salicylic acid	CA	2TMS	18	0.009	GC-MS
Glucose	S		5.5	0.030	LC-MS
Mannose	S	1MEOX, 5TMS	6.8	0.031	GC-MS
Mannitol	P	6TMS	5.0	0.031	GC-MS
2C-methyl-*D*-erythritol	P		43	0.015	LC-MS
Fructose 6-phosphate	SP		5.3	0.010	LC-MS
Glucose 6-phosphate	SP		5.0	0.011	LC-MS
Sedoheptulose 7-phosphate	SP		28	0.002	LC-MS
2C-methylerythritol 4-phosphate	SP		31	<0.001	LC-MS
2-keto-3-deoxy-6-phosphogluconate	SAP		5.1	0.015	LC-MS
Tyrosine	AA	2TMS	7.2	0.027	GC-MS
Argininosuccinate	AA		28	0.002	LC-MS
Ureidosuccinic acid	AA		37	0.008	LC-MS
Nonadecan-1-ol	FAl	1TMS	5.7	0.007	GC-MS
Metabolites with lower relative content in heterocyst-forming compared to non-heterocystous strains					
2-phosphoglycolic acid	CAP		14	<0.001	LC-MS
2-phosphoglyceric acid	CAP		39	0.009	LC-MS
3-phosphoglyceric acid	CAP		24	0.005	LC-MS
Phosphoenolpyruvic acid	CAP		25	0.024	LC-MS
α,α-Trehalose	S	8TMS	65	0.017	GC-MS
RI2255 Glucosylglycerol	S		>100	<0.001	GC-MS
Fructose-1,6-diphosphate	SP		27	0.001	LC-MS
Sedoheptulose-1,7-biphosphate	SP		9.4	0.001	LC-MS
Adenylosuccinic acid	Nuc		6.5	0.004	LC-MS
GTP	Nuc		17	0.002	LC-MS
ATP	Nuc		22	0.001	LC-MS
CTP	Nuc		5.6	<0.001	LC-MS
UTP	Nuc		7.2	0.004	LC-MS
GDP	Nuc		5.0	0.001	LC-MS
ADP	Nuc		6.6	0.002	LC-MS
dTTP	Nuc		5.5	0.013	LC-MS
ADP-glucose	NucS		7.4	0.019	LC-MS
S-acetyl CoA	CoAt		10	0.016	LC-MS
trans-9-Hexadecenoic acid	FA	1TMS	5.7	0.008	GC-MS
(2*E*)-4-hydroxy-3-methylbut-2-en-1-yl diphosphate (HMBDP)	PP		8.9	0.003	LC-MS
Δ^3-isopentenyl pyrophosphate	PP		10	0.001	LC-MS

* GC-MS- and LC-MS-detected polar metabolites from extracts of cyanobacteria cultivated under storage conditions showing significant differences (FC ≥ 5, $p \leq 0.05$) in relative content in heterocyst-forming (B-1520, B-1519, B-1213, B-1533, B-1535, B-1529; $n = 24$) compared to non-heterocystous (B-353, B-2050, B-2037, B-1526, B-287, B-256; $n = 24$) extremophilic strains studied. [a] Only structurally annotated metabolites exhibiting a considerable difference ($\geq$5-fold) in abundance between compared groups are presented and arranged according to their chemical structure: [b] carboxylic acids (CAs) and their phosphate derivatives (CAPs), sugars (Ss), polyols (Ps), sugar phosphates (SPs), sugar-derived acids (SAs) and their phosphate derivatives (SAPs), amino acids (AAs), nucleotides and nucleosides (Nucs), sugar nucleotides (NucSs), CoA thioesters (CoAts), fatty acids (FAs) and their alcohols (FAls), prenyl phosphates (PPs). [c] Metabolite features representing each metabolite are specified with characteristic TMS and MEOX derivatization groups. The whole pattern including all compounds (annotated and nonannotated) demonstrating $\geq$ 2-fold difference (Student's *t*-test *p* (FDR-adjusted) ≤ 0.05) in abundance is presented in Supplementary S2, Table S2-7. [d] The Student's *t*-test *p*-value was calculated with the FDR obtained according to the Benjamini–Hochberg approach [42].

2.6. Pathway Analysis

In view of the considerable range of metabolites accumulated by the haloalkaliphilic strains B-1529, B-287 and B-256, it is important to identify the molecular mechanisms that are most significant in maintaining the life processes of these cyanobacterial strains and their dependence on high-Na^+-concentration and high-pH conditions. For this purpose, the identified primary metabolites from observed strain-specific metabolite patterns were merged into a matrix and underwent pathway analysis. This analysis combines pathway enrichment (global test) and pathway topology (relative centrality test) analyses to highlight the pathways via which the selected metabolites might support the life of haloalkaliphilic cyanobacteria strains. The pathway analysis was based on the *Synechococcus elongatus* PCC7942 pathway library available from the KEGG web resource. It is important to note that several identified substances, which are characteristic of B-1529 (glucosylglycerol and methylmalonyl-CoA) and of B-287 (*D*-erythronic acid), were not included in the results of the metabolic pathway analysis. The reason is that these substances are not available or have not been identified in the KEGG database. A further discussion of their role in cyanobacterial metabolism will be based on an analysis of relevant literature.

Many significant ($p \leq 0.05$) pathways were identified for each of the haloalkaliphilic strains (B-1529, B-287 and B-256) (Supplementary S2, Figure S2-14). Among them, two pathways with the highest PI values as 1 were found to be common in at least two of the three strains: galactose metabolism for strains B-1529 and B-256, inositol phosphate metabolism for strains B-287 and B-256. This information on common important metabolomic pathways found points to possible common adaptations that evolved in similar extreme habitats. However, the most important pathways as indicated by the highest values of PI = 1 and $-\log10(p) > 5$ were found to be unique in the indicated strains as follows: alanine, aspartate and glutamate metabolism in B-1529; butanoate metabolism in B-287; and pentose and glucuronate interconversions in B-256 (Supplementary S2, Table S2-8, Figure S2-14). Each metabolic pathway determined to be highly reliable is associated with a group of crucial metabolites that play a significant role in these pathways (Supplementary S2, Table S2-8).

3. Discussion

Twelve cyanobacterial strains investigated in this work were isolated from ecologically various extreme environments. These cyanobacteria are extremophilic or extremotolerant depending on the location of their optimal growth range and how wide it is in relation to environmental parameters (temperature, water availability, salinity and alkalinity) [4]. Namely, out of the twelve strains, six were haloalkaliphilic (*Sodalinema* sps (B-2037, B-353), *Limnospira* sps (B-1526, B-287, B-256), *Nodularia* sp. B-1529) being isolated from salt alkaline and soda lakes and one was halophilic as being a marine strain (*Sodalinema stalii*, B-2050). Having evolved under conditions of high salinity, these organisms formed salt tolerance which made them salt dependent [43]. Indeed, these strains cannot survive without high Na^+ concentrations as their proteins lose stability [44]. The marine strain originating from coastal shoals, an intertidal environment with moderately fluctuating salinity, is capable of tolerating this range of salinities. In addition, this strain, like all marine strains, requires not only Na, but also $C1^-$, Mg^{2+} and Ca^{2+} [39]. Furthermore, salt alkaline and soda lakes are characterized by not only various salinity fluctuations, but also high concentrations of HCO_3 and $CO_3{}^{2-}$ and pH 9–10 [31]. Therefore, the haloalkaliphilic strains are restricted to narrow ranges of alkalinity fluctuations. Given these considerations, the cultivation media for the halo(alkali)philic strains were selected to provide optimal conditions for photoautotrophic growth. However, to maintain long-term storage, the cyanobacteria were kept in low light conditions and without additional CO_2 supply. Thus, this group of cyanobacteria presents Na^+- and alkalinity (except marine strain *S. stalii*)-requiring strains of various adaptational types.

Five other cyanobacterial strains considered as extremotolerant are representatives of terrestrial macrocolony *Nostoc commune* (B-1520, B-1519), able to sustain desiccation, thermotolerant *Dolichospermum* sp. B-1213 from a hot spring environment, and cold-tolerant cyanobacteria *Anabaena* sp. B-1535 and *Anabaena* cf. pirinica B-1533, able to sustain low positive temperature and high-rate water flows. These soil and freshwater cyanobacteria strains can survive over a wide range of fluctuations in one or more environmental factors. However, they are unable to develop under such wide factor fluctuations. For the majority of the strains, optimal conditions for growth are not extremely harsh. Nevertheless, their high adaptability potential allows them to withstand long periods of desiccation, heat and cold. This remarkable adaptability has sparked scientific interest in utilizing these strains in biotechnological applications [45,46]. Similar to halo(alkali)philic strains, the extremotolerant cyanobacteria were maintained under reduced light and atmospheric CO_2 to ensure long-term storage conditions. It is noteworthy that all the cyanobacterial strains were unialgal but non-axenic as they had been co-isolated with other bacteria from the same habitat. Despite the fact that cyanobacteria constituted the dominant component in each culture, their primary metabolome that was studied could also be influenced by the bacterial cohabitants.

Further experimental design was thoughtfully considered to address the following primary questions of this exploration work: Are there constitutive metabolic differences between strains in the absence of the pressure of environmental specialization factors under culture storage conditions? If such differences are identified, could they be related to the extreme conditions of origin of the strains? What is the pattern of strain-specific metabolites that might indicate the strains' extreme origin after their prolonged storage?

The responses to the first two posed queries were derived through an examination of cyanobacterial GC-MS- and LC-MS-based metabolomic data, employing multivariate statistics and hierarchical clustering (Figures 1 and 2). All strains studied here exhibited distinct clustering in the PCA and heatmap models, indicating the presence of strain-specific metabolomic characteristics. These latter reflect constitutive (inherited) genetic differences between the strains. Their occurrence can be explained by the position of 'plasticity-led evolution'. This suggests that environmentally induced alterations in an organism's phenotype may facilitate the emergence of new adaptations if they are consistently advantageous to the organism in the new environment. In this case, selection may lead to an evolutionary loss of plasticity and, consequently, to a reduced sensitivity to the original environmental trigger. Finally, this could culminate in the constitutive expression of the trait that was previously induced by the environment [47].

Furthermore, PCA results reflected inherent similarities in the metabolomes of isolates from similar environments and general differences in the metabolomes of representatives of different ecological groups. Indeed, the similarities in metabolomes found in organisms that evolved under similar environments can reflect the evolution of their common acclimation strategies which were traced to changes in their genome [11,15,48]. The genetic basis of cyanobacterial adaptation to specific habitats has been investigated in a comparative genomic analysis of 650 cyanobacterial genomes from various ecosystems [48]. The general strategy of marine strains was to shrink genomes adapting to nutrient-poor conditions. Terrestrial strains expanded genomes partly due to horizontal gene transfer to cope with environmental fluctuations.

In addition, the present work has yielded several noteworthy outcomes concerning halo(alkali)philic and natronophilic strains. Firstly, as would be expected, some phylogenetically close strains, e.g., *Sodalinema* strains, have been shown to share certain metabolome characteristics under long-term storage, despite their initially different ecological environments (Table 1). This is evidenced by the separation of the strains in a cluster in the PCA model (Figure 1). Thus, *Sodalinema stalii* B-2050, a strain that is only halophilic, but not alkaliphilic and natronophilic, appeared to be in the same cluster as other haloalkaliphilic and natronophilic *Sodalinema* strains (Figure 1, Cluster 3.1, Supplementary S1, Figure S1-2). In contrast, other closely related strains, such as *Limnospira* strains (Figure S1-2), which were isolated from ecologically similar but geographically disparate environments (Table 1), have been observed to be distributed by PCA across different metabolomic clusters (Figure 1). One of the strains (B-1526) was located in close proximity to the *Sodalinema* strains (Figure 1, Cluster 3.1). Two other *Limnospira* strains (B-256 and B-287) along with a heterocystous *Nodularia* strain formed a separate group (Figure 1, Cluster 3.2) that differed from the others by considerable intergroup metabolome variability. This group of strains (B-287, B-256, and B-1529) was distinguished by the highest number of various accumulated metabolites, as demonstrated by the heatmap (Figure 2). Consequently, the strains of the genus *Sodalinema* exhibit minimal variability in their metabolomes, while strains of the genus *Limnospira*, despite their close evolutionary relationships, display considerable variability in their metabolomes. This finding might be explained by at least one of the following reasons: (1) A heightened sensitivity of *Limnospira* strains to the influence of diverse extreme habitat conditions on the formation of strain-specific adaptations, as compared to *Sodalinema* strains; (2) more variable natural environments in the case of *Limnospira* than *Sodalinema* strains, which can reflect the evolution of more various acclimation strategies in the former; (3) different capacities of *Limnospira* and *Sodalinema* strains to withstand long-term storage cultivation.

It is also important to note that PCA enabled the identification of sample metabolomes representing diazotrophic heterocystous strains from the entire metabolomic dataset (Figure 1). All six diazotrophic heterocystous strains examined in this study (B-1520, B-1519, B-1213, B-1533, B-1535, and B-1529) are classified within the order *Nostocales* and are distinguished by their capacity to form heterocysts for effective atmospheric nitrogen fixation [49]. Remarkably, these strains, despite being isolated from ecologically disparate extreme environments, exhibited a certain degree of similarity in their metabolomes, as revealed by PCA. This metabolomic similarity may be attributable to their shared capacity for heterocyst formation for nitrogen fixation. It is noteworthy that, in this case, this similarity between the metabolomes of distantly related strains is indicative of a shared, ancient genetic heritage rather than the evolution of analogous strategies for adaptation to extreme environmental conditions. This functionality (heterocyst-forming ability) is likely to be evolutionarily advantageous due to its persistence in cyanobacteria under diverse extreme environmental conditions. Therefore, it can be hypothesized that the distinguishing pattern of metabolites that were revealed between the metabolomes of heterocystous and non-heterocystous strains is indicative of alterations in the primary metabolism of cyanobacteria associated with heterocyst formation and N_2 fixation. Interestingly, in this pattern of metabolites, among the nitrogen-containing metabolites, a variety of nucleoside di- and triphosphates exhibited notable abundance decreases, whilst only a small number of amino acids (in particular, argininosuccinate and ureidosuccinate) displayed significant increases in heterocystous cyanobacteria (Table 6). The former observation suggests a high level of energy consumption in heterocystous strains, as maintaining heterocysts (which ensures N_2 fixation and subsequent distribution of the fixed nitrogen in cell metabolism) requires a high energy investment [50,51]. The two discovered metabolites, whose contents were many times higher in heterocystous strains, may indicate two mutually active pathways involved in the distribution of fixed nitrogen within the metabolism of these cyanobacteria. Specifically, the argininosuccinate, a precursor of arginine (a nitrogen-rich amino acid), suggests the involvement of the arginine–ammonia cycle [51], which, in combination with aspartate, may further contribute to the formation of a cyanophycin co-polymer, serving as a nitrogen reservoir. Conversely, the enhanced abundance of argininosuccinate in heterocystous cyanobacteria may also be attributable to the inhibition of arginine biosynthesis by a feedback mechanism [52] under strain storage conditions. The enhanced abundance of ureidosuccinate (also known as carbomoyl-*L*-aspartate) indicates the transfer of fixed nitrogen in the pyridine biosynthetic pathway. Another interesting outcome of the comparison is that heterocystous cyanobacteria exhibit elevated levels of sugar monophosphates, while non-heterocystous cyanobacteria display higher levels of sugar diphosphates. This observation may be indicative of the differential contribution of the pentose phosphate pathway and the Calvin cycle in the metabolism of the respective cyanobacterial groups.

To identify the pattern of strain-specifically accumulated metabolites, we compared the metabolome of each strain with the average levels of metabolic characteristics of the other strains studied. Interestingly, few (up to four) strain-specific metabolites were identified in desiccation-tolerant terrestrial strains (Supplementary S2, Table S2-3). Their numbers were higher (8–15) in hot- and cold-extremotolerant strains (Table S2-4) and varied considerably (6–67) among halo(alkali)philic strains (Tables S2-5 and S2-6). This difference in the number of constitutively up-regulated metabolites between extremotolerant and extremophilic strains may reflect differences in their acclimation strategies. Indeed, terrestrial and freshwater extremotolerant strains isolated from turbulent ecosystems involve, on average, more inducible genes of more functional categories in shaping adaptation to environmental fluctuations [48]. Therefore, in the absence of the influence of such environmental fluctuations, no end products (metabolites) of these genes are formed in the strains under storage. In this context finding the patterns of strain-specifically accumulated metabolites indicates some originally environment-inducible biosynthetic mechanisms could lose sensitivity to the fluctuating environmental factor and, finally, result in the constitutive expression of the metabolic traits. On the other hand, organisms (extremophiles) adapted to thrive in less variable environments have a greater number of constitutively expressed genes enabling mechanisms of evolved acclimatization [53]. Their final metabolite products are expected to be synthesized in the strains during their storage cultivation.

3.1. Constitutive Patterns of Metabolites and Their Adaptive Potential for Extremotolerant Cyanobacteria

Nostoc commune strains B-1520 and B-1519 were isolated from terrestrial macrocolonies subjected to desiccation in natural habitat regularly and are tolerant to drying–rehydration cycles. Although desiccated cyanobacteria are incapable of photosynthesis, rehydration rapidly restores (in just 5 min!) their photosynthetic activity [54]. Extracellular polysaccharides (EPSs) produced in response to stress play a crucial role in their tolerance mechanisms [54–57].

An important metabolite found in significant amounts in B-1520, but not in B-1519, was salicylic acid (Table 2). This key signaling metabolite is known to activate defense responses in plants [58–61]. The accumulation of salicylic acid in the *N. commune* strain B-1520 agrees with the data of Toribio et al. [62]. They indicated the production of salicylic acid by 28 cyanobacteria strains and demonstrated that some strains of the genus *Nostoc* exhibited the most significant level of salicylic acid production. It is possible that salicylic acid acts as a trigger to enhance the EPS production in B-1520 under storage conditions. This can be explained by the fact that the production of EPSs (especially capsule EPSs bound to the cell surface) in *N. flagelliforme* [63] could be enhanced by salicylic acid added to the culture medium. Salicylic acid first increased the NO level which in turn promoted EPS biosynthesis. The latter occurred via the nitrosylation of EPS biosynthetic enzymes [64]. Additionally, the cyanobacteria's ability to synthesize enhanced levels of salicylic acid can be used in agrobiotechnologies to promote plant disease resistance [65,66]. It also cannot be excluded that salicylic acid could at least partly be produced by bacteria accompanying the strain culture. It is known that soil strains of *N. commune* may form associations with heterotrophic bacteria, of the Actinobacteria phylum [67], with some members of this group such as *Streptomyces* having been shown to produce salicylic acid [68].

Another important metabolite that was accumulated in both B-1520 and B-1519 under storage conditions is erythritol (Table 2). Erythritol was found in the composition of

EPSs from *N. commune* [69,70]. In addition, some biotechnological strategies have created genetically modified cyanobacteria that produce erythritol [71]. Erythritol is produced from erythrose, previously dephosphorylated from erythrose-4-phosphate (E4P). It is noteworthy that E4P, an intermediate of the Calvin and pentose phosphate oxidation (PPO) pathways, is the common precursor for two metabolites, erythritol and salicylic acid, found in this work as constitutively up-regulated in at least one desiccation-tolerant strain. These data suggest that carbon flux in those strains might be directed via E4P to two routes. The first is the shikimate pathway producing phenolic compounds, which mainly are known as antioxidants neutralizing free radicals formed during desiccation. The second is a branch reaction producing erythritol which might serve as an EPS constituent or an osmoprotectant, protecting cellular membrane integrity and proteins from denaturation under water-deficient conditions [72].

Dolichospermum sp. B-1213 is an isolate from a hot spring habitat (Table 1). In this work, storage cultivation of the strain was carried out at room temperature (22 °C), whereas the optimal growth conditions are in the range of 35–38 °C [32]. Therefore, in addition to reducing the light and CO_2 supply to limit cyanobacteria growth under storage, the temperature shifting from optimal to low values could act as an additional stress factor for B-1213 and cause specific metabolic responses of the strain cells. Indeed, the highest accumulation of ADP-ribose (27-fold) and NADH (4-fold, Table 3) may indirectly indicate the presence of this stress factor. High levels of ADP-ribose in cells are toxic because of its ability to cause nonenzymatic ADP-ribosylation inactivating proteins [73,74]. Elevated NADH levels can indicate that its oxidated state NAD^+ (the dominant NAD form in unstressed cells) can be depleted in glycolysis. An important role of glycolysis in the B-1213 is also evidenced by the accumulation of glucose 1-phosphate, a product of glycogen degradation, which enters glycolysis via conversation to glucose 6-phosphate (G-6-P). Furthermore, the accumulation of glutamate in the strain may indicate that the carbon flux derived from glycogen consumption is redirected for amino acid synthesis. This is consistent with the literature data [75] showing that redirection of carbon flow from glycogen synthesis in *Synechococcus elongatus* led to enhanced glutamate production.

The cold-tolerant *Anabaena* sp. B-1535 and *A.* cf. pirinica B-1533 (Table 1) were maintained at 22 °C during their storage. This meant that the growth of the strains was not limited by temperature fluctuations that the strains had adapted to, but only by the light and CO_2 supply. Under these conditions, both strains exhibited constitutive up-regulation in a range of metabolites (Table 3). Many of the metabolites can be related to EPS and glycolipid production during heterocyst formation in an environment lacking a fixed nitrogen source [38,76]. Heterocysts, N_2-fixing cells, sustain an O_2-limiting micro-environment for oxygen-sensitive nitrogenase by forming a dense envelope of glycolipids and EPSs [77,78]. The final step of glycolipid synthesis includes the reaction of glycolipid synthase transferring glucose to the fatty alcohol [77]. Thus, we proposed that the elevated levels of hexoses, their phosphates and fatty alcohol (nonadecan-1-ol) observed in at least one of the B-1533 and B-1535 strains may be involved in the synthesis of heterocyst glycolipids. Meanwhile, hexoses, their derivatives and digalacturonic acid may be involved in the synthesis of EPSs including heterocyst polysaccharides.

B-1533 shows more up-regulated metabolites than B-1535, indicating a more active metabolic status. This may be due to the different capacities of the strains to cope with photorespiration. The enhanced glycolate level in B-1535 suggests enhanced RUBISCO oxygenase activity in this strain.

B-1533 showed elevated levels of metabolites, members of all three principal metabolic pathways of glucose catabolism: the Embden–Meyerhoff–Parnas (EMP or glycolysis),

OPP and Entner–Dudoroff (ED) pathways, as well as the TCA cycle (Table 3) [79]. These pathways may provide energy and carbon flux towards N_2 fixation and biosynthesis of fatty acids, alcohols, EPSs, and various secondary metabolites. Indeed, the accumulation of compounds such as 2C-methyl-*D*-erythritol and 2C-methyl-*D*-erythritol-4-phosphate (MEP) in B-1533 and/or B-1535 may indicate active isoprenoid biosynthesis [80]. In addition, the potential contribution of bacteria associated with these cyanobacterial cultures to the elevated levels of these two metabolites cannot be dismissed [81]. Additionally, increased levels of 3-dehydroshikimic acid, an intermediate of the shikimate pathway, indicate active biosynthesis of aromatic amino acids and diverse phenolic metabolites. Both strains also exert high levels of ureidosuccinate, an intermediate of de novo pyrimidine biosynthesis. This indicates the direction of nitrogen flux fixed by nitrogenase from amino acids to pyrimidine nucleotides [82].

3.2. Constitutive Patterns of Metabolites and Their Adaptive Potential for Haloalkaliphiles and Natronophiles

The marine *S. stalii* (B-2050), the haloalkaliphiles and natronophiles *S. gerasimenkoae* (B-353) and *S. orleanskyi* (B-2037) from salt alkaline lakes, and *Limnospira* sp. (B-1526) from a soda lake habitat exhibited less pronounced metabolic differences among themselves under storage conditions compared to three other strains from soda lake environments: *Limnospira* sps B-287 and B-256 and *Nodularia* sp. (B-1529) (Figure 1). This finding enabled us to distribute all seven halo(alkali)philic strains between two groups according to the relative similarity in their metabolomes.

Despite the general metabolic similarity (Figure 1), the cyanobacteria strains from the first group exhibited distinctive metabolite levels showing the strains' unique adaptations to specific habitats. B-2037 demonstrated the accumulation of glucosylglycerol (Table 4) which acts as compatible solute; i.e., by increasing cellular osmolality, it contributes to the osmotic acclimatization of cells [53]. This compound also protects membrane stability and proteins from denaturation under high salinity and alkalinity. In addition, elevated related abundance of phosphate and ADP in B-2037 can at least partially be related to glucosylglycerol biosynthesis [83]. Three other metabolites from the pattern of B-2037-distinguished metabolites, up-regulated sedoheptulose-1,7-biphosphate and 2-phosphoglycolate and down-regulated 3-hydroxypyruvate, indicate activities of the Calvin cycle and photorespiration. The latter pathway may be stimulated by the limited light supply to maintain the strain storage conditions. To support CO_2 fixation by RUBISCO (and therefore to decrease photorespiration) under high-alkalinity conditions and to utilize the available HCO_3^-, haloalkaliphilic cyanobacteria have evolved carbon concentration mechanisms [84]. These mechanisms enable the active transport of HCO_3^- into the cell where it is converted to CO_2 in specialized protein carboxysomes containing RUBISCO [85]. The elevated ADP levels observed in this strain may indicate energy consumption associated with the described process. The elevated Cl^- level in B-2037 may have toxic effects on cellular metabolism, potentially due to its effects on protein structure [53]. The *trans*-isomer of 9-hexadecenoic acid, which was identified as accumulating in the strain, may have been produced by bacteria associated with the culture of the strain rather than by the strain itself. This is because *S. orleanskyi*, like other cyanobacterial strains, is characterized by a high content of only the *cis* but not the *trans* isomer of 9-hexadecenoic acid [86].

For B-353, the elevated level of shikimate may be related to the carbon flux directed towards the biosynthesis of aromatic secondary metabolites [87]. Additionally, this strain exhibited the accumulation of adenosine (41-fold higher than the mean level of this metabolite across all strains studied, Table 4). This was unexpected, and further investigation is needed into the role of this metabolite in tolerance mechanisms. However, this result is in line with other studies also indicating that cyanobacteria are capable of producing enhanced level of adenosine [88,89]. Fatima et al. showed a genetically modeled strain of *Synechococcus elongatus* which responded to salt stress by the accumulation of various nucleosides, including adenosine [88]. Moreover, adenosine produced by cyanobacteria has promising biotechnological potential, as it exhibits cytotoxicity and is used as an anti-leukemic agent [89,90].

In the marine strain B-2050, up-regulation of glycerol, galacturonic acid and several other sugar derivatives may be associated with EPS biosynthesis, which according to some data may be associated with increased NaCl tolerance [91]. B-2050 also exhibited the highest levels of orotic acid and NADPH among the other strains studied. A high level of orotic acid may reflect constitutive activity of de novo pyrimidine synthesis. Elevated levels of NADPH may indicate low activity of the Calvin cycle, in the reactions of which NADPH can be utilized.

The soda lake strain B-1526 has demonstrated the accumulation of a considerable quantity of glucosylglycerol, functioning as a compatible solute, similar to that observed in B-2037. B-1526 also displayed the highest level of 3-hydroxybutyric acid (Table 4), indicating that it is the most effective producer of poly-3-hydroxybutyrate (PHB) among the strains examined in this study. PHB is a non-water-soluble energy-rich polymer that serves as a specific carbon and energy reserve for withstanding nitrogen starvation under stress conditions [92,93].

The second group, comprising the three haloalkaliphiles B-1529, B-287 and B-256 derived from soda lake habitats, displayed the most distinctive metabolomes (Figure 1) and the richest patterns of diverse accumulated metabolites (Figure 2, Supplementary S2, Table S2-6). The *Nodularia* sp. B-1529, which is the only diazotrophic heterocystous strain among other haloalkaliphiles studied here, exhibited the highest number of accumulated metabolites in its primary metabolome (Table 5). The metabolites were participants of the major catabolic (EMP, ED, OPP, TCA), biosynthetic (e.g., de novo pyrimidine and purine nucleotide, shikimate, EPS, MEP, fatty acid pathways) and remodeling routes (fatty acid desaturation). Additionally, hexoses, sucrose and glucosylglycerol are compatible solutes ensuring high cellular osmolality and homeostasis under high-Na^+-concentration and high-pH conditions. It is noteworthy that in B-1529, the levels of the majority of amino acids were higher than those in the other heterocystous diazotrophs studied in this work (Supplementary S2, Table S2-6). It might be speculated that amino acids such as arginine and aspartate may be involved in the metabolism of cyanophycin, serving as a nitrogen reserve in cyanobacteria [94]. On the whole, the elevated levels of many nitrogen-containing metabolites (amino acids, nucleosides, nucleotides) point to a high cellular nitrogen status [95,96]. This suggests that the N_2-fixation process is highly efficient in B-1529. It is notable that *Nodularia* is notorious for its production of toxic non-ribosomal pentapeptides, nodularins [95,97–99]. These toxic compounds may act as liver tumor promotors [100]. The amino acids produced by this strain might be used for nodularin biosynthesis. The production of the nodularin is not constitutive, but rather inducible [95]. However, the conditions that influence its biosynthesis as well as the biological role of the compounds have not been well elucidated [101,102]. As nodularin is a nitrogen-rich compound, its biosynthesis depends on nitrogen availability and also requires active photosynthesis [103]. Nodularin binds to proteins. The functionality of the binding has been discussed in the context of ROS protective stabilization of proteins by nodularins or as a means for nitrogen storage [95].

Moreover, pathway analysis was employed to elucidate the most efficient pathways among those discussed herein. This enabled the identification of the most efficient pathway in the strain under storage conditions, based on its $-\log_{10}(p)$ and PI values, as alanine, aspartate and glutamate metabolism (Supplementary S2, Table S2-8). This result was anticipated, as this pathway reflected the aforementioned metabolism of amino acids resulting from nitrogen fixation and their potential utilization for nodularin biosynthesis.

The genus *Limnospira* to which the two strains B-256 and B-287 (and also previously described B-1526) belong, in contrast to *Nodularia,* is not a toxic cyanobacteria genus [104]. *Limnospira,* also designated as Spirulina, is acknowledged as a rich source of diverse nutrients [105]. It usually exhibits optimal growth and a high quality of other nutrient parameters when cultivated under 30 ± 5 °C [106]. Indeed, these facts are consistent with our observations, which revealed a marked abundance of various important primary metabolites in both strains even under storage conditions (Supplementary S2, Table S2-6).

Strain B-287 is characterized by a high energetic status, as can be concluded from its highest relative content of ATP and GTP (Table 5, Supplementary S2, Table S2-6). The high ATP and 2- and 3-phosphoglycerate levels may indicate that carbon flow is directed not to energy production but rather towards biosynthetic processes. This is evidenced by the relatively high content of glycerol 3-phosphate, acetyl-CoA and malonyl-CoA in B-287, which may be involved in polyketide synthesis [107,108], fatty acid metabolism, and PHB production [109]. Furthermore, pathway analysis confirmed the importance of these metabolic pathways. It identified 'butanoate metabolism' as the most crucial for B-287 cultivated under storage conditions (Supplementary S2, Table S2-8, Figure S2-14). The butanoate metabolism encompasses processes involving short-chain fatty acids, thereby establishing a connection between the fatty acid, polyketide and PHB metabolisms. Additionally, B-287 exhibited elevated levels of important intermediates involved in isoprenoid biosynthesis (Table 5, Supplementary S2, Table S2-6) [110]. B-287 also has the highest trehalose accumulation among the strains studied. Trehalose and sucrose, along with glucosylglycerol, provide cellular osmoprotection as compatible solutes [111]. Trehalose synthesis is supported by elevated levels of UDP- and ADP-glucose (Table 5), starts from glycogen and is induced by high salt concentrations [53].

B-256 accumulated deoxy-pentoses and deoxyribonucleotides, compounds (adenylosuccinic acid, xanthosine-5′-phosphate, inosinic acid) involved in de novo purine nucleotide synthesis and metabolites (UDP-*N*-acetylglucosamine and glucosamine 6-phosphate) associated with cyanobacterial cell wall biogenesis (Supplementary S2, Table S2-6). Pathway analysis identified 'pentose and glucuronate interconversions' as the most significant metabolic route in this strain under storage conditions (Table S2-8). This indicates the importance of sugar-phosphate substrates in nucleotide synthesis, cell wall biosynthesis and, overall, the active growth of this strain.

An overview of the aforementioned metabolic processes that may occur in extremotolerant and extremophilic cyanobacteria to maintain their viability under storage is given in Table 7.

Table 7. Primary metabolites accumulated in specific cyanobacterial strains and the main metabolic processes in which they may be involved to maintain the life of extremotolerant and extremophilic strains under long-term storage conditions *.

Tolerance Group	Species Name, IPPAS ID, Extreme Environment	Storage Medium, t °C, Storage Period	Strain-Specifically Accumulated Metabolites	Metabolic Processes
Desiccation-tolerant	*Nostoc commune* B-1520, heterocystous diazotroph, terrestrial macrocolony	BG-11 without nitrogen, 22 °C, 3 months	Salicylic acid; erythritol	Production of EPS, osmoprotection, component of EPS
	Nostoc commune B-1519, heterocystous diazotroph, terrestrial macrocolony	BG-11 without nitrogen, 22 °C, 3 months	Erythritol	Osmoprotection, component of EPS

Table 7. *Cont.*

Tolerance Group	Species Name, IPPAS ID, Extreme Environment	Storage Medium, t °C, Storage Period	Strain-Specifically Accumulated Metabolites	Metabolic Processes
High-temperature-tolerant	*Dolichospermum* sp. B-1213, heterocystous diazotroph, hot springs	BG-11 without nitrogen, 22 °C, 8 months	ADP-ribose, NADH; glucose-1-P, glutamate	Protein ADP-ribosylation; glycogen degradation and redirected carbon flux via glycolysis towards amino acid synthesis
Low-temperature-tolerant up to 10–11 °C	*Anabaena* cf. pirinica B-1533, heterocystous diazotroph, high-rate cold water flow	No. 6 without nitrogen, 22°, 3 months	Hexoses, hexose phosphates, fatty alcohol; glycolysis and pentose phosphate intermediates; ureidosuccinate; 2C-methyl-*D*-erythritol, MEP	Production of EPS and glycolipids; enhanced carbon metabolism via glycolysis and pentose phosphate pathways; de novo pyrimidine biosynthesis; isoprenoid biosynthesis
	Anabaena sp. B-1535, heterocystous diazotroph, high-rate cold water flow	No. 6 without nitrogen, 22 °C, 3 months	Glycolate; MEP	Photorespiration; isoprenoid biosynthesis
Haloalkaliphilic and natronophilic	*Sodalinema orleanskyi* B-2037, saline–alkaline lake	S (pH 9.0–9.5) 32 °C, 3 weeks, 22 °C, 2 months	Glucosylglycerols; glycolate-2P	Osmoregulation; photorespiration
	Sodalinema gerasimenkoae B-353, saline–alkaline lake	S (pH 9.0–9.5) 27 °C, 3 weeks, 22 °C, 2 months	Shikimate; adenosine	Aromatic amino acids, phenolic compounds; not well understood
	Sodalinema stalii B-2050, halophilic, coastal shoals	ASNIII, (pH 7.5) 27 °C, 3 weeks, 22 °C, 2 months	Glycerol, galacturonic acid; orotic acid; NADPH	EPS production; de novo pyrimidine biosynthesis; low Calvin cycle activity
	Limnospira sp. B-1526, soda lake	Zarrouk, (pH 9) 32 °C, 6 weeks	Glucosylglycerol; 3-hydroxybutyrate	Osmoprotection; PHB synthesis
	Nodularia sp. B-1529, heterocystous diazotroph, soda lake	Zarrouk, 27 °C, 3 weeks, 22 °C, 3 weeks	Amino acids; compatible solutes (hexoses, sucrose, glucosylglycerol)	Nodularin biosynthesis (protein protection from oxidative damage or N-storage polymer?); osmoprotection
	Limnospira sp. B-287, origin is not known, haloalkaliphile, natronophile	Zarrouk, (pH 9) 32 °C, 6 weeks	ATP, CoA-thioesters, isopentenyl-PP; trehalose	Biosynthetic processes (polyketide, fatty acid, PHB, isopropanolds); osmoprotection
	Limnospira sp. B-256, soda lake	Zarrouk, (pH 9) 32 °C, 6 weeks	Deoxypentoses; deoxy-ribonucleotides; UDP-*N*-acetylglucosamine	De novo nucleotide synthesis; cell wall biogenesis

* Reduced light supply up to light intensity 50 µmol photons m^{-2} s^{-1} and no additional CO_2 supply.

4. Materials and Methods

4.1. Reagents

The reagents were obtained from various manufacturers as follows: *L*-aspartic acid and 2-oxoglutaric acid from Reanal (Budapest, Hungary); hexane (puriss p.a.) from Conlac GmbH (Leipzig, Germany); *N*-methyl-*N*-(trimethylsilyl)trifluoroacetamide (MSTFA, MS grade) from Macherey-Nagel GmbH and Co KG (Düren, Germany); methanol (LC grade) from Vekton (Saint-Petersburg, Russia). Other chemicals were sourced from Sigma-Aldrich Chemie GmbH (Taufkirchen, Germany). Water purification was performed using a Millipore Milli-Q Gradient A10 system (resistance 18 mΩ/cm, Merck Millipore, Darmstadt, Germany).

4.2. Cyanobacterial Strains' Characterization and Cultivation

Twelve non-axenic cyanobacterial strains were obtained from the Collection of Microalgae and Cyanobacteria IPPAS (K.A. Timiryazev Institute of Plant Physiology RAS, Moscow, Russia) (Table 1). For the experiments, the strains were grown in 350 mL Erlenmeyer flasks containing 150–200 mL of appropriate liquid medium under conditions (reduced light and an absence of additional CO_2 supply) which are used in the IPPAS collection for long-term maintenance of the studied strains (Table 1).

For the analysis, cyanobacterial biomass was collected in 50 mL tubes by centrifugation (4500× *g*, 15 °C, 10 min), the supernatant was thoroughly removed and the obtained pellets were washed twice with deionized water, frozen at −80 °C and freeze-dried in a Labconco FreeZone 6 L lyophilizer (Labconco, Kansas City, MO, USA) at a condenser temperature of −40 °C and a pressure of 100 Pa for 2–3 days.

4.3. GC-MS-Based Analysis of Thermally Stable Polar Metabolites

Sample preparation procedure (metabolite extraction and derivatization) and GC-MS analysis were performed according to the protocol described by Bilova et al. [112, 113] with minor changes. In detail, to extract thermally stable metabolites, lyophilizates of cyanobacteria were processed in a two-step procedure. First, cyanobacteria material (approximately 10 mg) was supplemented with 800 µL ice-cold methanol containing the internal standard (IS) adonitol at a final concentration of 50 µmol/L. After intensive vortexing (30 s) and subsequent centrifugation of the resulting suspension (12,000× *g*, 4 °C, 10 min), a portion (500 µL) of the obtained supernatant was transferred to a new tube. Water (400 µL) was added to the remaining supernatant, and cyanobacteria residues were resuspended by vortexing to obtain suspension. After repeated centrifugation, the obtained supernatant was collected and combined with the first extract portion. The total volume (1100 µL) of the combined aqueous (aq.) methanolic (1:3, *v*/*v*) extract was mixed with 300 µL hexane to purify the extract from low-polarity and high-molecular-weight compounds such as chlorophylls, carotenoids and lipids which are not suitable for GC-MS. The resulting mixture was vortexed and briefly centrifugated (2 min, 5000× *g*) to separate the nonpolar hexane phase from the aq. methanolic phase containing polar metabolites. Aliquots (50 µL) from the sample aq. methanolic extract were taken for further analysis. The indicated volume of sample aliquots was selected as the optimal one in a series of preliminary experiments directed to optimize the performance of subsequent derivatization and GC-MS analysis.

The aliquots were evaporated to dryness at +4 °C using a Labconco CentriVap vacuum concentrator (Labconco Corporation, Kansas City, MO, USA). The resulting dry residues were derivatized through a two-step procedure by subsequently using methoxyamine hydrochloride (MEOX) and *N*-methyl-*N*-(trimethylsilyl)trifluoroacetamide (MSTFA), following a previously established protocol [114]. The obtained mixtures of thermally stable metabolite derivatives were analyzed by gas chromatography–mass spectrometry (GC-MS). For that 1 µL from each sample was injected with a CTC GC PAL liquid injector (Shimadzu Deutschland GmbH, Duisburg, Germany) in a GC2010 gas chromatograph. The separation of analytes was carried out in the mode of programmable temperature gradient from 40 to 320 °C on a ZB-5MS nonpolar capillary column (30 m, 0.25 mm, 0.25 µm, Phenomenex Inc., Torrance, CA, USA) coupled to a Shimadzu GCMS QP2010 quadrupole mass-selective detector equipped with electron ionization ion source and operating at the settings summarized in Supplementary S2, Table S2-9.

A sample sequence (batch) organized for the GC-MS analysis involved various types of samples arranged in the following order: (i) hexane, (ii) a mixture of C_{10}–C_{40} alkanes dissolved in hexane, (iii) MSTFA, (iv) a derivatization blank (containing only derivatization agents), (v) an extraction blank (containing only IS and derivatization agents), (vi) experimental samples in a randomized sequence, (vii) quality controls (QCs) obtained by mixing equal volumes (200 µL) of all analyzed samples and prepared from aliquots (50 µL) of the resulted pooled extracts and (viii) 28 mixes of 5–6 authentic standards (a final concentration of each component 50 µmol/L) serving for reliable metabolite identification. QCs were inserted into the sample batch in every 5–6 experimental samples to evaluate method performance.

The processing of the obtained GC-MS data (chromatograms) employed an untargeted approach aimed at unbiased annotation of all detected total ion current (TIC) chromatographic peaks with a signal-to-noise ratio (S/N) ≥ 3. For that, the following pieces of software were used: the Automated Mass Spectral Deconvolution and Identification System (AMDIS, version 2.66 from 8 August 2008, www.amdis.net, accessed on 25 September 2023); Xcalibur, version 3.0.63 from 5 August 2013 (TermoFisher Scientific Inc., Bremen, Germany); and MS-DIAL, version 4.9 from 1 January 2022 (RIKEN Center for Sustainable Resource Science, Kanagawa, Japan). The quality of the obtained GCMS data was evaluated by the baseline position, retention time, shape and height of chromatographic peaks and the level of noise background. The annotation of TIC peaks to specific analytes (also called features) relied on a search of experimentally obtained retention indices (RIs) and electron ionization mass spectra (EI-MS) against established reference mass spectral libraries. For each analyte, RI was calculated by retention time (t_R) and retention indices of linear alkanes C_{10}–C_{40} (Supplementary S2, Table S2-10) according to the following formula indicated in the AMDIS manual instruction: $RI_{comp} = RI_b + ((RI_a - RI_b) \times (t_{Rcomp} - t_{Rb})/(t_{Ra} - t_{Rb}))$, where t_{Rb} and t_{Ra} are the retention times of alkanes which are the closest before and after the actual retention time (t_{Rcomp}) of the target compound, and RI_b and RI_a represent the retention indices associated with t_{Rb} and t_{Ra}, respectively. The following libraries were used for the metabolite structural annotation: National Institute of Standards and Technology (NIST), Golm Metabolome Database (GMD) and an in-house authentic standard library (IhASL) (Supplementary S2, Table S2-1). The identification of analytes was achieved by matching the obtained GC-MS data (t_R, RI and EI-MS) with those of authentic standards from IhASL coeluted with the experimental samples. The quantitation of analyte relative abundances was conducted by integrating the corresponding extracted ion chromatograms (XICs) built for the analyte characteristic fragment ions ($m/z \pm 0.5$) at the analyte specific retention times.

4.4. LC-MS Analysis of Thermally Labile Polar Metabolites

Analysis of thermally labile anionic polar metabolites was performed based on reversed-phase ion pair ultrahigh-performance liquid chromatography and tandem mass spectrometry accomplished with electrospray ionization and a triple quadrupole mass analyzer (RP-IP-UHPLC-ESI-QqQ-MS/MS), following a method described by Balcke et al. [115] with minor adaptations.

The lyophilized material of cyanobacteria was weighed, and 25 ± 3 mg DW was added to polypropylene 2 mL microtubes containing 200 mg glass beads (0.75–1 mm diameter), 3 stainless steel beads (3 mm diameter) and 1 stainless steel bead (5 mm diameter). The cyanobacteria material was extracted through a two-step procedure. The first extraction was performed with 900 µL of an ice-cold (−80 °C) dichloromethane/ethanol mixture (2:1, *v*/*v*) and 100 µL of ice-cold HCl/water (1:200, *v*/*v*). The mixture was intensively homogenized (5.0 m/s, 3 × 20 s, FastPrep-24™, MP Biomedicals, Eschwege, Germany). After subsequent centrifugation (4 °C, 10,000× *g*, 5 min), the polar supernatant fraction (300 µL) was collected into a new 1.5 mL microtube. During the second extraction step, the plant material residue was resuspended in an additional portion (50 µL) of ice-cold HCl/water (1:200, *v*/*v*). After the sample mixing and centrifugation, as described above, the obtained polar supernatant fraction (120 µL) was combined with the first polar extract portion. The combined extract was evaporated to dryness at +4 °C using a Labconco CentriVap vacuum concentrator

(Labconco Corporation, Kansas City, MO, USA) and stored at −20 °C. For LCMS analysis, the dried extract residue was solubilized in 180 μL of a water–ethanol mixture (1:3, v/v), and after additional centrifugation (12,000× g, 4 °C, 5 min) it was transferred into a glass insert of a chromatographic vial, capped with membrane-secured plastic septa. Aliquots (5 μL) of the extracts were analyzed by a Waters ACQUITY H-Class UPLC System (Waters GmbH, Eschborn, Germany), online coupled to AB Sciex QTRAP 6500 MS/MS System (AB Sciex, Darmstadt, Germany) under the chromatographic and mass spectrometric settings presented in the Supplementary Materials S2 (Table S2-2). A sequence for the LC-MS analysis included samples of different types (experimental samples in randomized order, QC samples, samples of a multicomponent standard solution, extraction blanks). Targeted acquisition of LC-MS data was performed in the multiple reaction monitoring (MRM) mode and relied on MRM ion pairs (m/z of molecular ion and m/z of characteristic fragment ion) of 183 individual authentic standards (target compounds). For the quantitative analysis of the target compounds in the experimental samples, the extracted MRM peaks were integrated using MultiQuan^{TM} (version 3.0.2) tool (AB Sciex, Darmstadt, Germany).

4.5. Statistical Analysis

The integrated areas of all analyte peaks recorded by GC-MS and LC-MS methods were combined into a digital matrix. The dataset after normalization to the dry weight of samples was filtrated to remote analytes with a high percentage of missing data (i.e., analytes that were absent in ≥20% of the samples) and those exhibiting significant in-group variation (relative standard deviation (RSD) ≥ 70%). The resulting dataset was statistically processed using the online platform MetaboAnalyst 5.0 (www.metaboanalyst.ca, accessed on 1 December 2023). The methods of multivariate statistics such as principal component analysis (PCA) and partial least squares discriminant analysis (PLS-DA) were applied. Additionally, hierarchical clustering with heatmap representations and a Volcano plot were used to visualize differences in relative abundances of annotated analytes across all experimental groups and to highlight analytes displaying specific behavior (accumulation or sharp decrease in abundance and disappearance) for each analyzed cyanobacteria strain, respectively.

Univariate statistics (Volcano plot) was applied to compare the abundances of individual metabolites in each of the twelve strains studied here with a pool of samples compiled from eleven other strains. The comparison results were presented as the fold difference (FC) in the mean abundances of individual metabolites and the significance of the differences (p-value) between compared groups estimated by Student's t-test. For FC and t-test p presented in tables and figures, the threshold values were as follows: the minimal value of FC was 2 and the maximal p-value was 0.05 as typically recommended for metabolomics studies [116,117]. The reliability of the observed FCs was assessed by adjusting the p-value with the false discovery rate (FDR) estimated with the Benjamini–Hochberg method [42]. As the variance of the groups compared was different and the data may not be normally distributed, the significance of the observed differences in the abundances of metabolites detected by the Volcano analysis was additionally assessed using Welch's t-test and the Mann–Whitney U-test [116,117]. The threshold p-values for the tests reported in the tables were as follows: the maximum p-value was 0.05 for the Welch's t-test and 0.01 for the Mann–Whitney U-test.

4.6. Metabolomic Pathway Analysis

Pathway analysis, available via the web resource Metaboanalyst 5.0, was used to evaluate the metabolic pathways of the analyzed cyanobacterial strains. This analysis combined pathway enrichment (global test) and pathway topology (relative betweenness centrality) methods. The former is used to highlight pathways that include metabolites displaying statistically significant alternations in their relative abundances between compared groups. The latter assesses metabolic pathways by measuring the contribution of the identified compounds to the corresponding metabolic pathway. The assignment of metabolites from the input list to metabolic pathways relied on the cyanobacterium *Synechococcus elongatus* PCC7942 metabolic pathway library available at the Kyoto Encyclopedia of Genes and Genomes (KEGG) database (https://www.genome.jp/kegg/, accessed on 15 December 2023). The results of the pathway analysis were visualized as a scatter plot built in coordinates of $-\log_{10}$ (p-value) (Y-axis) and the impact value reflecting the metabolite's contribution to a given metabolic pathway (pathway impact, X-axis).

5. Conclusions and Perspectives

The results of this exploration work suggest that all the extremotolerant and extremophilic cyanobacterial strains studied exhibited distinctions in their primary metabolomes under storage conditions. The metabolic differences between strains may be associated with their origin from extreme habitats and were more pronounced between strains from different extreme environments. Conversely, a group of strains representing heterocystous diazotrophs exhibited certain shared metabolomic characteristics, irrespective of their extreme habitats. This finding suggests that the metabolomic similarities observed among these strains may be indicative of an ancestral heritage, rather than the development of analogous strategies to cope with a stressful environment. Nevertheless, the extremotolerant heterocystous diazotrophic terrestrial and freshwater strains cultured under similar storage conditions showed lower amounts of strain-specifically accumulated primary metabolites than extremophiles from habitats with high salinity and alkalinity. The latter group was highly diverse in the amount of specific primary metabolites. This may indicate the diversity of adaptive mechanisms in which metabolites may be involved in sustaining microbial life in saline and alkaline environments.

The results of this study also highlighted three strains of the genus *Limnospira* which, despite their close evolutionary relationship, displayed considerable metabolome variability and a *Nodularia* strain which among all studied strains was distinguished by the majority of accumulated compounds. Despite the analogous habitats in which the four strains evolved, they may form disparate adaptations to adjust their metabolism to thrive in high-Na^+ and high-pH conditions. Therefore, these four strains might be a subject worthy of further investigation, specifically with regard to the molecular mechanisms underpinning their high metabolic plasticity. In addition, the following directions are suggested as potential areas for further research in relation to this study: (1) assessing the effect of storage conditions on the metabolism of cyanobacteria and determining the sensitivity of their adaptations to storage conditions, which is of particular interest when considering groups of terrestrial and freshwater strains, especially those that are tolerant to desiccation and exhibit depleted strain-specific metabolite profiles under storage conditions; (2) identification of major strain-specific secondary metabolites in the stored strains, as combining this information with primary metabolomics data will allow the linkages between primary and secondary pathways to be established and their contribution to strain-specific adaptations to be assessed and will also allow the biotechnological potential of each strain to be assessed.

Supplementary Materials: The following supporting information can be downloaded at: https://www.mdpi.com/article/10.3390/ijms26052201/s1.

Author Contributions: D.L., M.S. and A.F. proposed the idea and the design of the research. M.S. maintained cultures of cyanobacterial strains and collected their material. A.O., S.S., A.K. and A.S. performed sample preparation for metabolomics and performed GC-MS and LC-MS method optimization and measurements. T.B., N.G. and N.F. accomplished processing and post-processing of metabolomics data. T.B. and N.G. contributed in writing the manuscript draft. T.B, N.G., N.F., M.S., A.F. and D.L. contributed to critical reading and discussion and prepared the final version of the manuscript. D.L. acquired the funding. All authors have read and agreed to the published version of the manuscript.

Funding: This research was funded by the Russian Science Foundation, grant No. 23-44-00101.

Institutional Review Board Statement: Not applicable.

Informed Consent Statement: Not applicable.

Data Availability Statement: All relevant data are available within the article and Supplementary Materials.

Acknowledgments: We thank Aditi Basnet (Leibniz Institute of Plant Biochemistry, Halle, Germany) for technical support in sample measurement. The authors acknowledge Saint-Petersburg State University for a research project 118201839 and the Chemical Analysis and Materials Research Centre and Center of Molecular and Cell Technologies for technical support under agreement C-111/30(01/1-69-46-SPbU). The infrastructural support from the Ministry of Science and Higher Education of the Russian Federation (theme #122042700045-3) is acknowledged.

Conflicts of Interest: The authors declare no conflicts of interest.

Abbreviations

AMDIS: Automated Mass Spectral Deconvolution and Identification System; ASNIII, Artificial Seawater Nutrient Medium (III); BG-11, Blue-Green Medium 11; E4P, erythrose-4-phosphate; ED, Entner–Doudoroff pathway; EI-MS, electron ionization mass spectrum; EMP pathway, Embden–Meyerhof–Parnas pathway; EPS, extracellular polymeric substance; FDR, false discovery rate; FC, fold change; G-6-P, glucose-6-phosphate; GC-MS, gas chromatography–mass spectrometry; GMD, Golm Metabolome Database; HMBPP, (*E*)-4-hydroxy-3-methyl-but-2-enyl pyrophosphate; ihASL, in-house authentic standard library; IS, internal standard; KEGG, Kyoto Encyclopedia of Genes and Genomes; LC-MS, liquid chromatography–mass spectrometry; MEOX, methoxyamine hydrochloride derivative; MEP, 2C-methyl-*D*-erythritol 4-phosphate; MS-DIAL: Mass Spectrometry Data Independent Analysis; MSTFA, *N*-methyl-*N*-(trimethylsilyl)trifluoroacetamide; MRM, multiple reaction monitoring; NIST, National Institute of Standards and Technology; PPO, pentose phosphate oxidation pathway; PCA, principal component analysis; PHB, poly-3-hydroxybutyric acid; PI, pathway impact, PLS-DA, partial least squares discriminant analysis; QC, quality control; RP-IP-UHPLC-ESI-QqQ-MS/MS: reversed-phase ion pair ultrahigh-performance liquid chromatography–electrospray ionization triple quadrupole tandem mass spectrometry; RI, retention index; ROS, reactive oxygen species; RSD, relative standard deviation; RuBisCO, ribulose-1,5-bisphosphate carboxylase/oxygenase; S medium, Spirulina medium, S/N: signal-to-noise ratio; TCA, tricarboxylic acid cycle; TIC, total ion current; TMS, trimethylsilyl group; XIC, extracted ion chromatogram.

References

1. Mehner, T. *Encyclopedia of Inland Waters*; Academic Press: Cambridge, MA, USA, 2009.
2. Demoulin, C.F.; Lara, Y.J.; Cornet, L.; François, C.; Baurain, D.; Wilmotte, A.; Javaux, E.J. Cyanobacteria evolution: Insight from the fossil record. *Free Radic. Biol. Med.* **2019**, *140*, 206–223. [CrossRef] [PubMed]
3. Waditee-Sirisattha, R.; Kageyama, H. Extremophilic cyanobacteria. In *Cyanobacterial Physiology*; Elsevier: Amsterdam, The Netherlands, 2022; pp. 85–99. [CrossRef]

4. Merino, N.; Aronson, H.S.; Bojanova, D.P.; Feyhl-Buska, J.; Wong, M.L.; Zhang, S.; Giovannelli, D. Living at the extremes: Extremophiles and the limits of life in a planetary context. *Front. Microbiol.* **2019**, *10*, 780. [CrossRef]
5. Pagels, F.; Vasconcelos, V.; Guedes, A.C. Carotenoids from cyanobacteria: Biotechnological potential and optimization strategies. *Biomolecules* **2021**, *11*, 735. [CrossRef] [PubMed]
6. Jain, S.; Prajapat, G.; Abrar, M.; Ledwani, L.; Singh, A.; Agrawal, A. Cyanobacteria as efficient producers of mycosporine-like amino acids. *J. Basic Microbiol.* **2017**, *57*, 715–727. [CrossRef]
7. Singh, D.P.; Prabha, R.; Verma, S.; Meena, K.K.; Yandigeri, M. Antioxidant properties and polyphenolic content in terrestrial cyanobacteria. *3 Biotech.* **2017**, *7*, 134. [CrossRef] [PubMed]
8. Latifi, A.; Ruiz, M.; Zhang, C.-C. Oxidative stress in cyanobacteria. *FEMS Microbiol. Rev.* **2009**, *33*, 258–278. [CrossRef]
9. Harada, M.; Akiyama, A.; Furukawa, R.; Yokobori, S.-i.; Tajika, E.; Yamagishi, A. Evolution of superoxide dismutases and catalases in cyanobacteria: Occurrence of the antioxidant enzyme genes before the rise of atmospheric oxygen. *J. Mol. Evol.* **2021**, *89*, 527–543. [CrossRef]
10. Singh, V.K.; Jha, S.; Rana, P.; Mishra, S.; Kumari, N.; Singh, S.C.; Anand, S.; Upadhye, V.; Sinha, R.P. Resilience and mitigation strategies of cyanobacteria under ultraviolet radiation stress. *Int. J. Mol. Sci.* **2023**, *24*, 12381. [CrossRef]
11. Dabravolski, S.A.; Isayenkov, S.V. Metabolites facilitating adaptation of desert cyanobacteria to extremely arid environments. *Plants* **2022**, *11*, 3225. [CrossRef]
12. Potts, M. Mechanisms of desiccation tolerance in cyanobacteria. *Eur. J. Phycol.* **1999**, *34*, 319–328. [CrossRef]
13. Oren, A. Diversity of Organic Osmotic Compounds and Osmotic Adaptation in Cyanobacteria and Algae. In *Algae and Cyanobacteria in Extreme Environments*; Springer: Berlin, Germany, 2007; pp. 639–655. [CrossRef]
14. Kirst, G.; Thiel, C.; Wolff, H.; Nothnagel, J.; Wanzek, M.; Ulmke, R. Dimethylsulfoniopropionate (DMSP) in icealgae and its possible biological role. *Mar. Chem.* **1991**, *35*, 381–388. [CrossRef]
15. Dasauni, K.; Divya; Nailwal, T.K. Cyanobacteria in cold ecosystem: Tolerance and adaptation. In *Survival Strategies in Cold-adapted Microorganisms*; Springer: Berlin, Germany, 2022; pp. 1–29. [CrossRef]
16. Joset, F.; Jeanjean, R.; Hagemann, M. Dynamics of the response of cyanobacteria to salt stress: Deciphering the molecular events. *Physiol. Plant.* **1996**, *96*, 738–744. [CrossRef]
17. Kageyama, H.; Waditee-Sirisattha, R. Osmoprotectant molecules in cyanobacteria: Their basic features, biosynthetic regulations, and potential applications. In *Cyanobacterial Physiology*; Elsevier: Amsterdam, The Netherlands, 2022; pp. 113–123. [CrossRef]
18. Kageyama, H.; Waditee-Sirisattha, R. Halotolerance mechanisms in salt-tolerant cyanobacteria. In *Advances in Applied Microbiology*; Elsevier: Amsterdam, The Netherlands, 2023; Volume 124, pp. 55–117.
19. Yadav, P.; Singh, R.P.; Rana, S.; Joshi, D.; Kumar, D.; Bhardwaj, N.; Gupta, R.K.; Kumar, A. Mechanisms of stress tolerance in cyanobacteria under extreme conditions. *Stresses* **2022**, *2*, 531–549. [CrossRef]
20. Prause, M.; Schulz, H.J.; Wagler, D. Rechnergestützte Führung von Fermentationsprozessen, Teil 2. *Acta Biotechnol.* **1984**, *4*, 143–151. [CrossRef]
21. Drobac-Čik, A.V.; Dulić, T.I.; Stojanović, D.B.; Svirčev, Z.B. The importance of extremophile cyanobacteria in the production of biologically active compounds. *Zb. Matice Srp. Prir. Nauk.* **2007**, *112*, 57–66. [CrossRef]
22. Bourgade, B.; Stensjö, K. Synthetic biology in marine cyanobacteria: Advances and challenges. *Front. Microbiol.* **2022**, *13*, 994365. [CrossRef]
23. Kochhar, N.; Shrivastava, S.; Ghosh, A.; Rawat, V.S.; Sodhi, K.K.; Kumar, M. Perspectives on the microorganism of extreme environments and their applications. *Curr. Res. Microb. Sci.* **2022**, *3*, 100134. [CrossRef]
24. Fiehn, O. Metabolomics by gas chromatography–mass spectrometry: Combined targeted and untargeted profiling. *Curr. Protoc. Mol. Biol.* **2016**, *114*, 30.4.1–30.4.32. [CrossRef]
25. Zhou, B.; Xiao, J.F.; Tuli, L.; Ressom, H.W. LC-MS-based metabolomics. *Mol. BioSyst.* **2012**, *8*, 470–481. [CrossRef]
26. t'Kindt, R.; Morreel, K.; Deforce, D.; Boerjan, W.; Van Bocxlaer, J. Joint GC–MS and LC–MS platforms for comprehensive plant metabolomics: Repeatability and sample pre-treatment. *J. Chromatogr. B* **2009**, *877*, 3572–3580. [CrossRef]
27. Schwarz, D.; Orf, I.; Kopka, J.; Hagemann, M. Recent applications of metabolomics toward cyanobacteria. *Metabolites* **2013**, *3*, 72–100. [CrossRef] [PubMed]
28. Ren, J.-L.; Zhang, A.-H.; Kong, L.; Wang, X.-J. Advances in mass spectrometry-based metabolomics for investigation of metabolites. *RSC Adv.* **2018**, *8*, 22335–22350. [CrossRef] [PubMed]
29. Lorenz, M.; Friedl, T.; Day, J.G. Perpetual Maintenance of Actively Metabolizing Microalgal Cultures. In *Algal Culturing Techniques*; Academic Press: New York, NY, USA, 2005; pp. 145–156. [CrossRef]
30. Yuorieva, N.; Sinetova, M.; Messineva, E.; Kulichenko, I.; Fomenkov, A.; Vysotskaya, O.; Osipova, E.; Baikalova, A.; Prudnikova, O.; Titova, M. Plants, cells, algae, and cyanobacteria in vitro and cryobank collections at the Institute of Plant Physiology, Russian Academy of Sciences—A platform for research and production center. *Biology* **2023**, *12*, 838. [CrossRef]

31. Samylina, O.S.; Sinetova, M.A.; Kupriyanova, E.V.; Starikov, A.Y.; Sukhacheva, M.V.; Dziuba, M.V.; Tourova, T.P. Ecology and biogeography of the 'marine *Geitlerinema*' cluster and a description of *Sodalinema orleanskyi* sp. nov., *Sodalinema gerasimenkoae* sp. nov., *Sodalinema stalii* sp. nov. and *Baaleninema simplex* gen. et sp. nov. (Oscillatoriales, Cyanobacteria). *FEMS Microbiol. Ecol.* **2021**, *97*, fiab104. [CrossRef]
32. Bozieva, A.M.; Khasimov, M.K.; Voloshin, R.A.; Sinetova, M.A.; Kupriyanova, E.V.; Zharmukhamedov, S.K.; Dunikov, D.O.; Tsygankov, A.A.; Tomo, T.; Allakhverdiev, S.I. New cyanobacterial strains for biohydrogen production. *Int. J. Hydrog. Energy* **2023**, *48*, 7569–7581. [CrossRef]
33. Anishchenko, O.V.; Gladyshev, M.I.; Kravchuk, E.S.; Ivanova, E.A.; Gribovskaya, I.V.; Sushchik, N.N. Seasonal variations of metal concentrations in periphyton and taxonomic composition of the algal community at a Yenisei River littoral site. *Cent. Eur. J. Biol.* **2010**, *5*, 125–134. [CrossRef]
34. Grant, W.; Mwatha, W.; Jones, B. Alkaliphiles: Ecology, diversity and applications. *FEMS Microbiol. Rev.* **1990**, *6*, 255–269. [CrossRef]
35. Jones, B.E.; Grant, W.D.; Duckworth, A.W.; Owenson, G.G. Microbial diversity of soda lakes. *Extremophiles* **1998**, *2*, 191–200. [CrossRef]
36. Samylina, O.S.; Kosyakova, A.I.; Krylov, A.A.; Sorokin, D.Y.; Pimenov, N.V. Salinity-induced succession of phototrophic communities in a southwestern Siberian soda lake during the solar activity cycle. *Heliyon* **2024**, *10*, e26120. [CrossRef]
37. Stanier, R.Y.; Kunisawa, R.; Mandel, M.; Cohen-Bazire, G. Purification and properties of unicellular blue-green algae (order Chroococcales). *Bacteriol. Rev.* **1971**, *35*, 171–205. [CrossRef] [PubMed]
38. Vladimirova, M.; Barcevič, E.; Žoldakov, I.; Epifanova, O.; Markelova, A.; Maslova, I.; Kupcova, E. IPPAS-коллекция культур микроводорослей Института физиологии растений им. КА Тимирязева АН СССР. In Каталог Культур Коллекций СССР, M; Nauka: Moscow, Russia, 1991; pp. 8–61.
39. Rippka, R.; Deruelles, J.; Waterbury, J.B.; Herdman, M.; Stanier, R.Y. Generic assignments, strain histories and properties of pure cultures of cyanobacteria. *Microbiology* **1979**, *111*, 1–61. [CrossRef]
40. Sarsekeyeva, F.K.; Usserbaeva, A.A.; Zayadan, B.K.; Mironov, K.S.; Sidorov, R.A.; Kozlova, A.Y.; Kupriyanova, E.V.; Sinetova, M.A.; Los, D.A. Isolation and characterization of a new cyanobacterial strain with a unique fatty acid composition. *Adv. Microbiol.* **2014**, *4*, 1033. [CrossRef]
41. Harvey, D.J.; Vouros, P. Mass spectrometric fragmentation of trimethylsilyl and related alkylsilyl derivatives. *Mass Spectrom. Rev.* **2020**, *39*, 105–211. [CrossRef] [PubMed]
42. Benjamini, Y.; Hochberg, Y. Controlling the false discovery rate: A practical and powerful approach to multiple testing. *J. R. Stat. Soc. Ser. B Methodol.* **1995**, *57*, 289–300. [CrossRef]
43. Golubic, S. Halophily and halotolerance in cyanophytes. *Orig. Life* **1980**, *10*, 169–183. [CrossRef]
44. Wydro, R.; Kogut, M.; Kushner, D. Salt response of ribosomes of a moderately halophilic bacterium. *FEBS Lett.* **1975**, *60*, 210–215. [CrossRef]
45. Gacitua, M.; Urrejola, C.; Carrasco, J.; Vicuña, R.; Srain, B.M.; Pantoja-Gutiérrez, S.; Leech, D.; Antiochia, R.; Tasca, F. Use of a thermophile desiccation-tolerant cyanobacterial culture and Os redox polymer for the preparation of photocurrent producing anodes. *Front. Bioeng. Biotechnol.* **2020**, *8*, 900. [CrossRef]
46. Rasul, F.; You, D.; Jiang, Y.; Liu, X.; Daroch, M. Thermophilic cyanobacteria—Exciting, yet challenging biotechnological chassis. *Appl. Microbiol. Biotechnol.* **2024**, *108*, 270. [CrossRef]
47. Levis, N.A.; Pfennig, D.W. Evolution: Ancestral plasticity promoted extreme temperature adaptation in thermophilic bacteria. *Curr. Biol.* **2020**, *30*, R68–R70. [CrossRef]
48. Chen, M.-Y.; Teng, W.-K.; Zhao, L.; Hu, C.-X.; Zhou, Y.-K.; Han, B.-P.; Song, L.-R.; Shu, W.-S. Comparative genomics reveals insights into cyanobacterial evolution and habitat adaptation. *ISME J.* **2021**, *15*, 211–227. [CrossRef]
49. Komárek, J.; Kaštovský, J.; Mareš, J.; Johansen, J.R. Taxonomic classification of cyanoprokaryotes (cyanobacterial genera) 2014, using a polyphasic approach. *Preslia* **2014**, *86*, 295–335.
50. Stal, L.J.; Paerl, H.W.; Bebout, B.; Villbrandt, M. Heterocystous versus non-heterocystous cyanobacteria in microbial mats. In Proceedings of the Microbial Mats: Structure, Development and Environmental Significance; Springer: Berlin, Heidelberg, 2014; pp. 403–414. [CrossRef]
51. Flores, E. Studies on the regulation of arginine metabolism in cyanobacteria should include mixotrophic conditions. *mBio* **2021**, *12*, e0143321. [CrossRef] [PubMed]
52. Katayama, N.; Osanai, T. Arginine inhibition of the argininosuccinate lyases is conserved among three orders in cyanobacteria. *Plant Mol. Biol.* **2022**, *110*, 13–22. [CrossRef]
53. Pade, N.; Hagemann, M. Salt acclimation of cyanobacteria and their application in biotechnology. *Life* **2014**, *5*, 25–49. [CrossRef] [PubMed]

54. Tamaru, Y.; Takani, Y.; Yoshida, T.; Sakamoto, T. Crucial role of extracellular polysaccharides in desiccation and freezing tolerance in the terrestrial cyanobacterium *Nostoc commune*. *Appl. Environ. Microbiol.* **2005**, *71*, 7327–7333. [CrossRef] [PubMed]
55. Bergman, B. Glyoxylate induced changes in the carbon and nitrogen metabolism of the cyanobacterium Anabaena cylindrica. *Plant Physiol.* **1986**, *80*, 698–701. [CrossRef]
56. Otero, A.; Vincenzini, M. *Nostoc* (cyanophyceae) goes nude: Extracellular polysaccharides serve as a sink for reducing power under unbalanced c/n metabolism 1. *J. Phycol.* **2004**, *40*, 74–81. [CrossRef]
57. Laroche, C. Exopolysaccharides from microalgae and cyanobacteria: Diversity of strains, production strategies, and applications. *Mar. Drugs* **2022**, *20*, 336. [CrossRef]
58. Rai, K.K.; Pandey, N.; Rai, S.P. Salicylic acid and nitric oxide signaling in plant heat stress. *Physiol. Plant.* **2020**, *168*, 241–255. [CrossRef]
59. Yadava, P. Salicylic acid alleviates methyl viologen induced oxidative stress through transcriptional modulation of antioxidant genes in *Zea mays* L. *Maydica* **2016**, *60*, 1–9.
60. Holuigue, L.; Salinas, P.; Blanco, F.; GarretÓn, V. Salicylic Acid and Reactive Oxygen Species in the Activation of Stress Defense Genes. In *Salicylic acid: A Plant Hormone*; Springer: Berlin, Germany, 2007; pp. 197–246. [CrossRef]
61. Wani, A.B.; Chadar, H.; Wani, A.H.; Singh, S.; Upadhyay, N. Salicylic acid to decrease plant stress. *Environ. Chem. Lett.* **2017**, *15*, 101–123. [CrossRef]
62. Toribio, A.; Suárez-Estrella, F.; Jurado, M.; López, M.; López-González, J.; Moreno, J. Prospection of cyanobacteria producing bioactive substances and their application as potential phytostimulating agents. *Biotechnol. Rep.* **2020**, *26*, e00449. [CrossRef]
63. Han, C.-F.; Liu, S.-T.; Yan, R.-R.; Li, J.; Chen, N.; Zhang, L.-L.; Jia, S.-R.; Han, P.-P. Salicylic Acid and Jasmonic Acid Increase the Polysaccharide Production of *Nostoc flagelliforme* via the Regulation of the Intracellular NO Level. *Foods* **2023**, *12*, 915. [CrossRef]
64. Li, M.Y.; Li, Y.R.; Han, C.F.; Zhang, J.; Zhu, R.Y.; Zhang, Y.; Li, J.; Jia, S.R.; Han, P.P. Nitric oxide mediates positive regulation of *Nostoc flagelliforme* polysaccharide yield via potential S-nitrosylation of G6PDH and UGDH. *BMC Biotechnol.* **2024**, *24*, 58. [CrossRef] [PubMed]
65. Nawaz, T.; Saud, S.; Gu, L.; Khan, I.; Fahad, S.; Zhou, R. Cyanobacteria: Harnessing the power of microorganisms for plant growth promotion, stress alleviation, and phytoremediation in the Era of sustainable agriculture. *Plant Stress* **2024**, *11*, 100399. [CrossRef]
66. Zhou, Y.; Bao, J.; Zhang, D.; Li, Y.; Li, H.; He, H. Effect of heterocystous nitrogen-fixing cyanobacteria against rice sheath blight and the underlying mechanism. *Appl. Soil Ecol.* **2020**, *153*, 103580. [CrossRef]
67. Inthasotti, T.; Pathom-aree, W. Diversity of actinobacteria associated with *Nostoc commune* Vaucher ex Bornet & Flahault macrocolonies. *Ann. Microbiol.* **2015**, *65*, 2229–2240. [CrossRef]
68. Malviya, M.K.; Pandey, A.; Trivedi, P.; Gupta, G.; Kumar, B. Chitinolytic activity of cold tolerant antagonistic species of Streptomyces isolated from glacial sites of Indian Himalaya. *Curr. Microbiol.* **2009**, *59*, 502–508. [CrossRef]
69. Helm, R.F.; Huang, Z.; Edwards, D.; Leeson, H.; Peery, W.; Potts, M. Structural characterization of the released polysaccharide of desiccation-tolerant *Nostoc commune* DRH-1. *J. Bacteriol.* **2000**, *182*, 974–982. [CrossRef]
70. Wang, X.; Yang, Z.; Liu, Y.; Wang, X.; Zhang, H.; Shang, R.; Laba, C.; Wujin, C.; Hao, B.; Wang, S. Structural characteristic of polysaccharide isolated from *Nostoc commune*, and their potential as radical scavenging and antidiabetic activities. *Sci. Rep.* **2022**, *12*, 22155. [CrossRef]
71. Van der Woude, A.D.; Perez Gallego, R.; Vreugdenhil, A.; Puthan Veetil, V.; Chroumpi, T.; Hellingwerf, K.J. Genetic engineering of *Synechocystis* PCC6803 for the photoautotrophic production of the sweetener erythritol. *Microb. Cell Factories* **2016**, *15*, 60. [CrossRef] [PubMed]
72. Yang, L.-B.; Dai, X.-M.; Zheng, Z.-Y.; Zhu, L.; Zhan, X.-B.; Lin, C.-C. Proteomic analysis of erythritol-producing *Yarrowia lipolytica* from glycerol in response to osmotic pressure. *J. Microbiol. Biotechnol.* **2015**, *25*, 1056–1069. [CrossRef]
73. Okuda, K.; Hayashi, H.; Nishiyama, Y. Systematic characterization of the ADP-ribose pyrophosphatase family in the Cyanobacterium *Synechocystis* sp. strain PCC 6803. *J. Bacteriol.* **2005**, *187*, 4984–4991. [CrossRef] [PubMed]
74. Jacobson, E.L.; Cervantes-Laurean, D.; Jacobson, M.K. Glycation of proteins by ADP-ribose. In *ADP-Ribosylation: Metabolic Effects and Regulatory Functions*; Springer: Berlin, Germany, 1994; pp. 207–212. [CrossRef]
75. Kato, Y.; Hidese, R.; Matsuda, M.; Ohbayashi, R.; Ashida, H.; Kondo, A.; Hasunuma, T. Glycogen deficiency enhances carbon partitioning into glutamate for an alternative extracellular metabolic sink in cyanobacteria. *Commun. Biol.* **2024**, *7*, 233. [CrossRef] [PubMed]
76. Golden, J.W.; Yoon, H.-S. Heterocyst development in *Anabaena*. *Curr. Opin. Microbiol.* **2003**, *6*, 557–563. [CrossRef]
77. Halimatul, H.S.M.; Ehira, S.; Awai, K. Fatty alcohols can complement functions of heterocyst specific glycolipids in *Anabaena* sp. PCC 7120. *Biochem. Biophys. Res. Commun.* **2014**, *450*, 178–183. [CrossRef]

78. Gambacorta, A.; Romano, I.; Trincone, A.; Soriente, A.; Giordano, M.; Sodano, G. Heterocyst glycolipids from five nitrogen-fixing cyanobacteria. *Gazz. Chim. Ital.* **1996**, *126*, 653–655.
79. Lucius, S.; Hagemann, M. The primary carbon metabolism in cyanobacteria and its regulation. *Front. Plant Sci.* **2024**, *15*, 1417680. [CrossRef]
80. Kudoh, K.; Hotta, S.; Sekine, M.; Fujii, R.; Uchida, A.; Kubota, G.; Kawano, Y.; Ihara, M. Overexpression of endogenous 1-deoxy-d-xylulose 5-phosphate synthase (DXS) in cyanobacterium *Synechocystis* sp. PCC6803 accelerates protein aggregation. *J. Biosci. Bioeng.* **2017**, *123*, 590–596. [CrossRef]
81. Clercin, N.A.; Koltsidou, I.; Picard, C.J.; Druschel, G.K. Prevalence of Actinobacteria in the production of 2-methylisoborneol and geosmin, over Cyanobacteria in a temperate eutrophic reservoir. *Chem. Eng. J. Adv.* **2022**, *9*, 100226. [CrossRef]
82. Mills, L.A.; McCormick, A.J.; Lea-Smith, D.J. Current knowledge and recent advances in understanding metabolism of the model cyanobacterium *Synechocystis* sp. PCC 6803. *Biosci. Rep.* **2020**, *40*, BSR20193325. [CrossRef]
83. Hagemann, M.; Erdmann, N. Activation and pathway of glucosylglycerol synthesis in the cyanobacterium Synechocystis sp. PCC 6803. *Microbiology* **1994**, *140*, 1427–1431. [CrossRef]
84. Kuprijanova, E.; Samylina, O. CO_2-concentrating mechanism and its traits in haloalkaliphilic cyanobacteria. *Microbiology* **2015**, *84*, 112–124. [CrossRef]
85. Yi, L.; Solanki, R.; Strous, M. In search of the pH limit of growth in halo-alkaliphilic cyanobacteria. *Environ. Microbiol. Rep.* **2024**, *16*, e13323. [CrossRef]
86. Sinetova, M.A.; Los, D.A. New insights in cyanobacterial cold stress responses: Genes, sensors, and molecular triggers. *Biochim. Biophys. Acta BBA-Gen. Subj.* **2016**, *1860*, 2391–2403. [CrossRef]
87. Qi, H.; Zhao, S.; Wen, J.; Chen, Y.; Jia, X. Analysis of ascomycin production enhanced by shikimic acid resistance and addition in Streptomyces hygroscopicus var. ascomyceticus. *Biochem. Eng. J.* **2014**, *82*, 124–133. [CrossRef]
88. Fathima, A.M.; Laviña, W.A.; Putri, S.P.; Fukusaki, E. Accumulation of sugars and nucleosides in response to high salt and butanol stress in 1-butanol producing Synechococcus elongatus. *J. Biosci. Bioeng.* **2020**, *129*, 177–183. [CrossRef] [PubMed]
89. Liu, L.; Herfindal, L.; Jokela, J.; Shishido, T.K.; Wahlsten, M.; Døskeland, S.O.; Sivonen, K. Cyanobacteria from terrestrial and marine sources contain apoptogens able to overcome chemoresistance in acute myeloid leukemia cells. *Mar. Drugs* **2014**, *12*, 2036–2053. [CrossRef]
90. Humisto, A.; Herfindal, L.; Jokela, J.; Karkman, A.; Bjørnstad, R.; Choudhury, R.R.; Sivonen, K. Cyanobacteria as a source for novel anti-leukemic compounds. *Curr. Pharm. Biotechnol.* **2016**, *17*, 78–91. [CrossRef]
91. Ozturk, S.; Aslim, B. Modification of exopolysaccharide composition and production by three cyanobacterial isolates under salt stress. *Environ. Sci. Pollut. Res.* **2010**, *17*, 595–602. [CrossRef] [PubMed]
92. Damrow, R.; Maldener, I.; Zilliges, Y. The multiple functions of common microbial carbon polymers, glycogen and PHB, during stress responses in the non-diazotrophic cyanobacterium *Synechocystis* sp. PCC 6803. *Front. Microbiol.* **2016**, *7*, 966. [CrossRef]
93. Koch, M.; Berendzen, K.W.; Forchhammer, K. On the role and production of polyhydroxybutyrate (PHB) in the cyanobacterium Synechocystis sp. PCC 6803. *Life* **2020**, *10*, 47. [CrossRef]
94. Flores, E.; Arévalo, S.; Burnat, M. Cyanophycin and arginine metabolism in cyanobacteria. *Algal Res.* **2019**, *42*, 101577. [CrossRef]
95. Gehringer, M.M.; Wannicke, N. Climate change and regulation of hepatotoxin production in Cyanobacteria. *FEMS Microbiol. Ecol.* **2014**, *88*, 1–25. [CrossRef] [PubMed]
96. Ermilova, E.; Forchhammer, K. PII signaling proteins of cyanobacteria and green algae. New features of conserved proteins. *Russ. J. Plant Physiol.* **2013**, *60*, 483–490. [CrossRef]
97. Carlsson, P.; Rita, D. Sedimentation of *Nodularia spumigena* and distribution of nodularin in the food web during transport of a cyanobacterial bloom from the Baltic Sea to the Kattegat. *Harmful Algae* **2019**, *86*, 74–83. [CrossRef]
98. Moffitt, M.C.; Neilan, B.A. Characterization of the nodularin synthetase gene cluster and proposed theory of the evolution of cyanobacterial hepatotoxins. *Appl. Environ. Microbiol.* **2004**, *70*, 6353–6362. [CrossRef]
99. Lage, S.; Mazur-Marzec, H.; Gorokhova, E. Competitive interactions as a mechanism for chemical diversity maintenance in Nodularia spumigena. *Sci. Rep.* **2021**, *11*, 8970. [CrossRef]
100. Lankoff, A.; Wojcik, A.; Fessard, V.; Meriluoto, J. Nodularin-induced genotoxicity following oxidative DNA damage and aneuploidy in HepG2 cells. *Toxicol. Lett.* **2006**, *164*, 239–248. [CrossRef]
101. Silveira, S.B.; Odebrecht, C. Effects of salinity and temperature on the growth, toxin production, and akinete germination of the cyanobacterium *Nodularia spumigena*. *Front. Mar. Sci.* **2019**, *6*, 339. [CrossRef]
102. Hobson, P.; Fallowfield, H. Effect of irradiance, temperature and salinity on growth and toxin production by *Nodularia spumigena*. *Hydrobiologia* **2003**, *493*, 7–15. [CrossRef]

103. Sevilla, E.; Martin-Luna, B.; Bes, M.T.; Fillat, M.F.; Peleato, M.L. An active photosynthetic electron transfer chain required for mcyD transcription and microcystin synthesis in *Microcystis aeruginosa* PCC7806. *Ecotoxicology* **2012**, *21*, 811–819. [CrossRef] [PubMed]
104. Pinchart, P.-E.; Marter, P.; Brinkmann, H.; Quilichini, Y.; Mysara, M.; Petersen, J.; Pasqualini, V.; Mastroleo, F. The genus *Limnospira* contains only two species both unable to produce microcystins: *L. maxima* and *L. platensis* comb. nov. *Iscience* **2024**, *27*, 110845. [CrossRef]
105. Kwei, C.K.; Lewis, D.; King, K.; Donohue, W.; Neilan, B.A. Molecular classification of commercial Spirulina strains and identification of their sulfolipid biosynthesis genes. *J. Microbiol. Biotechnol.* **2011**, *21*, 359–365. [CrossRef]
106. Sinetova, M.A.; Kupriyanova, E.V.; Los, D.A. Spirulina/Arthrospira/Limnospira—Three names of the single organism. *Foods* **2024**, *13*, 2762. [CrossRef]
107. Yuzawa, S.; Keasling, J.D.; Katz, L. Bio-based production of fuels and industrial chemicals by repurposing antibiotic-producing type I modular polyketide synthases: Opportunities and challenges. *J. Antibiot.* **2017**, *70*, 378–385. [CrossRef] [PubMed]
108. Roulet, J.; Taton, A.; Golden, J.W.; Arabolaza, A.; Burkart, M.D.; Gramajo, H. Development of a cyanobacterial heterologous polyketide production platform. *Metab. Eng.* **2018**, *49*, 94–104. [CrossRef]
109. Tharasirivat, V.; Jantaro, S. Increased biomass and polyhydroxybutyrate production by *Synechocystis* sp. PCC 6803 overexpressing RuBisCO genes. *Int. J. Mol. Sci.* **2023**, *24*, 6415. [CrossRef]
110. Pattanaik, B.; Lindberg, P. Terpenoids and their biosynthesis in cyanobacteria. *Life* **2015**, *5*, 269–293. [CrossRef]
111. Carninci, P.; Nishiyama, Y.; Westover, A.; Itoh, M.; Nagaoka, S.; Sasaki, N.; Okazaki, Y.; Muramatsu, M.; Hayashizaki, Y. Thermostabilization and thermoactivation of thermolabile enzymes by trehalose and its application for the synthesis of full length cDNA. *Proc. Natl. Acad. Sci. USA* **1998**, *95*, 520–524. [CrossRef] [PubMed]
112. Bilova, T.; Lukasheva, E.; Brauch, D.; Greifenhagen, U.; Paudel, G.; Tarakhovskaya, E.; Frolova, N.; Mittasch, J.; Balcke, G.U.; Tissier, A. A snapshot of the plant glycated proteome: Structural, functional, and mechanistic aspects. *J. Biol. Chem.* **2016**, *291*, 7621–7636. [CrossRef] [PubMed]
113. Bilova, T.; Paudel, G.; Shilyaev, N.; Schmidt, R.; Brauch, D.; Tarakhovskaya, E.; Milrud, S.; Smolikova, G.; Tissier, A.; Vogt, T. Global proteomic analysis of advanced glycation end products in the *Arabidopsis* proteome provides evidence for age-related glycation hot spots. *J. Biol. Chem.* **2017**, *292*, 15758–15776. [CrossRef]
114. Milkovska-Stamenova, S.; Schmidt, R.; Frolov, A.; Birkemeyer, C. GC-MS method for the quantitation of carbohydrate intermediates in glycation systems. *J. Agric. Food Chem.* **2015**, *63*, 5911–5919. [CrossRef] [PubMed]
115. Balcke, G.U.; Bennewitz, S.; Bergau, N.; Athmer, B.; Henning, A.; Majovsky, P.; Jiménez-Gómez, J.M.; Hoehenwarter, W.; Tissier, A. Multi-omics of tomato glandular trichomes reveals distinct features of central carbon metabolism supporting high productivity of specialized metabolites. *Plant Cell* **2017**, *29*, 960–983. [CrossRef]
116. Vinaixa, M.; Samino, S.; Saez, I.; Duran, J.; Guinovart, J.J.; Yanes, O. A guideline to univariate statistical analysis for LC/MS-based untargeted metabolomics-derived data. *Metabolites* **2012**, *2*, 775–795. [CrossRef]
117. Xi, B.; Gu, H.; Baniasadi, H.; Raftery, D. Statistical analysis and modeling of mass spectrometry-based metabolomics data. In *Mass Spectrometry in Metabolomics: Methods and Protocols*; Springer: Berlin, Germany, 2014; pp. 333–353. [CrossRef]

International Journal of
Molecular Sciences

MDPI

Article

Evaluating Drought Tolerance in *Codonopsis pilosula* Seedlings: Combining Growth, Physiology, Yield, and Tolerance Indices

Hongyan Wang, Yuan Chen *, Fengxia Guo, Di Wu, Wei Liang, Pengbin Dong and Jiali Cheng

College of Agronomy, College of Life Science and Technology, State Key Laboratory of Aridland Crop Science, Gansu Agricultural University, Lanzhou 730070, China; why8852@163.com (H.W.); guofx@gsau.edu.cn (F.G.); andywu173@163.com (D.W.); liangw@gsau.edu.cn (W.L.); dongpb@stumail.nwu.edu.cn (P.D.); chengjiali24@icloud.com (J.C.)

* Correspondence: chenyuan@gsau.edu.cn

Abstract: Drought stress during the *Codonopsis pilosula* (Campanulaceae) seedling stage significantly affects its growth, quality, and yield. The aim of this study was to identify drought-tolerant cultivars of *C. pilosula* by using the growth—physiology—yield composite index (GPYCI) and drought-tolerant indices. Nine *C. pilosula* cultivars were evaluated under normal-watered (black plastic film hole sowing, BF) and water-stressed (spread in the open field, SF) conditions in a design that adopted a two-factor paired experiment with three replications. The emergence rate was significantly influenced by the water treatment, while both the water treatment and the cultivar affected root length, proline content, APX activity, and chlorophyll levels. The G1 cultivar performed better than others in multiple aspects. Yields and their attributes varied among cultivars under different water levels. The average yield was 7350.76 kg/hm^2 under BF conditions and 4856.32 kg/hm^2 under SF conditions. Drought stress reduced the total root length, single root fresh weight, and yield by 18.33%, 28.4%, and 33.9%, respectively. Correlation analysis revealed unique physiological response mechanisms to water changes among cultivars. Drought tolerance indices and comprehensive factor analysis indicated varying levels of drought tolerance among cultivars. This study has provided valuable insights into the growth, physiology, and yield response of *C. pilosula* under drought conditions and laid the foundation for breeding drought-tolerant cultivars.

Keywords: *Codonopsis pilosula*; drought stress; growth-physiology-yield composite index; seedlings; hole sowing

1. Introduction

Climate change, a concern affecting everyone, has brought many challenging tasks to modern agriculture [1]. The rising temperature and fluctuating precipitation caused by climate change have been widely reported to intensify the frequency and severity of drought on a global scale [1,2]. Drought-induced agricultural losses account for approximately 70% of the global potential yield losses each year [3]. *Codonopsis pilosula*, a traditional herbal plant, and its dry roots, known as *Codonopsis radix* or Dangshen [4], can be used for both food and medicine [5] and are widely used in Asian countries.

Gansu Province is the regional production area of *C. pilosula* [6]. It features a dry climate, with 75% of its total area being arid or semi-arid regions. Seasonal drought and drought during the critical growth period severely hinder the quality and yield of *Codonopsis radix* [7]. Breeding drought-tolerant new *C. pilosula* cultivars is a crucial strategy to solve this issue. *C. pilosula* reproduces via seeds [8], and the yield and quality of *Codonopsis radix* are largely dependent on the quality of its seedlings [9]. *Codonopsis* seedlings demand more water than during the stage of medicinal value formation of *C. pilosula*, mainly because water deprivation causes a decrease in water potential, nutrient uptake, and photosynthesis, as well as induces oxidative damage from reactive oxygen species (ROS) and the disturbance of metabolism, finally resulting in a reduced production and yield quality of *Codonopsis* seedlings [10–12]. However, plant responses in terms of leaf water relations, stomatal regulation, photosynthesis, and other regulatory processes vary among species and cultivars [13]. For example, Zhang et al. [14] suggests that water stress reduces the stomatal conductance of field-grown maize through osmotic adjustment, thereby improving water use efficiency. Shi et al. [15] suggests that a higher rate of soluble sugar accumulation in the source is one of the reasons for triggering non-sequential leaf senescence. A higher rate of soluble sugar mobilization during non-sequential senescence contributes to high and stable wheat yield and drought tolerance. At the same time, there were significant differences and a negative correlation in stomatal density and size between indica and japonica rice [16]. The growth–physiology–yield comprehensive index (GPYCI) is an important reference for enhancing drought resistance as it is related to the plant's adaptive mechanisms under stress conditions [17].

Identifying traits that tolerate drought stress can help breeders develop drought-resistant plant types. Hafeez et al. [18] believe that in the drought-tolerant breeding of wheat, selecting plants with excellent traits such as efficient water use and deep root systems is of great significance for improving breeding strategies. Toulotte et al. [19] believe that utilizing the rich genetic diversity of wild crop ancestors is a promising way. These wild relatives can thrive in diverse environments and have many adaptive traits. Identifying and introducing these traits into staple cereals can cut yield losses under stress. Through the evaluation of 114 rice genotypes under water stress, Verma and Sarma [20] found significant variations in the root and shoot traits of each genotype. Eleven significant marker—trait associations and a new QTL related to root length associated with RM127 were detected. The marker alleles related to these can be used in marker-assisted selection to improve root and drought tolerance traits. Although *C. pilosula* exhibits high genetic diversity, studies assessing the genotypes of the plant under drought stress using the GPYCI and drought index are relatively scarce. Selecting the best-performing cultivars from under well-watered and water-stressed conditions is an effective method for identifying drought-resistant genotypes [21]. Several drought tolerance indices have been proposed in other research [1,22,23], most of which could be used in this study to select the best *C. pilosula* cultivars for drought tolerance. We hypothesized that the responses of *C. pilosula* cultivars vary significantly under different water levels, and that the best performing cultivar can be chosen using the growth-physiology-yield composite index (GPYCI) and drought tolerance indices. Therefore, the present work was aimed at evaluating the GPYCI of the seedlings in different *C. pilosula* cultivars under well-watered and water-stressed conditions and identifying the best drought-tolerant cultivar using drought tolerance indices.

2. Results

2.1. Variations in Stress-Resistant Physiological Indicators of Different *C. pilosula* Cultivar Seedlings

As shown in the soil measurements on July 30 (Figure 1a), the average soil moisture content in the open field plots (SF) for each treatment was 136.5 g/kg. Given the preference of *Codonopsis pilosula* seedlings for humidity, this moisture level might slow down their growth, potentially limiting plant height and root length. In contrast, the soil water content in the black plastic film hole-sown plots (BF) was approximately 176.1 g/kg, which could foster better growth and development of *C. pilosula*, potentially increasing yield and quality. Thus, the soil conditions in this study are suitable for assessing the drought resistance of various *C. pilosula* cultivars.

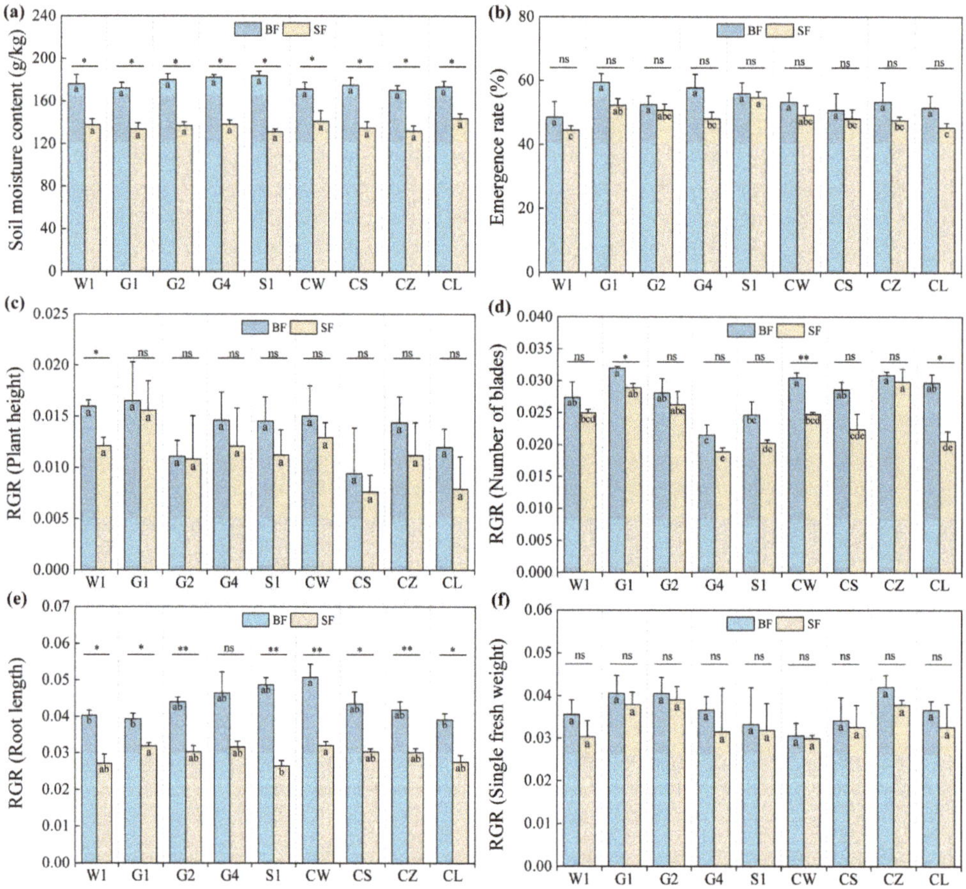

Figure 1. Growth indicators of C. pilosula cultivar seedlings. The data in the figure represent the $\overline{X} \pm$ SD (standard deviation) (r = 3). (**a**) Soil moisture content; (**b**) emergence rate; (**c**) RGR (plant height); (**d**) RGR (number of blades); (**e**) RGR (root length); (**f**) RGR (single fresh weight); Different lowercase letters mean a significant difference ($p < 0.05$); '*' indicates a significant difference ($p < 0.05$); '**' indicates a highly significant difference ($p < 0.01$); 'ns' indicates that the difference is not significant.

Figure 1 illustrates significant trait variations among *C. pilosula* cultivars under BF and SF conditions. Under the BF condition, the emergence rate did not differ significantly among the cultivars, with the G1 cultivar showing the highest rate. However, under the SF condition, significant differences were observed, particularly with the S1 and G1 cultivars having higher emergence rates than the W1 cultivar (Figure 1b). The analysis of variance indicated that the water treatment significantly affected the emergence rate, RGR (plant height), RGR (number of blades), and RGR (root length), but the cultivar only significantly affected RGR (number of blades) and has an interaction effect on RGR (single fresh weight)

(Table 1). For the relative growth rate of the number of leaves, both the water treatment and the cultivar had significant effects. Under the BF condition, the G1 cultivar had the highest rate, while the G4 cultivar was significantly lower than the W1 cultivar. Under the SF condition, the CZ cultivar exhibited the highest relative growth rate, with the G4 cultivar again being significantly lower than the W1 cultivar (Figure 1d). The water treatment significantly influenced the emergence rate, the relative growth rate of plant height, and the root length of *C. pilosula* seedlings. Both the water treatment and the cultivar significantly influenced the relative growth rate of the number of blades (Table 1). Under the BF condition, the CW cultivar had the highest rate, but water stress impeded root growth. Under the SF condition, the G1 cultivar showed the highest relative growth rate, indicating better water and nutrient uptake through root development, which is crucial for drought tolerance (Figure 1e).

Table 1. Analysis of variance for the main and interaction effects of growth indicators.

Source	df	MS/Sig.				
		Emergence Rate	RGR (Plant Height)	RGR (Number of Blades)	RGR (Root Length)	RGR (Single Fresh Weight)
W	1	305.422 **	8.138×10^{-5} *	0.000 **	0.002 **	0.000 ns
C	8	56.263 ns	3.240×10^{-5} ns	6.704×10^{-5} **	2.811×10^{-5} ns	7.693×10^{-5} ns
Block	2	19.454 ns	0.000 **	1.121×10^{-5} ns	0.000 **	0.000 ns
W × C	8	11.319 ns	2.572×10^{-6} ns	9.737×10^{-6} ns	6.083×10^{-5} ns	4.734×10^{-6} **
Error	34	36.978	1.647×10^{-5}	6.393×10^{-6}	2.754×10^{-5}	6.584×10^{-5}

'W' means water level; 'C' means cultivar; 'Block' means replications; 'Error' means experimental error; '*' indicates a significant difference ($p < 0.05$); '**' indicates a highly significant difference ($p < 0.01$); 'ns' indicates that the difference is not significant.

Water stress significantly impacts the rate and efficiency of various physiological and biochemical traits by disrupting physiological pathways. Significant alterations were observed in the relative water content (RWC) of leaves, proline content, antioxidant enzyme activities (such as superoxide dismutase (SOD), catalase (CAT), peroxidase (POD), and ascorbate peroxidase (APX)), as well as the chlorophyll content, which were influenced by both the cultivar and the water conditions (Figure 2, Table 2). The cultivars had no significant effect on the RWC of leaves (Figure 2a, Table 2), but they did influence other physiological indicators and had no interaction effect on RGR (Leaf), POD, SOD, and CAT. Under the SF condition, the proline content notably increased in the CZ, CS, CW, and G1 cultivars compared to the W1 cultivar (Figure 2b). The CW cultivar exhibited the highest activity of the APX enzyme before and after stress, while the G2 cultivar showed the most significant change in APX activity (Figure 2c). Under the SF condition, the G1 cultivar had the highest activities of CAT and POD, and there was no significant difference in SOD activity among the cultivars under both treatments (Figure 2d–f). The G1 cultivar had the highest levels of chlorophylls a and b, as well as carotenoid (Figure 2g–j), suggesting enhanced photoprotection under the SF condition. In contrast, the CZ cultivar showed the most significant changes in chlorophyll a, the chlorophyll a/b ratio, and carotenoid, indicating a strong physiological response to water stress, which affected pigment synthesis, degradation, or regulation. The a/b ratio increased in the G1 and G2 cultivars under the SF condition compared to the BF condition (Figure 2i), likely due to a slight rise in chlorophyll a, which could boost carbon fixation activity. This suggests that the drought stress was not severe enough to markedly inhibit photosynthesis in the G1 and G2 cultivars.

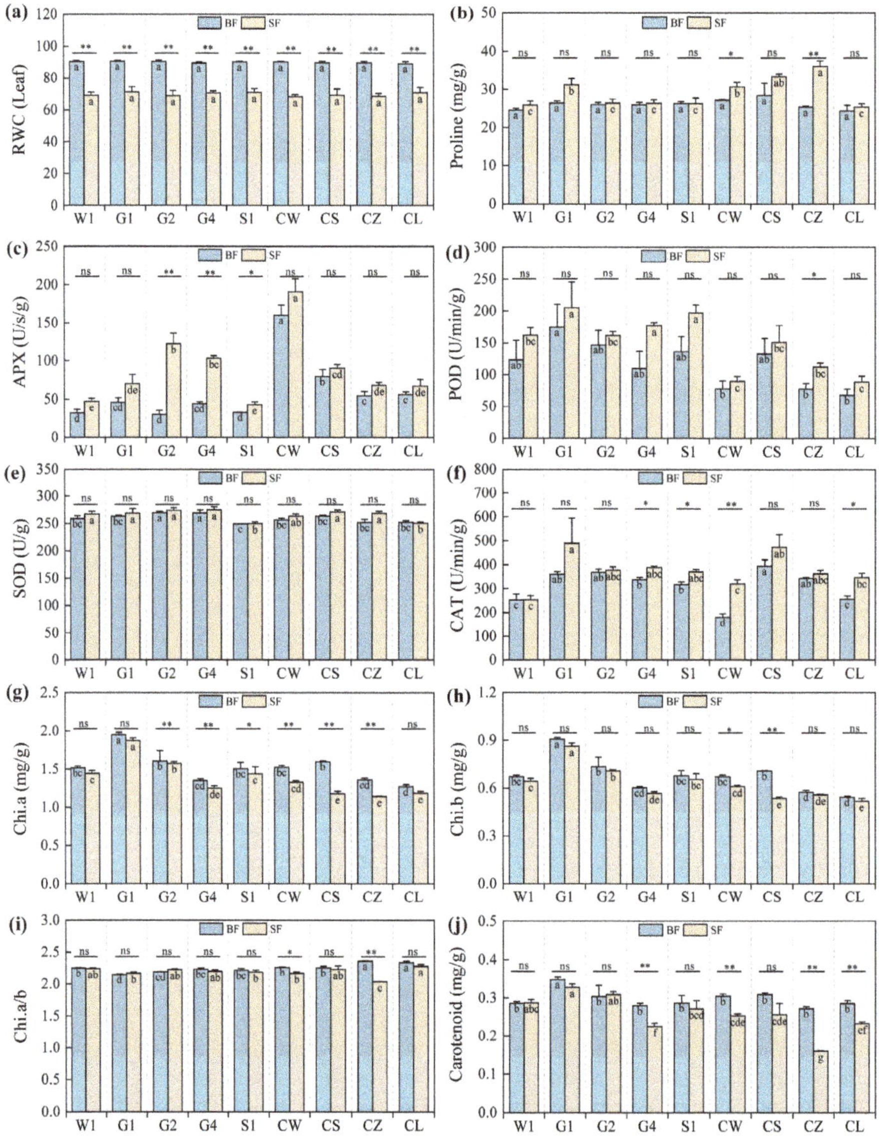

Figure 2. Physiological indicator responses of *C. pilosula* cultivar seedlings. The data in the figure represent the $\overline{X} \pm SD$ (r = 3). **(a)** Relative water content of leaf (RWC (leaf)); **(b)** leaf proline content; **(c)** leaf ascorbate peroxidase (APX) content; **(d)** leaf peroxidase (POD) content; **(e)** leaf superoxide dismutase (SOD) content; **(f)** leaf catalase (CAT) content; **(g)** content of chlorophyll a (Chi.a); **(h)** content of chlorophyll b (Chi.b); **(i)** content of chlorophyll a/b (Chi.a/b); **(j)** content of carotenoids; Different lowercase letters mean a significant difference ($p < 0.05$); '*' indicates a significant difference ($p < 0.05$); '**' indicates a highly significant difference ($p < 0.01$); 'ns' indicates that the difference is not significant.

Table 2. Analysis of variance for the main and interaction effects of physiological indicators.

Source	df	MS/Sig.									
		RWC (Leaf)	Proline	APX	POD	SOD	CAT	Chi.a	Chi.b	Chi.a/b	Carotenoid
W	1	5564.354 **	122.949 **	12,072.268 **	14,939.910 **	491.478 **	54,924.322 **	0.266 **	0.031 **	0.043 *	0.021 **
C	8	2.285 ns	31.844 **	10,082.467 **	9004.174 **	380.612 **	26,088.905 **	0.262 **	0.065 **	0.011 **	0.007 **
Block	2	0.635 ns	1.929 ns	3.508 ns	5157.167 *	68.423 ns	663.737 ns	0.000 ns	0.000 ns	0.001 ns	1.993×10^{-5} ns
W × C	8	2.577 ns	18.031 **	1199.301 **	600.412 ns	38.736 ns	3868.050 ns	0.022 *	0.004 *	0.017 **	0.002 **
Error	34	12.356	4.566	208.483	1105.257	52.773	3076.853	0.008	0.002	0.001	0.001

'W' means water level; 'C' means cultivar; 'Block' means replications; 'Error' means experimental error; '*' indicates a significant difference ($p < 0.05$); '**' indicates a highly significant difference ($p < 0.01$); 'ns' indicates that the difference is not significant.

2.2. Yield and Its Attributes of C. pilosula Cultivar Seedlings

Significant differences ($p < 0.05$, Figure 3, Table 3) were exhibited by *C. pilosula* cultivars in biomass production, yield, and yield-related traits under two water levels. The analysis of variance showed that the water treatment had a significant impact on the total root length, fresh weight per root, yield, and disease incidence. Meanwhile, the cultivar significantly influenced the root diameter, fresh weight per root, total number of plants, and disease incidence. Notably, there was an interaction effect between the water treatment and the cultivar on the root diameter, fresh weight per root, and disease incidence (Table 3). Under the BF and SF conditions, the mean total root lengths were 22.47 cm and 18.35 cm, respectively. The corresponding average single root fresh weights were 2.31 g and 1.65 g, and the average yields were 7350.76 kg/hm^2 and 4856.32 kg/hm^2. Compared to the BF condition, water stress reduced the total root length, single root fresh weight, and yield by 18.33%, 28.4%, and 33.9%, respectively. Under the BF and SF conditions, the G2 cultivar had the thickest root diameter, which was 5.64 mm and 5.15 mm, respectively (Figure 3a). Under the same conditions, the G1 cultivar had the longest total root length, measuring 23.36 cm and 20.46 cm, respectively (Figure 3b). The G2 cultivar also had the highest single root fresh weight, being 4.20 g and 2.06 g, respectively, under the BF and SF conditions, with extremely significant differences shown (Figure 3c). The yield of the G2 cultivar was the highest, reaching 8366.00 kg/hm^2 and 4834.68 kg/hm^2 under the BF and SF conditions, respectively (Figure 3d). Under the BF and SF conditions, the S1 cultivar had the highest plant count, which was 752.50×10^4 and 615.93×10^4 plants/hm^2, respectively (Figure 3e). Under the BF and SF conditions, the W1 cultivar had the highest incidence rate, being 10.4% and 7.0%, respectively, while the CZ cultivar had the lowest, with rates of 1.2% and 0.7%, respectively (Figure 3f).

Table 3. Analysis of variance for the main and interaction effects of yield and yield-related indicators.

Source	df	MS/Sig.					
		Root Diameter	Total Root Length	Fresh Weight per Root	Total Number of Plants	Yield	Incidence of Disease
W	1	0.215 ns	224.714 **	5.799 **	44.262 ns	83,999,917.780 **	66.168 **
C	8	1.320 **	10.012 ns	1.423 **	61,324.448 **	2,585,914.143 ns	32.410 **
Block	2	0.532 *	3.350 ns	0.263 ns	3484.338 ns	2,171,003.736 ns	3.077 ns

Table 3. *Cont.*

Source	df	MS/Sig.					
		Root Diameter	Total Root Length	Fresh Weight per Root	Total Number of Plants	Yield	Incidence of Disease
W × C	8	0.362 *	6.618 ns	0.570 *	17,607.675 ns	962,622.148 ns	6.691 **
Error	34	0.141	6.811	0.194	13,991.329	2,377,142.101	1.640

'W' means water level; 'C' means cultivar; 'Block' means replications; 'Error' means experimental error; '*' indicates a significant difference ($p < 0.05$); '**' indicates a highly significant difference ($p < 0.01$); 'ns' indicates that the difference is not significant.

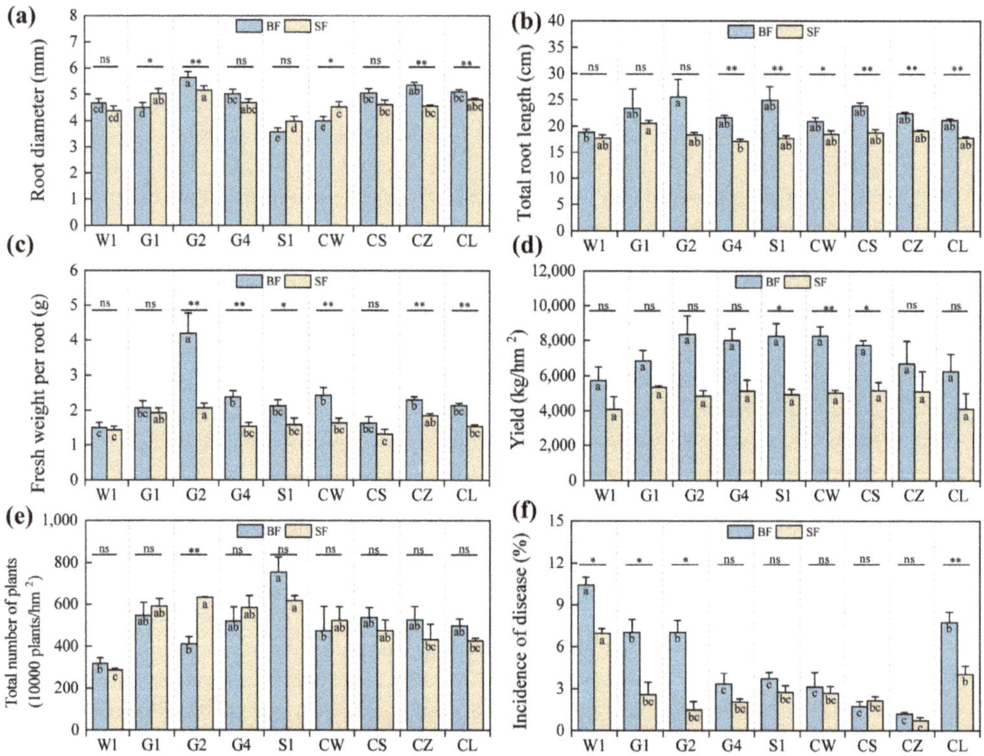

Figure 3. Yield, and yield-related indicators of *C. pilosula* cultivar seedlings. The data in the figure represent the $\overline{X} \pm$ SD (r = 3). (**a**) Root diameter; (**b**) total root length; (**c**) fresh weight per root; (**d**) fresh yield; (**e**) total number of plants; (**f**) incidence of disease; Different lowercase letters mean a significant difference ($p < 0.05$); '*' indicates a significant difference ($p < 0.05$); '**' indicates a highly significant difference ($p < 0.01$); 'ns' indicates that the difference is not significant.

2.3. Correlation Analysis of C. pilosula Cultivar Seedlings

From the distribution of the selected growth–physiology–yield composite index (GPYCI) and cultivars on the PC1 axis in the principal component biplot (Figure 4a), the growth-related indicators (e.g., relative growth rate-related indices) and some physiological indicators (e.g., antioxidant enzymes and chlorophyll) exhibit different distribution patterns on the PC1 axis, suggesting that the PC1 axis reflects a comprehensive performance trend of plants during growth and environmental adaptation. However, it can be observed that from the left to the right on the PC1 axis, the distribution of different cultivars is relatively dispersed. Cultivar CL is located at the far left of the PC1 axis, while cultivar G1 is at the far right. This implies that there are differences in the physiological responses of different cultivars to drought. It can also be seen from Figure 4b that different cultivars have differences in tolerance to different GPYCIs. Among them, the peroxidase (POD) activity of the G1 cultivar shows a significant positive correlation under the two water conditions; the relative water content (RWC) of leaves of the G2 cultivar shows a significant positive correlation under the two water conditions; the chlorophyll a (Chi.a) and carotenoids of the S1 cultivar show a significant positive correlation under the two water conditions, while the total number of plants per unit area shows a significant negative correlation; the fresh weight of a single root of the CW cultivar shows a significant positive correlation under the two water conditions; and the yield of the CS cultivar shows a significant negative correlation under the two water conditions. This indicates that the genetic characteristics of *C. pilosula* cultivars play a key role in responding to water changes. Each cultivar has its own unique physiological response mechanism, which determines their performance of various indices under different water environments.

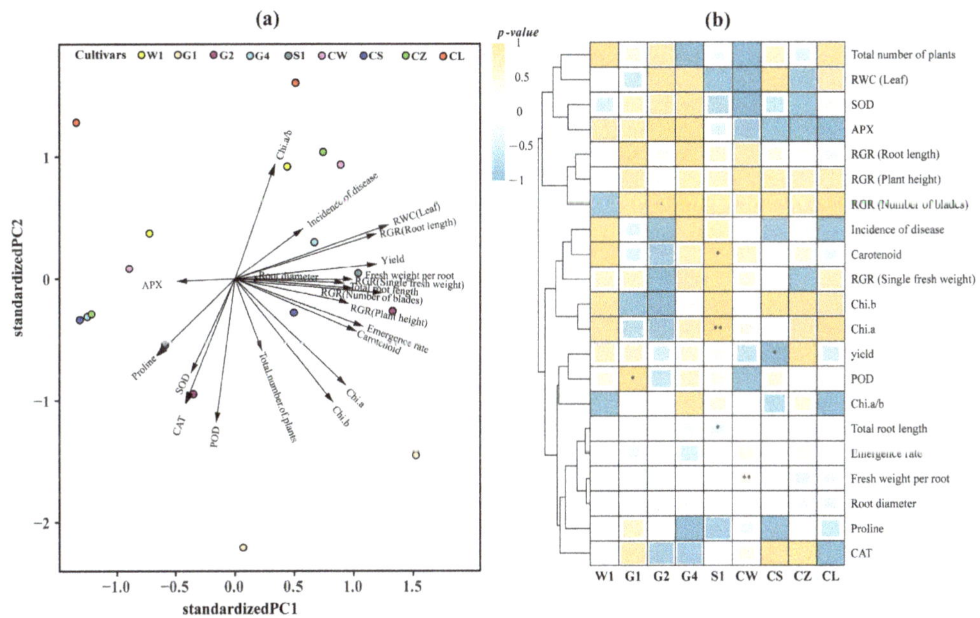

Figure 4. Biplot analysis and Pearson's correlation analysis of the GPYCI for *C. pilosula* seedlings under the BF and SF conditions. (**a**) Biplot analysis of indices for *C. pilosula* cultivars grown under the BF and SF conditions; The length of the variable vector reflects the contribution degree of the variable to the principal component, while the angle between the vectors indicates the correlation between the variables. (**b**) Pearson's correlation analysis of the GPYCI of *C. pilosula* seedlings under the BF and SF conditions. The size of each square represents the magnitude of the Pearson correlation coefficient between variables and cultivars under different treatments, and different colors represent the significance level. GPYCI is the growth–physiology–yield composite index. '*' indicates a significant difference ($p < 0.05$); '**' indicates a highly significant difference ($p < 0.01$).

2.4. Drought Tolerance Indices and Comprehensive Factor Analysis

Drought tolerance indices can be utilized to assess the differences in drought tolerance among cultivars, the sensitivity of cultivars to drought, as well as the stability and adaptability of cultivars under drought conditions. In this study, we calculated the drought tolerance indices through physiological and yield indicators under two water levels to evaluate the drought resistance of different *C. pilosula* cultivars at the seedling stage (Table 4). The higher the values of SSI% and SSI reduction, the more sensitive the cultivar is to stress. Meanwhile, larger values or values closer to one for other indices indicate that the cultivar is more physiologically or yield-stable under the SF condition and is more drought-tolerant. From the results of the drought tolerance indices, it can be observed that when different cultivars are evaluated for drought tolerance using different drought tolerance indices, the results of drought tolerance vary. Therefore, it is necessary to comprehensively consider various drought tolerance indicators for the evaluation of the drought resistance of *C. pilosula* seedling cultivars. Based on the principal component analysis of 22 drought resistance indices, the eigenvalues of the first five principal components were greater than one. The contribution rates were 38.76%, 23.52%, 11.72%, 10.39%, and 7.46%, respectively, and the cumulative contribution rate was 91.86%. Consequently, the eigenvalues and contribution rates of the first five principal components (Table A1) were extracted to calculate the weight values of each index (Table A2). According to the test indices' membership degree, weight value, and the addition and multiplication operations of all the indices, the comprehensive evaluation indices ranked the cultivars in the order of G1 > CZ > CS > G2 > S1 > CW > G4 > CL > W1 (Table 5).

Table 4. Drought tolerance index values of *C. pilosula* cultivar seedlings grown under two water conditions.

Drought Indices	*C. pilosula* Cultivars								
	W1	G1	G2	G4	S1	CW	CS	CZ	CL
ESI	0.92	0.88	0.97	0.83	0.98	0.92	0.95	0.89	0.88
RSI—Plant height	0.73	0.94	0.98	0.81	0.74	0.85	0.79	0.75	0.59
RSI—Number of blades	0.91	0.90	0.93	0.87	0.80	0.79	0.76	0.96	0.64
RSI—Root length	0.61	0.79	0.63	0.62	0.41	0.55	0.64	0.67	0.65
RSI—Single fresh weight	0.84	0.93	0.96	0.85	0.96	0.98	0.95	0.90	0.88
RWCSI	0.76	0.79	0.76	0.79	0.79	0.76	0.77	0.76	0.79
CSI	0.95	0.96	0.98	0.93	0.96	0.88	0.74	0.88	0.94
EASI-CAT	1.00	0.64	0.97	0.85	0.83	0.21	0.80	0.95	0.64
EASI-POD	0.68	0.82	0.90	0.39	0.55	0.86	0.86	0.55	0.69
EASI-APX	0.50	0.47	−2.09	−0.35	0.69	0.81	0.86	0.75	0.81
EASI-SOD	0.97	0.98	0.98	0.98	1.00	0.97	0.97	0.93	1.01
PSI	1.05	1.18	1.02	1.02	1.00	1.13	1.18	1.42	1.04
STI	0.43	0.68	0.75	0.76	0.75	0.77	0.74	0.63	0.48
MPI	4904.36	6089.67	6600.33	6568.97	6590.75	6648.67	6451.06	5895.11	5182.95
TOL	1645.61	1508.67	3531.33	2871.72	3324.17	3262.67	2579.22	1597.11	2129.43
YI	0.56	0.73	0.66	0.70	0.67	0.68	0.70	0.69	0.56
GMP	48,34.85	6042.77	6359.78	6410.12	6377.73	6445.43	6320.84	5840.77	5072.41
YSI	0.71	0.78	0.58	0.64	0.60	0.61	0.67	0.76	0.66
SSI	0.85	0.65	1.24	1.06	1.19	1.16	0.98	0.70	1.00
% reduction	28.73	22.04	42.21	35.87	40.28	39.40	33.32	23.86	34.08
YSSI	0.64	0.66	1.00	0.91	0.97	0.97	0.86	0.67	0.74
YPSI	1629.17	2290.51	1534.25	1848.48	1633.07	1692.80	1935.80	2148.96	1526.49

Table 5. Membership value and comprehensive evaluation index of all drought tolerance indices of *C. pilosula* seedlings.

Drought Indices	*C. pilosula* Cultivars								
	W1	G1	G2	G4	S1	CW	CS	CZ	CL
ESI	0.600	0.333	0.933	0.000	1.000	0.600	0.800	0.400	0.333
RSI—Plant height	0.359	0.897	1.000	0.564	0.385	0.667	0.513	0.410	0.000

Table 5. *Cont.*

Drought Indices	*C. pilosula* Cultivars								
	W1	G1	G2	G4	S1	CW	CS	CZ	CL
RSI—Number of blades	0.844	0.813	0.906	0.719	0.500	0.469	0.375	1.000	0.000
RSI—Root length	0.526	1.000	0.579	0.553	0.000	0.368	0.605	0.684	0.632
RSI—Single fresh weight	0.000	0.643	0.857	0.071	0.857	1.000	0.786	0.429	0.286
RWCSI	0.000	1.000	0.000	1.000	1.000	0.000	0.333	0.000	1.000
CSI	0.875	0.917	1.000	0.792	0.917	0.583	0.000	0.583	0.833
EASI-CAT	1.000	0.544	0.962	0.810	0.785	0.000	0.747	0.937	0.544
EASI-POD	0.569	0.843	1.000	0.000	0.314	0.922	0.922	0.314	0.588
EASI-APX	0.878	0.868	0.000	0.590	0.942	0.983	1.000	0.963	0.983
EASI-SOD	0.500	0.625	0.625	0.625	0.875	0.500	0.500	0.000	1.000
PSI	0.119	0.429	0.048	0.048	0.000	0.310	0.429	1.000	0.095
STI	0.000	0.735	0.941	0.971	0.941	1.000	0.912	0.588	0.147
MPI	0.000	0.680	0.972	0.954	0.967	1.000	0.887	0.568	0.160
TOL	0.932	1.000	0.000	0.326	0.102	0.133	0.471	0.956	0.693
YI	0.000	1.000	0.588	0.824	0.647	0.706	0.824	0.765	0.000
GMP	0.000	0.750	0.947	0.978	0.958	1.000	0.923	0.625	0.147
YSI	0.650	1.000	0.000	0.300	0.100	0.150	0.450	0.900	0.400
SSI	0.661	1.000	0.000	0.305	0.085	0.136	0.441	0.915	0.407
% reduction	0.668	1.000	0.000	0.314	0.096	0.139	0.441	0.910	0.403
YSSI	0.000	0.056	1.000	0.750	0.917	0.917	0.611	0.083	0.278
YPSI	0.134	1.000	0.010	0.421	0.139	0.218	0.536	0.815	0.000
Comprehensive evaluation index	0.413	0.783	0.598	0.553	0.597	0.558	0.616	0.603	0.420
Comprehensive sorting	9	1	4	7	5	6	3	2	8

3. Discussion

The growth of *Codonopsis pilosula*, especially in the seedling stage, is frequently hampered by various abiotic stress factors, such as temperature, precipitation, light intensity, and heavy metal [24,25], with drought being a major concern. Drought refers to a climatic phenomenon in which there is no rain for a long period or the precipitation is abnormally low, resulting in dry air and water shortage in the soil, thus having an adverse impact on plants [26]. Recently, researchers have cloned hundreds of genes related to crop drought resistance through genetic and molecular methods, but only a few have been used in breeding, showing that drought resistance regulation is complex [27]. Numerous studies [28,29] have also elucidated the drought resistance mechanism of plants. However, there remain challenges in applying these findings to large-scale breeding programs. For instance, how to effectively introduce such genetic modifications into plant varieties without compromising other important agronomic traits and ensure the stable inheritance and expression of these genes across different varieties. Consequently, identifying the key genotype and assessing its performance under different water stress conditions are crucial objectives for plant researchers [30]. Under natural conditions of water stress, evaluating various growth, physiological, and yield aspects of *C. pilosula* cultivars offers breeders an efficient approach. These trait evaluations can serve as the basis for selecting drought-tolerant materials, facilitating their promotion in drought-tolerant areas or the development of new drought-tolerant *C. pilosula* cultivars [31].

3.1. Drought Stress Affects the Growth Rate of C. pilosula Seedlings

Drought is a major abiotic stress that has numerous adverse effects on crop growth and development [32]. It mainly manifests itself by slowing down the growth rate of crop stems, leaves, and roots. While stems and leaves are crucial for sunlight absorption, competition for space, and biomass accumulation [33], leaves are the main site of photosynthesis in plants [34–36], and the root system facilitates the plant's water and nutrient uptake [37,38]. In this study, under the water-stress condition (SF), the emergence rate and the relative growth rates of plant height, as well as the number of leaves, root length, and the fresh weight of individual roots of *C. pilosula* seedlings generally showed a reduction compared to the normal-watered condition (BF). Makonya et al. [39] observed a decrease in growth rates and biomass accumulation in two chickpea cultivars subjected to drought during their vegetative and flowering stages, and this was consistent with the results of this study. The reduction in the relative growth rate of growth indicators can be attributed to the reallocation of resources towards stress tolerance mechanisms rather than growth and development [40]. Among them, cultivars G1, G2, and CZ exhibited relatively better growth under drought conditions, with higher emergence rates and growth rates in plant height, the number of leaves, root length, and the fresh weight of individual roots. This is because they were selected under extremely arid conditions, and drought hardening can improve the drought tolerance and adaptability of plants [32].

3.2. *Drought Stress Affects Physiological Changes in C. pilosula Seedlings*

Under drought stress, the physiological responses of different cultivars are the key indicators of how plants tolerate drought stress [41] as they reflect a plant's ability to maintain cellular homeostasis, prevent oxidative damage, and regulate water absorption and loss [42,43]. Maintaining a good water status is crucial for optimal physiological function and growth. Some studies have shown that a high relative water content (RWC) is closely related to drought resistance [44,45]. In this study, under the BF condition, most cultivars maintained relatively high RWC values. However, under the SF condition, the RWC significantly decreased across all cultivars, and the degree of reduction varied. Higher RWC indicates better maintenance of water balance and potentially greater drought tolerance, as plants can maintain turgor pressure and normal physiological functions [46]. Cultivar W1 showed the most significant reduction in RWC, suggesting that it is more susceptible to water stress when maintaining tissue water balance. Osmotic adjustment is regarded as an important component of drought tolerance. As an osmoregulatory substance, proline accumulates in large quantities when plants are subjected to environmental stresses such as drought. It can increase the solute concentration of the cell and reduce the osmotic potential of the cell, thus enabling the cell to absorb more water from the external environment, maintaining the turgor pressure of the cell, and ensuring the normal physiological function of the cell [47,48]. In this study, the proline content of all cultivars increased under the SF conditions, but the extent of the increase varied. Cultivars CW and CZ showed a significant increase in proline content, suggesting a stronger osmoprotective response. In addition, the initial proline content of different cultivars may affect the extent of their increase under the SF conditions. Cultivars with lower basal proline content may have more scope for up-regulation. Salsinha et al. [49] showed that proline content, which is usually elevated under drought stress, acts as an osmoprotectant, helping to maintain cellular turgor and protect cellular structures. Chlorophyll is a photosynthetic pigment that participates in light absorption and plays an important role in plant photosynthesis. Since drought stress can accelerate chlorophyll decomposition, chlorophyll content is one of the most commonly used indicators of the severity of drought stress [46,50]. We found that the chlorophyll content of all cultivars significantly decreased under the BF condition. However, cultivars W1 and G1 showed a relatively smaller decline, indicating a stronger ability to maintain their photosynthetic machinery under drought conditions. Bano et al. [51] concluded that a plant maintaining chlorophyll content under drought stress indicates the ability to sustain energy production. However, some studies have also shown that maintaining a lower chlorophyll content under severe drought stress may help plants reduce photo-oxidative damage [52]. This may be related to differences in the plant's ability to protect chlorophyll from degradation or maintain its synthesis under stress. Plants accumulate reactive oxygen species (ROS) under drought stress, which can impair chloroplast and mitochondrial functions and subject the plant cells to oxidative damage, including lipid peroxidation, protein oxidation, and DNA damage. H_2O_2 is one of the ROS closely associated with oxidative stress. It is derived from the disproportionation of superoxide anion, and the product of H_2O_2 has a strong oxidation ability [53,54]. Throughout evolution, plants have developed an enzymatic antioxidant system to remove excess ROS, primarily involving ascorbate peroxidase (APX), superoxide dismutase (SOD), peroxidase (POD), and catalase (CAT) [55,56]. Under the SF conditions, cultivars G2, G4, S1, CZ, and others showed a significant increase in antioxidant enzyme activity, indicating a more effective defense against oxidative stress. These cultivars may possess a more efficient antioxidant system, with genes encoding antioxidant enzymes being more sensitive to drought-induced ROS production. Hosseini et al. [57] found that gradual drought reduced the biomass and leaf relative water content of licorice and gradually increased the concentration of osmotic regulatory substances in leaves and the activity of antioxidant enzymes in leaf extracts, which was consistent with the results of this study.

3.3. *Drought Stress Affects Yield and Quality in C. pilosula Seedlings*

Drought stress is a major abiotic stress; the impacts of it on the yield and quality of plants during plant growth and development have been well studied [58–60], and they vary based on drought type, intensity, and duration [61,62]. In this study, the water treatment had an impact on the total root length, fresh weight of a single root, and yield of all *C. pilosula* cultivars' seedlings. In the early stage of drought stress, some plants increase water use efficiency by reducing the lateral growth rate of roots in order to utilize water resources more effectively. Plant roots are usually organs for storing water and nutrients, which directly affect the drought tolerance of plants and the later yield of crops [63]. Under the SF condition, cultivars G1, S1, and CW showed a greater diameter of root compared to the BF condition. This suggests that their root systems have greater water storage and water uptake capacity and will act as a compensatory mechanism when plants face water deficit. Previous studies have shown that efficient root systems have greater root vigor and greater rooting depth, which is a useful trait for improving drought tolerance in crops [64,65]. Meanwhile, cultivars G1, G2, and G4 were able to maintain relatively high yields under the SF condition. This is due to the fact that these cultivars may be more drought-resistant or have undergone long periods of natural selection under drought conditions, resulting in the development of traits adapted to such environments [66].

3.4. *Comprehensive Evaluation of Different Cultivars of C. pilosula Seedlings Under Drought Stress*

The correlation between cultivars and the growth–physiology–yield composite index (GPYCI) reveals that environmental factors have a relatively larger impact on their expression [1]. The drought resistance index is calculated using specific formulas, which take into account various growth and physiological parameters of plants under both the BF and SF conditions [22]. Drought tolerance indices provide a broader and more systematic analysis than analyzing a single growth or physiological indicator, and they contribute to a more in-depth understanding of the mechanisms of drought tolerance in plants as well as the differences in the adaptability of different *C. pilosula* cultivars under drought conditions. Comprehensive factor analysis is a multivariate statistical method. Based on the integrated performance of seedling growth, physiology, and yield of different cultivars under two water conditions, the drought tolerance of different *C. pilosula* cultivars can be assessed more comprehensively. Xiao et al. [67] conducted a comprehensive factor analysis based on six medicinal root trait indicators, early bolting rate, disease resistance, and individual medicinal yield under four fertilization treatments. They ultimately concluded that an organic fertilizer application of 2000 kg/hm^2 is optimal for the medicinal performance of *Angelica sinensis*. Cui et al. [68] determined through a comprehensive factor analysis of 16 traits of space-grown A. sinensis that the selection efficiency of transplanting space-grown *A. sinensis* 22 h is higher, providing scientific and technical support for the expansion of new cultivars of *A. sinensis*.

4. Materials and Methods

4.1. The Experimental Area

The experiment was conducted from April 2023 to March 2024 in Yuanyi Village, Gongchang Town, Longxi County, Gansu Province (35°01′8″ E, 104°37′11″ N), a traditional area for *C. pilosula* production. Longxi County, located in the Loess Plateau, has a temperate continental climate [69]. The experimental site, at an altitude of 1733.6 m, received an average annual rainfall of 226.3 mm. The temperature and rainfall during the experimental period (April 2023–March 2024) are presented in Figure 5. The previous crop was *Astragalus mongholicus* Bunge, and the field had been fertilized with organic fertilizer for five years, thus creating an organic microenvironment. A composite soil sample was collected using a soil drill with the multi-point composite sampling method from 0 to 20 cm before sowing to evaluate soil nutrients. The loose soil had a soil water content of 203.10 g/kg. Additionally, it showed an electrical conductivity of 192.97 μs/cm and a pH of 7.95 and was enriched with 29.89 g/kg of organic matter. It also contained 29.91 mg/kg of effective phosphorus, 148.23 mg/kg of available potassium, 1.14% of total nitrogen, and had a cation exchange capacity of 12.1 cmol/kg (+).

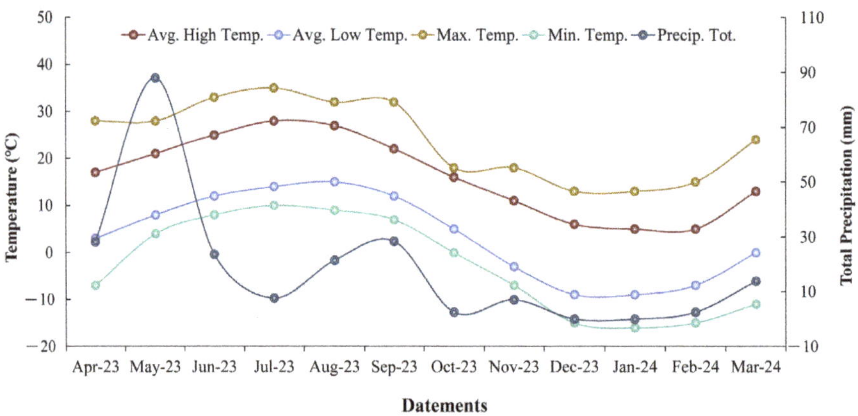

Figure 5. The temperature and rainfall conditions during the experimental period (April 2023–March 2024).

4.2. Experimental Design and Treatments

The field was prepared manually, and the weeds were removed. The experiment adopted a two-factor paired experiment with three replications for each of the nine *C. pilosula* cultivars under normal-watered (black plastic film hole sowing, BF) and water-stressed (Spread in the open field, SF) conditions. The field, covering an area of 224 m^2, was divided into 54 plots (each with an area of 2.5 m^2). The seeding density is 3.0 g/m^2. When broadcasting, the seeds were evenly spread. When sowing in holes on plastic film, the seeding amount per hole is approximately 60 seeds. After sowing, gently cover the seeds with soil, and then cover the entire experimental field with a shade net until the emergence is complete, after which the shade net is removed. The total seeding amount for each plot is 7.5 g, approximately 2500 *C. pilosula* seedlings. Path widths and buffer zones were established to minimize edge effects and facilitate field management and data collection.

The tested cultivars included Weidang No. 1, which is a variety promoted in Dingxi, Gansu Province (the seeds were sourced from the Dry Farming Research and Promotion Center of Dingxi, Gansu Province), and other *C. pilosula* cultivars were selected by the group of Yuan Chen and Feng-xia Guo from Gansu Agricultural University, as detailed in Table 6. The application rate of the organic fertilizer in the experimental field was 270 g/m^2, which was characterized by an organic matter content of >45.0%, N + P_2O_2 + $K_2O \geq 5.0\%$, pH within the range of 5.5–8.5, a moisture content of $\leq$30.0%, and was produced by Gansu Tianyao Biotechnology Co., Ltd., Dingxi City, China. All other field management practices were in line with local large-scale agricultural production.

Table 6. The details of the experimental cultivars.

Cultivar Number	Cultivar Name	Breeding Methods	Abbreviation
2006-92-02	Weidang No. 1	Pedigree selection	W1
2011-D07-3	Gandang No. 1	Mass selection	G1
2011-D07-1	Gandang No. 2	Mass selection	G2
2011-D07-7	Gandang No. 4	Mass selection	G4
2011-13-3	Space-bred No. 1	Mutation breeding	S1
2011-2-1	Wild patterned *Codonopsis*	Domestication breeding	CW
2011-2-2	Succulent patterned *Codonopsis*	Domestication breeding	CS
2011-1-1	CL	Pedigree Selection	CZ
2011-1-2	CZ	Pedigree Selection	CL

4.3. The Studied Traits

On 10 July 2023, the soil water content of each treatment was determined at intervals of 20 days. Nine cultivars of *Codonopsis pilosula* seedlings were randomly and destructively sampled from each plot to determine growth data. When measuring individual indicators, 20 samples were randomly measured in each replicate block. Leaves were collected on 30 July and 30 August and rapidly frozen with liquid nitrogen for the subsequent evaluation of physiological indexes. After excavation on 14 March 2024, the yield data were determined. According to the local climatic conditions (Figure 5), this study assumed that there was less rainfall (7.6 mm) and a higher temperature in July 2023, creating natural drought stress conditions. The soil water content of each treatment is shown in Figure 1a; there were significant differences in soil water content among the treatments ($p < 0.05$) but no significant differences among the cultivars ($p > 0.05$). The average soil moisture content in the open field plots (SF) for each treatment was 136.5 g/kg, while that in the black plastic film hole sown plots (BF) was approximately 176.1 g/kg. Therefore, the growth data on 10 July and 30 July, the physiological indexes, and the soil moisture content on 30 July were selected for this study to evaluate the drought resistance of *C. pilosula* seedlings of various cultivars.

Soil moisture was determined using the oven-drying method, and the plant height, the number of leaves, the root length, and weight were also measured. The relative growth rate (RGR) was calculated as RGR = $(\ln Ht_2 - \ln Ht_1)/(t_2 - t_1)$, where Ht_1 and Ht_2 are the plant heights at times t_1 and t_2, respectively [70]. Relative water content (RWC) was estimated according to L. González and M. González-Vilar [71]. The proline levels were assessed with a proline assay kit (A107-1-1, Jiancheng, Nanjing, China). Chlorophyll was determined via ethanol extraction. The antioxidant enzymatic activity was measured at 4 °C [72–74].

4.4. Drought Tolerance Indices

We identified, in Table 7, the drought tolerance indices used in this study by reviewing other crop-related studies, and we defined stability indices 12–17 based on other formulas.

Table 7. Drought tolerance indices.

	Drought Tolerance Indices	Formula	Reference
1	Stress tolerance index, *STI*	$STI = \frac{(Yi)_{SF} \times (Yi)_{BF}}{\overline{(Yi)_{BF}}^2}$	[1]
2	Mean productivity index, *MPI*	$MPI = \frac{(Yi)_{SF} + (Yi)_{BF}}{2}$	[75]
3	Stress tolerance, *TOL*	$TOL = (Yi)_{BF} - (Yi)_{SF}$	[23]
4	Yield index, *YI*	$YI = \frac{(Yi)_{SF}}{\overline{(Yi)_{BF}}}$	[76]
5	Geometric mean productivity, *GMP*	$GMP = \sqrt[2]{(Yi)_{BF} \times (Yi)_{SF}}$	[77]
6	Yield stability index, *YSI*	$YSI = \frac{(Yi)_{SF}}{(Yi)_{BF}}$	[78]
7	Stress intensity, *SI*	$SI = 1 - \frac{\overline{(Yi)_{SF}}}{\overline{(Yi)_{BF}}}$	[79]
8	Stress susceptibility index, *SSI*	$SSI = \frac{1 - \frac{(Yi)_{SF}}{(Yi)_{BF}}}{SI}$	[23]
9	% reduction	$\% \text{ reduction} = \frac{(Yi)_{BF} - (Yi)_{SF}}{(Yi)_{BF}} \times 100\%$	[80]
10	Yield potential score index, *YPSI*	$YPSI = 0.5\,(MPI + STI) - 0.5\,(SSI + TOL)$	[81]
11	Yield stress score index, *YSSI*	$YSSI = 0.5\,(STI + SSI)$	[81]
12	Emergence rate stability index, *ESI*	$ESI = \frac{(Eri)_{SF}}{(Eri)_{BF}}$	
13	RWC stability index, *RWCSI*	$RWCSI = \frac{(RWCi)_{SF}}{(RWCi)_{BF}}$	
14	Proline content stability index, *PSI*	$PSI = \frac{(Pro.i)_{SF}}{(Pro.i)_{BF}}$	
15	Chlorophyll stability index, *CSI*	$PSI = \frac{[(Chl.a+b)i]_{SF}}{[(Chl.a+b)i]_{BF}}$	
16	RGR stability index, *RSI-X*	$RSI - X = 1 - \frac{\lvert (RGRi)_{SF} - (RGRi)_{BF} \rvert}{0.5\,\lvert (RGRi)_{BF} + (RGRi)_{BF} \rvert}$	
17	Enzyme activity stability index, *EASI-X*	$EASI - X = 1 - \frac{(Ei-X)_{SF} - (Ei-X)_{BF} \rvert}{(Ei)_{BF}}$	

Where $\overline{(Yi)_{SF}}$ is the mean yield of each cultivar under the water-stressed (*SF*) condition, $(Eri)_{SF}$ is the emergence rate of each cultivar under the SF condition, $(RGRi)_{SF}$ is the relative growth rate of growth parameters of each cultivar under the SF condition, $(RWCi)_{SF}$ is the leaf relative water content of each cultivar under the SF condition, $[(Chl.a + b)i]_{SF}$ is the chlorophyll content of each cultivar under the SF condition, $(Ei - X)_{SF}$ is the enzyme activity (CAT, APX, POD, and SOD) of each cultivar under the SF condition, $(Pro.i)_{SF}$ is the proline content of each cultivar under the SF condition, $(Yi)_{SF}$ is the yield of each cultivar under the SF condition; $\overline{(Yi)_{BF}}$ is the mean yield of each cultivar under the well-watered (BF) condition, $(Eri)_{BF}$ is the emergence rate of each cultivar under the BF condition, $(RGRi)_{BF}$ is the relative growth rate of growth parameters of each cultivar under the BF condition, $(RWCi)_{BF}$ is the leaf relative water content of each cultivar under the BF condition, $[(Chl.a + b)i]_{BF}$ is the chlorophyll content of each cultivar under the BF condition, $(Ei - X)_{BF}$ is the enzyme activity (CAT, APX, POD, and SOD) of each cultivar under the BF condition, $(Pro.i)_{BF}$ is the proline content of each cultivar under the BF condition, and $(Yi)_{BF}$ is the yield of each cultivar under the BF condition.

4.5. Statistical Analysis

The results were analyzed via analyses of variance (ANOVAs) and principal component analysis using the SPSS 22.0 software (Chicago, IL, USA). The means were analyzed using the least significant difference (LSD) method at $p = 0.05$ (LSD 0.05). A comprehensive evaluation was conducted according to the methods by Wang et al. [82]. First, based on factor analysis, the principal component values of each indicator whose initial characteristic root was greater than 1 were extracted, and then, the weight (W_j) and positive and negative membership function values were calculated according to the relevant properties. Finally, the comprehensive index (*CI*) was estimated.

$$W_j = \sum(C_{l,j} \times VP_l)/\sum\sum(C_{l,j} \times VP_l)$$

$$R(X_{ij}) = (X_{ij} - X_{j\min})/(X_{j\max} - X_{j\min})$$

$$RR(X_{ij}) = 1 - (X_{ij} - X_{j\min})/(X_{j\max} - X_{j\min})$$

$$CI_j = \sum[R(X_{ij}) \times W_j]$$

where $C_{l,j}$ represents the l-th principal component of the j-th indicator, VP_l is the percentage of the variance of the first principal component, W_j is the weight value of the j-th indicator, i is different treatment, j is the measurement indicator, $R(X_{ij})$ is the membership function value of indicator j under i treatment, $RR(X_{ij})$ represents the value of the anti-membership function of i processing the j-th indicator, X_{ij} is the average observation value of indicator j of i treatment, X_{jmin} is the minimum value of the j-th indicator in all treatments, X_{jmax} is the maximum value of the j-th indicator in all treatments, CI_j is the cumulative composite indicator of the i-th indicator. CI_j is the cumulative composite indicator of the i-th processed j indicator.

5. Conclusions

In conclusion, this study evaluated the drought tolerance of different *C. pilosula* cultivars at the seedling stage. The soil environment created was conducive for studying drought resistance. Variations in growth-, physiological-, and yield-related indices were observed among cultivars under different water conditions. The water treatment and the cultivar had different effects on these indices. Cultivar G1 showed better performance in its emergence rate, RGR (number of blades), chlorophyll content, and antioxidant enzyme activities. Cultivar CZ showed better performance in proline content, the content of chlorophyll a/b, and the incidence of disease. Cultivar G2 showed better performance in root diameter, total root length, fresh weight per root, and fresh yield. Drought tolerance indices and comprehensive factor analysis showed that different cultivars had different drought tolerance levels, and a comprehensive evaluation considering multiple indicators was necessary. The comprehensive evaluation ranked the cultivars, with G1 being the most drought-tolerant, followed by CZ, CS, etc. These findings provide valuable information for selecting drought-tolerant *C. pilosula* cultivars in drought-prone areas.

Author Contributions: Conceptualization, Y.C.; methodology, H.W. and F.G.; validation, H.W. and F.G.; formal analysis, H.W. and W.L.; investigation, Y.C.; resources, Y.C. and F.G.; data curation, H.W.; writing—original draft preparation, H.W.; writing—review and editing, Y.C., F.G., D.W., W.L., P.D. and J.C.; visualization, H.W.; supervision, Y.C.; project administration, Y.C.; funding acquisition, Y.C. All authors have read and agreed to the published version of the manuscript.

Funding: This research was funded by the Ministry of Agriculture and Rural Affairs of the People's Republic of China Foundation for Traditional Chinese Medicinal Materials Advantaged Characteristic Industry Cluster Construction of Improved Variety Breeding (NO. 2022GSCYJQ07-33); Integrated Research on Standardized Cultivation Techniques of *Codonopsis pilosula* (NO. 2018YFC1706301-1); Research on the Breeding Technology of Seeds and Seedlings of *Codonopsis pilosula* (NO. 2018YFC1706301-2); A Grant for a Chief Expert of Traditional Chinese Medicinal Industry in Modern Agricultural Industry System & Safety (NO. GARS-ZYC-1); The Breeding and Base Construction of New Varieties of Chinese Medicinal Materials (NO. 7010722087).

Institutional Review Board Statement: Not applicable.

Informed Consent Statement: Not applicable.

Data Availability Statement: Data will be made available on request.

Acknowledgments: We gratefully acknowledge the help of the Guo Z. J., Mou X. H. and Guo T. M. in field cultivation and management. The authors are also very grateful to Li H. L. and Ruan Z. H. for their diligent scientific work.

Conflicts of Interest: The authors declare that the research was conducted in the absence of any commercial or financial relationships that could be construed as a potential conflict of interest.

Appendix A

Table A1. The result of principal component analysis of all drought tolerance indices of *C. pilosula* seedlings.

Principal Components	Eigenvalue	Contribution Rate (%)	Cumulative Contribution Rate (%)
1	8.529	38.768	38.768
2	5.175	23.523	62.290
3	2.579	11.722	74.013
4	2.285	10.385	84.398
5	1.642	7.464	91.862

Table A2. Capacity and weight of all drought tolerance indices of *C. pilosula* seedlings.

Drought Indices	Load					Weight Value
	1	2	3	4	5	
ESI	−0.547	−0.009	−0.168	−0.588	−0.026	0.037
RSI-Plant height	−0.334	0.707	−0.426	0.072	0.413	0.057
RSI-Number of blades	0.225	0.543	−0.743	0.100	−0.164	0.052
RSI-Root length	0.647	0.359	−0.108	0.084	0.526	0.048
RSI-Single fresh weight	−0.643	0.466	0.242	−0.370	0.229	0.054
RWCSI	0.000	−0.245	0.444	0.773	0.211	0.057
CSI	−0.039	−0.365	−0.520	0.443	0.317	0.056
EASI-CAT	0.242	−0.099	−0.673	0.138	−0.374	0.048
EASI-POD	−0.209	0.225	−0.008	−0.665	0.669	0.061
EASI-APX	0.445	−0.096	0.781	−0.191	−0.230	0.051
EASI-SOD	−0.406	−0.649	0.244	0.317	0.445	0.061
PSI	0.563	0.638	0.145	−0.292	−0.258	0.052
STI	−0.699	0.658	0.140	0.228	−0.075	0.045
MPI	−0.744	0.615	0.119	0.213	−0.088	0.044
TOL	0.991	0.017	0.037	−0.025	0.093	0.023
YI	−0.196	0.908	0.18	0.307	−0.022	0.046
GMP	−0.684	0.671	0.139	0.230	−0.089	0.046
YSI	0.934	0.306	0.066	0.034	0.092	0.031
SSI	0.941	0.295	0.051	0.038	0.087	0.030
% reduction	0.940	0.294	0.050	0.044	0.083	0.030
YSSI	−0.984	0.077	0.018	0.061	−0.103	0.025
YPSI	0.558	0.764	0.196	0.219	0.030	0.045

References

1. Ahmed, K.; Shabbir, G.; Ahmb, M.; Shah, K.N. Phenotyping for drought resistance in bread wheat using physiological and biochemical traits. *Sci. Total Environ.* **2020**, *729*, 139082. [CrossRef] [PubMed]
2. Mean, F.H.; Yusop, Z.; Yusof, F. Drought analysis and water resource availability using standardised precipitation evapotranspiration index. *Atmos. Res.* **2018**, *201*, 102–115. [CrossRef]
3. Min, H.; Chen, C.; Wei, S.; Shang, X.; Sun, M.; Xia, R.; Liu, X.; Hao, D.; Chen, H.; Xie, Q. Identification of drought tolerant mechanisms in maize seedlings based on transcriptome analysis of recombination inbred lines. *Front. Plant Sci.* **2016**, *26*, 1080. [CrossRef]
4. Chinese Pharmacopoeia Committee. *Edition of Chinese Pharmacopoeia*; Chemical Industry Press: Beijing, China, 2020; Volume 1, pp. 293–294.
5. Liang, Z.S. The development of the homology of medicine and food industry of exciting fire. *Econ. Trade* **2024**, *7*, 19–20.
6. He, J.Y.; Zhu, S.; Komatsu, K.; Goda, Y.; Cai, S.Q. Genetic polymorphism of medicinally-used *Codonopsis* species in an internal transcribed spacer sequence of nuclear ribosomal DNA and its application to authenticate *Codonopsis radix*. *J. Nat. Med.* **2014**, *68*, 112–124. [CrossRef]
7. Liang, Y.C.; Wei, G.F.; Ning, K.; Li, M.Z.; Zhang, G.Z.; Luo, L.; Zhao, G.H.; Wei, J.H.; Liu, Y.P.; Dong, L.L.; et al. Increase in carbohydrate content and variation in microbiome are related to the drought tolerance of *Codonopsis pilosula*. *Plant Physiol. Biochem.* **2021**, *165*, 19–35. [CrossRef]
8. Shin, J.; Kathuria, A.; Lee, Y.S. Effect of hydrophilic and hydrophobic cyclodextrins on the release of encapsulated allyl isothiocyanate (AITC) and their potential application for plastic film extrusion. *Appl. Polym.* **2019**, *136*, 48137. [CrossRef]
9. Xiao, S.X.; Lei, Z.H.; Song, C.X.; Wang, X.F.; Wang, Y.L.; Zhao, Y.; Feng, Z.; Meng, R.L. Effects of different seedling grades on yield and commercial quality of *Codonopsis pilosula*. *Agric. Biotech.* **2020**, *9*, 18–21.

10. Ahmed, M.; Aslam, M.; Hassan, E.; Hayat, R.; Ahmed, S. Biochemical, physiological and agronomic response of wheat to changing climate of rainfed Pakistan. *Pak. J. Bot.* **2018**, *51*, 535–551. [CrossRef] [PubMed]
11. Ahmed, H.G.M.; Sajjad, M.; Li, M.J.; Azmat, M.A.; Rizwan, M.; Maqsood, R.H.; Khan, S.H. Selection criteria for drought-tolerant bread wheat genotypes at seedling stage. *Sustainability* **2019**, *11*, 2584. [CrossRef]
12. Ahmed, K.; Shahid, S.; Chung, E.; Wang, X.; Harun, S. Climate change uncertainties in seasonal drought severity-area-frequency curves: Case of arid region of Pakistan. *J. Hydrol.* **2019**, *570*, 473–485. [CrossRef]
13. Wang, X.; Mao, Z.; Zhang, J.; Hemat, M.; Huang, M.; Cai, J.; Zhou, Q.; Dai, T.; Jiang, D. Osmolyte accumulation plays important roles in the drought priming induced tolerance to post-anthesis drought stress in winter wheat (*Triticum aestivum* L.). *Environ. Exp. Bot.* **2019**, *166*, 103804. [CrossRef]
14. Liao, Q.; Gu, S.; Kang, S.; Du, T.; Tong, L.; Wood, J.D.; Ding, R. Mild water and salt stress improve water use efficiency by decreasing stomatal conductance via osmotic adjustment in field maize. *Sci. Total Environ.* **2022**, *805*, 150364. [CrossRef]
15. Shi, H.R.; Wang, B.; Yang, P.J.; Li, Y.B.; Miao, F. Differences in sugar accumulation and mobilization between sequential and non-sequential senescence wheat cultivars under natural and drought conditions. *PLoS ONE* **2016**, *11*, e0166155. [CrossRef] [PubMed]
16. Chen, H.W.; Zhao, X.Q.; Zhai, L.Y.; Shao, K.T.; Jiang, K.W.; Shen, C.C.; Chen, K.; Wang, S.; Wang, Y.; Xu, J.L. Genetic bases of the stomata-related traits revealed by a genome-wide association analysis in rice (*Oryza sativa* L.). *Front. Genet.* **2020**, *11*, 00611. [CrossRef]
17. Sallam, A.; Alqudah, A.M.; Dawood, M.F.A.; Baenziger, P.S.; Börner, A. Drought stress tolerance in wheat and barley: Advances in physiology, breeding and genetics research. *Int. J. Mol. Sci.* **2019**, *20*, 3137. [CrossRef] [PubMed]
18. Hafeez, A.; Ali, S.; Javed, M.A.; Iqbal, R.; Khan, M.N.; Çiğ, F.; Sabagh, A.E.; Abujamel, T.; Harakeh, S.; Ercisli, S.; et al. Breeding for water-use efficiency in wheat: Progress, challenges and prospects. *Mol. Biol. Rep.* **2024**, *51*, 429. [CrossRef] [PubMed]
19. Toulotte, J.M.; Pantazopoulou, C.K.; Sanclemente, M.A.; Voesenek, L.A.C.J.; Sasidharan, R. Water stress resilient cereal crops: Lessons from wild relatives. *J. Integr. Plant Biol.* **2022**, *64*, 412–430. [CrossRef]
20. Verma, H.; Sarma, R.N. Identification of markers for root traits related to drought tolerance using traditional rice germplasm. *Mol. Biotechnol.* **2021**, *63*, 1280–1292. [CrossRef] [PubMed]
21. Pandey, B.B.; Ratnakumar, P.; Kiran, B.U.; Dudhe, M.Y.; Lakshmi, G.S.; Ramesh, K.; Guhey, A. Identifying traits associated with terminal drought tolerance in sesame (*Sesamum indicum* L.) genotypes. *Front. Plant Sci.*. [CrossRef]
22. Yemata, G.; Bekele, T. Evaluation of sesame (*Sesamum indicum* L.) varieties for drought tolerance using agromorphological traits and drought tolerance indices. *PeerJ* **2024**, *12*, e16840. [CrossRef]
23. Anwaar, H.A.; Perveen, R.; Mansha, M.Z.; Abid, M.; Sarwar, Z.M.; Aatif, H.M.; Umar, U.u.d.; Sajid, M.; Usman Aslam, H.M.; Alam, M.M.; et al. Assessment of grain yield indices in response to drought stress in wheat (*Triticum aestivum* L.). *Saudi J. Biol. Sci.* **2020**, *27*, 1818–1823. [CrossRef]
24. Meng, T.T.; Zhao, T.T.; Leng, F.F.; Chen, J.X.; Wang, Y.G. Changes of key soil factors, biochemistry and bacterial species composition during seasons in the rhizosphere and roots of *Codonopsis pilosula* (tangshen). *Agronomy* **2023**, *13*, 1545. [CrossRef]
25. Sperdouli, I. Heavy Metal Toxicity Effects on Plants. *Toxics* **2022**, *10*, 715. [CrossRef]
26. Orimoloye, I.R.; Belle, J.A.; Orimoloye, Y.M.; Olusola, A.O.; Ololade, O.O. Drought: A Common Environmental Disaster. *Atmosphere* **2022**, *13*, 111. [CrossRef]
27. He, Z.H.; Zhang, P.; Jia, H.T.; Zhang, S.L.; Nishawy, E.; Sun, X.P.; Dai, M.Q. Regulatory mechanisms and breeding strategies for crop drought resistance. *New Crops* **2024**, *1*, 100029. [CrossRef]
28. Li, J.H.; Huang, Y.; Yu, X.S.; Wu, Q.Q.; Man, X.X.; Diao, Z.H.; You, H.; Shen, J.B.; Cai, Y. Identification and Application of CLE Peptides for Drought Resistance in Solanaceae Crops. *J. Agric. Food Chem.* **2024**, *72*, 13869–13884. [CrossRef]
29. Deng, Y.R.; Yang, J.W.; Wei, J.J.; Liu, S.Y.; Yang, L.; Wang, X.F.; Zhang, N.; Si, H.J. stu-miR159a negatively regulates anthocyanin-specific MYB transcription factor to mediate drought stress tolerance in potato. *Hortic. Plant J.* **2024**. [CrossRef]
30. Toker, C.; Canci, H.; Yildirim, T. Evaluation of perennial wild *Cicer* species for drought resistance. *Genet. Resour. Crop. Evol.* **2007**, *54*, 1781–1786. [CrossRef]
31. Mwadzingeni, L.; Shimelis, H.; Dube, E.; Liang, M.D.; Tsilo, T.J. Breeding wheat for drought tolerance: Progress and technologies. *J. Integr. Agric.* **2016**, *15*, 935–943. [CrossRef]
32. Khan, R.; Ma, X.H.; Shah, S.; Wu, X.Y.; Shaheen, A.; Xiao, L.X.; Wu, Y.H.; Wang, S.S. Drought-hardening improves drought tolerance in Nicotiana tabacum at physiological, biochemical, and molecular levels. *BMC Plant Biol.* **2020**, *20*, 486. [CrossRef] [PubMed]
33. Guo, F.; Hou, E.; Ma, C.; Li, G.; Lin, R.; Zhao, Y.; Wang, X. Comparative transcriptome analysis of the peanut semi-dwarf mutant 1 reveals regulatory mechanism involved in plant height. *Gene* **2021**, *791*, 145722. [CrossRef]
34. Wang, Y.; Peng, X.; Salvat, F.; Wang, Y.; Yan, X.; Zhou, Z.; Lin, J. Salt-adaptive strategies in oil seed crop *Ricinus communis* early seedlings (cotyledon vs. true leaf) revealed from proteomics analysis. *Ecotoxicol. Environ. Saf.* **2019**, *171*, 12–25. [CrossRef]
35. Zhi, Y.; Cai, X.; Wang, J.; Jia, M.; Wang, Z. Photosynthesis promotion mechanisms of artificial humic acid depend on plant types: A hydroponic study on C3 and C4 plants. *Sci. Total Environ.* **2024**, *917*, 170404. [CrossRef]

36. Henry, R.J.; Furtado, A.; Rangan, P. Pathways of photosynthesis in non-leaf tissues. *Biology* **2020**, *9*, 438. [CrossRef] [PubMed]
37. Li, D.; Li, G.; Xi, B.; Gan, J.; Wen, D.; Cao, F.; Suo, F.; Li, J.; Ma, B.; Guo, B. Response mechanism of growth and gypenosides content for *Gynostemma longipes* cultivated at two altitude habitats to fine root morphological characteristics. *Front. Plant Sci.* **2023**, *14*, 1143745. [CrossRef]
38. Tripathi, S.; Tiwari, K.; Mahra, S.; Victoria, J.; Rana, S.; Tripathi, D.K.; Sharma, S. Nanoparticles and root traits: Mineral nutrition, stress tolerance and interaction with rhizosphere microbiota. *Planta* **2024**, *260*, 34. [CrossRef]
39. Makonya, G.M.; Ogola, J.B.O.; Muasya, A.M.; Crespo, O.; Maseko, S.; Valentine, A.J.; Ottosen, C.O.; Rosenqvist, E.; Chimphango, S.B.M. Intermittent moisture supply induces drought priming responses in some heat-tolerant chickpea genotypes. *Crop Sci.* **2020**, *60*, 2527–2542. [CrossRef]
40. Dong, J.; Zhang, H.; Ai, X.; Dong, Q.; Shi, X.; Zhao, X.; Zhong, C.; Yu, H. Improving chilling tolerance of peanut seedlings by enhancing antioxidant-modulated ROS scavenging ability, alleviating photosynthetic inhibition, and mobilizing nutrient absorption. *Plant Biol.* **2024**, *26*, 532–543. [CrossRef] [PubMed]
41. Mahmood, D.; Shahid Iqbal, M.; Li, H.; Faisal Nazir, M.; Khalid, S.; Sarraz, Z.; Hu, D.; Baojun, C.; Geng, X.; Tajo, S.; et al. Differential seedling growth and tolerance indices reflect drought tolerance in cotton. *BMC Plant Biol.* **2022**, *22*, 331. [CrossRef]
42. Liu, F.; Zhao, Y.; Wang, X.; Wang, B.; Xiao, F.; He, K. Physiological response and drought resistance evaluation of *Gleditsia sinensis* seedlings under drought-rehydration state. *Sci. Rep.* **2023**, *13*, 19963. [CrossRef]
43. Wang, J.; Zhang, X.; Han, Z.; Feng, H.; Wang, Y.; Kang, J.; Han, X.; Wang, L.; Wang, C.; Li, H.; et al. Analysis of physiological indicators associated with drought tolerance in wheat under drought and re-watering conditions. *Antioxidants* **2022**, *11*, 2266. [CrossRef] [PubMed]
44. Altinkut, A.; Kazan, K.; Ipekci, Z.; Gozukirmizi, N. Tolerance to paraquat is correlated with the traits associated with water stress tolerance in segregating F2 populations of barley and wheat. *Euphytica* **2001**, *121*, 81–86. [CrossRef]
45. Osakabe, Y.; Osakabe, K.; Shinozaki, K.; Tran, L.S.P. Response of plants to water stress. *Front. Plant Sci.* **2014**, *5*, 86. [CrossRef] [PubMed]
46. Ying, Y.Q.; Song, L.L.; Jacobs, D.F.; Mei, L.; Liu, P.; Jin, S.H.; Wu, J.S. Physiological response to drought stress in *Camptotheca acuminata* seedlings from two provenances. *Front. Plant Sci.* **2015**, *6*, 361. [CrossRef]
47. Wei, L.; Zhang, D.; Xiang, F.; Zhang, Z. Differentially expressed miRNAs potentially involved in the regulation of defense mechanism to drought stress in maize seedlings. *Int. J. Plant Sci.* **2009**, *170*, 605122. [CrossRef]
48. Fang, Y.; Xiong, L. General mechanisms of drought response and their application in drought resistance improvement in plants. *Cell. Mol. Life Sci.* **2015**, *72*, 673–689. [CrossRef]
49. Febrianti Salsinha, Y.C.; Rini, D.S.; Indradewa, D.; Rachmawati, D.; Alam, T.; Purwestri, Y.A. Exogenously applied *Casuarina equisetifolia* leaf extracts act as an osmoprotectant on proline accumulation under drought stress in local rice from Indonesia. *Front. Plant Sci.* **2023**, *14*, 1210241. [CrossRef]
50. Efeoglu, B.; Ekmekci, Y.; Cicek, N. Physiological responses of three maize cultivars to drought stress and recovery. *S. Afr. J. Bot.* **2009**, *75*, 34–42. [CrossRef]
51. Bano, Q.; Ilyas, N.; Bano, A.; Zafar, N.; Akram, A.; Hassan, F. Effect of Azospirillum inoculation on maize (*Zea mays* L.) under drought stress. *Pak. J. Bot.* **2013**, *45*, 13–20.
52. Aranjuelo, I.; Molero, G.; Erice, G.; Avice, J.C.; Nogués, S. Plant physiology and proteomics reveals the leaf response to drought in alfalfa (*Medicago sativa* L.). *J. Exp Bot.* **2011**, *62*, 111–123. [CrossRef]
53. Kissen, R.; Øverby, A.; Winge, P.; Bones, A.M. Allyl-isothiocyanate treatment induces a complex transcriptional reprogramming including heat stress, oxidative stress and plant defence responses in *Arabidopsis thaliana*. *BMC Genom.* **2016**, *17*, 740. [CrossRef] [PubMed]
54. Yin, Y.; Qin, K.; Song, X.; Zhang, Q.; Zhou, Y.; Xia, X.; Yu, J. BZR1 transcription factor regulates heat stress tolerance through FERONIA receptor-like kinase mediated reactive oxygen species signaling in tomato. *Plant Cell Physiol.* **2018**, *59*, 2239–2254. [CrossRef]
55. Prerostova, S.; Dobrev, P.; Kramna, B.; Gaudinova, A.; Knirsch, V.; Spichal, L.; Zatloukal, M.; Vankova, R. Heat acclimation and inhibition of cytokinin degradation positively affect heat stress tolerance of *Arabidopsis*. *Front. Plant Sci.* **2020**, *11*, 87. [CrossRef]
56. Medina, E.; Kim, S.H.; Yun, M.; Choi, W.G. Recapitulation of the function and role of ROS generated in response to heat stress in plants. *Plants* **2021**, *10*, 371–383. [CrossRef] [PubMed]
57. Hosseini, M.S.; Samsampour, D.; Ebrahimi, M.; Abadía, J.; Khanahmadi, M. Effect of drought stress on growth parameters, osmolyte contents, antioxidant enzymes and glycyrrhizin synthesis in licorice (*Glycyrrhiza glabra* L.) grown in the field. *Phytochemisty* **2018**, *156*, 124–134. [CrossRef]
58. Zhu, J.K. Abiotic stress signaling and responses in plants. *Cell* **2016**, *167*, 313–324. [CrossRef]
59. Blum, A. Crop responses to drought and the interpretation of adaptation. *Plant Growth Regul.* **1996**, *20*, 135–148. [CrossRef]

60. Chaves, M.M.; Maroco, J.P.; Pereira, J.S. Understanding plant responses to drought from genes to the whole plant. *Funct. Plant Biol.* **2003**, *30*, 239–264. [CrossRef] [PubMed]
61. Passioura, J. The drought environment: Physical, biological and agricultural perspectives. *J. Exp Bot.* **2007**, *58*, 113–117. [CrossRef] [PubMed]
62. Rao, I.M.; Beebe, S.E.; Polania, J.; Grajales, M.; Cajiao, C.; Ricaurte, J.; García, R.; Rivera, M. Evidence for genotypic differences among elite lines of common bean in the ability to remobilize photosynthate to increase yield under drought. *J. Agric. Sci.* **2017**, *155*, 857–875. [CrossRef]
63. Dinneny, J.R. Developmental responses to water and salinity in root systems. *Annu. Rev. Cell Dev. Biol.* **2019**, *35*, 239–257. [CrossRef] [PubMed]
64. Polania, J.; Poschenrieder, C.; Rao, I.; Beebe, S. Root traits and their potential links to plant ideotypes to improve drought resistance in common bean. *Theor. Exp. Plant Physiol.* **2017**, *29*, 143–154. [CrossRef]
65. Beebe, S.E.; Rao, I.M.; Blair, M.W.; Acosta-Gallegos, J.A. Phenotyping common beans for adaptation to drought. *Front. Physiol.* **2013**, *4*, 35. [CrossRef]
66. Basu, S.; Ramegowda, V.; Kumar, A.; Pereira, A. Plant adaptation to drought stress. *F1000Research* **2016**, *5*, 1554. [CrossRef] [PubMed]
67. Xiao, W.J.; Guo, F.X.; Chen, Y.; Liu, L.L.; Cheng, Y.Z.; Jiao, X.S.; Zhang, B.Q.; Bai, G.; Jin, J.Q. Effect of organic fertilizer application on the medicinal character, yield and disease resistance of *Angelica sinensis*. *Acta Pratacult. Sin.* **2021**, *30*, 189–199.
68. Cui, L.G.; Guo, F.X.; Chen, Y.; Xu, B.; Li, R.X. Effects of cultivation methods on the propagating characteristics of *Angelica sinensis* SP2 induced by space flight. *J. Nucl. Agric. Sci.* **2024**, *38*, 813–821.
69. Zhou, B.P.; Peng, Y.W.; Zhao, C.L.; Ding, C.W.; Miao, G.X.; Liu, X.J.; Wang, S. Analysis on climate change characteristics of Longxi in the upstream of Wei River in recent 61 years. *J. Anhui Agric. Sci.* **2018**, *46*, 125–128. [CrossRef]
70. Asmare, B.; Mekuriaw, Y. Assessment of livestock production system and feed balance in watersheds of North Achefer district, Ethiopia. *J. Agric. Environ. Int. Dev.* **2017**, *11*, 574. [CrossRef]
71. González, L.; González-Vilar, M. Determination of relative water content. In *Handbook of Plant Ecophysiology Techniques*; Springer: Dordrecht, The Netherlands, 2001; pp. 207–212, ISBN 978-0-7923-7053-6.
72. Guo, F.X.; Zhang, M.X.; Chen, Y.; Zhang, W.H.; Xu, S.J.; Wang, J.H.; An, L.Z. Relation of several antioxidant enzymes to rapid freezing resistance in suspension cultured cells from alpine *Chorispora bungeana*. *Cryobiology* **2006**, *52*, 241–250. [CrossRef]
73. Guo, F.X.; Xiao, W.J.; Chen, Y.; Zhang, Y.J.; Chen, Y.Z.; Liu, L.L.; Gao, X. Initiation of early bolting by pre-enhancing anthocyanin and catalase activity in *Angelica sinensis* tender leaf during medicine formation cultivation year. *Russ. J. Plant Physiol.* **2021**, *68*, 763–773. [CrossRef]
74. Niu, J.; Zhang, S.P.; Liu, S.D.; Ma, H.J.; Chen, J.; Shen, Q.; Ge, C.W.; Zhang, X.M.; Pang, C.Y.; Zhao, X.H. The compensation effects of physiology and yield in cotton after drought stress. *J. Plant Physiol.* **2018**, *224–225*, 30–48. [CrossRef] [PubMed]
75. Sun, F.; Chen, Q.; Chen, Q.; Jiang, M.; Qu, Y. Yield-based drought tolerance index evaluates the drought tolerance of cotton germplasm lines in the interaction of genotype-by-environment. *PeerJ* **2023**, *11*, e14367. [CrossRef] [PubMed]
76. Shahrokhi, M.; Khorasani, S.K.; Ebrahimi, A. Evaluation of drought tolerance indices for screening some of super sweet maize (*Zea mays* L. var. *saccharata*) inbred lines. *Agrivita J. Agric. Sci.* **2020**, *42*, 435–448. [CrossRef]
77. Adhikari, N.R.; Sharma, S.; Gairhe, J.; Bhandari, R.M.; Poudel, S. Evaluation of drought tolerant rice cultivars using drought tolerant indices under water stress and irrigated condition. *Am. J. Clim. Chang.* **2019**, *8*, 228–236. [CrossRef]
78. Kouighat, M.; Nabloussi, A.; Kettani, R.; Fakhour, S.; El Fechtali, M.; Hamdani, A. Drought-tolerant sesame mutant lines assessed by physiological traits and stress indices under water deficit conditions. *J. Agric. Food Res.* **2023**, *14*, 100842. [CrossRef]
79. Zare, M. Evaluation of drought tolerance indices for the selection of Iranian barley (*Hordeum vulgare*) cultivars. *Afr. J. Biotechnol.* **2012**, *11*, 15975–15981. [CrossRef]
80. Choukan, R.; Taherkhani, I.; Ghanadha, M.; Khodarahmi, M. Evaluation of drought tolerance in grain maize hybrids using drought tolerance indices. *Seed Plant Improvment J.* **2008**, *24*, 543–562. Available online: https://api.semanticscholar.org/CorpusID:111322998 (accessed on 8 September 2024).
81. Malinowska, M.; Donnison, I.; Robson, P. Morphological and physiological traits that explain yield response to drought stress in Miscanthus. *Agronomy* **2020**, *10*, 1194. [CrossRef]
82. Wang, H.Y.; Chen, Y.; Guo, F.X.; Dong, P.B.; Liang, W.; Cheng, J.L. Improvement in the quality and productivity of *Codonopsis pilosula* seedlings by dazomet soil fumigation. *Sci. Rep.* **2024**, *14*, 5407–5420. [CrossRef]

International Journal of
Molecular Sciences

MDPI

Article

Effect of Temperature on Polyamine Oxidase Genes in *Skeletonema dohrnii*

Wei Teng [1,2] and Jun Sun [1,2,*]

1 Research Centre for Indian Ocean Ecosystem, Tianjin University of Science and Technology, Tianjin 300457, China; tengwei990412@163.com
2 Institute for Advanced Marine Research, China University of Geosciences, Guangzhou 511462, China
* Correspondence: phytoplankton@163.com

Abstract: In our experiments, we investigated the effect of temperature on diatom polyamine metabolism using *Skeletonema dohrnii* as an experimental algal species. We set three different temperature conditions for incubation and selected *Skeletonema dohrnii* in the exponential growth period, and analyzed basic physiological parameters, polyamine composition and content, and polyamine oxidase (PAO) gene expression at different temperatures. The results showed that low temperatures led to a decrease in growth rate, an increase in biogenic silica content, an increase in the content of putrescine and spermine, a decrease in the concentration of spermidine, and a down-regulation of PAO gene expression. In addition, high temperature led to an increase in growth rate, a significant change in the concentration of putrescine and spermine, and an increase in spermidine. These findings suggest that changes in temperature affect the growth rate of algae, low temperature increases the biogenic silica content of diatoms, different temperature stresses lead to different kinds of polyamine changes in diatoms, and the PAO gene may play a role in regulating the response of algae to temperature changes. This study lays a foundation for further exploration of the function of the PAO gene in *Skeletonema dohrnii*.

Keywords: abiotic stress; diatom; polyamine metabolism

1. Introduction

According to the latest survey report, global temperatures will rise between 1 °C and 5 °C [1]. In this case, biomes around the world will be seriously affected [2–4]. Phytoplankton are thermophiles and cannot regulate their own temperature. Therefore, they are extremely sensitive to temperature, and changes in temperature have a great impact on them, and temperature controls cell metabolism and stability by regulating the activity of certain enzymes [5–7]. Phytoplankton play a very important role in the ecological environment, which is the basis of the ecosystem food chain and the main source of primary productivity [8,9]. Therefore, studying the effects of temperature on phytoplankton physiology and cellular metabolism is crucial for better predicting the evolution of ecosystems in changing environments and is likewise immensely helpful in exploring the mechanisms by which phytoplankton respond to temperature.

Diatoms are unicellular photosynthetic eukaryotes that are recognized as being of global significance in biogeochemical cycles and the functioning of aquatic food webs [10,11]. They constitute a significant portion of the aquatic biomass, especially during pronounced seasonal phytoplankton blooms, and are estimated to account for 20%

of the Earth's total primary plant production [8,12]. Due to their complex evolutionary history [13], diatoms have a "mixed-match genome" [14], which equips them with a variety of potentially advantageous traits, such as rigid silicified cell walls, the ability to rapidly respond to changes in ambient light, the formation of a stationary phase, and engagement in urea cycling [15]. Typically, planktonic diatoms are well suited to environments with intermittent light and nutrient availability. However, they are especially abundant in nutrient-rich regions, such as the polar zones, upwelling areas, and coastal environments [16], highlighting their remarkable adaptability to diverse ecological niches and biomes.

Skeletonema dohrnii was selected in this study as an important factor contributing to the occurrence of red tide [17]. It has been noted that in areas where red tides are frequent, there are periodic variations between diatom red tides and methanotrophic red tides due to environmental factors and ecological processes [18]. Studies of red tides have shown that environmental conditions are key in determining which species become dominant during red tide outbreaks, with temperature being a key factor in promoting diatom growth. In addition, during red tide outbreaks, the concentration of polyamines in seawater tends to increase with phytoplankton blooms [19,20].

Polyamines, which are organic polymeric cations commonly present in most organisms at physiological pH, serve as crucial growth regulators during the growth processes of phytoplankton [21–23]. In microalgae, polyamines are associated with both cell growth and the accumulation of metabolites [24–26]. Among the polyamines, the most extensively studied are putrescine (PUT), spermidine (SPD), and spermine (SPM), which are widely distributed and relatively abundant. In addition to these, less common polyamines such as diaminopropane (DAP), norspermidine (NSPD), and norspermine (NSPM) have recently garnered attention for their potential involvement in newly identified biochemical pathways. The dominant polyamine species differ across algae, potentially due to variations in their evolutionary histories [27,28]. The metabolic pathways of polyamines in microalgae remain poorly understood compared to those studied in higher plants. Nevertheless, higher plant polyamine metabolism can serve as a reference for modeling these pathways in microalgae (Figure 1). Interestingly, the chemical synthesis of polyamines in microalgae diverges from that of higher plants, with a key distinction lying in the starting point: polyamine synthesis in microalgae typically begins with arginine. This precursor undergoes decarboxylation and enzymatic reactions to form putrescine, which is further converted into H_2O_2, 4-aminobutanal, and, ultimately, 1-pyrroline through the catalytic action of diamine oxidase (DAO). Moreover, putrescine serves as a precursor for the synthesis of triamines and tetraamines via the activity of aminopropyl transferases. Specifically, spermidine synthase (SPDS) transfers an aminopropyl group from decarboxylated S-adenosyl methionine (dcSAM) to putrescine, producing SPD, which can then be converted into SPM through the addition of another aminopropyl group by spermine synthase (SPMS). Alternatively, TSPM synthase (TSPMS) can attach the aminopropyl group to the opposite end of the SPD molecule to produce thermospermine (TSPM), an isomer of SPM. A crucial step in these processes is the generation of dcSAM by SAM decarboxylase (SAMDC), which provides the aminopropyl donor for the reactions catalyzed by aminopropyl transferases.

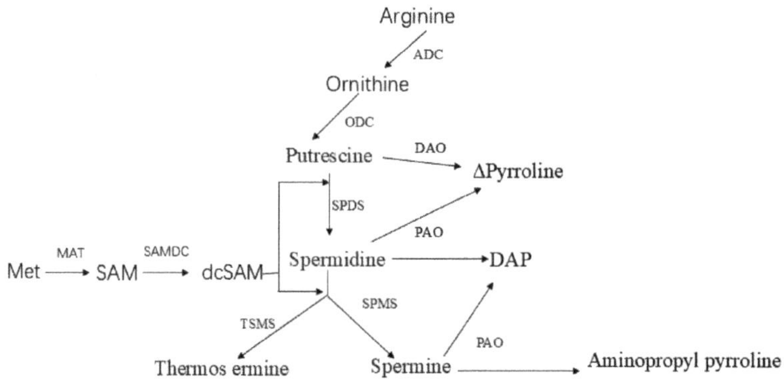

Figure 1. A simple pathway for polyamine anabolism in microalgae (adapted from Miguel A. Blázquez [29]).

Polyamines are widely found in animals, plants and bacteria. Current research on the role of polyamines has focused on plants and animals. Polyamines play an important role in plant growth, development and other physiological processes. Polyamines play a role in mitigating abiotic stress in plants [30–32]. Similar effects have been found in polyamine studies in marine microalgae. Previously, it was found that silicate concentration leads to changes in polyamine concentration and gene expression of polyamine oxidase in diatoms [33]. In general, there are few studies on the response of microalgal polyamines to environmental stresses, and the molecular regulatory mechanisms of polyamine metabolism are also rarely investigated.

Polyamine oxidases (PAOs) are FAD-dependent enzymes involved in the catabolism of polyamines [34]. They catalyze the oxidation of spermidine (Spd), spermine (Spm), and/or their acetylated derivatives at secondary amino groups [35]. The roles of plant PAOs in polyamine catabolism can be grouped into two primary categories. The first category includes PAOs that mediate the terminal catabolism of polyamines. These enzymes catalyze reactions that oxidize Spd or Spm, producing hydrogen peroxide, 1,3-diaminopropane (DAP), and 4-aminobutane (in the case of Spd oxidation) or N-(3-aminopropyl)-4-aminobutane (in the case of Spm oxidation). The second category includes PAOs that carry out the trans-conversion of polyamines, such as the conversion of Spm to Spd and Spd to putrescine (Put). These reactions also generate hydrogen peroxide as a by-product of their catalytic activity. The terminal catabolic pathway of polyamines is specifically activated in the extracellular compartment, whereas the trans-conversion pathway occurs within the intracellular cytoplasmic space, predominantly in the peroxisomes [36].

The purpose of this study is to analyze how *Skeletonema dohrnii* cells respond to different temperatures. At the same time, *Skeletonema dohrnii*, as a red tide algal species, can also be used to investigate the effect of temperature on red tide. This study will explore factors such as cell growth rate, Fv/Fm, chlorophyll content, biosilicon content and polyamine composition as well as content. In addition, we will evaluate the expression of polyamine oxidase genes.

2. Results

2.1. Physiological Property

The change in temperature significantly affected the growth of *Skeletonema dohrnii* (Figure 2A). The growth rate of *Skeletonema dohrnii* was significantly lower at 15 °C than at the normal temperature of 25 °C. In particular, the maximum growth rate of 1.58 times d^{-1} at 25 °C was significantly higher than the rate of 1.18 times d^{-1} at 15 °C. At 28 °C, the growth rate of *Skeletonema dohrnii* was significantly higher than the rate at 25 °C. The maximum growth rate of 2.01 times d^{-1} at 28 °C was higher than the maximum growth rate at 25 °C. From Figure 1, we can see that the growth rate at 15 °C was significantly lower than the rate at 25 °C, and the growth rate at 15 °C was significantly lower than the rate at 28 °C. Similarly, the growth rate at 25 °C was significantly lower than that at 28 °C, so it can be seen that the increase in temperature promoted the growth rate of *Skeletonema dohrnii*. In addition, all Fv/Fm values ranged from 0.61–0.70 with no significant difference among the three treatments, indicating that temperature did not significantly affect the photosynthetic performance of the algae during the experiment (Figure 2B). The content of chlorophyll *a* fluctuated with time in *Skeletonema dohrnii* under each temperature condition (Figure 3). However, the chlorophyll concentrations of *Skeletonema dohrnii* under the three temperature incubations showed no significant differences.

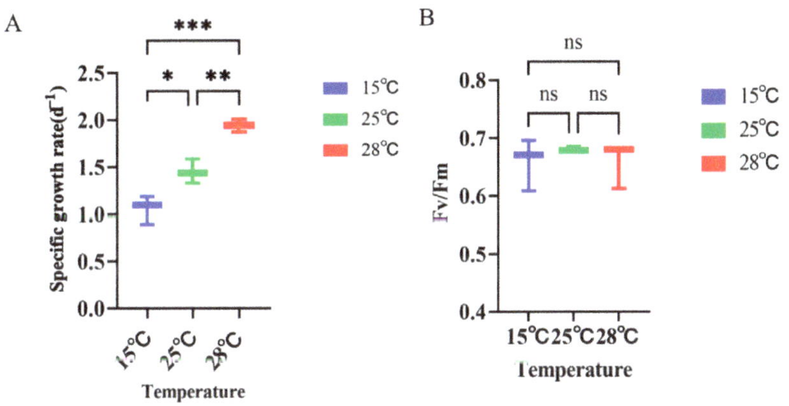

Figure 2. Physiological traits of *Skeletonema dohrnii* cultured under different temperature conditions. (**A**): specific growth rate of *Skeletonema dohrnii* at different temperatures; (**B**): Fv/Fm of *Skeletonema dohrnii* at different temperatures; (ns indicates no significance; * indicates $p < 0.05$; ** indicates $p < 0.01$; *** indicates $p < 0.001$).

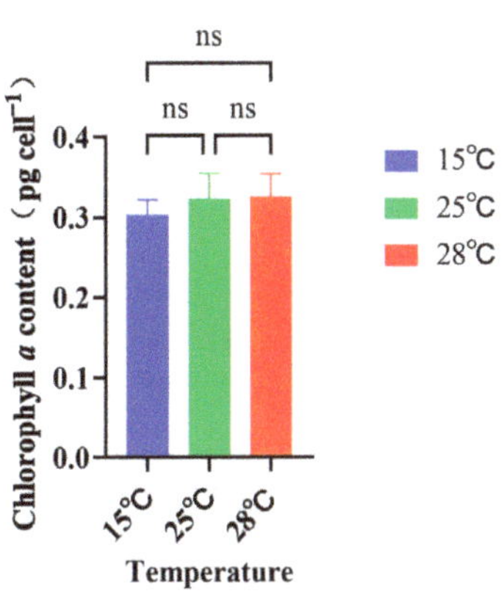

Figure 3. Chlorophyll *a* content of *Skeletonema dohrnii* cultured under different temperature conditions (ns indicates not significant).

2.2. Biosilicon Content

The biogenic silica content (pg $cell^{-1}$) of *Skeletonema dohrnii* cultured at different temperature culture conditions was different (Figure 4). In terms of intracellular silicon content (Figure 4A), the biosilicon content was 1.078 ± 0.082 at the 25 °C (normal) culture condition, while the biosilicon content at 15 °C (low temperature) was 1.51 times higher than that of the 25 °C culture condition, reaching 1.638 ± 0.291. The biosilicon content at 28 °C (high temperature) was 0.81 times higher than that of the 25 °C culture condition, which reached 0.876 ± 0.074. Changes in cell wall silica content were consistent with intracellular silica content (Figure 4B). The cell wall silica content at 25 °C (normal) incubation condition was 0.14 ± 0.033, whereas the biogenic silica content under the incubation condition at 15 °C (low temperature) was 1.67 times higher than that of the incubation condition at 25 °C, which reached 0.234 ± 0.029. The biogenic silica content under the incubation condition at 28 °C (high temperature) was 1.67 times higher than that of the incubation condition at 28 °C (high temperature). The biosilica content was 0.67 times higher than that of the 25 °C incubation condition, reaching 0.095 ± 0.002. However, it was determined that the change from normal in the high-temperature group was not significant.

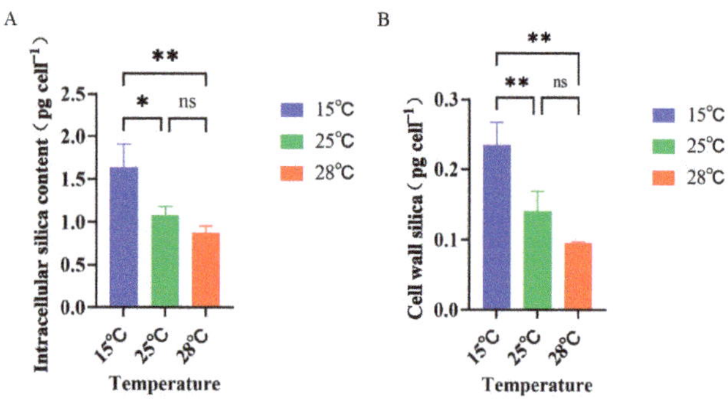

Figure 4. Biosilica content of *Skeletonema dohrnii* cultured under different temperature conditions. (**A**): Intracellular silica content of *Skeletonema dohrnii* at different temperatures; (**B**): Cell wall silica content of *Skeletonema dohrnii* at different temperatures (ns indicates non-significant; * indicates $p < 0.05$; ** indicates $p < 0.01$).

2.3. Polyamine Concentration

The concentration of polyamines differed in different temperature conditions of culture (Figure 5). There was a significant difference in the concentration of putrescine in *Skeletonema dohrnii* cultured at 15 °C (low temperature) compared to 25 °C (normal temperature), with the putrescine concentration in the low-temperature group being 2.53 times higher than that in the normal group. In contrast, there was no significant difference in the concentration of putrescine in *Skeletonema dohrnii* under the 28 °C (high temperature) culture condition (Figure 5A). The distribution of spermine concentration in different treatment groups was similar to that of putrescine (Figure 5B). There was a significant difference between the spermine concentration of *Skeletonema dohrnii* under low-temperature culture conditions and that of the normal-temperature group, with the spermidine concentration in the low-temperature group being 1.39 times higher than that of the normal group, whereas there was no significant difference between the high-temperature group and that of the normal-temperature group. The distribution of spermine content under different temperature conditions was different from both putrescine and spermidine (Figure 5C). Spermidine was not detected in the 15 °C culture condition, and there was also a significant difference in the content of spermidine between the 25 °C and 28 °C incubation conditions. The spermidine content in the 28 °C culture condition was 1.78 times higher than that in the 25 °C group.

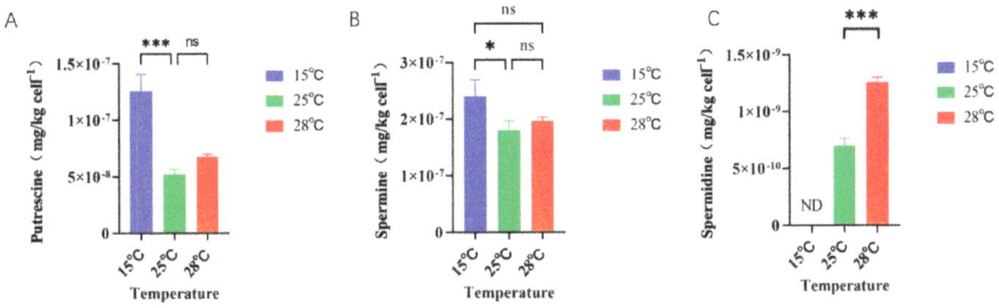

Figure 5. Concentration of polyamines in *Skeletonema dohrnii* cultured under different temperature conditions. (**A**): putrescine content of *Skeletonema dohrnii* at different temperatures; (**B**): spermine content of *Skeletonema dohrnii* at different temperatures; (**C**): spermidine content of *Skeletonema dohrnii* at different temperatures (ND indicates none detected; ns indicates non-significance; * indicates $p < 0.05$; *** indicates $p < 0.001$).

2.4. Polyamine Oxidase Gene Expression

Polyamine oxidase is an enzyme involved in the regulation of polyamine levels in vivo, and its expression was significantly different under different temperature conditions (Figure 6). The polyamine oxidase protein gene expression of *Skeletonema dohrnii* cultured at a low temperature of 15 °C was lower than that of *Skeletonema dohrnii* cultured at a normal temperature of 25 °C, and the polyamine oxidase protein gene expression of *Skeletonema dohrnii* cultured at a low temperature of 15 °C was 0.25 times higher than that of A. *Skeletonema dohrnii* cultured at a normal temperature of 25 °C. The polyamine oxidase protein gene expression of *Skeletonema dohrnii* cultured at a high temperature of 28 °C was lower than that of *Skeletonema dohrnii* cultured at a normal temperature of 25 °C, and the polyamine oxidase protein gene expression of *Skeletonema dohrnii* cultured at a high temperature of 28 °C was 0.3 times of that of *Skeletonema dohrnii* cultured at a normal temperature of 25 °C.

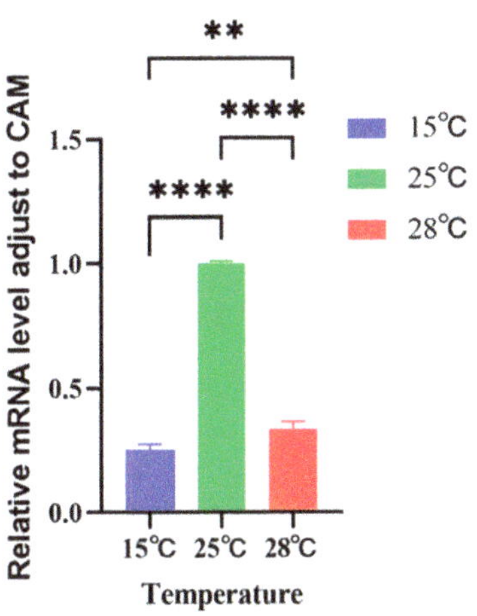

Figure 6. Expression of polyamine oxidase gene in *Skeletonema dohrnii* cultured at different temperatures (** indicates $p < 0.01$; **** indicates $p < 0.0001$).

3. Discussion

In recent years, research on the effects of ocean warming on phytoplankton has always been devoted to the study of the impacts of ocean warming on phytoplankton, and most of the studies have focused on the impacts on the dynamics of phytoplankton communities. Few studies have focused on the molecular mechanisms behind the phytoplankton response to ocean warming and the changes in differential gene expression. In this study, we investigated the response of an ecologically significant diatom in polyamine metabolism at different temperatures by combining the changes in specific gene expression and physiological traits and found differences in the expression of key genes of polyamine metabolism at different temperatures, so as to provide a reference for the study of molecular regulatory mechanisms in the response of marine diatoms to temperature stress.

3.1. Response of Physiological Traits to Temperature in *Skeletonema dohrnii*

Previous studies have shown that warmer temperatures have a positive or neutral effect on the growth of marine phytoplankton [37–40]. For example, cyanobacteria, when cultured individually in batches, have higher growth rates at higher temperatures than at normal temperatures [38]. It has also been shown that functional types of phytoplankton may have different temperature coefficients (Q10), maximum-temperature dependence of growth, and temperature ranges, which would lead to different responses to different degrees of temperature change. In response to global modeling of future temperatures, these differences lead to taxon-specific predictions of growth and geographic distribution, with low-latitude coccolithophores facing considerable reductions and cyanobacterial growth rates increasing substantially. A single effect of temperature change may alter the global community structure of phytoplankton due to significant differences in thermal response among functional phytoplankton types [41]. In the present study, it was observed that the growth rate of *Skeletonema dohrnii* algae incubated under different temperature conditions was directly proportional to the increase in temperature, which increased with the increase in temperature.

The Fv/Fm parameter is generally used to assess whether the cells are stressed by environmental stress, and the value of Fv/Fm is generally constant in the absence of stress from environmental stressors. In the present study, the change in temperature did not significantly change the Fv/Fm values of *Skeletonema dohrnii* throughout the experiment, indicating that the cells remained relatively active throughout the experiment. Similarly, the Fv/Fm values of *Skeletonema dohrnii* did not change significantly after 400 generations of selection in a high-temperature environment [42]. In conclusion, *Skeletonema dohrnii* under experimental temperature stress still maintains constant PSII functional efficiency. It is noteworthy that there was no significant change in the chlorophyll *a* content within the experimental temperature. Similarly, some species showed no significant change in chlorophyll content in response to changes in temperature. For example, the chlorophyll content remained unchanged in *Chaetoceros* sp., *Rhodomonas* sp., *Cryptomonas* sp., and *Isochrysis* sp. grown at 25 °C, 27 °C, 30 °C, 33 °C, and 35 °C [43].

Notably, we found that the biogenic silica content was higher in the low-temperature than in the normal- and high-temperature groups in the experimental temperatures. Some researchers have found that diatom cell size decreases with increasing temperature [44]. The size of the cell volume affects the biogenic silica content, which is similar to the results of the present study. Physically, diatoms with higher densities at lower temperatures appear to be better able to contribute to the "ballasting effect" (referring to the ability of BSi, as the major silicate biomineral, to largely promote POC transport from the surface to deeper ocean), whereas diatom blooms accelerate at higher temperatures and have lower silica content, which may be detrimental to carbon sequestration in the oceans [45].

3.2. Response of Polyamine Content to Temperature

Temperature is widely acknowledged as a critical environmental factor influencing algal growth. Polyamines (PAs), such as spermidine (SPD) and spermine (SPM), are considered significant plant growth regulators (PGRs) that play a role in controlling plant growth and senescence, including enhancing resistance to heat stress [46,47]. The composition of polyamines in microalgae varies in proportions, and changes in polyamine content in response to changes in the environment. It was found that spermine was the main polyamine altered in diatoms under silicate stress [33], whereas in the present study, putrescine and spermine were the main polyamines altered in response to the change of temperature conditions. A similar situation was observed in different salt stresses. Under high-salinity stress, *Skeletonema costatum* significantly increases its polyamine levels, particularly the content of free polyamines, whereas *Prorocentrum donghaiense* primarily regulates its bound spermidine and free spermine. Under low-salinity stress, *Skeletonema costatum* enhances the production of free putrescine, while *Prorocentrum donghaiense* exhibits an overall increase in the levels of all forms of polyamines [48]. According to our results, the main polyamine controlling the growth rate in *Skeletonema dohrnii* under temperature change was spermidine. The beneficial, general stimulatory effect of putrescine in plants has long been known. However, the specific role of putrescine in microalgae is not clearly understood. It has been found that a low temperature increases the putrescine content of maize [49], and a similar situation occurred in the low-temperature group in the present study. Changes in putrescine concentration appear to act on genes at the transcriptional level. For example, putrescine alters the level of heat-shock protein gene expression [50]. According to previous studies, spermine treatment also increased the expression levels of stress-related genes that protect seedlings from stress damage [51]. In our study, we found that the content of spermine increased at low temperatures compared with normal temperatures, suggesting that spermine is involved in stress alleviation during low-temperature stress. Interestingly, we did not find an increase in spermine content in the high-temperature group, which suggests that spermine is not involved in stress relief during high temperature stress. Polyamines can regulate the up- or down-regulation of gene expression either directly or by stimulating the phosphorylation of regulatory proteins, such as transcription factors. For example, a protein kinase in rice is regulated by spermidine at the transcriptional and translational levels [52]. Another example of the covalent binding of polyamines to proteins is observed in the biosynthesis of hypusine. In this process, the butylamino group of spermidine is utilized to facilitate the post-translational modification of the precursor of the eukaryotic translation initiation factor 5A (eIF5A), converting lysine into hypusine [53]. Hypusine-modified eIF5A is essential for the growth of all eukaryotic cells. This could potentially explain the higher growth rates observed under high-temperature conditions. Additionally, the content of spermidine may play a role in regulating diatom cell division.

3.3. Response of Polyamine Oxidase Gene Expression to Temperature

The metabolism of polyamines is very closely related to two enzymes. One is polyamine synthase, and the other is polyamine oxidase. Polyamines are involved in the process of linking stresses in algae facing environmental stresses. Based on the results of the study, we found that the expression of polyamine oxidase was down-regulated in both the high-temperature and low-temperature groups relative to the normal-temperature group (Figure 6). Previous studies have also found that diatom polyamine oxidase genes are down-regulated in response to silicate stress [33]. This suggests that the algal cells need a higher concentration of polyamine to maintain the physiological activity of the cells. Elevated PA levels were found to make plants stress-tolerant in the study [54]. It was found that AtPAO2 and AtPAO5 expression was up-regulated under saline stress; however, AtPAO1, AtPAO3, and AtPAO4 were not affected by salinity treatment [55]. Sagor et al. [56] generated a series of *A. thaliana* mutant lines in which the PAO gene was knocked out. The AtPAO1 and AtPAO 5 double mutants were tolerant to saline and dehydration stresses, while the double mutant AtPAO 2–AtPAO 4 was sensitive to saline and dehydration stresses, compared to the wild type (WT). It was further observed that PAO activity was reduced to 62% and Na^+ uptake was reduced to 75% in the former double mutant compared with the wild type. In addition, the level of spermine was higher in the mutant than in the wild type. Tolerance of the mutant phenotype may be related to the down-regulation of PAO metabolism leading to increased levels of polyamines. Our study revealed a decrease in PAO gene expression in the low- and high-temperature groups compared to the normal-temperature group, which tentatively suggests that polyamines are involved in the mitigation of temperature stress in diatoms.

4. Materials and Methods

4.1. Source and Cultivation of Algal Species

The experimental species *Skeletonema dohrnii* was collected from the Yellow Sea and preserved in f/2 culture medium. The algae were kept at 25 °C, with a light exposure of 100–120 μmol photons $m^{-2}{\cdot}s^{-1}$, and a light exposure to darkness ratio of 12 h:12 h. The experimental algae were cultivated in artificial seawater (ASW) under the same storage conditions as the original algal species. The specific composition of ASW is shown in Table 1; final salinity is 35 PSU.

Table 1. Composition of artificial seawater.

Composition	Final Concentration
NaCl	4.20×10^{-1} M
Na_2SO_4	2.88×10^{-2} M
KCl	9.39×10^{-3} M
$NaHCO_3$	2.38×10^{-3} M
KBr	8.40×10^{-4} M
H_3BO_3	4.85×10^{-5} M
NaF	7.15×10^{-5} M
$MgCl_2{\cdot}6H_2O$	5.46×10^{-2} M
$CaCl_2{\cdot}2H_2O$	1.05×10^{-2} M
$SrCl_2{\cdot}6H_2O$	6.38×10^{-5} M

4.2. Experimental Training Conditions

The experiment involved three temperatures: 15 °C, 25 °C, and 28 °C. These concentrations corresponded to low (15 °C), normal (25 °C), and high (28 °C) temperatures. Each concentration was mixed with 300 mL of culture medium in 3 replicates. The algal cells were cultured semi-continuously in a 500 mL polycarbonate bottle and diluted every 3 days to facilitate cell transfer during the exponential growth phase.

4.3. Determination of Growth Rate

We counted under a light microscope using a blood cell counting plate and then calculated the growth rate (d^{-1}) using Equation (1).

$$\mu = (\mathrm{Ln}N_1 - \mathrm{Ln}N_0)/t \quad (1)$$

where N_0 is the cell density at the beginning of the algae dilution cycle, N_1 is the cell density at the end of the algae dilution cycle, and t is the duration of the algae dilution cycle.

4.4. Determination of Fv/Fm

A hand-held algal fluorometer (cuvette type) (model: AquaPen AP-C100; PSI., Ltd., Prague, Czech Republic) with an optical glass cuvette and FluorPen system software (version number 1.1) was utilized to measure chlorophyll fluorescence parameters. The maximum photochemical quantum yield of PSII, represented by the chlorophyll fluorescence parameter F_v/F_m, is typically determined using Equation (2):

$$\frac{F_v}{F_m} = (F_m - F_0)/F_m \quad (2)$$

where F_0 and F_m are the minimum and maximum fluorescences measured when the test algal sample is dark-adapted, respectively [57]. Dark adaptation time is 10 min.

4.5. Determination of Chlorophyll Concentration

We introduced 5 mL of algal solution into a GF membrane, then transferred the membrane to a brown vial. After adding 5 mL of 90% acetone to the vial, we allowed the extraction process to continue at −20 °C for 24 h. Subsequently, we utilized a TriLog fluorescence analyzer (Turner, CA, USA) to determine the chlorophyll *a* content and compute the Chl-a concentration.

4.6. Determination of Biogenic Silicon Concentration

A 5 mL sample of algal solution was passed through a 0.6 μm PC membrane for filtration. The membrane, now containing algae, was transferred to a centrifuge tube filled with 10 mL of ultrapure water. The tube was agitated to ensure complete dispersion of algae from the membrane into the water. This mixture was then subjected to heat treatment in a 100 °C water bath for 10 min. Following this, the heated solution underwent centrifugation and filtration. The resulting filtrate and residue were subsequently analyzed to determine biosilica content, following the methodology outlined by Strickland and Parsons [58].

4.7. Determination of Polyamine Content

Using high-performance liquid chromatography (HPLC), polyamines were quantified in accordance with Aizi. Shanghai Fuda Detection Technology Group Co., Ltd. (Tianjin, China) conducted the analysis to measure the concentration of polyamines in this research.

4.8. Design of Primer Sequences

Based on unpublished transcriptome data from the literature [33], primer 5.0 was used to design primers for polyamine oxidase genes and housekeeping genes (Table 2). The CaM gene is a housekeeping gene, and CaM is a calmodulin protein that is widespread in eukaryotic cells and is highly evolutionarily conserved. Calmodulin itself has no enzymatic activity and is biologically inactive in the absence of Ca^{2+}.

Table 2. Primer sequences.

Primer Name	Sequence
PAO-U	GCGGCTGAGAGAGCAGCAAACTCCC
PAO-D	CAATAACACGTGGAGGACGTCG
CaM-U	TCGATGTTGATGGTGGTGGAA
CaM-D	TCGGGGTCGTCTTTGCTGTT

4.9. Reverse Transcription

A reagent kit from Takara was used for this operation. First, 1 μL Oligo dT Primer, 1 μL dNTP Mixture, 4 μL RNase Free dH_2O, and 4 μL RNA from diatoms were added to the PCR tube for mixing, and then the PCR tube containing the solution was heated in a water bath at 65 °C for 5 min; the PCR tube was placed on ice after the heating was completed, and 4 μL of 5× Primer Script II Buffer, 0.5 μL of RNase Inhibitor, 1 μL of Primer Script II, and 4.5 μL of RNase Free dH_2O were added to the tube for mixing. Then, the PCR tube containing the mixture was put into the PCR tube containing the mixture, put into the Eppendorf PCR instrument and run according to the set reaction program, and cDNA was obtained. The reaction program was as follows: 42 °C for 60 min, 95 °C for 5 min, 72 °C for 15 min, and storage at 4 °C.

4.10. Real Time Fluorescence Quantitative PCR

After conducting primer suitability PCR screening, it was determined that the optimal annealing temperature (Tm) for the polyamine oxidase gene and CaM gene primers was 60 °C, within a temperature gradient range of 53–60 °C. CaM was used as an internal reference gene for quantitative experiments using cDNA from *Skeletonema dohrnii* vulgaris cultured at various silicate concentrations, following a two-step method. Three parallel experiments were conducted, with negative controls included to minimize systematic errors from sampling and instruments. The mRNA levels of polyamine oxidase genes were measured using the Step One Plus real-time PCR system under culture conditions with different silicate concentrations.

4.11. Data Analysis

The physiological data of algal cells in this study were analyzed using GraphPad Prism 9.4 for statistical evaluation. A t-test was used as a post-test for one-way ANOVA with a significance level of $p < 0.05$, indicating a statistical difference in the data.

Following the completion of real-time fluorescence quantitative PCR, the data were analyzed using the ABI software (version number 2.3, Applied Biosystems, USA). The relative expression level of the polyamine oxidase gene mRNA under various environmental conditions was calculated using the $2^{-\Delta\Delta Ct}$ method based on the Ct values of the target gene and the CaM gene. A temperature of 25 °C was used as the control for gene expression levels.

5. Conclusions

In this study, we investigated the physiological response and molecular mechanism of *Skeletonema dohrnii* in diatoms to different temperature stresses. We focused on the growth rate, biogenic silica content, polyamine level and polyamine oxidase expression of *Skeletonema dohrnii* in response to temperature. It was shown that the growth rate of *Skeletonema dohrnii* was significantly enhanced by increasing temperature. The low temperature contributed to the increase in biosilicon content. The polyamine composition of *Skeletonema dohrnii* was different between low- and high-temperature stresses. Both low and high temperatures led to a decrease in the gene expression level of PAO. These results may provide data for the study of the molecular regulation mechanism of diatoms facing abiotic stress.

Author Contributions: W.T.: Investigation, Methodology, Visualization, Writing—original draft, Writing—review and editing. J.S.: Conceptualization, Methodology, Project administration, Resources, Supervision, Visualization, Review and editing. All authors have read and agreed to the published version of the manuscript.

Funding: This research was financially supported by the National Key R&D Program of China (2019YFC407800), the National Natural Science Foundation of China (41876134, 41676112, and 41276124), and the Changjiang Scholars Program of the Ministry of Education of China (T2014253) to Jun Sun.

Institutional Review Board Statement: Not applicable.

Informed Consent Statement: Not applicable.

Data Availability Statement: Data are contained within the article.

Conflicts of Interest: The authors declare no conflict of interest.

References

1. Rogelj, J.; Meinshausen, M.; Knutti, R. Global warming under old and new scenarios using IPCC climate sensitivity range estimates. *Nat. Clim. Change* **2012**, *2*, 248–253. [CrossRef]
2. Frey, S.D.; Lee, J.; Melillo, J.M.; Six, J. The temperature response of soil microbial efficiency and its feedback to climate. *Nat. Clim. Change* **2013**, *3*, 395–398. [CrossRef]
3. Vezzulli, L.; Brettar, I.; Pezzati, E.; Reid, P.C.; Colwell, R.R.; HöFle, M.G.; Pruzzo, C. Long-term effects of ocean warming on the prokaryotic community: Evidence from the vibrios. *ISME J.* **2011**, *6*, 21–30. [CrossRef] [PubMed]
4. Thomas, M.K.; Kremer, C.T.; Klausmeier, C.A.; Litchman, E. A Global Pattern of Thermal Adaptation in Marine Phytoplankton. *Science* **2012**, *338*, 1085. [CrossRef]
5. Privalov, P.L. Stability of Proteins Small Globular Proteins. *Adv. Protein Chem.* **1979**, *33*, 167–241. [PubMed]
6. Danson, M.J.; Hough, D.W.; Russell, R.J.M.; Taylor, G.L.; Laurence, P. Enzyme thermostability and thermoactivity. *Protein Eng. Des. Sel.* **1996**, *9*, 629–630. [CrossRef] [PubMed]
7. Kingsolver, J.G. The well-temperatured biologist. (American Society of Naturalists Presidential Address). *Am. Nat.* **2009**, *174*, 755–768. [CrossRef] [PubMed]
8. Field, C.B.; Behrenfeld, M.J.; Randerson, J.T.; Falkowski, P. Primary production of the biosphere: Integrating terrestrial and oceanic components. *Science* **1998**, *281*, 237–240. [CrossRef]
9. Fuhrman, J.A.; Cram, J.A.; Needham, D.M. Marine microbial community dynamics and their ecological interpretation. *Nat. Rev. Microbiol.* **2015**, *13*, 133–146. [CrossRef] [PubMed]
10. Smetacek, V. Diatoms and the silicate factor. *Nature* **1998**, *101*, 38–39. [CrossRef]
11. Falkowski, P.G. The ocean's invisible forest. *Sci Am.* **2002**, *287*, 54–61. [CrossRef]
12. Nelson, D.M.; Tréguer, P.; Brzezinski, M.A.; Leynaert, A.; Quéguiner, B. Production and dissolution of biogenic silica in the ocean: Revised global estimates, comparison with regional data and relationship to biogenic sedimentation. *Glob. Biogeochem. Cycles* **1995**, *9*, 359–372. [CrossRef]
13. Kooistra, W.H.C.F.; Gersonde, R.; Medlin, L.K.; Mann, D.G. The Origin and Evolution of the Diatoms: Their Adaptation to a Planktonic Existence. In *Evolution of Primary Producers in the Sea*; Falkowski, P.G., Knoll, A.H., Eds.; Academic Press: Burlington, MA, USA, 2007; pp. 207–249.
14. Armbrust, E.V. The life of diatoms in the world's oceans. *Nature* **2009**, *459*, 185–192. [CrossRef] [PubMed]
15. Bowler, C.; Vardi, A.; Allen, A.E. Oceanographic and Biogeochemical Insights from Diatom Genomes. *Annu. Rev. Mar. Sci.* **2010**, *2*, 333–365. [CrossRef] [PubMed]
16. Smetacek, V. Making sense of ocean biota: How evolution and biodiversity of land organisms differ from that of the plankton. *J. Biosci.* **2012**, *37*, 589–607. [CrossRef]
17. Zhou, M.J.; Yan, T.; Zhou, J. Preliminary analysis of the characteristics of red tide areas in Changjiang River estuary and its adjacent sea. *Chin. J. Appl. Ecol.* **2003**, *14*, 1031–1038.

18. Zhou, M.; Zhu, M. Progress of the Project "Ecology and Oceanography of Harmful Algal Blooms in China". *Adv. Earth Sci.* **2006**, *21*, 673–679.
19. Nishibori, N.; Matuyama, Y.; Uchida, T.; Moriyama, T.; Hirota, H. Spatial and temporal variations in free polyamine distributions in Uranouchi Inlet, Japan. *Mar. Chem.* **2003**, *82*, 307–314. [CrossRef]
20. Li, C.; Zhao, W.; Miao, H. Distributions of free 2-Phenylethylamine, Putrescine, Spermidine, and Spermine in the East China Sea in summer of 2010. *Mar. Sci.* **2012**, *36*, 68–74.
21. Moschou, P.N.; Paschalidis, K.A.; Roubelakis-Angelakis, K.A. Plant polyamine catabolism: The state of the art. *Plant Signal Behav* **2008**, *3*, 1061–1066. [CrossRef] [PubMed]
22. Sebela, M.; Frebort, I.; Galuszka, P.; Brauner, F.; Lamplot, Z.; Lemr, K.; Pec, P. A study on the reactions of plant copper amine oxidase with short-chain aliphatic diamines. *Inflamm. Res.* **2001**, *50* (Suppl. 2), 138–139. [CrossRef] [PubMed]
23. Gholami, M.; Fakhari, A.; Ghanati, F. Selective Regulation of Nicotine and Polyamines Biosynthesis in Tobacco Cells by Enantiomers of Ornithine. *Chirality* **2013**, *25*, 22–27. [CrossRef]
24. Kotzabasis, K.; Fotinou, C.; Roubelakis-Angelakis, K.A.; Ghanotakis, D. Polyamines in the photosynthetic apparatus. *Photosynth. Res.* **1993**, *38*, 83–88. [CrossRef]
25. Alcázar, R.; Cuevas, J.C.; Patron, M.; Altabella, T.; Tiburcio, A.F. Abscisic acid modulates polyamine metabolism under water stress in Arabidopsis thaliana. *Physiol. Plant.* **2006**, *128*, 448–455. [CrossRef]
26. Handa, A.K.; Mattoo, A.K. Differential and functional interactions emphasize the multiple roles of polyamines in plants. *Plant Physiol. Biochem.* **2010**, *48*, 540–546. [CrossRef] [PubMed]
27. Hamana, K.; Aizaki, T.; Arai, E.; Saito, A.; Uchikata, K.; Ohnishi, H. Distribution of norspermidine as a cellular polyamine within micro green algae including non-photosynthetic achlorophyllous Polytoma, Polytomella, Prototheca and Helicosporidium. *J. Gen. Appl. Microbiol.* **2004**, *50*, 289–295. [CrossRef] [PubMed]
28. Nishibori, N.; Nishijima, T. Changes in polyamine levels during growth of a red-tide causing phytoplankton Chattonella antiqua (Raphidophyceae). *Eur. J. Phycol.* **2004**, *39*, 51–55. [CrossRef]
29. Blazquez, M.A. Polyamines: Their Role in Plant Development and Stress. *Annu. Rev. Plant Biol.* **2024**, *75*, 95–117. [CrossRef] [PubMed]
30. Wu, J.; Liu, W.; Jahan, M.S.; Shu, S.; Sun, J.; Guo, S. Characterization of polyamine oxidase genes in cucumber and roles of CsPAO3 in response to salt stress. *Environ. Exp. Bot.* **2022**, *194*, 104696. [CrossRef]
31. Xi, Y.; Hu, W.J.; Zhou, Y.; Liu, X.; Qian, Y.X. Genome-Wide Identification and Functional Analysis of Polyamine Oxidase Genes in Maize Reveal Essential Roles in Abiotic Stress Tolerance. *Front. Plant Sci.* **2022**, *13*, 950064. [CrossRef] [PubMed]
32. Samanta, I.; Roy, P.C.; Das, E.; Mishra, S.; Chowdhary, G. Plant Peroxisomal Polyamine Oxidase: A Ubiquitous Enzyme Involved in Abiotic Stress Tolerance. *Plants* **2023**, *12*, 652. [CrossRef] [PubMed]
33. Teng, W.; Shang, X.; Sun, J. The effect of silicate on polyamine oxidase genes in Skeletonema dohrnii. *Mar. Environ. Res.* **2025**, *204*, 106860. [CrossRef] [PubMed]
34. Cona, A.; Rea, G.; Angelini, R.; Federico, R.; Tavladoraki, P. Functions of amine oxidases in plant development and defence. *Trends Plant Sci.* **2006**, *11*, 80–88. [CrossRef]
35. Wu, T.; Yankovskaya, V.; Mcintire, W.S. Cloning, Sequencing, and Heterologous Expression of the Murine Peroxisomal Flavoprotein, N1-Acetylated Polyamine Oxidase. *J. Biol. Chem.* **2003**, *278*, 20514–20525. [CrossRef]
36. Wang, W.; Paschalidis, K.; Feng, J.C.; Song, J.; Liu, J.H. Polyamine Catabolism in Plants: A Universal Process With Diverse Functions. *Front. Plant Sci.* **2019**, *10*, 561. [CrossRef] [PubMed]
37. Goldman, J.C. Temperature effects on steady-state growth, phosphorus uptake, and the chemical composition of a marine phytoplankter. *Microb. Ecol.* **1979**, *5*, 153–166. [CrossRef] [PubMed]
38. Mesquita, M.A.-O.; Prestes, A.C.C.; Gomes, A.M.A.; Marinho, M.M. Direct Effects of Temperature on Growth of Different Tropical Phytoplankton Species. *Microb. Ecol.* **2019**, *79*, 1–11. [CrossRef] [PubMed]
39. Padfield, D.; Yvon-Durocher, G.; Buckling, A.; Jennings, S.; Yvon-Durocher, G. Rapid evolution of metabolic traits explains thermal adaptation in phytoplankton. *Ecol. Lett.* **2015**, *19*, 133–142. [CrossRef]
40. Anderson, S.A.-O.; Barton, A.D.; Clayton, S.; Dutkiewicz, S.A.-O.; Rynearson, T.A.-O. Marine phytoplankton functional types exhibit diverse responses to thermal change. *Nat. Commun.* **2021**, *12*, 6413. [CrossRef] [PubMed]
41. Cheng, L.M.; Zhang, S.F.; Xie, Z.A.-O.X.; Li, D.X.; Lin, L.; Wang, M.A.-O.; Wang, D.A.-O. Metabolic Adaptation of a Globally Important Diatom following 700 Generations of Selection under a Warmer Temperature. *Environ. Sci. Technol.* **2022**, *56*, 5247–5255. [CrossRef]
42. Carvalho, A.P.; Monteiro, C.M.; Malcata, F.X. Simultaneous effect of irradiance and temperature on biochemical composition of the microalga Pavlova lutheri. *J. Appl. Phycol.* **2009**, *21*, 543–552. [CrossRef]

43. Montagnes, D.J.S.; Franklin, D.J. Effect of Temperature on Diatom Volume, Growth Rate, and Carbon and Nitrogen Content: Reconsidering Some Paradigms Author(s). *Limnol. Ocean.* **2001**, *46*, 2008–2018. [CrossRef]
44. Sheehan, C.E.; Baker, K.G.; Nielsen, D.A.; Petrou, K. Temperatures above thermal optimum reduce cell growth and silica production while increasing cell volume and protein content in the diatom Thalassiosira pseudonana. *Hydrobiologia* **2020**, *847*, 4233–4248. [CrossRef]
45. Savvides, A.; Ali, S.; Tester, M.; Fotopoulos, V. Chemical Priming of Plants Against Multiple Abiotic Stresses: Mission Possible? *Trends Plant Sci.* **2016**, *21*, 329–340. [CrossRef] [PubMed]
46. Hussain, A.; Nazir, F.; Fariduddin, Q. Polyamines (spermidine and putrescine) mitigate the adverse effects of manganese induced toxicity through improved antioxident system and photosynthetic attributes in Brassica juncea. *Chemosphere* **2019**, *236*, 124830. [CrossRef]
47. Zhu, X.-W.; Zhao, W.-H.; Miao, H. Relationship between growth and variation of endogenous polymine content under salinity stress in skeletonema costatum s. L. And prorocentrum donghaiense. *Oceanol. Limnol. Sin.* **2015**, *46*, 50–57.
48. Szalai, G.; Janda, T.; Bartók, T.; Páldi, E. Role of light in changes in free amino acid and polyamine contents at chilling temperature in maize (Zea mays). *Physiol. Plant.* **2010**, *101*, 434–438. [CrossRef]
49. Kumar, R.R.; Singh, G.P.; Sharma, S.K.; Singh, K.; Goswami, S.; Rai, R.D. Molecular cloning of HSP17 gene (sHSP) and their differential expression under exogenous putrescine and heat shock in wheat (*Triticum aestivum*). *Acad. J.* **2012**, *11*, 16800–16808.
50. Fu, X.Z.; Xing, F.; Wang, N.Q.; Peng, L.Z.; Chun, C.P.; Cao, L.; Ling, L.L.; Jiang, C.L. Exogenous spermine pretreatment confers tolerance to combined high-temperature and drought stress in vitro in trifoliate orange seedlings via modulation of antioxidative capacity and expression of stress-related genes. *Biotechnol. Biotechnol. Equip.* **2014**, *28*, 192–198. [CrossRef] [PubMed]
51. Gupta, B.; Gupta, K.; Sengupta, D.N. Spermidine-mediated in vitro phosphorylation of transcriptional regulator OSBZ8 by SNF1-type serine/threonine protein kinase SAPK4 homolog in indica rice. *Acta Physiol. Plant.* **2012**, *34*, 1321–1336. [CrossRef]
52. Park, M.H.; Nishimura, K.; Zanelli, C.F.; Valentini, S.R. Functional significance of eIF5A and its hypusine modification in eukaryotes. *Amino Acids* **2010**, *38*, 491–500. [CrossRef]
53. Liu, J.-H.; Wang, W.; Wu, H.; Gong, X.; Moriguchi, T. Polyamines function in stress tolerance: From synthesis to regulation. *Front. Plant Sci.* **2015**, *6*, 827. [CrossRef] [PubMed]
54. Zarza, X.; Atanasov, K.E.; Marco, F.; Arbona, V.; Alcazar, R. Polyamine Oxidase 5 loss-of-function mutations in Arabidopsis thaliana trigger metabolic and transcriptional reprogramming and promote salt stress tolerance. *Plant Cell Environ.* **2016**, *40*, 1703–1713. [CrossRef] [PubMed]
55. Sagor, G.H.M.; Siyuan, Z.; Seiji, K.; Stefan, S.; Thomas, B.; Tomonobu, K. Reducing Cytoplasmic Polyamine Oxidase Activity in Arabidopsis Increases Salt and Drought Tolerance by Reducing Reactive Oxygen Species Production and Increasing Defense Gene Expression. *Front. Plant Sci.* **2016**, *7*, 214. [CrossRef]
56. Dao, L.H.T.; Beardall, J. Effects of lead on two green microalgae Chlorella and Scenedesmus: Photosystem II activity and heterogeneity. *Algal Res.* **2016**, *16*, 150–159. [CrossRef]
57. Strickland, J.D.H.; Parsons, T.R. A Practical Handbook of Seawater Analysis. *Fish. Res. Bd. Canada Bull.* **1968**, *167*, 185–194.
58. Aziz, A.; Larher, F. Changes in polyamine titers associated with the proline response and osmotic adjustment of rape leaf discs submitted to osmotic stresses—ScienceDirect. *Plant Sci.* **1995**, *112*, 175–186. [CrossRef]

International Journal of
Molecular Sciences

MDPI

Article

Unraveling Effects of miRNAs Associated with APR Leaf Rust Resistance Genes in Hybrid Forms of Common Wheat (*Triticum aestivum* L.)

Julia Spychała [1,*], Aleksandra Noweiska [1,*], Agnieszka Tomkowiak [2], Roksana Bobrowska [2], Katarzyna Szewczyk [2] and Michał Tomasz Kwiatek [1,2]

1 Plant Breeding and Acclimatization Institute–National Research Institute in Radzików, 05-870 Błonie, Poland; m.kwiatek@ihar.edu.pl

2 Department of Genetics and Plant Breeding, Poznań University of Life Sciences, 60-637 Poznań, Poland; agnieszka.tomkowiak@up.poznan.pl (A.T.); roksana.bobrowska@up.poznan.pl (R.B.); katarzyna.szewczyk1@up.poznan.pl (K.S.)

* Correspondence: j.spychala@ihar.edu.pl (J.S.); a.noweiska@ihar.edu.pl (A.N.)

Abstract: The fungus *Puccinia triticina* Eriks (*Pt*) is the cause of leaf rust, one of the most damaging diseases, which significantly reduces common wheat yields. In *Pt*-resistant adult plants, an APR-type resistance is observed, which protects the plant against multiple pathogen races and is distinguished by its persistence under production conditions. With a more complete understanding of the molecular mechanisms underlying the function of APR genes, it will be possible to develop new strategies for resistance breeding in wheat. Currently, mainly APR genes, such as *Lr34*, *Lr46*, and *Lr67*, are principally involved in resistance breeding as they confer durable resistance to multiple fungal races occurring under different climatic and environmental conditions. However, the mechanisms underlying the defence against pathogens mediated by APR genes remain largely unknown. Our research aimed to shed light on the molecular mechanisms related to resistance genes and miRNAs expression, underlying APR resistance to leaf rust caused by *Pt*. Furthermore, the present study aimed to identify and functionally characterize the investigated miRNAs and their target genes in wheat in response to leaf rust inoculation. The plant material included hybrid forms of wheat from the F_2 and BC_1F_1 generations, obtained by crossing the resistance cultivar Glenlea (CItr 17272) with agriculturally important Polish wheat cultivars. Biotic stress was induced in adult plants via inoculation with *Pt* fungal spores under controlled conditions. The RT-qPCR method was used to analyze the expression profiles of selected APR genes at five time points (0, 6, 12, 24, and 48 hpi). The results presented here demonstrate the differential expression of APR genes and miRNAs at stages of leaf rust development at selected timepoints after inoculation. We analyzed the expression of three leaf rust resistance genes, using different genetic backgrounds in F_2 and BC_1F_1 segregation materials, in leaf tissues after *Pt* infection. Our goal was to investigate potential differences resulting from the genetic background found in different generations of hybrid forms of the same parental forms. Gene ontology analysis predicted 190 target genes for tae-miR5384-3p and 167 target genes for tae-miR9653b. Our findings revealed distinct expression profiles for genes, with the highest expression levels observed mainly at 6, 24, and 48 hpi. The candidate gene *Lr46-Glu2* displayed an upregulation, suggesting its potential involvement in the immune response against *Pt* infection.

Keywords: adult plant resistance; leaf rust; *Lr46-Glu2*; target genes; small RNAs

1. Introduction

The global area under common wheat cultivation is more than 215 million hectares per year. Moreover, the scale of global wheat trade is larger than that of maize and rice combined [1]. Wheat grains are one of the most crucial sources of protein, providing approximately 20% of the calories consumed by people worldwide [2–5]. A disruption of these proportions could lead to a major crisis, i.e., insufficient food availability and economic changes resulting from the increased prices of wheat-containing products [6,7]. The growth of the world's population is forcing suppliers to satisfy a growing demand for food. In order to meet these expectations, wheat production has to be increased by approximately 60% by 2050, which imposes a substantial challenge on breeders. Currently, the main directions of wheat breeding focus on high yields, high trade quality, and resistance to biotic and abiotic stresses [3,8,9]. To develop a sustainable and cost-effective strategy to control the spread of cereal diseases, it is therefore necessary to implement a multidisciplinary research approach involving breeders, farmers, geneticists, phytopathologists, and biotechnologists [10].

Worldwide wheat production is constantly exposed to diseases and pests that are evolving and spreading to new areas [11]. Climate change is favoring an increase in the incidence of diseases such as powdery mildew, stripe rust, leaf rust, and septoria tritici blotch, leading to an average of 20% damage to wheat yields each year [12]. Rust-causing fungi are biotrophic and are consequently entirely dependent on nutrient resources obtained from living host cells [13]. Rust fungi differ widely in the number of hosts they can infect; for example, wheat leaf rust (*Pt*) can only infect plants belonging to *Triticum* and *Aegilops* genera [14]. One of the current priorities in the industry is improved and climate-friendly agronomic practices, reducing the use of crop protection products [13]. Therefore, an understanding of the genetic variability of plant resistance, the evolutionary capacity of pathogens, and the choice of modern breeding methods are essential for successful resistance breeding. Common wheat (*Triticum aestivum* L.) is an allohexaploid species (2n = 6x = 42; AABBDD) derived from a dynamic domestication history [15]. The polyploidy of this species is the result of two rounds of hybridization: the first interspecific hybridization between the wheat-related *T. urartu* (2n = 2x = 14; AA) [16] and an undiscovered species related to *Aegilops speltoides* (genome BB), and the second between *T. turgidum* (2n = 4x = 28; AABB) and *A. tauschii* (2n = 2x = 14; DD) [17,18]. However, the origin of the B subgenome is still unclear. Currently, there is no diploid species whose chromosomes have high affinity for the B subgenome, and this suggests that the ancestral species is extinct or still unidentified [16]. Alternatively, there is a third possibility, which is the polyphyletic origin of subgenome B and its evolution at the polyploid level through multiple crosses with related species, such as *Ae. speltoides*, *Ae. bicornis*, *Ae. sharonensis*, *Ae. longissima*, and *Ae. searsii* [19,20]. In comparison to wheat's wild relatives and its old cultivars, modern cultivars show increasingly lower genetic diversity, thereby also having a decreasing potential for disease-resistant genotypes. One of the strategies to overcome low genetic diversity in wheat is to exploit the gene pools present in native cultivars and wild ancestors of wheat [21,22]. Gene banks represent a huge potential that can be used in research and breeding applications to increase wheat resistance to biotic and abiotic stresses. Thus, by using closely related wild varieties to cross with elite crop cultivars, it is possible to introduce new genetic sources of desirable traits into breeding, especially genes that determine disease resistance. These genes can then be retained in subsequent hybrid generations through molecular-marker-assisted backcrossing [23]. Thereby, gene banks can be used to diversify resistance determinants, although the introgression of new alleles is a long and difficult process. Therefore, to accelerate the discovery, introgression, and pyramiding of resistance genes in resistance breeding, it is crucial to integrate as many tools as possible, most importantly advanced molecular biology techniques [24].

According to recent research, over 80 *Lr* genes and 14 additional leaf rust resistance genes have been identified in wheat. The above 14 genes were not assigned a new number in the *Lr* collection because they were not evaluated for allelism with known *Lr* genes to determine novelty [25]. There are two groups of resistance genes used in breeding wheat resistant to rust diseases. The first group, known as resistance genes (R major genes), confer specific resistance against one particular race of pathogen. Most of the identified *Lr* genes belong to this group [26]. In contrast, the second group, classified as APR resistance genes, is race-unspecific and often manifests itself as a "slow rusting" effect, which is associated with slower infection development, a longer latency period, and the formation of fewer uredinia. APR genes provide partial but durable resistance to pathogens [27]. The APR genes encode different proteins, such as ABC transporters, protein kinases, or hexose transporters, and these include *Lr34/Yr18/Pm38/Sr57*, *Lr46/Yr29/Pm39/Sr58*, and *Lr67/Yr46/Pm46/Sr55*.

The regulation of gene expression is enormously important, being essential for knowing and understanding the molecular mechanisms of resistance. In addition, researchers are increasingly turning their attention to the role of small non-coding RNAs (miRNAs, siRNAs), which play a significant role in the control of molecular processes. A more thorough understanding of the regulation and expression profile of genes related to RNA interference (RNAi) mechanisms and tolerance to abiotic or biotic stresses could lead to improved agronomic traits in many crops worldwide [28]. In plants, miRNA molecules are encoded by *MIR* genes, which are located in non-coding regions of the genome and are normally transcribed by RNA polymerase II [29]. miRNA are small, non-coding endogenous RNAs of 20-to-24 nucleotides (nt) that play a key role in regulating cellular processes. These small molecules can downregulate the protein levels of their target genes by repressing translation, cleaving the transcript, or inhibiting transcription [30,31]. The aim of the presented study was a multilevel analysis of the molecular mechanisms of resistance related to APR genes and miRNAs expression in response to *Pt* inoculation. In the experiment, we analyzed the expression of three leaf rust resistance genes, using different genetic backgrounds in F_2 and BC_1F_1 segregation materials, in leaf tissues after *Pt* infection. Our goal was to investigate potential differences resulting from the genetic background found in different generations of hybrid forms of the same parental forms. Furthermore, the present study aimed to identify and functionally characterize the investigated miRNAs and their target genes in the common wheat genome using GO analysis.

2. Results

2.1. Identification of Molecular Markers Associated with APR Genes in Hybrid Forms

The use of molecular markers successfully identified selected *Lr* resistance genes in the wheat hybrid forms tested. The marker *csLV34*, anchored to the *Lr34* gene sequence, was used to verify the presence of the respective alleles (Table 1). According to the literature, the STS marker *csLV34* is highly reliable for the identification of *Lr34* [32]. If *Lr34* is present, a product of 150 bp will be obtained. The absence of the *Lr34* gene is indicated by the 229 bp product obtained. In a previous study, four hybrid forms were analyzed, a heterozygous system was obtained in the F_1 generation [33], and only in Harenda × Glenlea form was a homozygote (+) obtained. In the absence of *Lr34*, a PCR product of 229 bp (−) was obtained. Further crosses resulted in hybrid forms of the F_2 and BC_1F_1 generations, which were also subjected to selection using the above molecular markers in the presented work (Table 1). In these generations, the use of the *csLV34* marker allowed for the identification of 46% heterozygous (H) forms of the *Lr34* gene and 6% homozygous (+) forms (Table 1).

Table 1. Identification of molecular markers associated with the resistance genes studied in wheat hybrid forms of wheat from the F_2 and BC_1F_1 generations. The grey color indicates the hybrid forms of wheat that have been selected and included in further molecular analyses.

No.	Plant Name	Breeding Company	Hybrid form BC_1F_1/F_2	*Lr34* (*csLV34*)	*Lr46* (*csLV46G22*)	*Lr67* (*cfd23*)	*Lr67* (*cfd71*)
1.	H × G 1 I	Małopolska HR	(Harenda × Glenlea) × Harenda	H	+	+	-
2.	H × G 1 II	Małopolska HR	(Harenda × Glenlea) × Harenda	-	+	-	-
3.	H × G 1 III	Małopolska HR	(Harenda × Glenlea) × Harenda	H	+	-	+
4.	H × G 1 IV	Małopolska HR	(Harenda × Glenlea) × Harenda	H	+	-	-
5.	H × G 2 I	Małopolska HR	(Harenda × Glenlea) × Harenda	-	+	-	-
6.	H × G 2 II	Małopolska HR	(Harenda × Glenlea) × Harenda	-	+	-	+
7.	H × G 2 III	Małopolska HR	(Harenda × Glenlea) × Harenda	H	+	-	+
8.	H × G 2 IV	Małopolska HR	(Harenda × Glenlea) × Harenda	-	+	-	-
9.	H × G 3 I	Małopolska HR	(Harenda × Glenlea) × Harenda	-	+	-	-
10.	H × G 3 II	Małopolska HR	(Harenda × Glenlea) × Harenda	H	+	-	-
11.	H × G 3 III	Małopolska HR	(Harenda × Glenlea) × Harenda	-	+	+	-
12.	H × G 3 IV	Małopolska HR	(Harenda × Glenlea) × Harenda	H	+	-	+
13.	J × G 1 I	Małopolska HR	(Jutrzenka × Glenlea) × Jutrzenka	H	+	+	-
14.	J × G 1 II	Małopolska HR	(Jutrzenka × Glenlea) × Jutrzenka	H	+	+	-
15.	J × G 1 III	Małopolska HR	(Jutrzenka × Glenlea) × Jutrzenka	H	+	+	-
16.	J × G 2 I	Małopolska HR	(Jutrzenka × Glenlea) × Jutrzenka	-	+	-	-
17.	J × G 2 II	Małopolska HR	(Jutrzenka × Glenlea) × Jutrzenka	-	+	-	-
18.	J × G 2 III	Małopolska HR	(Jutrzenka × Glenlea) × Jutrzenka	-	+	-	-
19.	J × G 3 I	Małopolska HR	(Jutrzenka × Glenlea) × Jutrzenka	-	+	-	-
20.	J × G 3 II	Małopolska HR	(Jutrzenka × Glenlea) × Jutrzenka	H	+	+	+
21.	J × G 3 III	Małopolska HR	(Jutrzenka × Glenlea) × Jutrzenka	-	+	-	-
22.	J × G 3 IV	Małopolska HR	(Jutrzenka × Glenlea) × Jutrzenka	H	+	+	-
23.	17 I	Strzelce HR	(Aura × Glenlea) × Aura	-	+	+	+
24.	17 II	Strzelce HR	(Aura × Glenlea) × Aura	H	+	+	+
25.	17 III	Strzelce HR	(Aura × Glenlea) × Aura	H	+	+	+
26.	17 IV	Strzelce HR	(Aura × Glenlea) × Aura	-	+	+	+
27.	17 V	Strzelce HR	(Aura × Glenlea) × Aura	H	+	+	+
28.	18 I	Strzelce HR	(Aura × Glenlea) × Aura	-	+	+	+
29.	18 II	Strzelce HR	(Aura × Glenlea) × Aura	H	+	+	+
30.	18 III	Strzelce HR	(Aura × Glenlea) × Aura	H	+	+	+
31.	18 IV	Strzelce HR	(Aura × Glenlea) × Aura	H	+	+	+
32.	21 I	Strzelce HR	(Aura × Glenlea) × Aura	H	+	+	+
33.	21 II	Strzelce HR	(Aura × Glenlea) × Aura	H	+	+	+
34.	21 III	Strzelce HR	(Aura × Glenlea) × Aura	H	+	+	+
35.	21 IV	Strzelce HR	(Aura × Glenlea) × Aura	-	+	+	+
36.	21 V	Strzelce HR	(Aura × Glenlea) × Aura	-	+	-	+
37.	22 I	Strzelce HR	(Aura × Glenlea) × Aura	-	+	+	+
38.	22 II	Strzelce HR	(Aura × Glenlea) × Aura	-	+	+	+
39.	22 III	Strzelce HR	(Aura × Glenlea) × Aura	H	+	-	+
40.	22 IV	Strzelce HR	(Aura × Glenlea) × Aura	-	+	-	+
41.	22 V	Strzelce HR	(Aura × Glenlea) × Aura	H	+	-	+
42.	24 I	Strzelce HR	(Aura × Glenlea) × Aura	-	+	-	+
43.	24 II	Strzelce HR	(Aura × Glenlea) × Aura	-	+	-	+
44.	24 III	Strzelce HR	(Aura × Glenlea) × Aura	-	+	-	+
45.	24 IV	Strzelce HR	(Aura × Glenlea) × Aura	H	+	-	+
46.	24 V	Strzelce HR	(Aura × Glenlea) × Aura	-	+	-	+
47.	25 I	Strzelce HR	(Aura × Glenlea) × Aura	-	+	-	+
48.	25 II	Strzelce HR	(Aura × Glenlea) × Aura	-	+	-	+
49.	25 III	Strzelce HR	(Aura × Glenlea) × Aura	H	+	-	+
50.	25 IV	Strzelce HR	(Aura × Glenlea) × Aura	-	+	-	+

Table 1. *Cont.*

No.	Plant Name	Breeding Company	Hybrid form BC_1F_1/F_2	*Lr34* (*csLV34*)	*Lr46* (*csLV46G22*)	*Lr67* (*cfd23*)	*Lr67* (*cfd71*)
51.	25 V	Strzelce HR	(Aura × Glenlea) × Aura	H	+	+	+
52.	D1 I	Danko HR	Itaka × Glenlea	+	+	+	-
53.	D1 II	Danko HR	Itaka × Glenlea	+	+	+	-
54.	D1 III	Danko HR	Itaka × Glenlea	+	+	-	+
55.	D1 IV	Danko HR	Itaka × Glenlea	H	+	+	-
56.	D1 V	Danko HR	Itaka × Glenlea	-	+	+	+
57.	D2 I	Danko HR	Itaka × Glenlea	H	+	+	+
58.	D2 II	Danko HR	Itaka × Glenlea	+	+	-	-
59.	D2 III	Danko HR	Itaka × Glenlea	-	+	+	-
60.	D2 IV	Danko HR	Itaka × Glenlea	H	+	+	-
61.	D2 V	Danko HR	Itaka × Glenlea	+	+	+	-
62.	D3 I	Danko HR	Itaka × Glenlea	+	+	+	+
63.	D3 II	Danko HR	Itaka × Glenlea	-	+	+	-
64.	D3 III	Danko HR	Itaka × Glenlea	-	+	+	-
65.	D3 IV	Danko HR	Itaka × Glenlea	-	+	+	-
66.	D3 V	Danko HR	Itaka × Glenlea	H	+	+	+
67.	M × G I	Smolice HR	Merkawa × Glenlea	H	+	-	+
68.	M × G II	Smolice HR	Merkawa × Glenlea	H	+	+	+
69.	M × G III	Smolice HR	Merkawa × Glenlea	-	+	+	+
70.	M × G IV	Smolice HR	Merkawa × Glenlea	H	+	+	+
71.	M × G V	Smolice HR	Merkawa × Glenlea	H	+	+	+

+, presence; H, heterozygous.

The *csLV46G22* marker linked to the *Lr46/Yr29* locus (E. Lagudah, unpublished data) was identified in all hybrid wheat forms of the obtained generations tested (Table 1). For *cfd23*, an amplification product of 211 bp was obtained, while for *cfd71*, a product of 214 bp was obtained. When the *cfd23* marker was used, the presence of the *Lr67* resistance determinant allele was confirmed in 55% of the hybrid forms of the F_2 and BC_1F_1 (+) generations, while the *cfd71* marker allowed for identification of the gene in 62%. Amplicons of both flanking markers (*cfd23* and *cfd71*) [34] were identified in 35% of the hybrid forms of the F_2 and BC_1F_1 (+) generations. Obtaining the above results allowed for a preliminary characterization of the genotypes and the selection of the obtained hybrid forms for further experiments, which included analysis of the expression of the aforementioned resistance genes and the candidate gene.

2.2. Observation and Evaluation of Infection Development by *Pt*

Observations of the plants after inoculation at the adult plant stage showed that the hybrids tested had low levels of infection by *Pt*. Based on the results of molecular marker identification, three plants were selected from each hybrid combination in which the presence of molecular markers was confirmed. The selected plants constituted biological replicates in the experiment. The evaluation of infection was performed at the adult plant stage (BBCH 39) using a rating scale according to COBORU (Research Centre For Cultivar Testing, Słupia Wielka, Poland (www.coboru.gov.pl/index_en; accessed on 9 September 2024)). The results of the infestation assessment are presented in Table 2.

Table 2. Evaluation of the level of infection development of tested hybrid forms of wheat.

Plant Name	Hybrid Form from BC_1F_1/F_2 Generations	Level of Infection
H × G 1 I	(Harenda × Glenlea) × Harenda	8
H × G 2 III	(Harenda × Glenlea) × Harenda	6
H × G 3 IV	(Harenda × Glenlea) × Harenda	6

Table 2. *Cont.*

Plant Name	Hybrid Form from BC_1F_1/F_2 Generations	Level of Infection
J × G 1 I	(Jutrzenka × Glenlea) × Jutrzenka	7
J × G 1 III	(Jutrzenka × Glenlea) × Jutrzenka	8
J × G 3 II	(Jutrzenka × Glenlea) × Jutrzenka	8
17 II	(Aura × Glenlea) × Aura	6
18 II	(Aura × Glenlea) × Aura	5
21 II	(Aura × Glenlea) × Aura	5
D1 III	Itaka × Glenlea	7
D3 I	Itaka × Glenlea	6
D2 V	Itaka × Glenlea	5
M × G II	Merkawa × Glenlea	6
M × G IV	Merkawa × Glenlea	5
M × G V	Merkawa × Glenlea	7

COBORU scale: 1—complete infection; 9—no infection.

2.3. *Analysis of APR Gene Expression in BC_1F_1 and F_2 Generations*

The expression results of the four candidate genes for *Lr46*, presented in our previous studies [33], allowed us to select *Lr46-Glu2* as the gene with the highest expression after inoculation with *Pt*. Therefore, the gene expression analysis in hybrid wheat forms of generations BC_1F_1 and F_2 included the three genes studied: *Lr34*, *Lr46-Glu2*, and *Lr67* (Figure 1). For the *Lr34* gene, as in previous years, it was characterized by low expression at all time points. However, an increase in expression could be observed in most of the hybrid forms tested, except (Harenda × Glenlea) × Harenda, (Jutrzenka × Glenlea) × Jutrzenka, which showed low expression at all time points. Interestingly, in (Aura × Glenlea) × Aura, an initial high expression level can be observed at 0 hpi. Thereafter, expression levels fluctuate but do not reach levels higher than before pathogen inoculation (Figure 1). Interestingly, the gene in both forms of the F_2 generation, namely, Merkawa × Glenlea and Itaka × Glenlea, responded very similarly, reaching similar expression values (Figure 1).

The candidate gene *Lr46-Glu2* showed the highest expression in two forms of the F_2 generation, namely, Merkawa × Glenlea and Itaka × Glenlea, reaching the highest values at 6 hpi (Figure 1). There was a decrease in expression at subsequent time points, but the Itaka × Glenlea hybrid form showed a renewed increase in *Lr46-Glu2* expression levels at 48 hpi (Figure 1). In contrast, the gene expression profile of the BC_1F_1 hybrid forms (Harenda × Glenlea) × Harenda and (Jutrzenka × Glenlea) × Jutrzenka was quite comparable, with the highest increase in *Lr46-Glu2* expression levels occurring at 12 hpi (Figure 1).

Analysis of *Lr67* expression showed that the response to inoculation varied widely among the hybrid forms tested (Figure 1). The most dynamic expression pattern was observed only in (Harenda × Glenlea) × Harenda, with a transient decrease in expression at 12 hpi. An interesting expression profile of this gene can also be observed in the second hybrid form from the BC_1F_1 generation, namely, (Aura × Glenlea) × Aura, where we observe a high level of expression even before inoculation (0 hpi). Then, a decrease in expression can be observed up to 12 h after inoculation and a dynamic increase in expression at 24 hpi (Figure 1). We also observed high basal expression of the F_1 form of the Aura cultivar in our previous study [33], where high expression at 0 hpi was noted in all genes tested (*Lr34*, *Lr46-Glu2*, *Lr67*). Very similar expression patterns can be observed in F_1 hybrid forms [33], with the exception of Jutrzenka × Glenlea.

Statistical analysis of the test genes, namely, *Lr34*, *Lr46-Glu2*, and *Lr67*, was also performed. The Kolmogorov–Smirnov test was used to test the null hypothesis that the set of normalized expression data for a particular gene and lineage comes from a normal distribution. This test was passed for all combinations of hybrid forms and genes except *Lr46-Glu2* in the Itaka × Glenlea hybrid form (p-value = 0.02397) (Supplementary Tables S1–S3). Levene's Brown–Forsythe test was used to assess the equality of variance.

The result of this test was positive for all tested combinations of hybrid forms and time points. To compare means, a two-sided Student's *t*-test was performed (Supplementary Tables S1–S3). The stability of reference gene expression was statistically calculated, and cT values at each time point tested after inoculation were compared to cT values before pathogen inoculation (Supplementary Table S4).

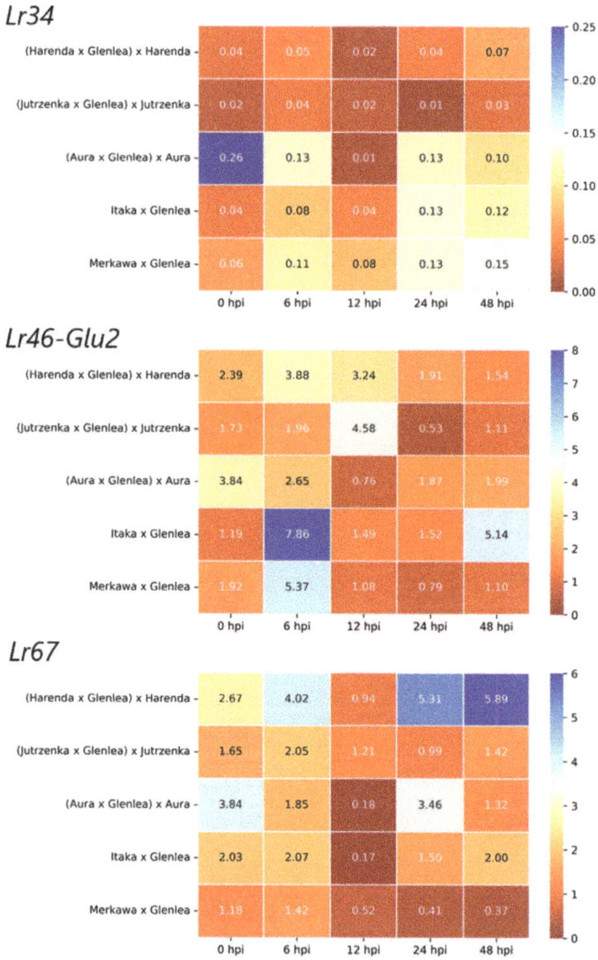

Figure 1. Heatmap plots showing the expression of *Lr34* and *Lr67* and the selected candidate gene *Lr46-Glu2* in leaf tissues after *Pt* infection. Expression analysis, using the RT-qPCR method, included wheat hybrid forms of BC_1F_1 and F_2 generations at five time points (0, 6, 12, 24, and 48 hpi). Red indicates lower normalized expression, and blue indicates higher normalized expression.

2.4. Analysis of miRNA Expression of Wheat Hybrid Forms of BC_1F_1 and F_2 Generations

The use of stem-loop RT-PCR and expression analysis via ddPCR was used to analyze the expression of miRNA molecules selected from databases as complementary to *Lr34* (tae-miR9653b; Table 3, Figure 2) and the candidate gene *Lr46-Glu2* (tae-miR5384-3p; Table 3, Figure 3). The obtained expression results of the selected genes were compared with the corresponding values for complementary miRNA molecules. Analysis of miRNA expression using ddPCR confirmed the potential role of tae-miR9653b in inhibiting the expression of the target gene—in this case, *Lr34*. Finally, in 24 and 48 hpi, when an increase in *Lr34* gene expression is observed, tae-miR9653b levels decrease. Such a tendency was observed in hybrid forms of the F_2 generation: Itaka × Glenlea and Merkawa × Glenlea (Figure 2). Interestingly, the *Lr34* gene identified in Merkawa × Glenlea was present in the homozygous form (Table 1). The highest expression of tae-miR9653b molecules was observed in hybrid forms of the BC_1F_1 generation: (Aura × Glenlea) × Aura, Merkawa × Glenlea, and (Jutrzenka × Glenlea) × Jutrzenka at 6, 24, and 48 hpi, respectively (Figure 2). An analysis of the tae-miR5384-3p gene, which is complementary to this gene, found that its expression level fluctuated after inoculation in the four hybrid forms, eventually adopting higher values than the initial values before inoculation. Therefore, this may suggest that tae-miR5384-3p did not specifically repress the *Lr46-Glu2* gene since the expression of this gene increased after inoculation. The highest expression of tae-miR5384-3p molecules was also observed in hybrid forms of BC_1F_1 generation—(Harenda × Glenlea) × Harenda, (Jutrzenka × Glenlea) × Jutrzenka, and (Aura × Glenlea) × Aura—at 24 hpi and 48 hpi (Figure 3). Table 3 shows the averaged values of the obtained expression levels (copies/µL) of the analyzed miRNA molecules using the ddPCR method.

Table 3. Average values of the obtained expression level (copies/µL) of the studied miRNAs, using the ddPCR method, for the studied genotypes at selected time points (hpi).

Hybrid Form	tae-miR9653b					tae-miR5384-3p				
	0 hpi	**6 hpi**	**12 hpi**	**24 hpi**	**48 hpi**	**0 hpi**	**6 hpi**	**12 hpi**	**24 hpi**	**48 hpi**
(Harenda × Glenlea) × Harenda	1149.51	1588.13	1340.06	1489.97	1726.36	89.77	84.14	146.39	167.93	154.10
(Jutrzenka × Glenlea) × Jutrzenka	425.41	434.56	475.90	836.50	362.28	129.60	62.57	120.24	147.38	130.92
(Aura × Glenlea) × Aura	741.59	1183.93	701.81	579.65	844.10	108.02	99.75	114.14	120.68	144.30
Itaka × Glenlea	1170.77	481.50	2169.09	1362.00	1332.23	102.19	86.76	103.48	131.42	160.96
Merkawa × Glenlea	774.80	738.21	2736.03	661.28	709.34	116.98	123.64	131.40	114.86	166.71

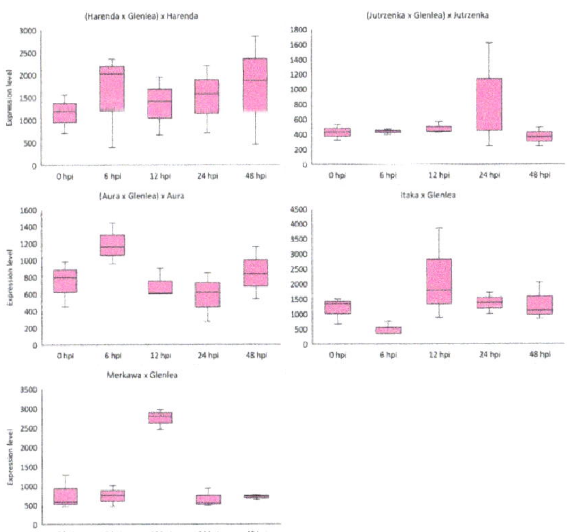

Figure 2. Box plots for tae-miR9653b expression at five timepoints after *Pt* inoculations. The ddPCR results show the numbers for the analyzed miRNAs during infection induced by *Pt*. Numbers 6, 12, 24, and 48 indicate hours after inoculation (hpi). The expression data shown were obtained from the analysis of three biological replicates, and each biological replicate had three technical replicates.

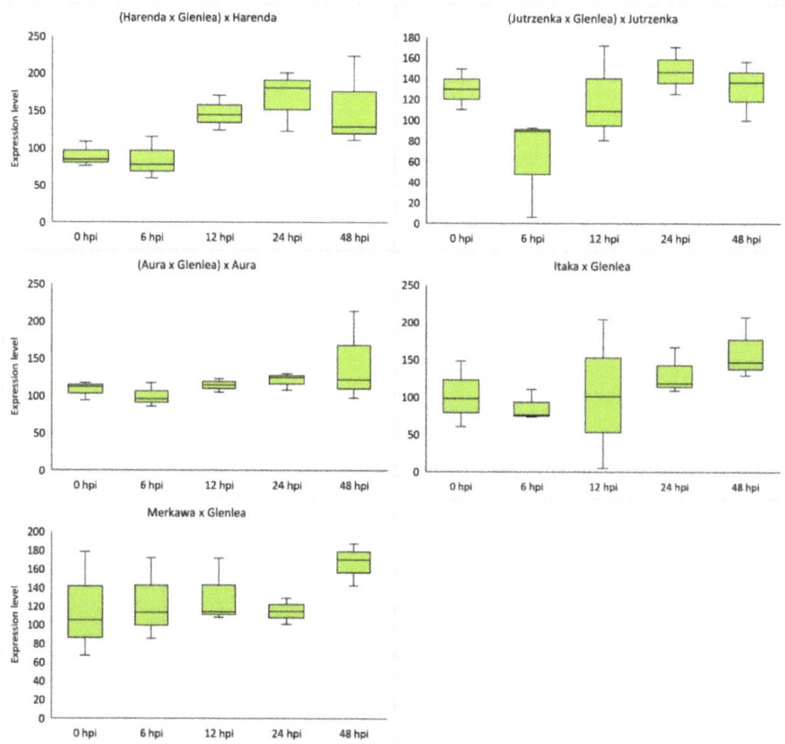

Figure 3. Box plots for tae-miR5384-3p expression at five timepoints after *Pt* inoculations. The ddPCR results show the droplet numbers for the analyzed miRNAs during infection induced by Pt. Numbers 6, 12, 24, and 48 indicate hours after inoculation (hpi). The expression data shown were obtained from the analysis of three biological replicates, and each biological replicate had three technical replicates.

2.5. GO Results for Tested miRNAs

A gene ontology (GO)-based enrichment analysis was conducted to further investigate the potential role of the miRNAs studied in the wheat response to leaf rust infection (Figure 4). For miRNA tae-miR9653b, the most genes (14) were located on chromosomes 7A and 7D (Figure 4). On chromosome 7A, they were located in the distal part of the short and long arms; on chromosome 7D, they were mostly located in the subtelomeric region of the long arm (Figure 5). For the analyzed miRNA tae-miR5384-3p, 16 genes were identified on chromosome 3B (Figure 4), mainly located in the distal part of the long arm (Figure 6). The sequence of studied mature miRNAs was analyzed using psRNATarget software.(www.zhaolab.org/psRNATarget/; accessed on 9 September 2024). The 190 target genes for tae-miR5384-3p and 167 target genes for tae-miR9653b have been predicted in the outcomes (Supplementary Materials File S5, Figures 5 and 6). GO analyses describe the functions, compartments, and biological processes in which genes are involved in a given organism, dividing them into three main domains: molecular function, biological process, and cellular component. The functional enrichment analysis of genes demonstrated the overall characteristics of candidate genes regarding GO functional entries (Figures 7 and 8). Primary GO (level 2) analysis of the tae-miR9653b miRNA for biological processes revealed 21 functions, with the largest number of genes located within the cellular process (GO:0009987), metabolic process (GO:0008152), and biosynthetic process

(GO:0009058) domains. For the cellular component domain, 17 functions were identified. In this domain, the greatest number of target genes were involved in the intracellular anatomical structure (GO:0005622), the membrane (GO:0016020), the nucleus (GO:0005634), and the cytoplasm (GO:0005737). For molecular process, 14 functions were observed, with most genes involved in binding (GO:0005488), catalytic activity (GO:0003824), nucleotide binding (GO:0000166), and transferase activity (GO:0016740) (Figure 7). In the GO analysis for tae-miR5384-3p in biological processes, the greatest number of genes were involved in the metabolic process (GO:0008152), the cellular process (GO:0009987), and the protein metabolic process (GO:0019538). For the cellular component aspect, the greatest number of genes were located in the membrane, the intracellular anatomical structure (GO:000562), the cytoplasm (GO:0005737), and the plasma membrane (GO:0005886). In biological processes, target genes mainly included catalytic activity (GO:0003824) and binding (GO:0005488) (Figure 8).

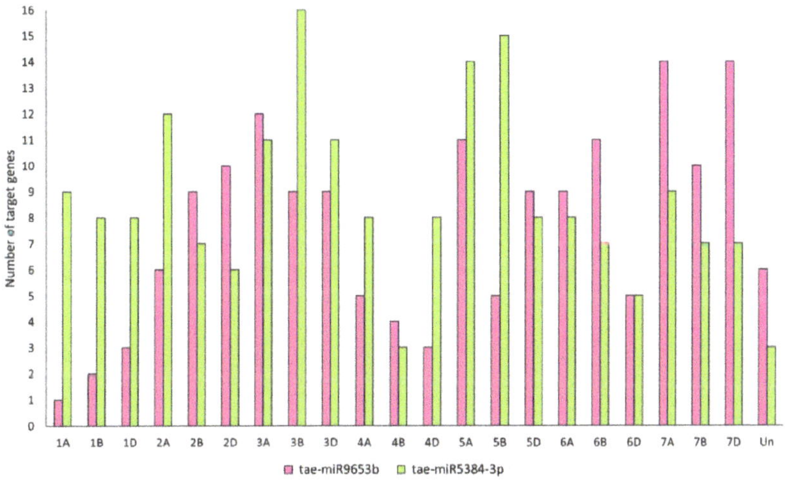

Figure 4. Number of predicted genes for specific wheat chromosomes for the miRNAs analyzed. "1A" and similar abbreviations indicate the name of the chromosomes of common wheat.

More in-depth GO analysis of the predicted target genes was performed for GO enrichment. For this analysis, the *p*-value was used as a selection criteria (*p*-value < 0.05) (Supplementary Materials File S6). The *p*-value test permitted us to find for the tae-miR9653b molecule 56 GO terms related to biological processes, 19 GO terms related to the cellular component, and 59 Go terms related to the molecular function. Figure 9 displays the top 15 significantly enriched GO elements for the target genes of tae-miR9653b. The in-depth GO analysis, which was performed on a *p*-value basis, also showed functions related to the expression of the *Lr34* gene we studied, and included in the top 15 functions are purine ribonucleotide binding (GO:0032555), ribonucleotide binding (GO:0032553), adenyl ribonucleotide binding (GO:0032559), and carbohydrate derivative binding (GO:0097367). The GO analysis for the tae-miR5384-3p molecule uncovered 122 GO terms relating to biological processes, 20 relating to cellular components, and 78 relating to molecular functions (Supplementary Materials File S6). The top 15 significantly enriched GO terms for the tae-miR5384-3p target genes are shown in Figure 10. Furthermore, our in-depth analysis for the miRNA—*Lr46-Glu2* (TraesCS1B02G454200.1) in these GO terms includes the presented *Lr46-Glu2* candidate gene as a function of hydrolase activity (GO:0016787) and catalytic activity (GO:0003824) (Figure 10).

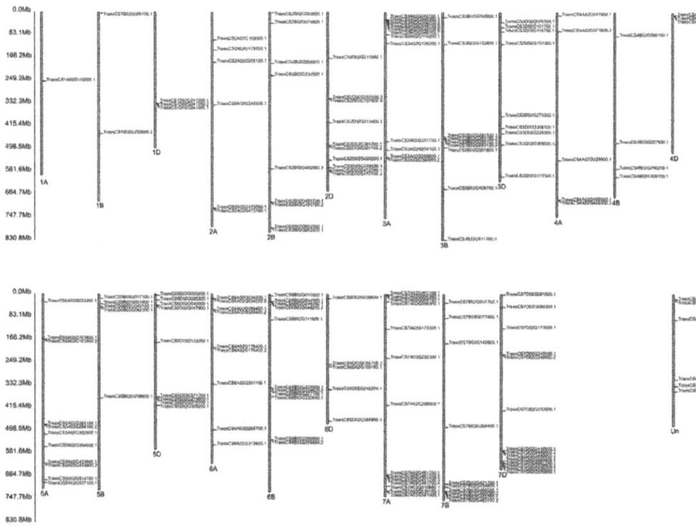

Figure 5. Localization of 167 target genes for tae-miR9653b on common wheat chromosomes. "1A" and similar indicate the name of the chromosome, and the abbreviation "Mb" next to the scale indicates the megabase unit referring to the physical length of the chromosomes.

Figure 6. Localization of 190 target genes for tae-miR5384-3p on common wheat chromosomes. "1A" and similar indicate the name of the chromosome, and the abbreviation "Mb" next to the scale indicates the megabase unit referring to the physical length of the chromosomes.

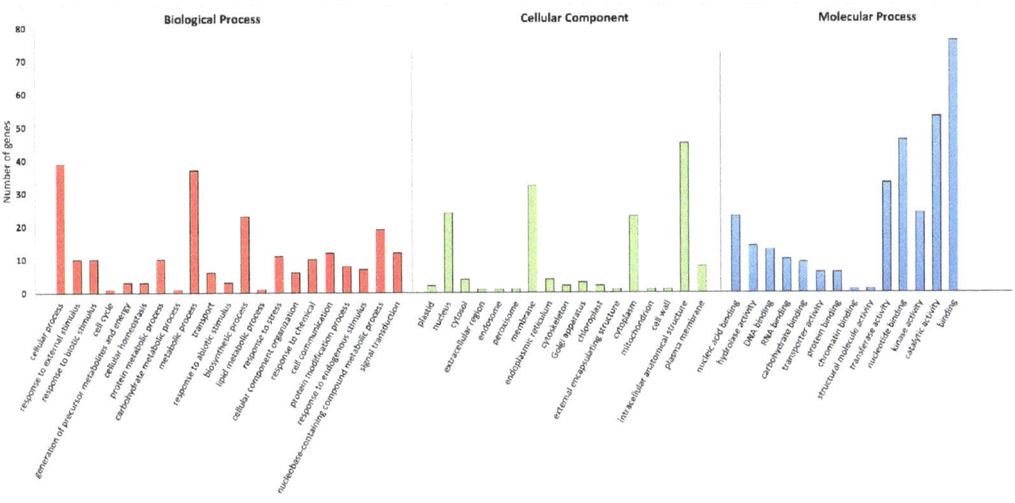

Figure 7. GO annotation of target genes for tae-miR9653b for three groups: biological process; cellular component; molecular function.

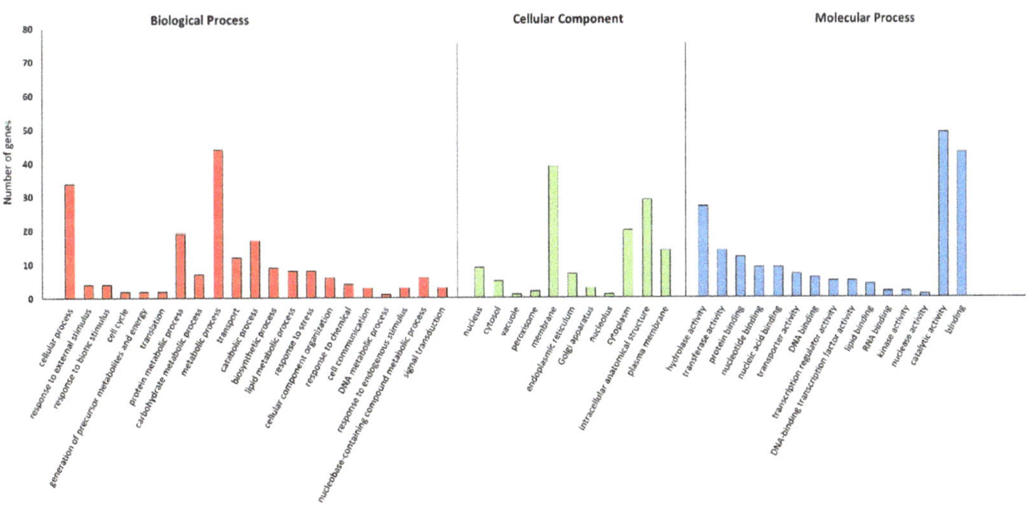

Figure 8. GO annotation of target genes for tae-miR5384-3p for three groups: biological process; cellular component; molecular function.

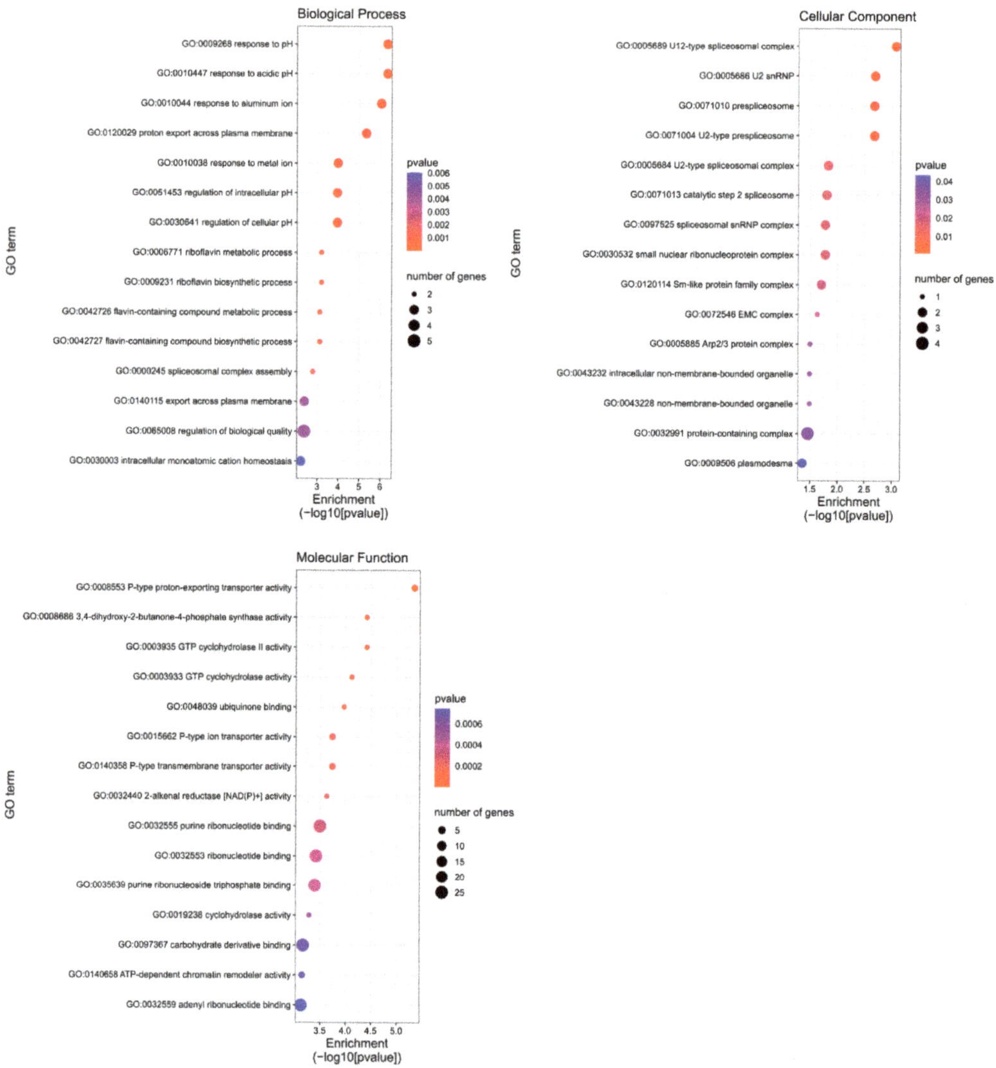

Figure 9. In–depth gene ontology (GO) pathway enrichment analysis for miRNA tae-miR9653b. The top 15 GO enrichment functions that were significantly enriched in biological processes, cellular components, and molecular functions were sorted by *p*-value.

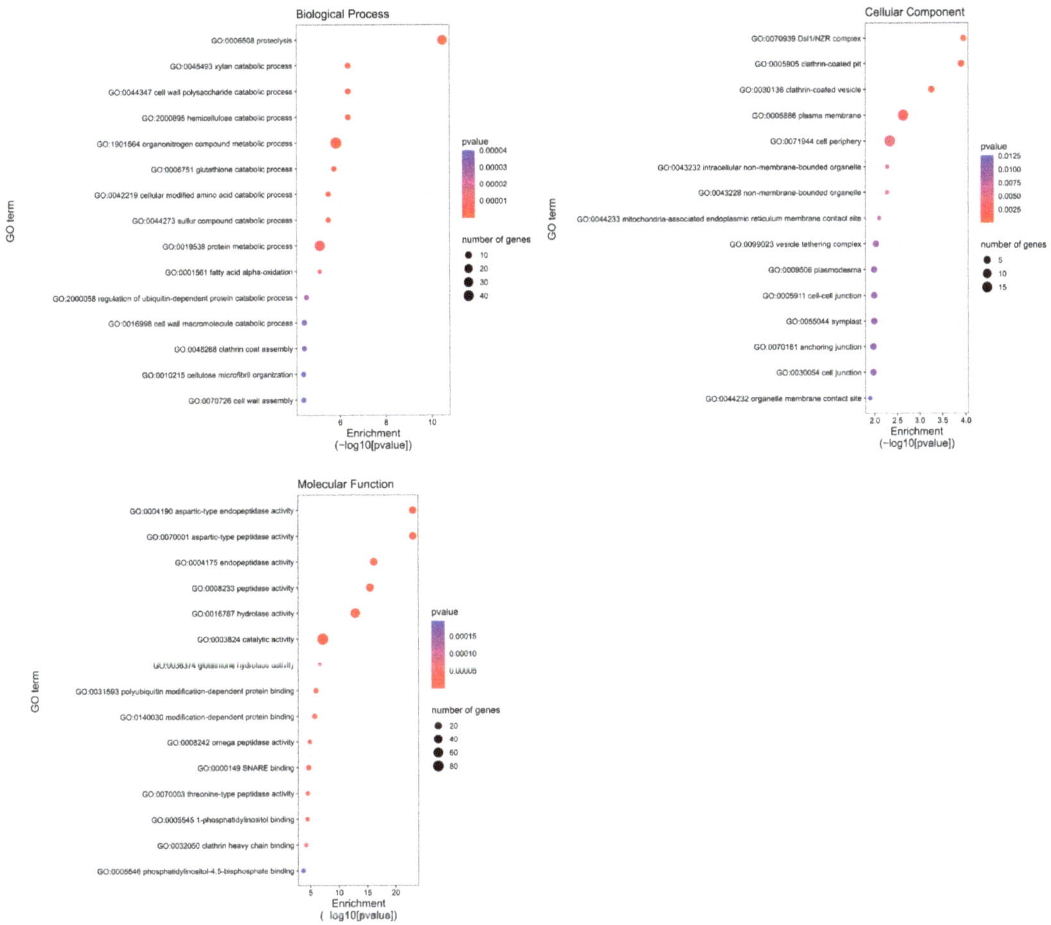

Figure 10. In–depth gene ontology (GO) pathway enrichment analysis for miRNA tae-miR5384-3p. The top 15 GO enrichment functions that were significantly enriched in biological processes, cellular components, and molecular functions were sorted by *p*-value.

3. Discussion

In this paper, we present the results of *Lr34* and *Lr67* expression analyses in hybrid forms of wheat obtained by crossing a selected Glenlea variety with high-yielding Polish wheat cultivars. In addition, we analyzed the expression of miRNAs complementary to the studied APR genes and extended the bioinformatics analysis by performing gene ontology enrichment. The obtained hybrid forms included the F_1 [33], F_2, and BC_1F_1 generations. As in the previous study [33], in the F_1 generation, the resistance genes *Lr34* and *Lr67* showed differential expression at selected time points. It should be noted that the *Lr34* gene is expressed at a relatively low level regardless of inoculation, while there is a rapid decrease in expression after inoculation (6 and 12 hpi). However, compared to the results described in our previous works, a similarity in the expression profile of *Lr34* in the donor form of Glenlea can be observed in the F_1 hybrid forms tested, where the highest expression was observed at 24 and 48 hpi [33]. In the BC_1F_1 generation, an increase in *Lr34* gene expression can be observed in most of the hybrid forms tested, except (Harenda × Glenlea)

× Harenda, (Jutrzenka × Glenlea) × Jutrzenka, which showed low expression at all time points. Notably, in the hybrid form (Aura × Glenlea) × Aura, we observed an initial high level of expression at 0 hpi. Thereafter, the expression level fluctuates but does not reach a level higher than before pathogen inoculation (Figure 1). Interestingly, the gene in both forms of the F_2 generation, namely, Merkawa × Glenlea and Itaka × Glenlea, responded very similarly, reaching similar expression values (Figure 1). In contrast, *Lr67* expression showed that the response to inoculation was again very different between the F_2 and BC_1F_1 hybrid forms tested (Figure 1). The most dynamic expression pattern was only observed in (Harenda × Glenlea) × Harenda, with a transient decrease in expression at 12 hpi. The interesting expression profile of this gene can also be observed in the second hybrid form of the BC_1F_1 generation, namely, (Aura × Glenlea) × Aura, where a high level of expression was observed even before inoculation (0 hpi), followed by a decrease in expression up to 12 hpi and a dynamic increase in expression at 24 hpi (Figure 1). Very similar expression patterns can be observed in F_1 hybrid forms [33], with the exception of Jutrzenka × Glenlea.

According to the literature, the *Lr34* and *Lr67* genes present several similarities in that these genes determine resistance to multiple diseases in a non-race-specific manner and are responsible for encoding cellular transporters. The *Lr34* gene plays a key role in wheat breeding for disease resistance largely because it interacts with other resistance genes, resulting in effective and durable resistance [35]. Considering that the *Lr34* gene from hexaploid wheat functions as a transgene in other cereals, such as rice and barley, these diploid species with small genomes and fully annotated gene sets are highly suitable for an extensive study of *Lr34* function [24]. According to another recent study, the *Lr34* gene and *TaCOMT-3B* synergistically increase resistance to stripe rust [36]. Researchers in this study showed that wheat lines carrying the *Lr34res* allele presented thicker cell walls and increased resistance to fungal penetration compared to lines without *Lr34res*. The aim of the study presented by McCallum and Hiebert (2022) was to determine whether the *Lr67* gene also interacts with other resistance genes in a similar manner to *Lr34* [35]. To this end, six different doubled haploid lines were developed, which were characterized by the presence of the *Lr67* or *Lr34* gene together with a second resistance gene: *Lr13*, *Lr16*, or *Lr32*. The presence of the *Lr67* and *Lr34* genes significantly reduced the level of intensity of leaf rust growth, and the *Lr34* gene (unlike gene *Lr67*) showed a significant interaction with *Lr13*. Both genes interacted with *Lr16* and *Lr67* and had a significant interaction with the *Lr32* gene. This analysis shows a similar effect of the *Lr67* gene to that of *Lr34* in interacting with other genes to provide better resistance than single genes. Scientists suggest that though *Lr67* is not widely implemented in crops, it may play an important role in disease resistance [35]. Despite the introduction of the *Lr67* gene into many wheat varieties created by CIMMYT since the 1950s [37], it has not been found in Canadian wheat varieties to date. The researchers highlight that exploiting the potential of the *Lr67* gene in leaf rust countries could improve resistance in future wheat varieties [35].

On the basis of our previous results [33,38], we decided to select the candidate gene *Lr46-Glu2* for further studies involving the F_2 and BC_1F_1 generations. The candidate gene *Lr46-Glu2* showed the highest expression in two forms of the F_2 generation, namely, Merkawa × Glenlea and Itaka × Glenlea, reaching the highest values at 6 hpi (Figure 1). There was a decrease in expression at subsequent time points, but the Itaka × Glenlea hybrid form showed a renewed increase in *Lr46-Glu2* expression levels at 48 hpi (Figure 1). A similar expression pattern of this candidate gene was observed in previous experiments in the Glenlea cultivar [38] and in hybrid forms of the F_1 generation [33]. In contrast, in the BC_1F_1 hybrid forms (Harenda × Glenlea) × Harenda and (Jutrzenka × Glenlea)

× Jutrzenka, the gene expression profile was very similar, and the greatest increase in *Lr46-Glu2* expression levels occurred at 12 hpi (Figure 1).

The miRNAs regulate gene expression by binding to mRNA. The seed sequence is essential for miRNA binding to the mRNA transcript. The seed sequence or seed region is a conserved heptametric sequence that is mainly located at positions 2–7 from the 5′-end of the miRNA [39]. Hence, even if the base pairing of the miRNA and its target mRNA is not perfect, the "seed sequence" must be perfectly complementary. In summary, a miRNA can regulate gene transcription or translation in different ways through a seed sequence, a non-seed, or the entire sequence, thus providing the possibility of dual or multiple synergism in the regulation of one target gene by a single miRNA. As is known, endogenous miRNAs usually have a mild effect on their targets, but dual regulation of miRNAs can form complex networks and ultimately produce additive effects at the cellular and physiological levels. It is worth emphasizing that accurate miRNA modeling is essential, and the technologies for identifying miRNA targets have been greatly improved in recent years [39]. Currently, one of the goals of research in this field should be to determine the function of miRNA molecules in common wheat and its close relatives or ancestors, which may increase the chances of understanding species-specific miRNA molecules. Furthermore, the study of resistance mechanisms using molecular tools should find application in the breeding practice of key crop species [39].

In our series of studies of the tae-miR9653b molecule, we showed an expression profile suggesting its involvement in the *Pt*-induced biotic stress response [40,41]. In a study presented by Tomkowiak et al. (2023), the inoculation of wheat cultivars with *Pt* spores resulted in altered expression of tae-miR9653b in cultivars showing the presence of the resistance-determining *Lr34* allele [40]. The expression level of tae-miR9653b varied by cultivar and time point. At 6 hpi, there was a slight increase in the expression level of tae-miR9653b observed in all wheat reference cultivars, except for the control cultivar HN ROD. Most cultivars showed increased expression of this miRNA at 12, 24, and 48 h after inoculation. The fastest response of tae-miR9653 to inoculation was observed in the cultivar Myna'S' [40]. In the present study, our analysis of primary GO (level 2) for the tae-miR9653b for biological processes revealed 21 functions, with the largest number of genes located within the cellular process (GO:0009987), metabolic process (GO:0008152), and biosynthetic process (GO:0009058) domains. For the cellular component domain, 17 functions were identified. The in-depth GO analysis, which was performed on a *p*-value basis, also showed that the functions related to the expression of the *Lr34* gene we studied included in the top 15 functions purine ribonucleotide binding (GO:0032555), ribonucleotide binding (GO:0032553), adenyl ribonucleotide binding (GO:0032559), and carbohydrate derivative binding (GO:0097367).

Previous studies have shown that tae-miR9653b and tae-miR5384-3p may be involved in immune responses to both biotic and abiotic stresses. A study presented by Li et al. (2017) examined the expression of a number of different miRNAs, including tae-miR9653b, after wheat exposure to phenanthrene [42], and it was shown that the expression of the tae-miR9653b studied was decreased after the treatment of plants with phenanthrene [42]. In contrast, in a study on drought stress in common wheat [43], the authors considered the role of miRNA molecules in drought response. One of the three miRNAs repressed by drought in the drought-tolerant variety was tae-miR9653 (including miR9653a and miR9653b) [43]. In contrast, in the susceptible genotype, tae-miR9653b expression was increased, while tae-miR9653a expression levels remained unchanged. Unfortunately, knowledge of the role of tae-miR9653b in plant responses to biotic stresses is still limited, although tae-miR9653b has been shown to respond positively to *Fusarium graminearum* infection in common wheat lines carrying the FHB resistance locus in chromosome 2DL [44].

The tae-miR5384-3p molecule, the expression of which was studied in this paper, was found to fluctuate in its expression level after inoculation in the four hybrid forms, eventually adopting values lower or equal to the initial values before inoculation. Therefore, as gene expression increased after inoculation, this may suggest that tae-miR5384-3p did not specifically repress the *Lr46-Glu2* gene at the time points tested. Analyzing the tae-miR5384-3p molecule complementary to this gene, its expression level was found to increase after inoculation in four hybrid forms of the BC_1F_1 generation—(Harenda × Glenlea) × Harenda, (Jutrzenka × Glenlea) × Jutrzenka, and (Aura × Glenlea) × Aura—and an increase in tae-miR5338-3p was observed at 24 and 48 hpi. The presented GO analysis for the tae-miR5384-3p molecule uncovered 122 GO terms relating to biological processes, 20 relating to cellular components, and 78 relating to molecular functions. Furthermore, our in-depth analysis for the miRNA—*Lr46-Glu2* (TraesCS1B02G454200.1) in these GO terms includes the presented *Lr46-Glu2* candidate gene as a function of hydrolase activity (GO:0016787) and catalytic activity (GO:0003824). Literature reports have determined that the tae-miR5384-3p molecule can inhibit a number of genes in the *TaFAB2* subfamily, which are associated with the formation of unsaturated fatty acids that play a role in plant development and in response to biotic and abiotic stresses [45]. In another study, the tae-miR5384-3p molecule showed a very high increase in expression after treating wheat seedlings with a chitosan suspension [42]. Such a phenomenon may relate to changing defense mechanisms in the early and late stages of plant responses to stresses. On the other hand, Gidhi's team (2022) conducted sequence analysis, revealing the involvement of tae-miR5384-3p (as well as tae-miR164 and tae-miR9679-5p) in targeting the *TaAFB6* gene (TraesCS5A02G281100), which encodes the *AUXIN SIGNALING F-BOX* (*AFB*) protein [46]. This suggested the possibility of post-transcriptional regulation of *TaAFB*-type gene expression. The researchers analyzed the expression of selected *AFB* genes in response to *Pt* spore inoculation by taking wheat leaf sections at six time points (0, 12, 48, 120, and 168 hpi). Some of the *AFB* genes analyzed showed highest expression at 12 and 48 hpi in both susceptible and resistant wheat lines. All known *TaAFB*-type genes are involved in the auxin-activated signaling pathway that leads to the ubiquitination process [46].

It is worth highlighting that the expression analyses presented here addressed gene expression patterns in various wheat hybrid forms obtained in F_2 and BC_1F_1 generations, while their genetic background and specific resistance mechanisms are not precisely known. Investigating the evolutionary history of these varieties could provide a context for interpreting the observed expression patterns of the resistance and candidate genes studied. Future research should address the functional validation of the candidate genes mentioned, including, but not limited to, gene knockout and overexpression analysis. In addition, the integration of transcriptomic, proteomic, and metabolomic data can provide a more comprehensive understanding of the molecular mechanisms underlying *Pt* resistance. This multi-omics approach enables the identification of key pathways and regulatory networks involved in plant immune responses [47].

4. Materials and Methods

4.1. Crossbreeding Between Wheat Cultivars and Conducting Experiments in a Controlled Environment

The plant material consisted of five hybrid combinations from the F_2 and BC_1F_1 generations obtained by crossing a Glenlea cultivar, obtained from the National Small Grains Collection of the United States Department of Agriculture—Agricultural Research Service (USDA-ARS) (Aberdeen, WA, USA), with economically important Polish spring wheat cultivars from plant breeding companies (Figure 11). The experiment was set up in a phytotron chamber with a system for maintaining and controlling constant temperature and humidity at the Wielkopolskie Centrum Zaanwansowanych Technologii (WCZT), 10 University of Poznań Street, 61-614 Poznań, Poland. Seedlings of hybrid forms were grown in the phytotron chamber at 22 °C and 17 °C (day and night, respectively), with a relative humidity of 60–70%, a 16 h photoperiod, and an illumination intensity of 250–300 μmol m^{-2} s^{-1}.

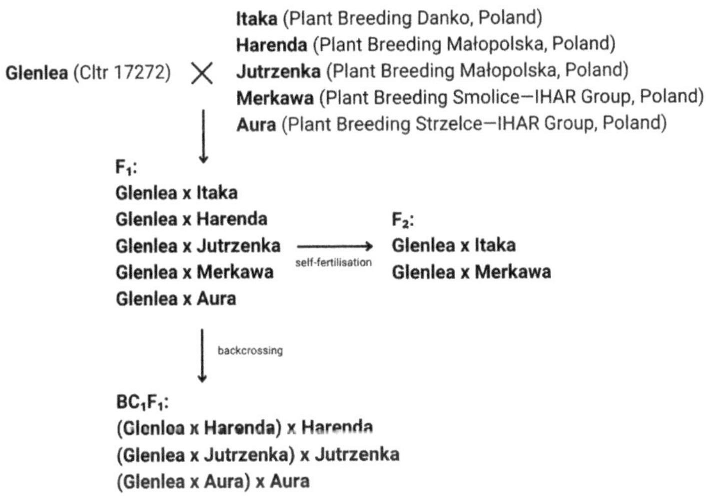

Figure 11. Diagram of crossbreeding conducted and the obtained hybrid forms of F_1, F_2, and BC_1F_1 generations.

4.2. Isolation of gDNA from Wheat Leaf Tissue

To confirm the presence of resistance-conditioning alleles in the hybrid forms tested, genomic DNA was extracted, and a PCR reaction was performed. DNA was isolated from leaves of 10-day-old seedlings (BBCH 13) using the GeneMATRIX Plant and Fungi DNA Purification Kit (EURx Ltd., Gdańsk, Poland) according to the procedure provided by the manufacturer. DNA concentration and quality were determined using a NanoDrop spectrophotometer at 260 nm. Samples were diluted with Elution buffer (EURx Ltd., Gdańsk, Poland) to a uniform concentration of 50 ng/μL.

4.3. Identification of Molecular Markers Associated with APR Genes

Polymerase chain reaction (PCR) was used to confirm the presence of alleles associated with the *Lr34*, *Lr46*, and *Lr67* genes. The marker *csLV34*, anchored to the sequence of the *Lr34* gene, was used to verify the presence of the respective alleles [48]. The marker *csLV46G22* linked to the *Lr46/Yr29* locus (E. Lagudah, unpublished data) was used to identify *Lr46*. Molecular flanking SSR markers, *cfd23* and *cfd71* [34], were used to confirm the presence of the *Lr67* gene. The PCR reaction was performed in a volume of 20 μL of reaction mixture, consisting of 1 μL of two primers (10 μM) (Merck-Sigma Aldrich, Burlington, MA, USA) and 12.5 μL of FastGene Optima HotStart ReadyMix (NIPPON Genetics Europe GmbH, Düren, Germany), which contained FastGene Optima DNA polymerase mix (0.2 U per 1 μL reaction), FastGene Optima buffer, dNTPs (0.4 mM), $MgCl_2$ (4 mM), and stabilizers. The temperature profile of the reaction was as follows: initial denaturation at 94 °C for 5 min, followed by 35 cycles (denaturation at 94 °C for 45 s; primer attachment at 60 °C for

30 s; strand extension at 72 °C for 1 min), followed by the final strand extension for 7 min at 72 °C and storage at 4 °C. Amplification products using the *csLV46G22* marker were digested with the restriction enzyme *BspEI* (Thermo Fisher Scientific, Waltham, MA, USA) at 37 °C for 1 h (Lagudah, private correspondence, 2020). A Labcycler thermocycler (SensoQuest GmbH, Göttingen, Germany) was used to perform the digestion reaction. The resulting PCR products were analyzed using a 2% agarose gel (Bioshop, Canada Inc., Burlington, ON, Canada) in 1x TBE buffer (Bioshop, Canada Inc., Burlington, ON, Canada) with the addition of 7 µL Midori Green Advanced DNA Stain (Nippon Genetics Europe, Düren, Germany). The results of the electrophoresis were photographed under UV light using a GelDoc XR UV Molecular Imager system with Image Lab software v5.1 (Bio-rad Laboratories, Inc., Hercules, CA, USA).

4.4. Inoculation of Adult Plants with Pt Spores

After reaching the adult plant stage (flag leaf stage, BBCH 39), biotic stress in hybrid forms was induced via inoculation with spores of the fungus *P. triticina*. The inoculation material was a mixture of *Pt* isolates collected from infected field experiments owned by Polish breeding companies. Before inoculation, the viability of *Pt* spores was assessed under a microscope. Hybrid wheat plants were inoculated by spraying whole plants with urediniospores at a concentration of approximately 5×10^5 spores/mL, suspended in sterile, nuclease-free water with Tween 20.

4.5. Observation and Evaluation of Pt Infection

The evaluation of infection included only selected plants in which the presence of molecular markers was confirmed. The selected plants constituted biological repeats in the experiment. Infection progression was assessed 10 days after *Pt* inoculation. To assess the stage of infection of the plants, a scale according to COBORU (Research Centre For Cultivar Testing, Słupia Wielka, Poland; www.coboru.gov.pl/index_en; accessed on 9 September 2024) was used: 1—complete infection; 9—no infection.

4.6. Total RNA Isolation and Reverse Transcription Reaction (cDNA Synthesis)

Isolation of total RNA from different leaf tissue samples collected at three biological replicates and time points (0, 6, 12, 24, and 48 hpi) was performed using the Maxwell RSC Plant RNA Isolation Kit (Promega, Madison, WI, USA) harnessing automated station Maxwell® RSC 48 Instrument (Promega). During RNA isolation, samples were treated with DNAase according to the manufacturer's recommendations. The concentration and purity of isolated total RNA was measured using a NanoDrop spectrophotometer in the A260/A280 absorbance range. Synthesis of cDNA was performed with 1 µg RNA using the iScript Reverse Transcription Supermix for RT-qPCR kit (Bio-Rad, Hercules, CA, USA) according to the manufacturer's protocol. The temperature profile of the cDNA synthesis reaction was as follows: pre-incubation for 5 min at 25 °C, reverse transcription for 60 min at 46 °C, inactivation of reverse transcriptase for 1 min at 95 °C, cooling to 4 °C, and storage at −20 °C.

4.7. Quantitative Gene Expression Analysis Using RT-qPCR

A two-step PCR protocol was exploited using iTaq Universal SYBR Green Supermix (Bio-Rad, Hercules, CA, USA) and the CFX Connect Real-Time PCR Detection System (Bio-Rad, Hercules, CA, USA). Each RT-qPCR experiment performed consisted of three biological repeats and three technical repeats, the results of which were averaged. In addition, a no template control (NTC), i.e., a negative control without a cDNA template, was also performed for each 96-well plate tested. The composition of the RT-qPCR reaction mixture was as follows: iTaq supermix—5 μL; forward and reverse primers (10 μM)—0.5 μL, each 3 μL of nuclease-free water; and cDNA template—1 μL. Following MIQE recommendations for the validity of RT-qPCR analyses, three biological repeats were used for each gene type at each time point for a given wheat cultivar or hybrid form, and three technical repeats were prepared for each [49]. Standard curves and reaction efficiency (%E) for tested and referenced genes were included according to earlier protocols [38,41]. Calculations of ΔΔCq included both reference genes using CFX Maestro (Bio-Rad, Hercules, CA, USA).

4.8. miRNA Isolation and Reverse Transcription Reaction (Stem Loop RT-PCR)

In the present study, we analyzed the expression of miRNAs associated with the *Lr34* gene and the candidate gene for the *Lr46* locus. For the *Lr34* gene, primers for tae-miR9653b were used. For the *Lr46-Glu2* candidate gene, the expression of tae-miR5384-3p was analyzed. Available databases for common wheat and literature data did not show miRNA molecules complementary to the *Lr67* gene. The coding sequences of the miRNA target genes studied were downloaded from the Ensembl Plants database and analyzed in the psRNATarget online database. The sequences of the miRNA molecules analyzed were found in the miRBase database and downloaded in FASTA format. Primers for miRNA reverse transcription and ddPCR reactions were designed using the OligoAnalyzer™ Tool from the Integrated DNA Technologies website, following early protocols [38,41].

Leaf tissue for miRNA isolation was collected analogously to the RNA isolation protocol. The mirVana miRNA Isolation Kit from ThermoFisher Scientific and a phenol–chloroform mixture were used for miRNA isolation. The miRNA isolation was carried out according to the protocol provided by the manufacturer. The obtained samples were stored in a cryofreezer (−80 °C) to avoid possible degradation. The miRNA fraction was then subjected to reverse transcription reaction using SuperScript IV Reverse Transcriptase (Thermo Fisher Scientific, Waltham, MA, USA), employing specific designed primers according to the protocol developed by Varkonyu-Gasic et al. (2007) [50]: incubation at 16 °C for 30 min; 60 cycles of 30 °C for 30 s, 42 °C for 30 s, and 50 °C for 1 s; incubation at 85 °C for 5 min.

4.9. ddPCR Expression Analysis

To quantify the number of miRNA molecules in plant samples, a ddPCR mixture, consisting of 10 μL ddPCR SuperMix Eva Green, primers (200 nM), template (reverse-transcribed elongated miRNA), and RNase-free water, was used. A 20 μL reaction mixture was used to generate reaction droplets in an 8-well cassette using a QX100 droplet generator (Bio-Rad Laboratories, Inc., Hercules, CA, USA). The droplets were carefully transferred to a 96-well ddPCR plate and heat-sealed using plastic heat-film coverslips (Bio-Rad Laboratories, Inc., Hercules, CA, USA). Amplification was then performed in a T100 PCR thermocycler (Bio-Rad Laboratories, Inc., Hercules, CA, USA) according to the previous protocols [38,41]. The resulting PCR products were quantitatively analyzed in a QX100 droplet reader (Bio-Rad Laboratories, Inc., Hercules, CA, USA). Data analysis was performed using QuantaSoft software version 1.7 (Bio-Rad Laboratories, Inc., Hercules, CA, USA). Positive droplets containing amplification products were distinguished from negative droplets by setting the fluorescence amplitude threshold to the lowest value of the positive droplet cluster.

4.10. Statistical Analysis

The Kolmogorov–Smirnov test was used to test the null hypothesis that a set of normalized expression values for a particular gene and cultivar indicates a distribution close to normal. This test was performed for all gene × genotype combinations. For verification with respect to homogeneity of variance, the Levene's variant Brown–Forsythe test was performed. For comparison of means, the Student's *t*-test (two-sided) was performed. For each gene tested, expression values at each time point tested were compared to expression before inoculation with the spore mixture. Normalized gene expression was presented using heat maps. To test the stability of the expression of the reference genes, comparisons were made between the cT values at each tested time point after inoculation relative to the cT values before pathogen inoculation. Statistical analyses were performed using Excel 2021 and Python 3.11.7.

4.11. Target Gene Prediction for Selected miRNAs

To analyze the miRNA target genes, an online tool psRNATarget (www.zhaolab.org/psRNATarget/; accessed on 9 September 2024) was used [51]. The software applies a predefined score scheme based on the "seed" region, examining complementary pairing and evaluating miRNA target accessibility based on the energy required to open the secondary structures. Target prediction was carried out by querying mature wheat miRNA sequences towards "*Triticum aestivum* (bread wheat), cDNA, EnsemblPlant, release 43" under default-parameters Schema V2 (release 2017). Gene ontology (GO) analyses of the predicted target genes were conducted using GO (www.geneontology.org (accessed on 9 September 2024)). An analytical tool from the PANTHER Classification System (https://pantherdb.org/ (accessed on 9 September 2024)) was used, which is continuously updated with GO annotations. GO enrichment plots and gene expression heat maps were created using Phyton 3.11.7. and TBtools v2.121 [52]. An in-depth enrichment analysis bubble chart was constructed for visualization by SRPLOT website (www.bioinformatics.com.cn/srplot; accessed on 9 September 2024).

5. Conclusions

Our findings revealed distinct expression profiles for different genes, with the highest expression levels observed at 6, 24, and 48 hpi. Candidate gene *Lr46-Glu2* displayed an upregulation, suggesting its potential involvement in the immune response against *Pt* infection. As a result of the analyses of 10 candidate genes for *Lr46* locus obtained in previous generations, we decided to select *Lr46-Glu2* for the present study. In the two BC_1F_1 forms (Harenda and Jutrzenka), the gene expression profiles were very similar, with the increase in *Lr46-Glu2* expression levels at 12 hpi. The *Lr46-Glu2* gene showed the highest expression (6 hpi) in two F_2 forms (Merkawa and Itaka). Analysis of *Lr67* expression showed that the response to inoculation varied widely. The *Lr34* gene was characterized by low expression at all time points. However, an increase in expression was observed in most of the hybrid forms tested, except Harenda and Jutrzenka. Furthermore, we observed potential intriguing interactions between APR genes expression and the associated miRNAs, which highlight the complex regulatory networks underlying *Pt* resistance. This study confirmed the potential role of tae-miR9653b in inhibiting *Lr34* target gene expression. Gene ontology analysis predicted 190 target genes for tae-miR5384-3p and 167 target genes for tae-miR9653b. The in-depth GO analysis, which was performed on a *p*-value basis, also showed functions related to the expression of the *Lr34* and *Lr46-Glu2* genes we studied. For tae-miR9653b, the most dynamic expression pattern was observed in Harenda hybrid form with the highest expression at 48 hpi and a temporary decrease at 12 hpi. As a result of the analysis of tae-miR5384-3p, it was found that its expression level fluctuated in the four hybrid forms, eventually adopting values lower than or equal to the initial values before inoculation. In contrast, tae-miR9653b levels decrease at 24 hpi to 48 hpi, when an increase in *Lr34* gene expression is observed. However, the results may suggest that tae-miR5384-3p did not specifically repress the *Lr46-Glu2* gene. Our presented study underscores the importance of further investigations to unravel the intricacies of wheat defense mechanisms against *Pt* and to facilitate the development of novel strategies for *Pt* resistance breeding.

Supplementary Materials: The following supporting information can be downloaded at https://www.mdpi.com/article/10.3390/ijms26020665/s1.

Author Contributions: Conceptualization, J.S. and M.T.K.; data curation, J.S. and A.N.; formal analysis, J.S. and A.N.; investigation, J.S., A.N. and K.S.; methodology, J.S., A.N. and A.T.; supervision, A.T.; visualization, A.N.; writing—original draft, J.S.; writing—review and editing, J.S., A.N., A.T., R.B., K.S. and M.T.K. All authors have read and agreed to the published version of the manuscript.

Funding: This research was financed by the framework of Ministry of Agriculture and Rural Development (Poland) program: "Biological Progress in Plant Production" in years 2021–2027, task No. 5: "A molecular analysis of an adult plant slow rusting genes conferring resistance to rusts caused by *Puccinia* sp." (DHR.hm.802.11.2024); project leader: Michał Tomasz Kwiatek. Publication costs were covered by funding from internal resources of IHAR—Director's Ordinance 12/D/2023.

Institutional Review Board Statement: Not applicable.

Informed Consent Statement: Not applicable.

Data Availability Statement: The data presented in this study are available on request from the corresponding author.

Conflicts of Interest: The authors declare no conflicts of interest.

References

1. FAO. The state of food security and nutrition in the world. In *Building Climate Resilience for Food Security and Nutrition*; FAO: Rome, Italy, 2018; ISBN 978-92-5-130571-3.
2. Karaman, M.; Barutçular, C.; Mennan, H.; Caglayan, K.; Erman, M.; Yildirim, M.; Gazel, M.; Erdemci, İ.; Kendal, E.; Pala, F.; et al. *Theoretical and Practical New Approaches in Cereal Science and Technology Theoretical and Practical New Approaches in Cereal Science and Technology Edited by Authors*; Iksad Publishing House: Tremblay-en-France, France, 2021; ISBN 978-625-7562-37-9.
3. Walkowiak, S.; Gao, L.; Monat, C.; Haberer, G.; Kassa, M.T.; Brinton, J.; Ramirez-Gonzalez, R.H.; Kolodziej, M.C.; Delorean, E.; Thambugala, D.; et al. Multiple Wheat Genomes Reveal Global Variation in Modern Breeding. *Nature* **2020**, *588*, 277–283. [CrossRef] [PubMed]
4. Shiferaw, B.; Smale, M.; Braun, H.-J.; Duveiller, E.; Reynolds, M.; Muricho, G. Crops That Feed the World 10. Past Successes and Future Challenges to the Role Played by Wheat in Global Food Security. *Food Secur.* **2013**, *5*, 291–317. [CrossRef]
5. Zhou, Y.; Zhao, X.; Li, Y.; Xu, J.; Bi, A.; Kang, L.; Xu, D.; Chen, H.; Wang, Y.; Wang, Y.; et al. Triticum Population Sequencing Provides Insights into Wheat Adaptation. *Nat. Genet.* **2020**, *52*, 1412–1422. [CrossRef] [PubMed]
6. Anderson, J. Food, Farming and Freedom: Sowing the Arab Spring. *Aust. J. Int. Aff.* **2012**, *66*, 403–404. [CrossRef]
7. Bentley, A. Broken Bread—Avert Global Wheat Crisis Caused by Invasion of Ukraine. *Nature* **2022**, *603*, 551. [CrossRef]
8. Ahmar, S.; Gill, R.A.; Jung, K.-H.; Faheem, A.; Qasim, M.U.; Mubeen, M.; Zhou, W. Conventional and Molecular Techniques from Simple Breeding to Speed Breeding in Crop Plants: Recent Advances and Future Outlook. *Int. J. Mol. Sci.* **2020**, *21*, 2590. [CrossRef] [PubMed]
9. Bokore, F.E.; Knox, R.E.; Hiebert, C.W.; Cuthbert, R.D.; DePauw, R.M.; Meyer, B.; N'Diaye, A.; Pozniak, C.J.; McCallum, B.D. A Combination of Leaf Rust Resistance Genes, Including *Lr34* and *Lr46*, Is the Key to the Durable Resistance of the Canadian Wheat Cultivar, Carberry. *Front. Plant Sci.* **2022**, *12*, 775383. [CrossRef] [PubMed]

10. Figlan, S.; Ntushelo, K.; Mwadzingeni, L.; Terefe, T.; Tsilo, T.; Shimelis, H. Breeding Wheat for Durable Leaf Rust Resistance in Southern Africa: Variability, Distribution, Current Control Strategies, Challenges and Future Prospects. *Front. Plant Sci.* **2020**, *11*, 549. [CrossRef]
11. Pequeno, D.N.L.; Ferreira, T.B.; Fernandes, J.M.C.; Singh, P.K.; Pavan, W.; Sonder, K.; Robertson, R.; Krupnik, T.J.; Erenstein, O.; Asseng, S. Production Vulnerability to Wheat Blast Disease under Climate Change. *Nat. Clim. Change* **2024**, *14*, 178–183. [CrossRef]
12. Singh, J.; Chhabra, B.; Raza, A.; Yang, S.H.; Sandhu, K.S. Important Wheat Diseases in the US and Their Management in the 21st Century. *Front. Plant Sci.* **2023**, *13*, 1010191. [CrossRef] [PubMed]
13. Kashyap, P.L.; Gupta, V.; Prakash Gupta, O.; Sendhil, R.; Gopalareddy, K.; Jasrotia, P.; Singh, G.P. (Eds.) *New Horizons in Wheat and Barley Research: Crop Protection and Resource Management*; Springer Nature: Singapore, 2022; ISBN 978-981-16-4133-6.
14. Lorrain, C.; Gonçalves dos Santos, K.C.; Germain, H.; Hecker, A.; Duplessis, S. Advances in Understanding Obligate Biotrophy in Rust Fungi. *New Phytol.* **2019**, *222*, 1190–1206. [CrossRef]
15. Petersen, G.; Seberg, O.; Yde, M.; Berthelsen, K. Phylogenetic Relationships of *Triticum* and *Aegilops* and Evidence for the Origin of the A, B, and D Genomes of Common Wheat (*Triticum aestivum*). *Mol. Phylogenet. Evol.* **2006**, *39*, 70–82. [CrossRef]
16. Katamadze, A.; Vergara-Díaz, O.; Uberegui, E.; Yoldi-Achalandabaso, A.; Araus, J.L.; Vicente, R. Evolution of Wheat Architecture, Physiology, and Metabolism during Domestication and Further Cultivation: Lessons for Crop Improvement. *Crop J.* **2023**, *11*, 1080–1096. [CrossRef]
17. Alonge, M.; Shumate, A.; Puiu, D.; Zimin, A.V.; Salzberg, S.L. Chromosome-Scale Assembly of the Bread Wheat Genome Reveals Thousands of Additional Gene Copies. *Genetics* **2020**, *216*, 599–608. [CrossRef] [PubMed]
18. Juery, C.; Concia, L.; De Oliveira, R.; Papon, N.; Ramírez-González, R.; Benhamed, M.; Uauy, C.; Choulet, F.; Paux, E. New Insights into Homoeologous Copy Number Variations in the Hexaploid Wheat Genome. *Plant Genome* **2021**, *14*, e20069. [CrossRef] [PubMed]
19. Levy, A.A.; Feldman, M. Evolution and Origin of Bread Wheat. *Plant Cell* **2022**, *34*, 2549–2567. [CrossRef] [PubMed]
20. Li, L.-F.; Zhang, Z.-B.; Wang, Z.-H.; Li, N.; Sha, Y.; Wang, X.-F.; Ding, N.; Li, Y.; Zhao, J.; Wu, Y.; et al. Genome Sequences of Five Sitopsis Species of Aegilops and the Origin of Polyploid Wheat B Subgenome. *Mol. Plant* **2022**, *15*, 488–503. [CrossRef]
21. Cavalet-Giorsa, E.; González-Muñoz, A.; Athiyannan, N.; Holden, S.; Salhi, A.; Gardener, C.; Quiroz-Chávez, J.; Rustamova, S.M.; Elkot, A.F.; Patpour, M.; et al. Origin and Evolution of the Bread Wheat D Genome. *Nature* **2024**, *633*, 848–855. [CrossRef]
22. Reynolds, M.; Dreccer, F.; Trethowan, R. Drought-Adaptive Traits Derived from Wheat Wild Relatives and Landraces. *J. Exp. Bot.* **2007**, *58*, 177–186. [CrossRef] [PubMed]
23. Müller, T.; Schierscher-Viret, B.; Fossati, D.; Brabant, C.; Schori, A.; Keller, B.; Krattinger, S.G. Unlocking the Diversity of Genebanks: Whole-Genome Marker Analysis of Swiss Bread Wheat and Spelt. *Theor. Appl. Genet.* **2018**, *131*, 407–416. [CrossRef]
24. Dinglasan, E.; Periyannan, S.; Hickey, L.T. Harnessing Adult-Plant Resistance Genes to Deploy Durable Disease Resistance in Crops. *Essays Biochem.* **2022**, *66*, 571–580. [CrossRef] [PubMed]
25. Kumar, K.; Jan, I.; Saripalli, G.; Sharma, P.K.; Mir, R.R.; Balyan, H.S.; Gupta, P.K. An Update on Resistance Genes and Their Use in the Development of Leaf Rust Resistant Cultivars in Wheat. *Front. Genet.* **2022**, *13*, 816057. [CrossRef] [PubMed]
26. Bhardwaj, S.C.; Gangwar, O.P.; Prasad, P.; Kumar, S.; Pal, D. Immunity to Rusts in Wheat: Theory, Fact and Practice. *Indian Phytopathol.* **2021**, *74*, 355–363. [CrossRef]
27. Haq, I.U.; Ijaz, S. *Plant Disease Management Strategies for Sustainable Agriculture Through Traditional and Modern Approaches*; Springer Nature: Berlin/Heidelberg, Germany, 2020; ISBN 978-3-030-35955-3.
28. Bazzini, A.A.; Hopp, H.E.; Beachy, R.N.; Asurmendi, S. Infection and Coaccumulation of Tobacco Mosaic Virus Proteins Alter microRNA Levels, Correlating with Symptom and Plant Development. *Proc. Natl. Acad. Sci. USA* **2007**, *104*, 12157–12162. [CrossRef]
29. Kumar, R.R.; Pathak, H.; Sharma, S.K.; Kala, Y.K.; Nirjal, M.K.; Singh, G.P.; Goswami, S.; Rai, R.D. Novel and Conserved Heat-Responsive microRNAs in Wheat (*Triticum aestivum* L.). *Funct. Integr. Genom.* **2015**, *15*, 323–348. [CrossRef] [PubMed]
30. Kamthan, A.; Chaudhuri, A.; Kamthan, M.; Datta, A. Small RNAs in Plants: Recent Development and Application for Crop Improvement. *Front. Plant Sci.* **2015**, *6*, 208. [CrossRef]
31. Summanwar, A.; Basu, U.; Rahman, H.; Kav, N.N.V. Non-Coding RNAs as Emerging Targets for Crop Improvement. *Plant Sci.* **2020**, *297*, 110521. [CrossRef] [PubMed]
32. Lagudah, E.S.; Krattinger, S.G.; Herrera-Foessel, S.; Singh, R.P.; Huerta-Espino, J.; Spielmeyer, W.; Brown-Guedira, G.; Selter, L.L.; Keller, B. Gene-Specific Markers for the Wheat Gene *Lr34/Yr18/Pm38* Which Confers Resistance to Multiple Fungal Pathogens. *TAG Theor. Appl. Genet. Theor. Angew. Genet.* **2009**, *119*, 889–898. [CrossRef]

33. Spychała, J.; Tomkowiak, A.; Noweiska, A.; Bobrowska, R.; Bocianowski, J.; Sobiech, A.; Kwiatek, M.T. Diversity of Expression Patterns of *Lr34*, *Lr67*, and Candidate Genes towards *Lr46* with Analysis of Associated miRNAs in Common Wheat Hybrids in Response to *Puccinia Triticina* Fungus. *Curr. Issues Mol. Biol.* **2024**, *46*, 5511–5529. [CrossRef]
34. Hiebert, C.W.; Thomas, J.B.; McCallum, B.D.; Humphreys, D.G.; DePauw, R.M.; Hayden, M.J.; Mago, R.; Schnippenkoetter, W.; Spielmeyer, W. An Introgression on Wheat Chromosome 4DL in RL6077 (Thatcher*6/PI 250413) Confers Adult Plant Resistance to Stripe Rust and Leaf Rust (*Lr67*). *Theor. Appl. Genet.* **2010**, *121*, 1083–1091. [CrossRef] [PubMed]
35. McCallum, B.D.; Hiebert, C.W. Interactions Between *Lr67* or *Lr34* and Other Leaf Rust Resistance Genes in Wheat (*Triticum aestivum*). *Front. Plant Sci.* **2022**, *13*, 871970. [CrossRef] [PubMed]
36. Zhang, Y.; Chen, G.; Zang, Y.; Bhavani, S.; Bai, B.; Liu, W.; Zhao, M.; Cheng, Y.; Li, S.; Chen, W.; et al. *Lr34/Yr18/Sr57/Pm38* Confers Broad-Spectrum Resistance to Fungal Diseases via Sinapyl Alcohol Transport for Cell Wall Lignification in Wheat. *Plant Commun.* **2024**, *5*, 101077. [CrossRef]
37. Huerta-Espino, J.; Singh, R.; Crespo-Herrera, L.A.; Villaseñor-Mir, H.E.; Rodriguez-Garcia, M.F.; Dreisigacker, S.; Barcenas-Santana, D.; Lagudah, E. Adult Plant Slow Rusting Genes Confer High Levels of Resistance to Rusts in Bread Wheat Cultivars From Mexico. *Front. Plant Sci.* **2020**, *11*, 824. [CrossRef] [PubMed]
38. Spychała, J.; Tomkowiak, A.; Noweiska, A.; Bobrowska, R.; Rychel-Bielska, S.; Bocianowski, J.; Wolko, Ł.; Kowalczewski, P.Ł.; Nowicki, M.; Kwiatek, M.T. Expression Patterns of Candidate Genes for the *Lr46/Yr29* "Slow Rust" Locus in Common Wheat (*Triticum aestivum* L.) and Associated miRNAs Inform of the Gene Conferring the *Puccinia triticina* Resistance Trait. *PLoS ONE* **2024**, *19*, e0309944. [CrossRef] [PubMed]
39. Pu, M.; Chen, J.; Tao, Z.; Miao, L.; Qi, X.; Wang, Y.; Ren, J. Regulatory Network of miRNA on Its Target: Coordination between Transcriptional and Post-Transcriptional Regulation of Gene Expression. *Cell Mol. Life Sci.* **2019**, *76*, 441–451. [CrossRef] [PubMed]
40. Tomkowiak, A.; Bobrowska, R.; Kwiatek, M.; Spychała, J.; Kuczyński, J.; Tyczewska, A.; Kowalczewski, P.; Weigt, D.; Kosiada, T. Analysis of miRNA Expression Associated with Gene *Lr34* Responsible for Resistance Mechanisms to Wheat Leaf Rust. *Pak. J. Bot.* **2023**, *55*, 379–385. [CrossRef]
41. Spychała, J.; Tomkowiak, A.; Noweiska, A.; Bobrowska, R.; Bocianowski, J.; Książkiewicz, M.; Sobiech, A.; Kwiatek, M.T. Expression Profiling of the Slow Rusting Resistance Genes *Lr34/Yr18* and *Lr67/Yr46* in Common Wheat (*Triticum aestivum* L.) and Associated miRNAs Patterns. *Genes* **2023**, *14*, 1376. [CrossRef] [PubMed]
42. Li, J.; Yue, L.; Shen, Y.; Sheng, Y.; Zhan, X.; Xu, G.; Xing, B. Phenanthrene-Responsive microRNAs and Their Targets in Wheat Roots. *Chemosphere* **2017**, *186*, 588–598. [CrossRef] [PubMed]
43. Bakhshi, B.; Fard, E.M.; Gharechahi, J.; Safarzadeh, M.; Nikpay, N.; Fotovat, R.; Azimi, M.R.; Salekdeh, G.H. The Contrasting microRNA Content of a Drought Tolerant and a Drought Susceptible Wheat Cultivar. *J. Plant Physiol.* **2017**, *216*, 35–43. [CrossRef] [PubMed]
44. Biselli, C.; Bagnaresi, P.; Faccioli, P.; Hu, X.; Balcerzak, M.; Mattera, M.G.; Yan, Z.; Ouellet, T.; Cattivelli, L.; Valè, G. Comparative Transcriptome Profiles of Near-Isogenic Hexaploid Wheat Lines Differing for Effective Alleles at the 2DL FHB Resistance QTL. *Front. Plant Sci.* **2018**, *9*, 37. [CrossRef] [PubMed]
45. Hajiahmadi, Z.; Abedi, A.; Wei, H.; Sun, W.; Ruan, H.; Zhuge, Q.; Movahedi, A. Identification, Evolution, Expression, and Docking Studies of Fatty Acid Desaturase Genes in Wheat (*Triticum aestivum* L.). *BMC Genom.* **2020**, *21*, 778. [CrossRef]
46. Gidhi, A.; Mohapatra, A.; Fatima, M.; Jha, S.K.; Kumar, M.; Mukhopadhyay, K. Insights of Auxin Signaling F-Box Genes in Wheat (*Triticum aestivum* L.) and Their Dynamic Expression during the Leaf Rust Infection. *Protoplasma* **2022**, *260*, 723–739. [CrossRef] [PubMed]
47. Shah, T.; Xu, J.; Zou, X.; Cheng, Y.; Nasir, M.; Zhang, X. Omics Approaches for Engineering Wheat Production under Abiotic Stresses. *Int. J. Mol. Sci.* **2018**, *19*, 2390. [CrossRef]
48. Lagudah, E.S.; McFadden, H.; Singh, R.P.; Huerta-Espino, J.; Bariana, H.S.; Spielmeyer, W. Molecular Genetic Characterization of the *Lr34/Yr18* Slow Rusting Resistance Gene Region in Wheat. *Theor. Appl. Genet.* **2006**, *114*, 21–30. [CrossRef]
49. Bustin, S.A.; Benes, V.; Garson, J.A.; Hellemans, J.; Huggett, J.; Kubista, M.; Mueller, R.; Nolan, T.; Pfaffl, M.W.; Shipley, G.L.; et al. The MIQE Guidelines: Minimum Information for Publication of Quantitative Real-Time PCR Experiments. *Clin. Chem.* **2009**, *55*, 611–622. [CrossRef]
50. Varkonyi-Gasic, E.; Wu, R.; Wood, M.; Walton, E.F.; Hellens, R.P. Protocol: A Highly Sensitive RT-PCR Method for Detection and Quantification of microRNAs. *Plant Methods* **2007**, *3*, 12. [CrossRef]
51. Dai, X.; Zhao, P.X. psRNATarget: A Plant Small RNA Target Analysis Server. *Nucleic Acids Res.* **2011**, *39*, W155–W159. [CrossRef]
52. Chen, C.; Chen, H.; Zhang, Y.; Thomas, H.R.; Frank, M.H.; He, Y.; Xia, R. TBtools: An Integrative Toolkit Developed for Interactive Analyses of Big Biological Data. *Mol. Plant* **2020**, *13*, 1194–1202. [CrossRef]

International Journal of
Molecular Sciences

MDPI

Article

Mitigation Effect of Exogenous Nano-Silicon on Salt Stress Damage of Rice Seedlings

Jian Xiong [1,2], Xiaohui Yang [1,2], Minmin Sun [1,2], Jianqin Zhang [1,2], Linchong Ding [1,2], Zhiyuan Sun [1], Naijie Feng [1,2], Dianfeng Zheng [1,2], Liming Zhao [1] and Xuefeng Shen [1,2,*]

1 College of Coastal Agricultural Sciences, Guangdong Ocean University, Zhanjiang 524008, China; 2112204010@stu.gdou.edu.cn (J.X.); 2112204049@stu.gdou.edu.cn (X.Y.); 2112204037@stu.gdou.edu.cn (M.S.); 21122040151@stu.gdou.edu.cn (J.Z.); dinglinchong0405@163.com (L.D.); sunzhiyuan@hndx2.wecom.work (Z.S.); fengnj@gdou.edu.cn (N.F.); zhengdf@gdou.edu.cn (D.Z.); nxyzlm@gdou.edu.cn (L.Z.)

2 National Saline-Tolerant Rice Technology Innovation South China Center, Zhanjiang 524008, China

* Correspondence: shenxuefeng@gdou.edu.cn

Abstract: Salt stress represents a significant abiotic stress factor that impedes the growth of rice. Nano-silicon has the potential to enhance rice growth and salt tolerance. In this experiment, the rice variety 9311 was employed as the test material to simulate salt stress via hydroponics, with the objective of investigating the mitigation effect of foliar application of nano-silicon on rice seedlings. The results demonstrated that NaCl stress markedly impeded the growth of rice seedlings after seven days of NaCl treatment. The foliar application of nano-silicon followed by NaCl stress alleviated the growth of rice seedlings, markedly improved the spatial conformation of the root system, and enhanced photosynthesis compared with that of NaCl stress alone. The activities of antioxidant enzymes were improved. The contents of antioxidants were increased, and the over-accumulation of ROS was reduced. Furthermore, the foliar application of nano-silicon was found to enhance the uptake of Si^{4+}, K^+, and Ca^{2+} in plants, while simultaneously reducing Na^+ and Cl^- accumulation. Additionally, the content of IAA, CTK, GA, JA, and SA was increased, and ABA was decreased. In conclusion, the foliar application of nano-silicon has been demonstrated to alleviate salt stress injury and improve the growth of rice seedlings.

Keywords: rice; salt stress; nano-silicon; antioxidant system; endogenous hormones; ion balance

1. Introduction

Soil salinization is one of the major abiotic stresses limiting food production. Saline soil areas in China are reaching 1.00×10^8 hectares, over half of which is suited for rice (*Oryza sativa* L.) cultivation [1]. Rice, a moderately salt-sensitive crop, plays a significant role in China's food crop production [2]. The seedling period is the stage of reproduction during which the plant is most susceptible to salt damage, with the greatest impact on morphogenesis and, consequently, on rice yield and quality [3]. Additionally, pot and field experiments have demonstrated that salt stress impairs seedling density and seedling biomass in rice [4]. The development and implementation of salt-tolerant rice cultivation strategies can improve the growth and development of rice seedlings, enhance rice yield, and increase quality; further, nanobiotechnology has great developmental potential in this field [5].

It has been found that salt stress can lead to crop yield reductions of up to 40% due to various environmental factors [6], with sodium chloride (NaCl) being one of the primary components affecting salt stress [7]. Salt stress disrupts osmotic balance and ionic homeostasis in plants, inducing secondary stress that leads to the accumulation of reactive oxygen species (ROS), such as hydrogen peroxide (H_2O_2) and superoxide anion ($O_2^{\cdot-}$). This accumulation results in oxidative damage to plant cells, affecting the growth and development of rice seedlings, which manifests as stunted plant growth, yellowing and wilting of leaves, and inhibited root development [8]. Malondialdehyde (MDA), a product of lipid peroxidation in cell membranes, can be measured to assess the degree of oxidative damage to plants [9]. Plants have evolved an antioxidant defense system to scavenge the excessive accumulation of ROS. This system is divided into two types: antioxidant enzymes and non-enzymatic antioxidants. The enzymatic antioxidant system comprises superoxide dismutase (SOD), peroxidase (POD), catalase (CAT), and ascorbate–glutathione cycle-related enzymes, including ascorbate reductase (APX), glutathione reductase (GR), dehydroascorbate reductase (DHAR), and monodehydroascorbate reductase (MDHAR) [10]. Non-enzymatic antioxidants include reduced glutathione (GSH), oxidized glutathione (GSSG), ascorbic acid (AsA), and dehydroascorbic acid (DHA) [11]. As the duration of the stress period increases, the transpiration rate declines, the net photosynthetic rate is suppressed, and photosynthetic pigment degradation may even occur [12]. An excessive concentration of salt ions in the growth environment will rapidly reduce the osmotic potential of the plant growth environment, thereby causing a physiological water deficit in the plant [13]. NaCl is one of the major constituents of salt stress, and excess Na^+ in the environment competes for K^+ uptake sites in the root system, accumulating excess Na^+ in the plant. This results in a Na^{+}/K^+ imbalance, thus affecting the plant's growth and metabolism [14].

Silicon (Si) is regarded as one of the most significant beneficial elements for plant life activities, exhibiting no detrimental effects at elevated concentrations [15], and studies have shown that it can enhance the salt tolerance of various crops, including soybean [16], corn [17], and rice [18]. Silicon supplementation helps plants retain water, enhances cell strength, can make stems more erect, and increases chlorophyll content and the rate of photosynthesis. The amount of effective silicon in the environment that can be absorbed and utilized by plants is generally very low, so silicon research is mostly limited to silica-loving crops [19]. Silicon is a key element for rice plants, which are often used as a benchmark in silicon nutrition studies. New research suggests that using silicon fertilizers can mitigate the negative impacts of salt stress on the development of rice seedlings [20]. Silica nanoparticles (NPs-Si) represent one of the applications of nanobiotechnology, and researchers have conducted numerous studies due to their unique physicochemical properties, eco-friendliness, and non-toxicity. Because of its small size and strong penetrating ability, nano-silicon can more effectively alleviate environmental stresses on plants compared to bulk silicon, such as by improving plant growth under salt stress [21]. Badawy demonstrated that the exogenous application of nano-silicon to rice plants subjected to salt stress resulted in enhanced water content and ion selectivity within the plants [22]. Riaz showed that the exogenous application of nano-silicon could enhance the resistance of rice seedlings by increasing the activity of antioxidant enzymes (SOD, CAT, and POD) and increasing the content of antioxidants (AsA and GSH) [23]. Ijaz observed that the application of exogenous nano-silicon resulted in an increase in the content of photosynthetic pigments and a reduction in the adverse effects of salt stress on rice seedlings [24]. The exogenous application of nano-silicon has been demonstrated to increase the content of osmoregulatory substances in rice, which has resulted in an improvement in growth status and antioxidant enzyme system activity [25]. The exogenous application of nano-silicon has been demonstrated to maintain beneficial element uptake and improve plant growth and development under salt stress conditions by regulating Na^+/Cl^- and Na^+/K^+ ratios [26]. Furthermore, Li et al. showed that nano-silicon has the ability to protect the ultrastructural integrity of the cell, thereby maintaining the normal functioning of the organelles [27]. These findings suggest that nano-silicon has the potential to improve the growth of rice seedlings under salt stress conditions.

The mechanism of action of exogenously applied nano-silicon to improve rice salt tolerance and promote the growth of rice seedlings under salt stress is currently poorly understood, particularly regarding the response mechanism of rice seedlings subjected to salt stress within a short period of time after nano-silicon treatment. The objective of this study is to gain further insight into the mechanisms by which the foliar application of nano-silicon mitigates the adverse effects of salt stress on rice seedlings and enhances morphological growth, photosynthesis, antioxidant regulation, and the reduction in oxidative stress in rice seedlings, providing new insights to explore the regulatory effects of exogenous nano-silicon on salt tolerance in rice.

2. Results

2.1. Effects of Nano-Silicon on the Above-Ground Morphological Parameters of Rice Seedlings Under NaCl Stress

NaCl stress significantly inhibited the above-ground growth of rice seedlings, and foliar application of nano-silicon could significantly alleviate the inhibitory effect of NaCl. From Figure S1 and Table 1, it can be seen that in comparison to the control (CK), the S treatment notably reduced the plant height by 11.46%, stem diameter at the base by 21.28%, leaf area by 34.04%, shoot fresh weight by 31.50% and shoot dry weight by 23.16% of the above-ground parts of rice seedlings. Compared to the S treatment, the SN treatment resulted in a notable increase in several morphological parameters of the seedlings, including plant height by 4.73%, stem basal width by 13.13%, leaf area by 19.63%, shoot fresh weight by 14.20% and shoot dry weight by 15.07%. The results showed that nano-silicon promoted the above-ground growth of rice seedlings undergoing NaCl stress.

Table 1. Effects of nano-silicon on morphological indicators of above-ground parts of rice seedlings under NaCl stress.

Treatment	Morphological Indicators				
	Shoot Length (cm)	**Stem Diameter (mm)**	**Leaf Area (mm^2)**	**Shoot Fresh Weight ($\times 10^{-2}$ g)**	**Shoot Dry Weight ($\times 10^{-2}$ g)**
CK	32.46 ± 0.10 a	6.58 ± 0.05 a	872.47 ± 28.09 a	94.80 ± 1.82 a	17.96 ± 0.58 b
N	32.88 ± 0.58 a	6.74 ± 0.02 a	794.41 ± 29.65 a	99.18 ± 3.24 a	19.60 ± 0.72 a
S	28.74 ± 0.10 c	5.18 ± 0.10 c	575.51 ± 7.31 c	64.94 ± 0.84 c	13.80 ± 0.39 d
SN	30.10 ± 0.04 b	5.86 ± 0.14 b	688.46 ± 31.34 b	74.16 ± 1.10 b	15.88 ± 0.31 c

Note: Values represent mean ± SEM ($n \geq 3$). Control (CK), nano-silica treatment (N), salt treatment (S), and nano-silica + salt treatment (SN). CK vs. S indicates increase/decrease in S treatment compared to CK. S vs. SN indicates increase/decrease in SN treatment compared to S treatment. Different lowercase letters indicate significant difference at the 0.05 level among different treatments based on Duncan's multiple range test.

2.2. Effects of Nano-Silicon on the Root Parameters of Rice Seedlings Under NaCl Stress

NaCl stress limits the root growth of rice seedlings, and foliar application of nano-silicon promotes root growth under NaCl stress. Figure S1 and Table 2 illustrate that, relative to CK, the S treatment resulted in a marked decrease in the root length by 11.89%, fresh weight by 30.35%, and dry weight by 35.68% of the seedlings. Compared to the

S treatment, the SN treatment significantly increased these root parameters: root length by 10.12%, fresh weight by 19.42%, and dry weight by 25.78%. After 7 d of NaCl stress, compared to CK, the S treatment significantly decreased the total root length by 47.91%, root surface area by 55.47%, root volume by 68.09%, root diameter by 25.99%, root bifurcation number by 57.28%, and root tip number by 51.91%. Relative to the S treatment, the SN treatment markedly increased these root structural parameters: total root length by 68.63%, root surface area by 34.17%, root volume by 38.75%, root diameter by 30.12%, root bifurcation number by 45.81%, and root tip number by 45.91%. Compared to CK, the S treatment led to a considerable reduction in root activity, of 42.86%. Conversely, the SN treatment caused a significant increase in root activity in comparison to the S treatment, of 51.65%, as shown in Table 3 and Figure S2. The results showed that nano-silicon could help rice seedlings by promoting root spatial configuration building and coordinate root growth under NaCl stress (Figure 1).

Table 2. Effects of nano-silicon on root morphological indexes of rice seedlings under NaCl stress.

Treatment	Morphological Indicators		
	Root Length (cm)	**Root Fresh Weight ($\times 10^{-2}$ g)**	**Root Dry Weight ($\times 10^{-2}$ g)**
CK	20.18 ± 0.34 a	32.82 ± 1.32 a	3.98 ± 0.13 a
N	20.70 ± 0.58 a	33.74 ± 1.53 a	4.18 ± 0.10 a
S	17.78 ± 0.43 b	22.86 ± 0.59 c	2.56 ± 0.07 c
SN	19.58 ± 0.13 a	27.30 ± 0.89 b	3.22 ± 0.02 b

Different lowercase letters indicate significant difference at the 0.05 level among different treatments based on Duncan's multiple range test.

Table 3. Effects of nano-silicon on root architecture of rice seedlings under NaCl stress.

Treatment	Morphological Indicators					
	Total Root Length (cm)	**Root Surface (cm^2)**	**Root Volume (cm^3)**	**Root Diameter ($\times 10^{-2}$ mm)**	**Root Bifurcation Number**	**Root Tips Number**
CK	288.27 ± 49.07 ab	29.03 ± 3.23 b	0.55 ± 0.04 ab	43.34 ± 0.31 a	481.00 ± 26.15 b	634.33 ± 42.19 b
N	361.63 ± 31.15 a	39.44 ± 3.38 a	0.71 ± 0.08 a	43.97 ± 1.16 a	732.67 ± 58.95 a	927.00 ± 90.28 a
S	150.16 ± 4.62 c	12.93 ± 1.01 c	0.17 ± 0.06 c	32.08 ± 2.21 b	231.33 ± 35.15 c	271.00 ± 16.52 c
SN	253.22 ± 6.59 b	25.96 ± 0.50 b	0.43 ± 0.04 b	41.74 ± 1.60 a	396.33 ± 31.47 b	502.33 ± 17.53 b

Different lowercase letters indicate significant difference at the 0.05 level among different treatments based on Duncan's multiple range test.

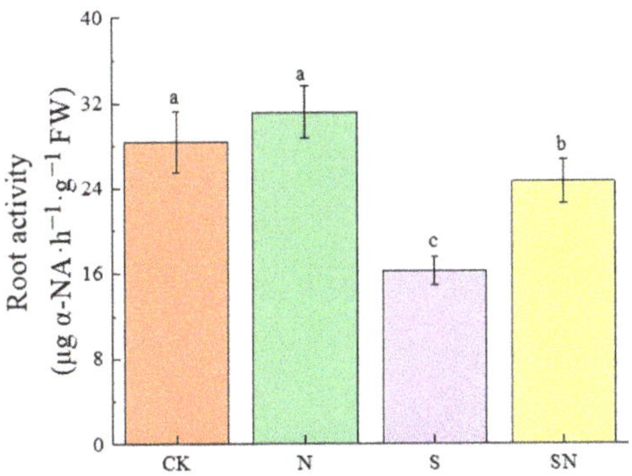

Figure 1. Effects of nano-silicon on root activity of rice seedlings under NaCl stress. (Control (CK), nano-silica treatment (N), salt treatment (S), and nano-silica + salt treatment (SN)). Different lowercase letters indicate significant difference at the 0.05 level among different treatments based on Duncan's multiple range test.

2.3. Effects of Nano-Silica on RWC of Leaves, R/S, and Seedling Index of Rice Seedlings Under NaCl Stress

NaCl stress severely limits the overall growth of rice seedlings, and foliar application of nano-silicon improves the water-holding capacity of leaves and induces coordinated above-ground and below-ground growth of rice seedlings. As shown in Figure 2, relative to CK, the S treatment significantly reduced the RWC of leaves by 14.44%, R/S by 15.26%, and seedling index by 34.35%. Compared to the S treatment, the SN treatment significantly increased the RWC of leaf by 6.09%, R/S by 11.06%, and seedling index by 24.83%.

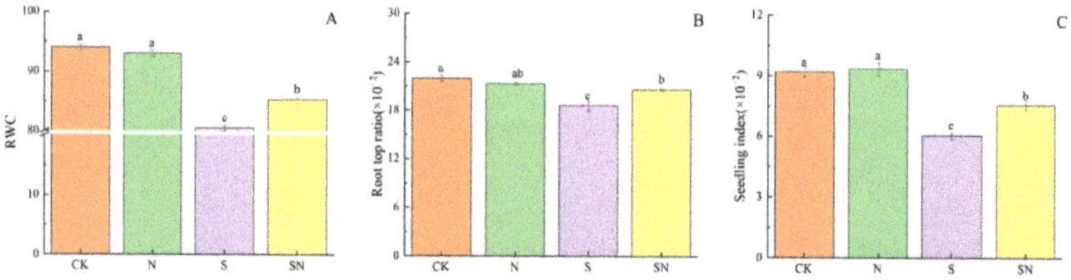

Figure 2. Effects of nano-silicon on RWC, R/S and seedling index of rice seedlings under NaCl stress. (Control (CK), nano-silica treatment (N), salt treatment (S), and nano-silica + salt treatment (SN), RWC (**A**), Root top ratio (**B**), Seedling index (**C**)). Different lowercase letters indicate significant difference at the 0.05 level among different treatments based on Duncan's multiple range test.

2.4. Effects of Nano-Silicon on the Photosynthetic System of Rice Seedlings Under NaCl Stress

NaCl stress resulted in reduced photosynthetic pigment content in rice seedling leaves, and foliar application of nano-silicon significantly alleviated photosynthetic pigment degradation due to NaCl stress. When compared to CK, the contents of Chl a, Chl b, total chl, and Car in the S treatment were all significantly reduced, by 29.19%, 39.12%, 30.84%, and 24.25%, respectively (Figure 3). In contrast to the S treatment, the SN treatment resulted in a significant increase in the content of Chl a, Chl b, total chl, and Car, with improvements of 27.14%, 38.17%, 29.42%, and 18.68%, respectively.

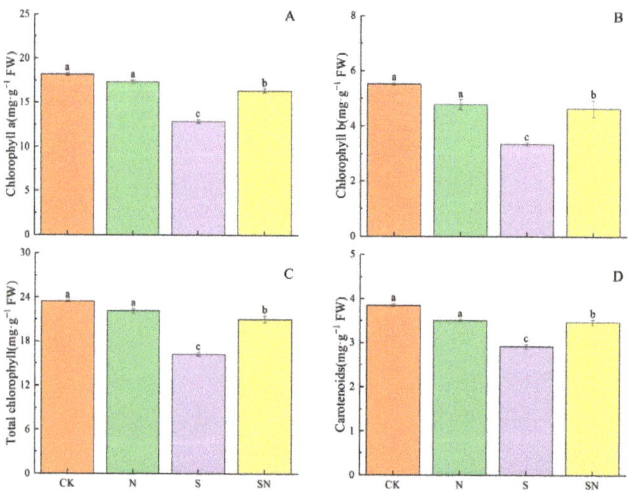

Figure 3. Effects of nano-silicon on photosynthetic pigment content in leaves of rice seedlings under NaCl stress. (Control (CK), nano-silica treatment (N), salt treatment (S), and nano-silica + salt treatment (SN). Chl a (**A**), Chl b (**B**), total chl (**C**), Car (**D**) in rice seeding). Different lowercase letters indicate significant difference at the 0.05 level among different treatments based on Duncan's multiple range test.

NaCl stress significantly inhibited photosynthesis in rice seedlings, and foliar application of nano-silicon improved the gas exchange parameters of leaves under NaCl stress and significantly enhanced photosynthesis in rice seedlings. As shown in Table 4, compared to CK, the S treatment resulted in a decrease in Pn, Tr, Ci, and Gs by 39.21%, 57.57%, 14.76%, and 72.70%, respectively. Compared to the S treatment, the SN treatment showed a significant increase in Pn, Tr, Ci, and Gs, with increases of 27.20%, 40.85%, 9.14%, and 90.00%, respectively. In comparison to CK, the S treatment led to a decrease in WUE and AMC by 31.48% and 20.35%, respectively, while the Ls increased by 58.45% compared to CK (Table 4). Compared to the S treatment, the SN treatment significantly increased WUE and AMC, by 27.86% and 9.89%, respectively, and significantly decreased Ls, by 7.87%.

Table 4. Effects of nano-silicon on photosynthetic gas exchange parameters of rice seedlings under NaCl stress.

Treatment	Photosynthetic Gas Exchange Parameter						
	Pn ($mmol·m^{-2}s^{-1}$)	Tr ($\mu mol·^{-2}s^{-1}$)	Ci ($\mu mol·mol^{-1}$)	Gs ($mmol·m^{-2}s^{-1}$)	WUE ($mg·kg^{-1}$)	Ls ($\times 10^{-2}$)	AMC ($\times 10^{-2}$)
CK	17.42 ± 1.24 a	5.02 ± 0.28 a	302.94 ± 4.11 b	346.94 ± 13.02 a	628.99 ± 19.96 ab	22.14 ± 0.97 c	4.57 ± 0.15 a
N	17.64 ± 0.70 a	4.12 ± 0.28 b	316.75 ± 4.83 a	332.98 ± 38.39 a	683.45 ± 62.59 a	24.21 ± 0.93 c	4.26 ± 0.05 b
S	10.59 ± 0.28 c	2.13 ± 0.12 d	258.24 ± 2.21 d	94.70 ± 2.64 c	430.99 ± 11.41 c	35.08 ± 0.56 a	3.64 ± 0.03 c
SN	13.47 ± 0.88 b	3.00 ± 0.17 c	281.84 ± 2.76 c	179.93 ± 5.03 b	551.05 ± 7.64 b	32.32 ± 0.29 b	4.00 ± 0.04 b

Note: Pn, net photosynthetic rate. Tr, transpiration rate. Ci, intercellular CO_2 concentration. Gs, stomatal conductivity. WUE, water use efficiency. Ls, stomatal limitation. AMC, apparent leaf mesophyll conductance. Different lowercase letters indicate significant difference at the 0.05 level among different treatments based on Duncan's multiple range test.

2.5. Effects of Nano-Silicon on Membrane Damage Indices and Reactive Oxygen Species Accumulation in Rice Seedlings Under NaCl Stress

NaCl stress caused rice seedlings to accumulate excessive ROS, resulting in cell membrane lipid peroxidation, leading to increased electrolyte extravasation. Foliar application of nano-silicon significantly reduced NaCl-induced ROS accumulation and attenuated cellular electrolyte extravasation. In comparison to CK, the S treatment significantly increased REC of leaves by 164.25% and roots by 14.41% (Figure 4). Compared to the S treatment, the SN treatment significantly reduced the leaf electrical conductivity, by 11.82%, while the change in root conductivity was not significant. As shown in Figures 5 and 6, after 3, 5, and 7 d of NaCl treatment, there was a significant reduction in leaf and root MDA content in both the S and SN treatments. Under NaCl stress for 3 to 7 d, the S treatment increased MDA in leaves by 23.20% to 52.23% and in roots by 21.65% to 33.16%, respectively, when compared to CK. Compared to CK, the H_2O_2 content in leaves under the S treatment increased by 7.11% to 27.12%. When compared to the S treatment, the SN treatment significantly reduced the MDA content in leaves and roots by 9.43% to 24.62% and 3.74% to 11.37%, respectively. Additionally, after 3, 5, and 7 d of NaCl treatment, the H_2O_2 content of S-treated leaves increased with progressive time under stress, whereas the H_2O_2 content of the SN-treated leaves decreased significantly. Compared to the S treatment, the H_2O_2 content under the SN treatment was significantly lower, by 18.12% and 11.43%, on days 5 to 7 of stress.

To visualize the distribution of $O_2^{\cdot -}$, histochemical staining was performed on the leaf of rice seedlings. $O_2^{\cdot -}$ appeared as dark blue spots in the leaf (Figure S3). Compared to CK, the leaf of rice seedlings in the S treatment exhibited a significant increase in the number of dark blue spots, which were densely distributed. In contrast to the S treatment, the SN treatment resulted in a significant reduction in the number of dark blue spots on the leaf. The results showed that foliar application of nano-silicon effectively resisted the oxidative damage induced by NaCl stress.

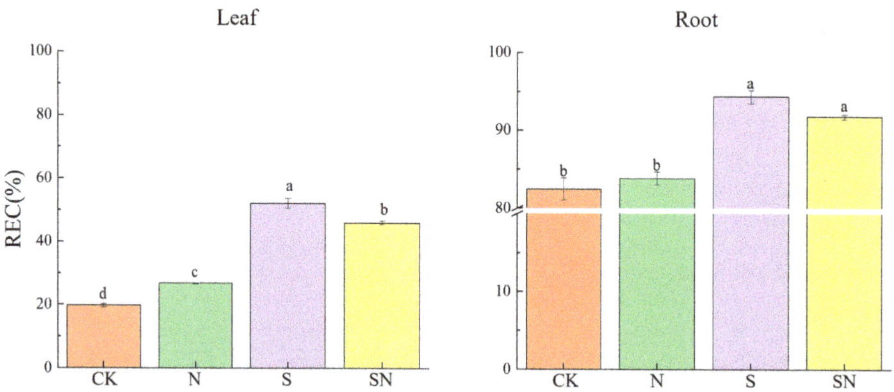

Figure 4. Effects of nano-silicon on REC in rice seedlings under NaCl stress. (Control (CK), nano-silica treatment (N), salt treatment (S), and nano-silica + salt treatment (SN)). Different lowercase letters indicate significant difference at the 0.05 level among different treatments based on Duncan's multiple range test.

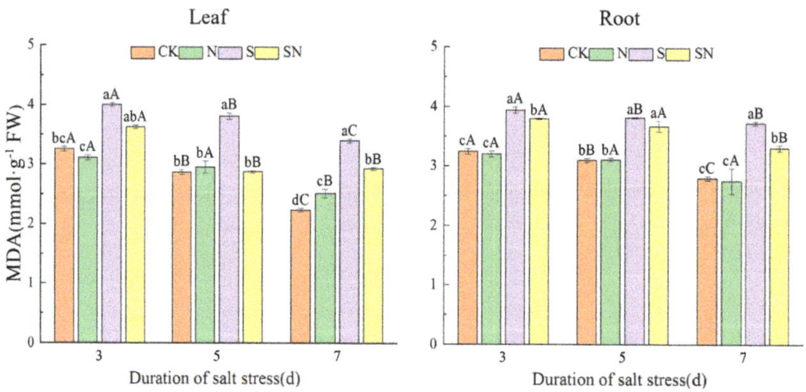

Figure 5. Effects of nano-silicon on MDA content in rice seedlings under NaCl stress. (Control (CK), nano-silica treatment (N), salt treatment (S), and nano-silica + salt treatment (SN). Note: 3, 5, 7 d indicates the duration of NaCl treatment. Lower case letters indicate differences between treatments at the same sampling time, upper case letters indicate differences between the same treatments at different sampling times ($p < 0.05$).

2.6. *Effects of Nano-Silicon on Antioxidant Enzymes in Rice Seedlings Under NaCl Stress*

Figure 7 demonstrates the changes in the activities of antioxidant enzymes (SOD, POD, and CAT) in leaves and roots of rice seedlings under different treatments. After NaCl treatment, the SOD activities of both the leaves and roots of S-treated rice seedlings decreased with increasing time under stress, while the SOD activities of those undergoing SN treatment were significantly higher (Figure 7A,B). After 7 d of NaCl treatment, compared with CK, the SOD activities in the leaves and roots of S-treated rice seedlings were significantly reduced, by 31.98% and 46.49%, respectively. Compared with the S treatment, the SOD activities of the leaves and roots of SN-treated rice seedlings were significantly increased by 127.60% and 125.12%, respectively. After NaCl treatment, the

POD activities of leaves and roots of both S-treated and SN-treated rice seedlings increased with increasing time under stress, but the elevated POD activities in the leaves and roots of SN-treated seedling were more significant (Figure 7C,D). After 7 d of NaCl treatment, compared to CK, the POD activities in S-treated leaves and roots increased by 12.80% and 17.14%, respectively. Compared to the S treatment, the POD activities of SN-treated leaves and roots were significantly increased by 25.39% and 23.66%, respectively. After NaCl treatment, CAT activity in the S-treated leaves decreased with increasing time under stress, and there was an increase in CAT activity in SN-treated leaves with increasing time (Figure 7E,F). After 7 d of NaCl treatment, compared with CK, the CAT activities of the S-treated leaves and roots were significantly reduced by 21.67% and 22.13%, respectively. Compared with the S treatment, the CAT activities of the leaves and roots undergoing the SN treatment were significantly increased by 105.47% and 63.79%, respectively. The results showed that foliar application of nano-silicon increased the antioxidant enzyme activity of rice seedlings and, at the same time, accelerated the response rate of antioxidant enzyme activity and enhanced the antioxidant capacity of the plants.

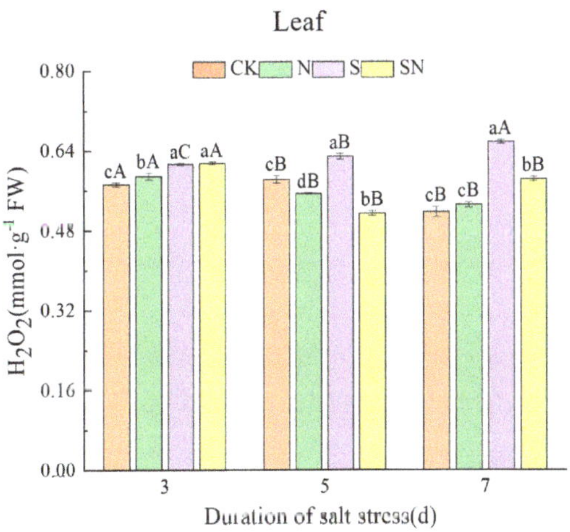

Figure 6. Effects of nano silicon on H_2O_2 content in leaf of rice seedlings under NaCl stress. (Control (CK), nano-silica treatment (N), salt treatment (S), and nano-silica + salt treatment (SN)). Lower case letters indicate differences between treatments at the same sampling time, upper case letters indicate differences between the same treatments at different sampling times ($p < 0.05$).

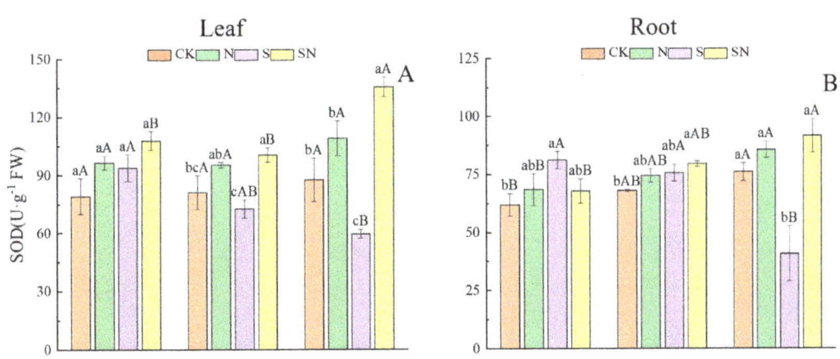

Figure 7. *Cont.*

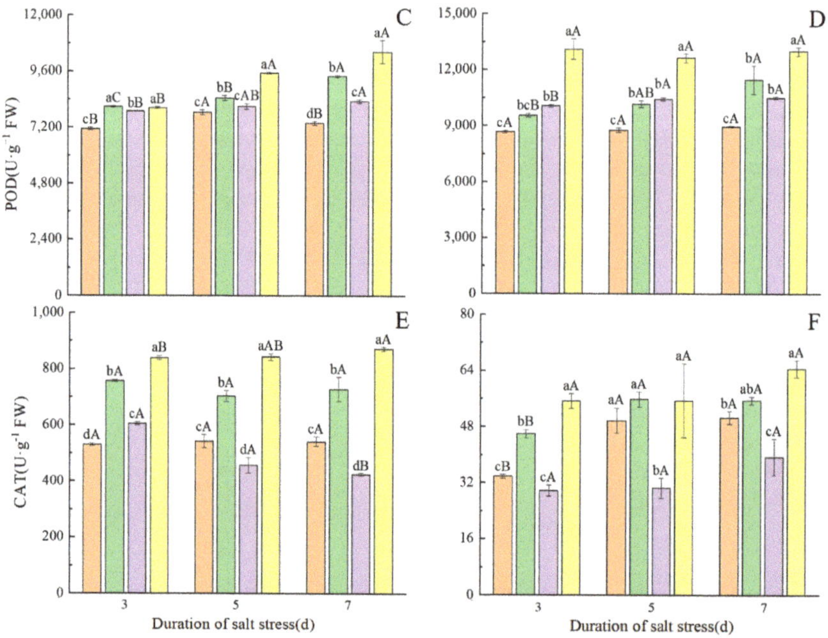

Figure 7. Of nano-silicon on antioxidant enzyme activities of rice seedlings under NaCl stress. (Control (CK), nano-silica treatment (N), salt treatment (S), and nano-silica + salt treatment (SN). SOD activity in leaf (**A**) and root (**B**), POD activity in leaf (**C**) and root (**D**), CAT activity in leaf (**E**) and root (**F**) of rice seeding). Lower case letters indicate differences between treatments at the same sampling time, upper case letters indicate differences between the same treatments at different sampling times ($p < 0.05$).

2.7. Effect of Nano-Silicon on Ascorbate-Glutathione Cycle in Rice Seedlings Under NaCl Stress

The AsA-GSH cycle-related enzyme activities of rice seedlings in each treatment can be seen in Figure 8, where NaCl stress increased APX activity in leaves and roots, but caused a decrease in MDHAR, DHAR, and GR activities. After NaCl treatment, APX activity in S-treated leaves and roots did not change with increasing time under stress, and APX activity in SN-treated leaves and roots was not affected by time under NaCl stress (Figure 8A,B). Compared with CK, APX activity was significantly elevated by 7.48–8.51% in S-treated leaves and 10.03–16.79% in roots at d 3, 5, and 7 after NaCl treatment. Compared to the S treatment, the APX activity was significantly higher in the SN treatment, by 13.36 to 17.75% in the leaves and 13.42 to 16.68% in the roots. After NaCl treatment, the MDHAR activity of S-treated leaves decreased with increasing time under stress, while the MDHAR activity of the root system increased, and SN-treated leaf and root MDHAR activities did not differ significantly across the three sampling occasions (Figure 8C,D). MDHAR activity in S-treated leaves was 16–36% lower than in CK at 3, 5, and 7 days after NaCl treatment. MDHAR activity in the root system was 21–36% lower. MDHAR activities of SN-treated leaves were higher by 24.98–34.68%, and activities in roots were higher by 20.29–41.46%, than those of the S treatment. After NaCl treatment, the DHAR activity of S-treated leaves significantly decreased with increasing time under stress, and a decrease in SN-treated leaf and root DHAR activity was also observed across the three sampling occasions (Figure 8E,F). Compared with CK, the DHAR activity of S-treated leaves was significantly reduced by 9.83 to 32.24%, while the DHAR activity of the root system was reduced by 18.40 to 19.38%, at 3, 5, and 7 days after S treatment. In comparison to the S treatment, the DHAR activity of SN-treated leaves was significantly increased, by 17.70

to 64.41%, and that of the root system was increased by 29.21 to 43.87%. After NaCl treatment, the GR activity of S-treated and SN-treated leaves showed an increasing trend at 3, 5, and 7 d after NaCl treatment, while the GR activity of S-treated roots gradually decreased, and there was no significant change in the GR activity of SN-treated roots (Figure 8G,H). At 3, 5, and 7 days after S treatment in comparison to CK, the GR activity of S-treated leaves was significantly reduced, by 8.03 to 27.36%, and that of the root system was decreased by 15.32 to 28.73%. Compared with the S treatment, the GR activity of SN-treated leaves was significantly increased by 16.73 to 21.44%, and that of the root system by 12.31 to 50.00%. The results showed that foliar application of nano-silicon enhanced the activities of enzymes related to the AsA-GSH cycle and helped to improve the antioxidant capacity of rice seedlings.

Figure 9 demonstrates the changes in the contents of AsA-GSH cycle-related substances in rice seedlings at each sampling time under different treatments. NaCl stress significantly increased the contents of DHA and GSSG in the leaves and roots of rice seedlings, and increased the AsA content of leaves, but the AsA content of roots showed a tendency of increasing and then decreasing. After NaCl treatment, the AsA content of S-treated leaves and roots decreased significantly with increasing time under stress, and the AsA content of SN-treated leaves and roots was relatively stable. At 3, 5, and 7 days after the S treatment compared with CK, the AsA content of S-treated leaves was significantly elevated, by 6.71 to 7.58%, and that of the root system was reduced by 1.30 to 6.48%. In comparison to the S treatment, the AsA content of leaves in the SN treatment was significantly higher, by 9.16–10.88%, and the AsA content of the root system was significantly lower, by 3.51–13.83%. After undergoing NaCl stress, the DHA content of S-treated leaves increased and then decreased, while the DHA content of roots gradually increased but was not significantly different, and that of SN-treated leaves decreased significantly. At 3, 5, and 7 days after the S treatment compared to CK, the DHA content of S-treated leaves was significantly elevated, by 21.29 to 36.36%, and that of the root system was significantly reduced, by 45.99 to 59.99%. In comparison to the S treatment, DHA content was significantly reduced, by 14.89 to 17.14% in SN-treated leaves, and significantly elevated, by 12.57 to 18.75%, in the root system. As a result, the AsA/DHA rate of S-treated leaves and roots significantly decreased at 3, 5, and 7 d of NaCl stress, while the AsA/DHA rate of SN-treated leaves decreased and then increased, and there was no significant change in the AsA/DHA rate of roots. Compared to CK, AsA/DHA was significantly reduced, by 11.57 to 21.16% in S-treated leaves and 35.58 to 41.41% in the root system. AsA/DHA was significantly increased, by 28.26 to 33.82% in leaves and 18.39 to 40.10% in roots of the SN treatment in comparison to the S treatment (Figure 10). After 3, 5, and 7 d of NaCl treatment, the GSH content of S-treated leaves and roots gradually decreased, while the GSH content of SN-treated leaves and roots was relatively stable but not significantly different. At 3, 5, and 7 days after NaCl treatment, in comparison to CK, the GSH content of S-treated leaves was significantly reduced, by 6.57 to 14.01%, while that of the root system was reduced by 7.26 to 11.56%. Compared to the S treatment, the GSH content of SN-treated leaves was significantly increased, by 7.89 to 16.57%, and that of the root system was significantly increased, by 7.83 to 17.51%. The GSSG content of S-treated leaves and roots gradually decreased at 3, 5, and 7 d after NaCl treatment, and the GSSG content of SN-treated leaves and roots significantly decreased with increasing time under stress. Compared with CK, the GSSG content of S-treated leaves was significantly increased, by 10.22 to 17.02%, and that of the root system was increased by 9.76 to 17.21% at 3, 5, and 7 days after the S treatment. In comparison to the S treatment, the GSSG content was significantly reduced, by 8.51 to 16.99% in the leaves and 4.30 to 13.07% in the root system of the SN treatment. As a result, GSH/GSSG gradually increased in S-treated leaves and roots at 3, 5, and 7 d of NaCl stress, and significantly increased in SN-treated leaves and roots with increasing time of nano-silicon treatment. Compared to CK, GSH/GSSG was significantly reduced, by 16.02 to 35.65% in S-treated leaves and 17.95 to 22.65% in the root system. Compared to the S treatment, GSH/GSSG was significantly increased by 19.16 to 29.88% in leaves and 15.36 to 35.17% in the root system in the SN treatment. The results showed that foliar application of nano-silicon could regulate the AsA/DHA and GSH/GSSG ratios in leaves and roots of rice seedlings under NaCl stress to enhance the antioxidant capacity of the plants.

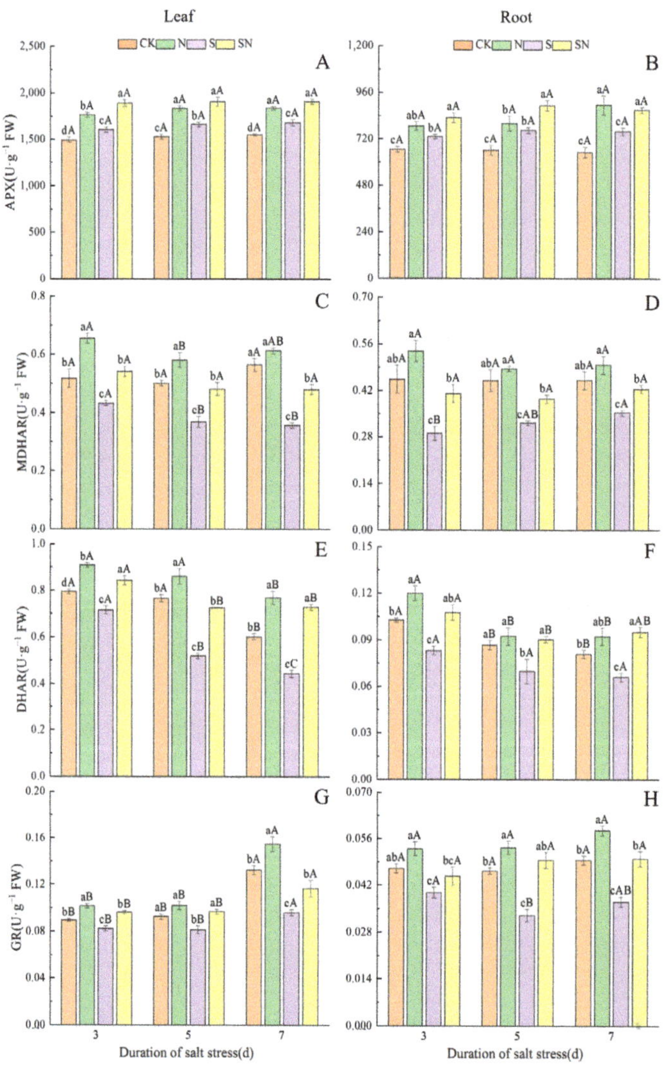

Figure 8. Effects of nano-silicon on the activities of AsA-GSH cycle-related enzymes of rice seedlings under NaCl stress. (Control (CK), nano-silica treatment (N), salt treatment (S), and nano-silica + salt treatment (SN). APX activity in leaf (**A**) and root (**B**), MDHAR activity in leaf (**C**) and root (**D**), DHAR activity in leaf (**E**) and root (**F**), GR activity in leaf (**G**) and root (**H**) of rice seeding). Lower case letters indicate differences between treatments at the same sampling time, upper case letters indicate differences between the same treatments at different sampling times ($p < 0.05$).

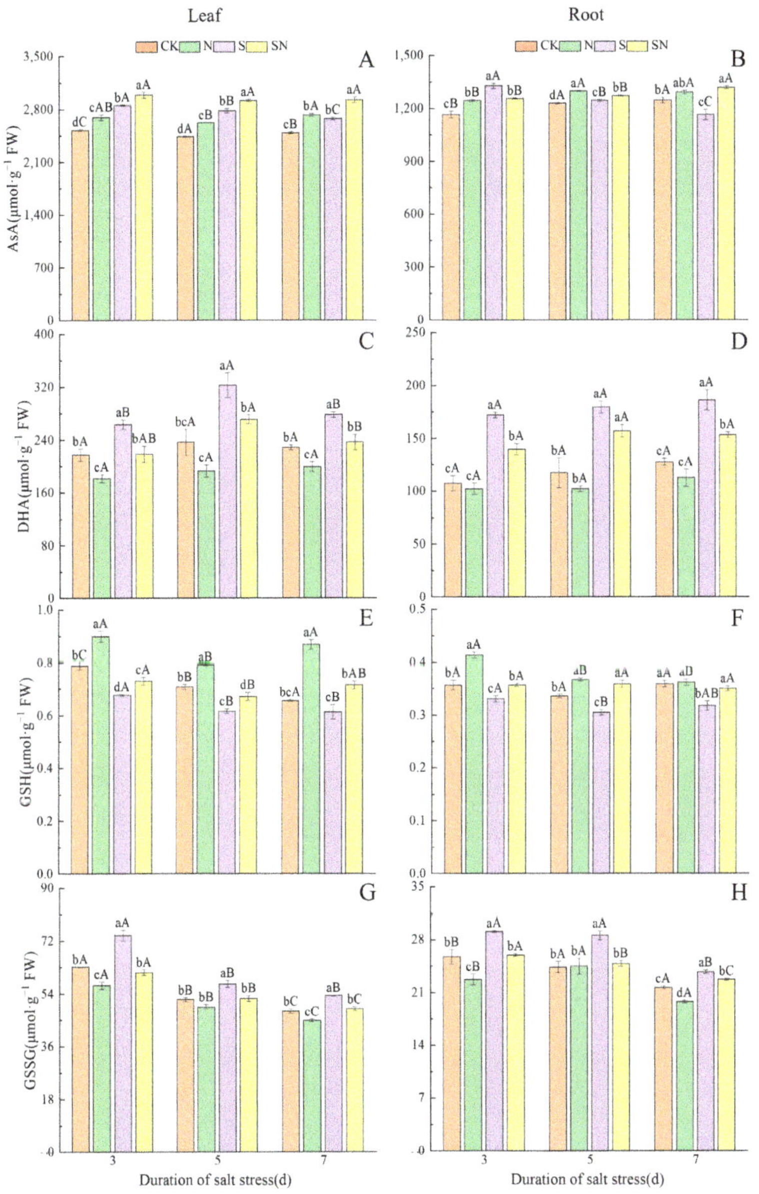

Figure 9. Effect of nano-silicon on the content of AsA-GSH cycle-related substances in rice seedlings under NaCl stress. (Control (CK), nano-silica treatment (N), salt treatment (S), and nano-silica + salt treatment (SN). AsA content in leaf (**A**) and root (**B**), DHA content in leaf (**C**) and root (**D**), GSH content in leaf (**E**) and root (**F**), GSSG content in leaf (**G**) and root (**H**) of rice seeding). Lower case letters indicate differences between treatments at the same sampling time, upper case letters indicate differences between the same treatments at different sampling times ($p < 0.05$).

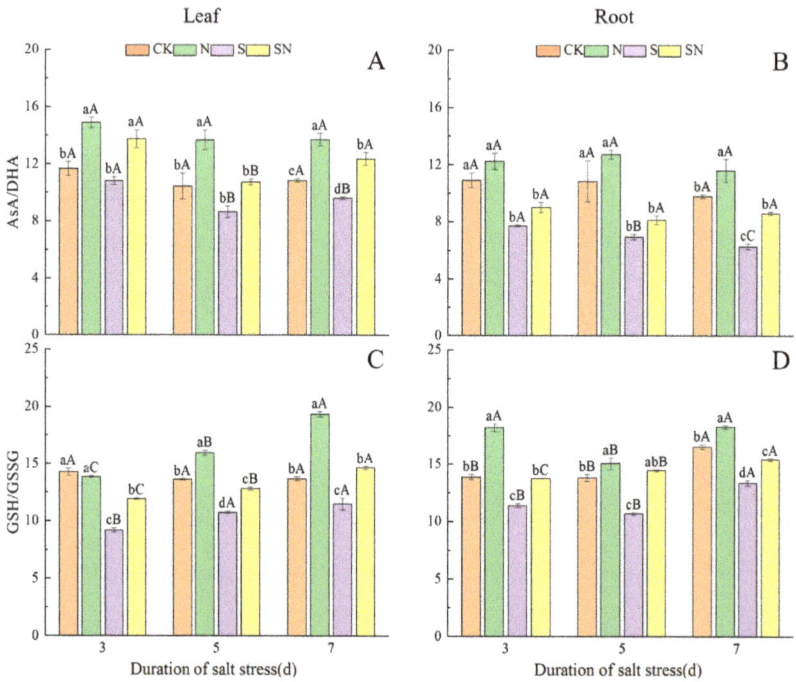

Figure 10. Effect of nano-silicon on AsA/DHA and GSH/GSSG ratios in rice seedlings under NaCl stress. (Control (CK), nano-silica treatment (N), salt treatment (S), and nano-silica + salt treatment (SN). AsA/DHA in leaf (**A**) and root (**B**), GSH/GSSG in leaf (**C**) and root (**D**) of rice seeding). Lower case letters indicate differences between treatments at the same sampling time, upper case letters indicate differences between the same treatments at different sampling times ($p < 0.05$).

2.8. Effects of Nano-Silicon on the Pro and SP Contents in Rice Seedlings Under NaCl Stress

NaCl stress caused the accumulation of osmoregulatory substances in the leaves and roots of rice seedlings, and the Pro content was higher than that of CK in both S-treated leaves and roots, while the SP content of S-treated leaves was also higher than that of CK, but the SP content of roots was significantly lower. After 3, 5, and 7 d of NaCl treatment, the Pro content of leaves and roots under both S and SN treatments increased significantly with the increase in time under stress, and there was no significant difference between the SP content of leaves in the S and SN treatments. In comparison to CK, the SP and Pro contents in the leaves of the S treatment group were significantly increased, by 24.07% and 20.17%, respectively. Furthermore, the Pro content in the roots of the S treatment showed a significant increase, of 47.37%, while the SP content was significantly reduced, by 13.02% (Figure 11). As opposed to the S treatment, the SN treatment led to notably higher levels of both SP and Pro in the leaf and root systems, with respective augmentations of 3.95%, 4.04%, 24.42%, and 47.37%. The results showed that foliar application of nano-silicon increased Pro and SP contents in leaves and roots of rice seedlings and improved osmotic potential to withstand NaCl stress.

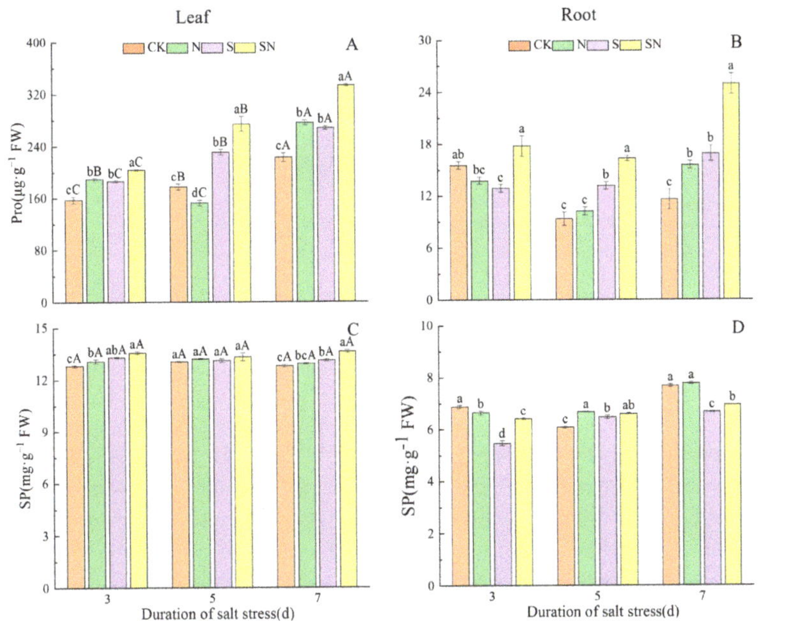

Figure 11. Effects of nano-silicon on soluble protein and proline contents in rice seedlings under NaCl stress. (Control (CK), nano-silica treatment (N), salt treatment (S), and nano-silica + salt treatment (SN). Pro content in leaf (**A**) and root (**B**), SP content in leaf (**C**) and root (**D**) of rice seeding). Lower case letters indicate differences between treatments at the same sampling time, upper case letters indicate differences between the same treatments at different sampling times ($p < 0.05$).

2.9. *Effects of Nano-Silicon on Growth Hormone Content of Rice Seedlings Under NaCl Stress*

As shown in Table 5, S stress significantly affected the endogenous hormone levels of each plant, and foliar application of nano-silicon could effectively reverse the effects of NaCl. In comparison to CK, the S treated leaves showed a significant reduction of 34.70% in IAA, 28.51% in CTK, 34.97% in GA, 47.11% in ABA, 18.08% in SA, and 25.75% in JA, and the root system showed a significant reduction of 32.61% in IAA, 38.89% in CTK, 17.00% in GA, while ABA significantly increased by 30.69%, SA significantly increased by 27.73%, and JA significantly increased by 9.14%. In comparison to the S treatment, the SN treatment showed a significant increase of 22.08% in IAA, 30.87% in CTK, 36.70% in GA, 15.45% in ABA, 32.17% in SA, and 22.21% in JA in the leaves, and in the root system it showed a significant increase in IAA by 56.14%, GA by 2.65%, ABA by 13.23%, SA by 27.71%, and JA by 24.62%.

Table 5. Phytoendogenous hormone contents in leaves and roots of rice seedlings.

Treatment		Endogenous Auxin					
		IAA ($\times 10^{-3}$ nmol·g^{-1} FW)	CTK (ng·g^{-1} FW)	GA (pmol·g^{-1} FW)	ABA (ng·g^{-1} FW)	SA (ng·g^{-1} FW)	JA (pmol·g^{-1} FW)
Leaf	CK	404.30 ± 11.04 b	319.89 ± 9.50 b	1.12 ± 0.02 b	545.81 ± 9.17 c	44.44 ± 0.43 d	12.96 ± 0.16 d
	N	457.10 ± 1.96 a	429.41 ± 9.18 a	1.32 ± 0.02 a	456.89 ± 15.10 d	52.47 ± 0.71 c	14.89 ± 0.31 c
	S	264.00 ± 4.45 d	228.70 ± 7.97 c	0.73 ± 0.02 d	802.96 ± 10.51 a	59.06 ± 1.13 b	16.30 ± 0.19 b
	SN	322.30 ± 6.11 c	299.30 ± 7.61 b	1.00 ± 0.02 c	678.13 ± 15.43 b	69.35 ± 1.31 a	19.92 ± 0.44 a

Table 5. *Cont.*

Treatment		Endogenous Auxin					
		IAA ($\times 10^{-3}$ nmol·g^{-1} FW)	CTK (ng·g^{-1} FW)	GA (pmol·g^{-1} FW)	ABA (ng·g^{-1} FW)	SA (ng·g^{-1} FW)	JA (pmol·g^{-1} FW)
Root	CK	485.40 ± 4.29 b	298.12 ± 6.54 b	0.91 ± 0.02 b	695.15 ± 12.660 c	37.45 ± 0.76 d	11.63 ± 0.28 b
	N	514.00 ± 3.42 a	359.22 ± 8.84 a	1.04 ± 0.03 a	548.13 ± 6.77 d	47.84 ± 0.98 c	15.41 ± 0.13 a
	S	327.10 ± 8.57 d	182.19 ± 2.10 c	0.75 ± 0.02 c	908.46 ± 8.73 a	55.21 ± 0.74 b	10.66 ± 0.39 b
	SN	406.00 ± 6.84 c	284.46 ± 8.18 b	0.77 ± 0.03 c	788.26 ± 11.86 b	61.10 ± 0.85 a	14.50 ± 0.38 a

Different lowercase letters indicate significant difference at the 0.05 level among different treatments based on Duncan's multiple range test.

2.10. Effect of Nano-Silicon on Ion Content of Rice Seedlings Under NaCl Stress

The ion content of seedling leaves and roots systems varied significantly and differently under different treatments (Table 6). Compared to CK, Na^+ was significantly elevated by 133.20% and 419.06%, in the leaves and roots of the S treatment. Compared to the S treatment, Na^+ was significantly reduced, by 13.29% and 7.12%, in the leaves and roots of the SN treatment, respectively. Compared to CK, Cl^- was significantly elevated, by 241.34% and 165.69%, in the leaves and roots of the S treatment, respectively. Compared to the S treatment, Cl^- was significantly reduced by 23.24% and 4.04% in the leaves and roots of the SN treatment, respectively. Compared to CK, Si^{4+} was significantly reduced by 33.98% and 23.17% in S-treated leaves and roots, respectively. Compared to the S treatment, Si^{4+} was significantly elevated, by 73.84% and 59.28%, in leaves and roots of the SN treatment, respectively. Compared to CK, K^+ was significantly reduced, by 14.06% and 41.30%, in S-treated leaves and roots, respectively. Compared to the S treatment, the K^+ of leaves and roots of the SN treatment was significantly higher, by 6.89% and 9.15%, respectively. Compared to CK, Ca^{2+} was significantly reduced, by 5.49% and 47.12%, in the leaves and roots of the S treatment, respectively. Compared to the S treatment, Ca^{2+} was significantly higher in the leaves and roots of the SN treatment, by 7.91% and 82.20%, respectively. As a result, in comparison to CK, Na^+/K^+ was significantly reduced, by 181.00% and 785.11%, in S-treated leaves and roots, respectively. Compared to the S treatment, Na^+/K^+ of leaves and roots in the SN treatment was significantly increased, by 18.89% and 14.94%, respectively. The results showed that foliar application of nano-silicon could reverse the ionic imbalance caused by NaCl stress, reduce Na^+ and Cl^- uptake, and promote K^+ and Ca^{2+} uptake.

Table 6. Ion content in leaves and roots of rice seedlings.

Treatment		Rice Seedling Ion Content					
		Na^+ (μg/g)	Cl^- (μg/g)	Si^{4+} (μg/g)	K^+ (μg/g)	Ca^{2+} (μg/g)	Na^+/K^+ ($\times 10^{-3}$)
Leaf	CK	321.20 ± 0.75 c	1462.02 ± 64.01 c	188.65 ± 3.30 c	8970.17 ± 14.25 a	1501.88 ± 1.07 b	35.80 ± 0.14 c
	N	127.06 ± 0.23 d	433.62 ± 25.34 d	367.92 ± 6.96 a	8666.60 ± 304.47 a	1833.36 ± 30.44 a	14.70 ± 0.54 d
	S	749.05 ± 2.60 a	4990.51 ± 73.55 a	124.54 ± 1.45 d	7448.48 ± 129.32 b	1419.38 ± 10.13 c	100.60 ± 1.40 a
	SN	650.04 ± 24.06 b	3830.46 ± 329.07 b	216.51 ± 10.45 b	7961.65 ± 143.74 b	1531.60 ± 7.31 b	81.60 ± 1.55 b
Root	CK	161.38 ± 0.55 c	428.27 ± 9.55 c	83.21 ± 0.16 c	2331.08 ± 10.65 a	214.88 ± 1.78 b	69.20 ± 0.08 c
	N	150.70 ± 0.16 c	262.43 ± 8.97 d	109.57 ± 1.30 a	2362.18 ± 94.98 a	239.54 ± 5.45 a	64.00 ± 2.51 c
	S	837.68 ± 0.56 a	1137.89 ± 20.41 a	63.93 ± 1.07 d	1368.44 ± 22.56 b	113.64 ± 3.65 c	612.50 ± 10.51 a
	SN	778.04 ± 11.89 b	1091.88 ± 3.95 b	101.83 ± 0.23 b	1493.62 ± 5.25 b	207.05 ± 1.30 b	521.00 ± 9.79 b

Different lowercase letters indicate significant difference at the 0.05 level among different treatments based on Duncan's multiple range test.

2.11. Changes in Stomata and Chloroplast Ultrastructure of Rice Seedling Leaves

Scanning electron microscopy observations revealed significant changes in the number of trichomes (Tr) and the degree of stomatal closure in rice seedling leaves after foliar application of nano-silicon. The number of Tr on the leaf surface of rice seedlings was significantly increased in the SN treatment compared to the S treatment (Figure S4C,D). Meanwhile, the stomata of rice leaves were almost entirely closed in the S treatment and closed significantly in the SN treatment compared to CK (Figure S4E,G,H). Transmission electron microscopy observations showed that there was a large amount of flocculent material in the leaf cells, impaired structural integrity of chloroplasts, ruptured chloroplast periplasm, and dried and thin starch grains in the S treatment (Figure S5C) compared to CK (Figure S5A). As shown in Figure S5D, in comparison to the S treatment, there was less flocculent in the leaf cells, chloroplasts were significantly enlarged and structurally intact, and starch granules were full and rounded in the SN treatment in rice.

2.12. Two-Factor Analysis and Correlation Analysis

Data on MDA, H_2O_2, osmoregulatory substances, and antioxidant enzymes in Table S1 showed that the treatments had significant effects on MDA, H_2O_2, Pro, SP, SOD, POD, CAT, APX, MDHAR, DHAR, and GR at the sampling times. The results indicated that both NaCl stress and foliar application of nano-silicon had a more stable stress/mitigation effect on rice seedlings throughout the treatments.

The experiment screened rice seedlings for morphological growth, photosynthetic gas exchange parameters, endogenous plant hormones, ion content, and AsA-GSH cycle-associated antioxidant enzyme activities with non-enzymatic antioxidants for correlation analysis (Figure S6). The results showed that IAA and GA were significantly and positively correlated with the biomass, seedling index, root spatial configuration, and photosynthetic gas exchange parameters of rice seedlings, and CTK content was significantly and positively correlated with root spatial configuration-related parameters; moreover, the increase in IAA, CTK and GA content could reduce the accumulation of Na+ in plants, and also reduce the level of the ABA hormone to break its morphological growth and photosynthesis inhibitory effects. Meanwhile, it was found that the contents of DHA and GSSG in the AsA-GSH cycle were all significantly negatively correlated with the morphological growth parameters of rice seedlings, while the activities of MDHAR and GR were inversely correlated, and there was a significant positive correlation between the activities of MDHAR and GR and the levels of CTK and GA hormones, respectively.

3. Discussion

3.1. Effect of Nano-Silicon on Morphological Growth Parameters of Rice Seedlings Under NaCl Stress

Soil salinization is a factor affecting food security due to its persistence and toxicity in environmental systems [28]. In general, salt stress, primarily NaCl stress, inhibits plant growth and development [29], and NaCl concentrations above the tolerance limit have severe negative effects on crop production. Furthermore, high Na^+ concentrations in the growth environment can cause cellular osmotic pressure imbalances, leading to water uptake difficulties, and can also cause excessive uptake of intracellular salt ions, which can lead to oxidative stress in cells. This results in the denaturation of cell membrane lipids and the disruption of normal cellular organelle functioning, ultimately manifesting as abnormal cell growth or even death [30]. Studies have reported [31] that salt stress reduces plant growth, leaf area index, and photosynthetic pigment content, while also leading to the greening of leaf in rice. In this study, we found that NaCl stress significantly inhibited the normal above- and below-ground growth of rice seedlings, which manifested as a decrease in growth indices such as plant height, leaf area, root length, and biomass. This, in turn, significantly reduced the relative water content of leaves, root crown ratio, and firmness index;

furthermore, NaCl stress limited the spatial conformation in the root systems of seedlings. Numerous studies [32] have shown that silica nanomaterials can significantly increase the stress tolerance of plants and improve the growth status of plants under NaCl stress. The foliar application of 2.00 mmol·L^{-1} nano-silicon significantly increased the biomass of rice seedlings subjected to saline stress, and their root system spatial configuration was significantly improved, with better promotion of the total root length, root volume, root thickness, and number of root forks and apices (Table 3). Yetgin et al. reported that the promotion of root elongation is crucial for enhancing a plant's ability to absorb water and nutrients, thereby improving overall growth conditions [33]. The total chlorophyll and carotenoid contents of NaCl-treated plants were significantly reduced compared to the control, and the application of nano-silicon significantly increased the net photosynthetic rate, transpiration rate, and stomatal conductance of rice seedlings under NaCl stress treatment (Table 4), which was similar to the results of a previous study [34]. Furthermore, the study revealed that NaCl stress not only resulted in stomatal closure in rice seedling leaves (Figures S4 and S5), but also led to damage to chloroplasts and disruption of cellular structure; the foliar application of nano-silicon maintained the stomatal opening of the leaf, effectively protected the integrity of chloroplasts, and promoted starch accumulation (Figure S5). Correlation analyses revealed that the spatial configuration of the root system of rice seedlings was significantly positively correlated with the gas exchange parameters of the leaf, and that the protection and enhancement of photosynthesis of the leaf could provide sufficient energy for the growth of the root system [35]. Furthermore, coordinated root growth could provide sufficient water and nutrient elements for the leaf [33]. Therefore, we concluded that the application of exogenous nano-silicon can promote the growth of rice seedlings under NaCl stress, coordinate the growth of both the above-ground and underground systems, and to a certain extent maintain their normal photosynthesis, which has a significant mitigating effect on salt damage.

3.2. Effect of Nano-Silicon on Antioxidant System and Reactive Oxygen Species Accumulation in Rice Seedlings Under NaCl Stress

Salt stress causes irreversible damage to the electron transfer process in plants, and this damage generates a large amount of reactive oxygen species (ROS), which, if not scavenged in time, can accumulate in excess, causing cell membrane lipid peroxidation and disrupting normal physiological and metabolic processes in plants [8]. When the crop growth process is subjected to salt stress, effective antioxidant defense systems, such as antioxidant enzymes, can play an important role in scavenging ROS and enhancing plant antioxidant activity [36]. In this study, the antioxidant enzyme activities of seedlings exhibited different performances with increases in time under NaCl stress, and the SOD and CAT activities of NaCl-treated leaves first increased and then decreased compared with the control treatment, which was assumed to be the up-regulation of enzyme activities stimulated by NaCl stress; however, as the exposure time increased, the accumulation of ROS was too high and ROS could not be efficiently scavenged in time, causing oxidative damage to the cells [9]. This is similar to the trend of antioxidant enzyme changes in rice under NaCl stress found in a previous study [37]. After the foliar application of nano-silicon, it was found that nano-silicon alleviated the adverse effects of NaCl stress on rice seedlings by promoting the increase in the activities of antioxidant enzymes (SOD, POD, and CAT) and scavenging the excess ROS (H_2O_2, $O_2^{\cdot-}$), thus reducing the degree of membrane damage (MDA) in seedlings. Badawy et al. also found that nano-silicon and nano-zinc mitigated the negative effects of NaCl stress by increasing the antioxidant enzyme activity and biomass of rice [22]. Riaz indicated that nano-silicon mitigated the toxicity of Cd to rice by enhancing antioxidant defenses [23]. Faten observed that the electrolyte permeability of the plant was significantly increased under NaCl stress conditions and membrane damage intensified, whereas exogenous nano-silicon could enhance the antioxidant capacity of the crop by increasing the antioxidant enzyme activity and promoting plant growth [38].

The antioxidant defense mechanism in higher plants consists of antioxidant enzymes and non-enzymatic antioxidants, of which the ascorbate–glutathione cycle is an important component, with AsA and GSH being key non-enzymatic constituents. The AsA-GSH cycle is able to effectively hydrolyze the H_2O_2 produced by physiological metabolic processes and plays a key role in the scavenging of ROS [39]. AsA, also known as vitamin C, is an electron donor in redox reactions and is able to mitigate salt stress damage to plant cells. The involvement of APX, DHAR, and MDHAR in the oxidation and synthesis cycle of AsA improves the antioxidant capacity of plants [40]. In this study, NaCl stress caused an increase in the overall AsA content of rice seedlings, suggesting that the synthesis of AsA was stimulated after the plants were subjected to stress, followed by the production of DHA from AsA catalyzed by APX, a process that contributes to H_2O_2 scavenging. This can be verified from the results where an increase in the activity of APX was observed (Figure 8), in conjunction with an increase in the DHA content of the seedlings (Figure 9). After foliar application of nano-silicon, we found that the APX activity and AsA content were further increased and the DHA content was decreased compared with NaCl treatment, accompanied by the enhancement of DHAR activity and MDHAR activity, and the reduction in MDA and ROS (H_2O_2, $O_2^{\cdot -}$), suggesting that nano-silicon could further enhance the antioxidant enzyme activity and accelerate the oxidative reduction of AsA as a means to reduce the excessive ROS accumulation brought about by oxidative stress [11]. The AsA/DHA ratio is an important indicator of antioxidant status; as shown in Figure 10, rice seedlings after foliar application of nano-silicon had a higher AsA/DHA ratio than in NaCl treatment, suggesting that nano-silicon treatment was able to maintain higher levels of AsA in the reduced state, which helped to maintain strong antioxidant capacity, as reported by Fan et al. [41].

GSH, as an important non-enzymatic antioxidant in the AsA-GSH cycle, is oxidized by DHAR catalyzing the oxidation of GSH to produce GSSG and the scavenging of H_2O_2, followed by GR-mediated reduction to GSH to complete the cycle; the rate at which the cycle is completed greatly influences the potency of AsA [42]. The content and ratio of GSH and GSSG are other important indicators of cellular antioxidant status. In this study, the GSH content of rice seedlings significantly decreased after being subjected to NaCl stress (Figure 9), presumably in order to scavenge the large amount of accumulated ROS as soon as possible; however, we observed that the GR activity was also inhibited by NaCl stress, resulting in elevated GSSG content, which could not be reduced to GSH in a timely manner, and the rate of redox cycling was restricted, resulting in an excessive accumulation of ROS and causing damage to the cell membrane [43] (the MDA content was elevated). The GSH/GSSG ratio is an important indicator of plant antioxidant capacity, and after the foliar application of nano-silicon, we found that the GSG/GSSG ratio of rice seedlings increased significantly compared with NaCl treatment alone, and enhanced GR and DHAR activities could also be observed, which provided a sufficient pool of GSH for the plant to fight against the accumulation of ROS. This is consistent with a previous study [39], which found that nano-silicon alleviates oilseeds' oxidative damage. Correlation analyses showed that the oxidation state of the AsA-GSH cycle was significantly negatively correlated with the growth of rice seedlings (Figure S6), while the Si^{4+} content was positively correlated with the GSH content and the GSH/GSSG ratio, and that increasing the level of the reduced state of the AsA-GSH cycle significantly improved the root conformation and leaf photosynthesis of rice seedlings. Therefore, we believe that the foliar application of nano-silicon can effectively enhance the activity of enzymes related to the AsA-GSH cycle, accelerate the redox cycle of the antioxidant substances AsA and GSH, and improve the antioxidant capacity of rice seedlings more comprehensively, to counteract the growth inhibitory effects of NaCl stress.

3.3. *Effect of Nano-Silicon on the Accumulation of Osmoregulatory Substances in Rice Seedlings Under NaCl Stress*

The osmotic stress experienced by plants during growth can be alleviated by increasing osmoregulatory substances in the plant to balance the cytoplasmic and external water potentials [44]. SP and Pro serve as key osmolytes; SP, being integral to cellular composition, can initiate responses to external stress and engage in osmotic adjustment, while Pro, acting as an osmoprotectant, is essential for plant recovery from environmental stressors [26]. In this study, it was found that the SP content of rice leaves increased under NaCl stress (Figure 11C), and the SP content of roots first increased and then decreased (Figure 11D). This may be due to the protective mechanism of rice in the early stage of NaCl stress, which caused the protein content to increase first; the content of SP then decreased in the later stages owing to the degradation of proteins under salt stress. NaCl stress in rice seedlings yielded an increasing trend in Pro content, indicating that plants increased the accumulation of osmoregulatory substances to adapt to the osmotic environment. Research has indicated that salt stress can lead to a suppression in protein synthesis or an acceleration in protein degradation [45], as the breakdown of SP can generate a substantial amount of free amino acids. Pro is among the first amino acids to rapidly increase in a variety of crops when exposed to stress [46]. In this study, the application of nano-silicon via foliar spray enhanced the levels of SP and Pro in both the leaves and roots of rice seedlings (refer to Figure 11). This enhancement helps maintain a lower osmotic potential within plant cells; increases the leaf relative water content, as shown in Figure 2A; and supports the normal development of plants even under the stress of saline conditions. In summary, nano-silicon has the capacity to modulate the water balance in rice seedlings and preserve leaf integrity by stimulating the accumulation of osmoprotectants, thereby alleviating the negative impacts of salt stress, a result consistent with Yan's research [47].

3.4. *Effect of Nano-Silicon on Endogenous Hormones in Rice Seedlings Under NaCl Stress*

Endogenous hormones play a direct or indirect regulatory role in plant development, and plant responses to environmental changes are often reflected in changes in the levels of many endogenous hormones; when subjected to external stress, the dynamic balance of hormones in the plant is disrupted, including synthesis and catabolism [48]. In this experiment, after treatment with foliar application of nano-silicon, the endogenous hormone levels of rice seedling leaves and roots changed significantly, with IAA, CTK, GA, SA, and JA contents significantly increasing and ABA contents decreasing. Nanomaterials can mitigate the adverse effects of salt stress by regulating hormone levels (ABA, CTK) and affecting cell division and root growth [49]. Yadav et al. found that plants decrease ABA when they increase the levels of GA and CTK [50]. Weria et al. found that nano-silicon can activate signaling pathways mediated by SA and JA, which modulate the plant's stress response mechanism [21]. Tripathi et al. suggested that plants increase JA levels under stress conditions and that JA participates in silica-regulated membrane signaling pathways [51]. In this experiment, endogenous hormone changes were significant in the leaves and roots of rice seedlings under NaCl stress, with significant increases in ABA, SA, and JA contents and decreases in the IAA, CTK, and GA contents. In comparison to salt stress, the endogenous hormone contents of seedlings changed significantly after the foliar application of nano-silicon, with a significant decrease in ABA content; a significant increase in IAA, CTK, and GA (Table 5); and a further increase in SA and JA content. Abouelsaad found that JA could improve plant salt tolerance by maintaining ROS homeostasis [52], and Wang reported that exogenous nano-silicon could enhance the expression of SA- and JA-related genes and increase the starting hormone levels to regulate the growth and antioxidant capacity of rice [53]. Correlation analysis determined that root conformation-related indexes showed a significant positive correlation with IAA and CTK (Figure S6), and a significant negative correlation with ABA. IAA and GA also showed a significant positive correlation with photosynthetic gas exchange parameters, and SA showed a positive correlation with antioxidant enzymes of the AsA-GSH cycle. This indicated that foliar application of nano-silicon modulated the level of endogenous hormones, and then each of the hormones acted to improve the root system by promoting its coordinated growth, which fed back to the above-ground part to improve the stoutness index of rice seedlings (Figure 2), guaranteeing the normal photosynthesis of leaves and increasing the overall biomass accumulation of the plant [54]. Therefore, we believe that the foliar application of nano-silicon can regulate the hormone levels of rice seedlings to improve salt tolerance and maintain normal physiological and biochemical functions.

3.5. Foliar Application of Nano-Silicon Modulates Ionic Homeostasis in Rice Seedlings

The dynamic balance of sodium and potassium ions is crucial for plant growth under salt stress, and high external concentrations of salt ions (e.g., Na^+ and Cl^-) can lead to physiological plant water deficits through osmotic stress; the root system passively accumulates a large amount of Na^+ and Cl^-, which hinders nutrient uptake and induces stress injuries, such as membrane lipid peroxidation and the disruption of the photosynthetic system [31]. Salt stress inhibits plant growth and development, and biomass accumulation is commonly used to characterize the severity of salt damage; it is a direct indicator of plant salt tolerance [55]. In this study, the large-scale accumulation of Na^+ and Cl^- in NaCl-treated rice seedlings hindered the uptake and utilization of nutrients such as Si^{4+}, K^+, and Ca^{2+}, causing nutrient imbalance in the plants and resulting in significant growth inhibition (Tables 1 and 2). The foliar application of nano-silicon yielded a significant reduction in the Na^+ and Cl^- contents of the seedlings, and a significant increase in the contents of Si^{4+} and Ca^{2+}, which effectively alleviated the ion toxicity produced by NaCl stress (Table 6). Si^{4+} promotes participation in phenolic accumulation, enhances plant antioxidant capacity, and is also a well-known beneficial element for growth [56]. Ca^{2+}, as an intracellular second messenger, not only interlinks with plant sensing in the face of salt stress, but also ensures subsequent signal transduction [57]. Previous studies have shown that rice can maintain the sodium–potassium balance by regulating K^+ uptake, Na^+ efflux, and compartmentalization [58]. In this study, we found that NaCl stress significantly reduced the K^+ content of rice seedlings, and the foliar application of nano-silicon improved this but did not make a significant difference; rather, it served to reduce Na^+/K^+ by controlling the accumulation of Na^+ uptake. In addition, after the foliar application of nano-silicon, the Na^+ content of leaves decreased more than that of roots, and in combination with its promotional effect on the spatial conformation of roots (Table 3 and Figure S2), we speculate that nano-silicon is more biased in regulating Na^+ uptake transporter-related genes, effectively reducing the accumulation of Na^+ while potentially restricting its translocation to the above-ground region. Bosnic showed that exogenous silicon was able to increase the exocytosis of Na^+ from the below-ground part of rice by increasing the expression of the Na^+ exocytosis gene, *OsSOS1*, under salt stress conditions, thus decreasing its translocation to the above-ground part [59]. We also noted that the application of nano-silicon elevated the content of JA, IAA, and CTK in the root system higher than that in the leaves, which contributed to the growth and development of the root system, accumulating more Na^+ while maintaining a certain level of Na^+/K^+. Gong reported a similar role of silicon in reducing the uptake and transport of Na^+ [60]. In this study, we found that Na^+, Cl^-, and Na^+/K^+ were all significantly positively correlated with ABA (Figure S6), and significantly negatively correlated with GA and IAA, whereas the contents of Si^{4+} and Ca^{2+} were significantly positively correlated with CTK. An increase in the content of Si^{4+} also enhanced the reduced state of the GSH cycle and encouraged a greater antioxidant capacity in the rice seedlings, improving NaCl and ultimately improving the growth status of plants under NaCl stress.

4. Materials and Methods

4.1. Plant Material and Growth Conditions

The rice variety used in this experiment was 9311 (Yangdao 6, National Approval No. 2001002), a salt-sensitive cultivar, which was bred by the Jiangsu Lixiahe Regional Agricultural Science Institute. The seeds were chosen based on their fullness and disease-free status, then sanitized in a 3.00% (*v*/*v*) hydrogen peroxide solution for a duration of 15 min, and subsequently washed extensively with deionized water. They were then transferred into beakers containing distilled water and soaked under dark conditions for 24 h. Subsequently, the seeds were evenly spread on double-layer grid germination trays (32.00 cm × 25.50 cm × 12.00 cm) and placed in a 30 °C constant-temperature incubator (JIDI-PX1100, JIDI, Shenzhen, China) for germination under dark conditions; they were misted with ultrapure water once in the morning and evening to maintain seed moisture. Once the rice germ had reached a length of approximately 1.00 cm, it was transferred to a black hydroponic box containing 900 mL of distilled water and allowed to acclimatize for 12 h (48 plants per box). Subsequently, the water was replaced with half-strength Hoagland nutrient solution (pH 6.70 to 6.90) [61], and the seedlings were put in a growth chamber (with a photosynthetic photon flux density of 30,000 lux, a photoperiod of 14 h light/10 h dark, and temperatures of 28 °C/23 °C day/night) for cultivation, with the nutrient solution being renewed every 3 d.

4.2. Characterization of Nano-Silica Materials

The original concentration of the nano-silica material used in the experiment was 2.00 $mol \cdot L^{-1}$ (purchased from Beijing Suo Laibao Science & Technology Co., Ltd., Beijing, China). It was diluted with deionized water and dispersed in an ultrasonic bath for 30 min before further processing. Transmission electron microscopy (TEM, HT7700 Exalens, Hitachi, Tokyo, Japan) was used to observe the morphology of the nano-silica (Figure S7).

4.3. Experimental Design

The experiment was conducted from September 2022 to July 2024 at the South China Center for National Salt-Tolerant Rice Technology Innovation, Institute of Agricultural Biotechnology, Guangdong Ocean University.

Once the seedlings reached the 3-leaf and 1-heart stage, the foliar application of nano-silica solution was administered (10.00 mL per box at a concentration of 2.00 $mmol \cdot L^{-1}$, equivalent to an actual silica application rate of 1.20 mg per box, The concentrations used in the experiment were derived from the results of the previous concentration screening test (Table S2, with the optimal concentration of 2.00 $mmol \cdot L^{-1}$ in bold), with deionized water used as a control (the front and back sides of leaves were sprayed with a high-pressure spray bottle until wet, but not dripping). Twenty-four hours post-treatment,

NaCl was added to simulate salt stress, achieving a final concentration of 60.00 mmol·L^{-1}. Four treatments were established: control (CK), nano-silica treatment (N), salt treatment (S), and nano-silica + salt treatment (SN), with nine replicates for each treatment. The experimental design is schematically illustrated in Figure S8. Samples were collected at 3, 5, and 7 d post-salt treatment, flash-frozen with liquid nitrogen, and then transferred to a −40 °C freezer for storage and subsequent analysis. Each morphological indicator was measured 7 d after NaCl treatment.

4.4. Determination of Indices and Methods

4.4.1. Morphological Indicators

In each replicate of each treatment, 20 uniform seedlings were randomly selected. Seedlings were divided into the shoot and root at the base of the stem, and their fresh weights (FWs) were measured separately. The leaves were removed from the leaf sheath, soaked in ultrapure water for 24 h, then quickly blotted dry with absorbent paper to measure the weight at full saturation (TW). After killing green matter at 105 °C for 30 min, they were dried to a constant weight at 80 °C, and the DWs of the leaves, stems, and roots were measured. The root systems of the seedlings were scanned using a desktop scanner (Epson CORP, Nagano, Japan), and the total root length, total root surface area, total root volume, and root tip numbers were obtained using Win RHIZO root analysis software (DJ-GX02, Regent Instruments, Inc., Quebec, QC, Canada). The root-to-shoot ratio (R/S), seedling index, and relative water content (RWC) [62] were calculated using the following formulas:

$$\text{R/S} = \text{DW of root/shoot} \quad (1)$$

$$\text{Seedling index} = (\text{R/S} + \text{Stem base width/Plant height}) \times (\text{DW of root + shoot}) \quad (2)$$

$$\text{RWC (\%)} = (\text{FW} - \text{DW}) \times 100/(\text{TW} - \text{DW}) \quad (3)$$

4.4.2. Determination of Root Oxidative Activity

The α-naphthylamine oxidation method [63] was employed to determine the root activity. One gram of fresh root tissue was placed in a conical flask containing 50.00 mL of α-naphthylamine (α-NA) solution at 20 μg·L^{-1}, and then incubated on a shaker at 25 °C for 3 h. After filtration, 2.00 mL of the filtrate was added to a test tube containing 1.00 mL of sodium nitrite, mixed well, and the absorbance at OD_{510} was measured. The oxidation amount of α-naphthylamine per gram of fresh root tissue per hour (μg·h^{-1}·g^{-1} FW) was then calculated.

4.4.3. Determination of Photosynthetic Pigment Content and Photosynthetic Gas Exchange Parameters

The extraction method [64] was employed for this phase of testing. Freshly chopped rice seedling leaves (0.10 g) were transferred into test tubes containing 10 mL of anhydrous ethanol and allowed to soak in the dark until the leaves were completely bleached, with five replications for each treatment. Using anhydrous ethanol as a blank, the absorbance at OD_{470}, OD_{649}, and OD_{665} was measured. The content of chlorophyll (Chl a, Chl b) and carotenoids (Car) (mg·g^{-1} FW) was calculated using the following formulas:

$$\text{Chl a} = 13.95 \times OD_{665} - 6.88 \times OD_{649} \quad (4)$$

$$\text{Chl b} = 24.96 \times OD_{649} - 7.32 \times OD_{665} \quad (5)$$

$$\text{Total Chl} = \text{Chl a} + \text{Chl b} \quad (6)$$

$$\text{Car} = (1000 \times OD_{470} - 2.05 \times \text{Chl a} - 114.8 \times \text{Chl b})/245 \quad (7)$$

Between 9:00 and 11:30 on the day of sampling, the net photosynthetic rate (Pn), stomatal conductance (gsw), intracellular carbon dioxide concentration (Ci), and transpiration rate (Tr) were measured using an LI-6800 portable photosynthesis system (LI-6800, LI-COR, Lincoln, NE, USA) [65]. The water use efficiency (WUE), stomatal limitation (Ls), and apparent leaf mesophyll conductance (AMC) were calculated using the following formulas:

$$\text{WUE} = \text{Pn}/\text{Tr} \quad (8)$$

$$\text{Ls} = (\text{Ca} - \text{Ci})/\text{Ca} \times 100 \quad (9)$$

$$\text{AMC} = \text{Pn}/\text{Ci} \quad (10)$$

4.4.4. Determination of Reactive Oxygen Species and Membrane Damage Indices

The nitro-blue tetrazolium (NBT) method [66] was used for the histochemical staining of superoxide anion radicals ($O_2^{\cdot-}$) in rice leaf tissues. From each experimental group, the initial fully developed leaf at the base was consistently harvested, with an analogous portion excised and submerged in phosphate-buffered saline (PBS, pH 7.80) containing 1% NBT (m/v). After vacuum infiltration, the samples were kept in the dark for 24 h until blue spots appeared. The samples were then transferred into a container with 85% ethanol and decolorized in a water bath at 80 °C until the leaves became bleached, after which they were observed and photographed.

The H_2O_2 content of the samples was determined using potassium iodide spectrophotometry [64], with 0.50 g of fresh samples weighed, ground in 5 mL of 0.1% TCA solution, and then subjected to centrifugation at 19,000× g for 20 min. The resulting supernatant was then added to a 10 mmol L^{-1} PBS solution with 1 mol L^{-1} potassium iodide solution, and a dark reaction was conducted at 28 °C for 1 h. The H_2O_2 content (μmol·g^{-1} FW) was then measured. The OD_{390} was employed to calculate the H_2O_2 content (μmol·g^{-1} FW). The MDA content was determined using the thiobarbituric acid (TBA) method, as described in a previous study [64]. Fresh samples (0.50 g) were weighed, ground in 10 mL of a 50 mmol L^{-1} PBS solution, and then centrifuged at 6000× g for 20 min. The resulting supernatant was transferred to a new tube and 0.6% TBA solution was added. The samples were then boiled in a water bath for 15 min. Following centrifugation at 4000× g for 20 min, the supernatant was collected to measure OD_{450}, OD_{532}, and OD_{600}, with the aim of calculating the MDA content (mmol·g^{-1} FW).

Freshly cut rice leaves (0.10 g) were accurately weighed and placed in a beaker containing 10.00 mL of ultrapure water, and then fully soaked for 24 h. The electrical conductivity of the solution (EC1) was measured using a conductivity meter. Subsequently, the solution was boiled for 30 min, cooled to room temperature, and the electrical conductivity (EC2) was measured again. The relative electrical conductivity (REC) was calculated using the formula [64]:

$$REC (\%) = (EC1/EC2) \times 100. \quad (11)$$

4.4.5. Determination of Antioxidant System

A quantity of 0.50 g of fresh plant samples was weighed and put into an ice bath. A quantity of 10.00 mL PBS (pH 7.80) was added, grinding was conducted at 4 °C for homogenization, and 10,000× *g* centrifugation was carried out for 20 min to obtain the crude enzyme solution to be measured. The SOD enzyme activity was determined using the NBT method, as described in a previous study [67]. The enzyme extract was prepared, and riboflavin and NBT were added. The OD_{560} was then measured after illumination. The total SOD activity ($U^{-1} \cdot g^{-1}$ FW) was calculated by taking 50% inhibition of NBT photochemical reduction by SOD activity units, with one enzyme activity unit (U) defined as the amount of enzyme required to inhibit 50% of the photochemical reduction of NBT. The POD enzyme activity was determined by the guaiacol method [68], whereby the catalysis by POD of the oxidation of guaiacol by H_2O_2 results in a product with maximum light absorption at 470 nm. The POD activity ($U^{-1} \cdot g^{-1}$ FW) was calculated by measuring OD_{470} with a change of 0. CAT enzyme activity was determined by the UV-absorbent method [64], whereby H_2O_2 had maximum light absorption at 240 nm. CAT was thus able to decompose H_2O_2, which resulted in a decrease in the solution's absorbance. CAT activity ($U^{-1} \cdot g^{-1}$ FW) was calculated by measuring the change in OD_{240} per unit time [64].

The contents of AsA, DHA, GSH, and GSSG were determined by referring to the method proposed by Zhao [39], with minor modifications. AsA and red phenanthroline can form a chelate with a strong absorption peak at 534 nm, and the content of AsA can be calculated by determining the OD_{534} ($\mu mol \cdot g^{-1}$ FW). DHA is reduced to AsA by dithiothreitol (DTT), and the content of DHA can be calculated by determining the AsA production rate ($\mu mol \cdot g^{-1}$ FW). GSH and 2-nitrobenzoic acid (DTNB) can form compounds with strong absorption peaks at 412 nm, and the content of GSH can be calculated by determining the OD_{412} ($\mu mol \cdot g^{-1}$ FW). GSSG can be reduced to GSH by GR, and then reacted with DTNB, which allows for the determination of the OD_{412}. APX catalyzes the oxidation of AsA. To calculate APX enzyme activity ($U^{-1} \cdot g^{-1}$ FW), a decrease of 0.01 in OD_{290} in 1 min was taken as 1 U. The action of GR consumes reduced coenzyme II (NADPH), which has an absorption peak at 340 nm. Therefore, GR enzyme activity ($U^{-1} \cdot g^{-1}$ FW) can be calculated by determining the change in OD_{340} ($U^{-1} \cdot g^{-1}$ FW). The absorption peak of NADPH occurs at 340 nm, thus enabling the calculation of GR activity through the measurement of the change in OD_{340} ($U^{-1} \cdot g^{-1}$ FW). DHAR is responsible for the production of 2-nitro-5-mercaptobenzoic acid (TNB) from GSH; thus, the activity of DHAR can be calculated by measuring OD_{412} ($U^{-1} \cdot g^{-1}$ FW). The action of MDHAR results in a depletion of the NADH content. The MDHAR enzyme activity can be calculated by determining the rate of change in OD_{340} ($U^{-1} \cdot g^{-1}$ FW).

4.4.6. Determination of Osmotic Adjustment Substances

Proline (Pro) levels were assayed employing the sulfosalicylic acid technique [69]. Pro and sulfosalicylic acid can form a compound with a strong absorption peak at 520 nm, and the Pro content ($\mu g \cdot g^{-1}$ FW) can be calculated by measuring OD_{520}. The determination of soluble protein (SP) content was carried out by using the Coomassie brilliant blue G-250 staining method [70]; SP and Coomassie brilliant blue form a compound with a strong absorption peak at 595 nm, and the SP content can be calculated by measuring OD_{595} ($mg \cdot g^{-1}$ FW).

4.4.7. Determination of Endogenous Hormone Contents

The determination of endogenous hormones abscisic acid (ABA), cytokinin (CTK), gibberellin (GA), indole-3-acetic acid (IAA), jasmonic acid (JA), and salicylic acid (SA) was performed using assay kits developed by Shanghai Enzyme-Linked Biotech Co., Ltd. (Shanghai, China) The samples were analyzed using an enzyme-linked immunosorbent assay (ELISA) with a double-antibody one-step sandwich method, and the OD_{450} value was determined by an enzyme labeler (RT-6100) to calculate the sample concentration.

4.4.8. Ion Content Determination

Fresh root and leaf samples were subjected to a 105 °C heat treatment for the purpose of killing any microorganisms. They were then transferred to a constant temperature oven set at 80 °C until a constant weight was achieved. The samples were ground into a homogeneous powder and weighed with an accuracy of 0.20 g for pre-treatment procedures such as disintegration and ashing. The concentrations of Na^+, Cl^-, K^+, Ca^{2+}, and Si^{4+} were determined through inductively coupled plasma mass spectrometry (ICP-OES, Thermo Scientific ICAP 6000 Series, Waltham, MA, USA).

4.4.9. Observation of Rice Leaf Cell and Chloroplast Ultrastructure by Scanning Electron Microscopy and Transmission Electron Microscopy

Samples were taken after 7 d of NaCl treatment, the topmost fully expanded leaves of rice seedlings from each treatment group were excised and observed for surface changes using SEM (MIRA3, TESCAN, Europe, Brno, Czech Republic). A transmission electron microscope was used to further observe and analyze the ultrastructural changes in chloroplasts induced by nano-silica under NaCl stress conditions [71]. The collected leaf samples were fixed in a 0.10 mol·L^{-1} PBS (pH 7.00) solution containing 2.50% glutaraldehyde and left to soak for 24 h. The leaves were then rinsed with buffer solution and further fixed in a PBS solution containing 1% osmium tetroxide (m/v, OsO_4). Following this, the specimens underwent dehydration via a serial dilution of ethanol (30%, 50%, 70%, 90%, 100%), and some samples were freeze-dried for 24 h. These samples were then observed using SEM (gold sputtering for 30 s, acceleration voltage of 5.00 kV), and images were captured. The remaining samples were embedded in epoxy resin, and ultra-thin sections of 70 to 90 nm thickness were cut using an ultramicrotome. These sections were then placed onto copper grids, stained, and observed using a TEM to analyze and collect images.

4.5. Statistical Analyses

Each parameter was analyzed using three technical replicates ($n \geq 3$), with three technical replicates conducted for each experiment. The results are expressed as mean ± standard error (SEM), and the presence of different lowercase letters indicates a significant difference ($p < 0.05$). Data were organized using Microsoft Excel 2016 software. Statistical analysis, including one-way ANOVA, was performed using IBM SPSS Statistics 24 software. Graphs and charts were created using Origin 2019 and Microsoft PowerPoint 2016 software.

5. Conclusions

In summary, the foliar application of nano-silicon can improve the coordinated growth, biomass accumulation, and photosynthetic efficiency of the root and above-ground parts of rice seedlings under NaCl stress. Nano-silicon also enhances the antioxidant defense system, balances Pro and SP levels, maintains the dynamic balance of Na^+/K^+, increases the uptake of nutrients, mitigates NaCl-induced oxidative damage, maintains the integrity of the chloroplasts and cell membranes, and jointly regulates the levels of endogenous hormones to alleviate the salt damage of rice seedlings and improve salt tolerance. The existence of complex but efficient regulatory mechanisms of nano-silicon in mitigating the negative effects of NaCl stress indicates the great potential of exploiting it in practical production, which could advance the development of environmentally friendly agricultural production (Figure 12). In addition, a deeper understanding of the molecular mechanisms behind the exogenous nano-silicon mitigation of salt damage in rice is more helpful in protecting and promoting practical production.

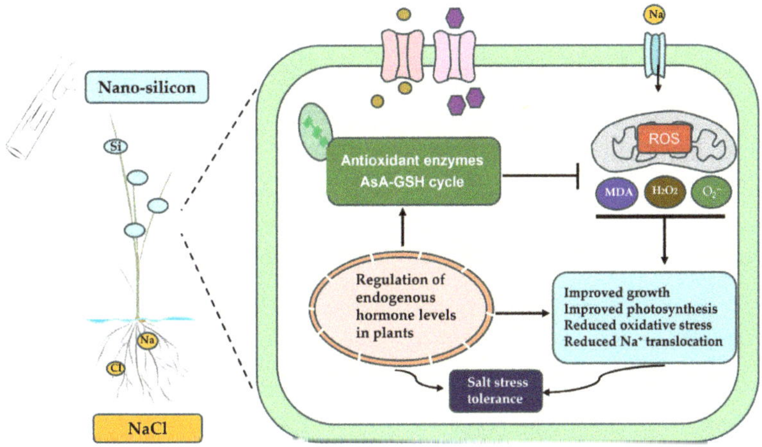

Figure 12. Modeling the response mechanism of nano-silicon to mitigate salt damage in rice seedlings.

Supplementary Materials: The following supporting information can be downloaded at: https://www.mdpi.com/article/10.3390/ijms26010085/s1.

Author Contributions: Conceptualization, X.S.; methodology, J.X., J.Z. and M.S.; validation, X.S. and D.Z.; investigation, J.X., J.Z. and X.S.; resources, X.S. and D.Z.; data curation, J.X. and X.S.; writing—original draft preparation, J.X. and X.Y.; writing—review and editing, X.S., N.F., L.Z. and Z.S.; visualization, J.X., L.D. and X.Y.; supervision, X.S.; project administration, X.S.; funding acquisition, X.S. All authors have read and agreed to the published version of the manuscript.

Funding: This research was funded by Guangdong Ocean University Introduced Talents Scientific Research Start-up Project, grant number (R20062). The APC was funded by Zhanjiang Science and Technology Plan Project (2022A01044).

Institutional Review Board Statement: Not applicable.

Informed Consent Statement: Not applicable.

Data Availability Statement: Data are contained within the article or Supplementary Materials.

Acknowledgments: We would like to thank the Analytical and Testing Center of Guangdong Ocean University for detection.

Conflicts of Interest: The authors declare no conflicts of interest.

References

1. He, L.; Luo, Y. Effects of long-term sand mixing ameliorating measures on soil salinity characteristics of saline-alkali land in northern Shaanxi. *IOP Conf. Ser. Earth Environ. Sci.* **2021**, *781*, 22084–22089. [CrossRef]
2. Qin, H.; Li, Y.; Huang, R. Advances and challenges in the breeding of salt-tolerant rice. *Int. J. Mol. Sci.* **2020**, *21*, 8385. [CrossRef] [PubMed]

3. Li, Q.; Zhu, P.; Yu, X.; Xu, J.; Liu, G. Physiological and molecular mechanisms of rice tolerance to salt and drought stress: Advances and future directions. *Int. J. Mol. Sci.* **2024**, *25*, 9404. [CrossRef]
4. Hussain, S.; Zhang, J.-H.; Zhong, C.; Zhu, L.-F.; Cao, X.-C.; Yu, S.-M.; Bohr, J.A.; Hu, J.-J.; Jin, Q.-Y. Effects of salt stress on rice growth, development characteristics, and the regulating ways: A review. *J. Integr. Agric.* **2017**, *16*, 2357–2374. [CrossRef]
5. Wang, L.; Ning, C.; Pan, T.; Cai, K. Role of silica nanoparticles in abiotic and biotic stress tolerance in plants: A review. *Int. J. Mol. Sci.* **2022**, *23*, 1947. [CrossRef] [PubMed]
6. Liu, L.; Wang, B. Protection of halophytes and their uses for cultivation of saline-alkali soil in China. *Biology* **2021**, *10*, 353. [CrossRef] [PubMed]
7. Hasanuzzaman, M.; Raihan, M.R.H.; Masud, A.A.C.; Rahman, K.; Nowroz, F.; Rahman, M.; Nahar, K.; Fujita, M. Regulation of reactive oxygen species and antioxidant defense in plants under salinity. *Int. J. Mol. Sci.* **2021**, *22*, 9326. [CrossRef]
8. Chung, Y.S.; Kim, K.S.; Hamayun, M.; Kim, Y. Silicon confers soybean resistance to salinity stress through regulation of reactive oxygen and reactive nitrogen species. *Front. Plant Sci.* **2019**, *10*, 1725–1732. [CrossRef]
9. Aslam, M.M.; Raja, S.; Saeed, S.; Farhat, F.; Tariq, A.; Rai, H.M.; Javaid, A.; Shahzadi, I.; Asim, M.; Zulfiqar, S.; et al. Revisiting the crucial role of reactive oxygen species and antioxidant defense in plant under abiotic stress. *Antioxidants* **2022**, *9*, 681–690.
10. Rahman, M.; Ghosh, P.K.; Akter, M.; Al Noor, M.; Rahman, A.; Keya, S.S.; Roni, S.; Biswas, A.; Bulle, M. Green vanguards: Harnessing the power of plant antioxidants, signal catalysts, and genetic engineering to combat reactive oxygen species under multiple abiotic stresses. *Plant Stress* **2024**, *13*, 100547–100554. [CrossRef]
11. Hasanuzzaman, M.; Bhuyan, M.H.M.B.; Anee, T.I.; Parvin, K.; Nahar, K.; Mahmud, J.A.; Fujita, M. Regulation of ascorbate-glutathione pathway in mitigating oxidative damage in plants under abiotic stress. *Antioxidants* **2019**, *8*, 384. [CrossRef] [PubMed]
12. Zahra, N.; Al Hinai, M.S.; Hafeez, M.B.; Rehman, A.; Wahid, A.; Siddique, K.H.; Farooq, M. Regulation of photosynthesis under salt stress and associated tolerance mechanisms. *Plant Physiol.* **2022**, *178*, 55–69. [CrossRef]
13. Lutts, S.; Kinet, J.M.; Bouharmont, J. Changes in plant response to NaCl during development of rice (*Oryza sativa* L.) varieties differing in salinity resistance. *J. Exp. Bot.* **1995**, *46*, 1843–1852. [CrossRef]
14. Chen, T.; Shabala, S.; Niu, Y.; Chen, Z.-H.; Shabala, L.; Meinke, H.; Venkataraman, G.; Pareek, A.; Xu, J.; Zhou, M. Molecular mechanisms of salinity tolerance in rice. *Crop. J.* **2021**, *9*, 506–520. [CrossRef]
15. Pavlovic, J.; Kostic, L.; Bosnic, P.; Kirkby, E.A.; Nikolic, M. Interactions of silicon with essential and beneficial elements in plants. *Front. Plant Sci.* **2021**, *12*, 6975–6992. [CrossRef] [PubMed]
16. Wang, P.; Zhang, H.; Hu, X.; Xu, L.; An, X.; Jin, T.; Ma, R.; Li, Z.; Chen, S.; Du, S.; et al. Comparing the potential of silicon nanoparticles and conventional silicon for salinity stress alleviation in soybean (*Glycine max* L.): Growth and physiological traits and rhizosphere/endophytic bacterial communities. *J. Agric. Food Chem.* **2024**, *72*, 10781–10793. [CrossRef]
17. Shoukat, A.; Pitann, B.; Hossain, S.; Saqib, Z.A.; Nawaz, A.; Mühling, K.H. Zinc and silicon fertilizers in conventional and nano-forms: Mitigating salinity effects in maize (*Zea mays* L.). *J. Plant Nutr. Soil Sci.* **2024**, *3*, 232–239. [CrossRef]
18. Taha, R.S.; Seleiman, M.F.; Shami, A.; Alhammad, B.A.; Mahdi, A.H.A. Integrated application of selenium and silicon enhances growth and anatomical structure, antioxidant defense system and yield of wheat grown in salt-stressed soil. *Plants* **2021**, *10*, 1040. [CrossRef] [PubMed]
19. Mandlik, R.; Thakral, V.; Raturi, G.; Shinde, S.; Nikolić, M.; Tripathi, D.K.; Sonah, H.; Deshmukh, R. Significance of silicon uptake, transport, and deposition in plants. *J. Exp. Bot.* **2020**, *71*, 6703–6718. [CrossRef] [PubMed]
20. Naidu, S.; Pandey, J.; Mishra, L.C.; Chakraborty, A.; Roy, A.; Singh, I.K.; Singh, A. Silicon nanoparticles: Synthesis, uptake and their role in mitigation of biotic stress. *Ecotoxicol. Environ. Saf.* **2023**, *255*, 114783–114788. [CrossRef]
21. Weisany, W.; Razmi, J.; Eshaghabadi, A.H.; Pashang, D. Silicon nanoparticles (SiNP): A novel and sustainable strategy for mitigating environmental stresses in plants. *J. Soil Sci. Plant Nutr.* **2024**, *24*, 2167–2191. [CrossRef]
22. Badawy, S.A.; Zayed, B.A.; Bassiouni, S.M.A.; Mahdi, A.H.A.; Majrashi, A.; Ali, E.F.; Seleiman, M.F. Influence of nano silicon and nano selenium on root characters, growth, ion selectivity, yield, and yield components of rice (*Oryza sativa* L.) under salinity conditions. *Plants* **2021**, *10*, 1657. [CrossRef]
23. Riaz, M.; Kamran, M.; Fahad, S.; Wang, X. Nano-silicon mediated alleviation of Cd toxicity by cell wall adsorption and antioxidant defense system in rice seedlings. *Plant Soil* **2022**, *486*, 103–117. [CrossRef]
24. Ijaz, U.; Ahmed, T.; Rizwan, M.; Noman, M.; Shah, A.A.; Azeem, F.; Alharby, H.F.; Bamagoos, A.A.; Alharbi, B.M.; Ali, S. Rice straw based silicon nanoparticles improve morphological and nutrient profile of rice plants under salinity stress by triggering physiological and genetic repair mechanisms. *Plant Physiol. Biochem.* **2023**, *201*, 107788–107790. [CrossRef]
25. Abd-El-Aty, M.S.; Kamara, M.M.; Elgamal, W.H.; Mesbah, M.I.; Abomarzoka, E.A.; Alwutayd, K.M.; Mansour, E.; Ben Abdelmalek, I.; Behiry, S.I.; Almoshadak, A.S.; et al. Exogenous application of nano-silicon, potassium sulfate, or proline enhances physiological

parameters, antioxidant enzyme activities, and agronomic traits of diverse rice genotypes under water deficit conditions. *Heliyon* **2024**, *10*, e26077. [CrossRef] [PubMed]
26. Mahawar, L.; Ramasamy, K.P.; Suhel, M.; Prasad, S.M.; Živčák, M.; Brestic, M.; Rastogi, A.; Skalický, M. Silicon nanoparticles: Comprehensive review on biogenic synthesis and applications in agriculture. *Environ. Res.* **2023**, *232*, 116292. [CrossRef] [PubMed]
27. Li, T.Z.; Chen, A.T.; Li, C.; Yang, D.; He, N. Effects of silicon on growth and physiological responses of rice seedlings under cadmium stress. *J. Agro-Environ. Sci.* **2018**, *5*, 15–26.
28. Tarolli, P.; Luo, J.; Park, E.; Barcaccia, G.; Masin, R. Soil salinization in agriculture: Mitigation and adaptation strategies combining nature-based solutions and bioengineering. *iScience* **2024**, *27*, e108830. [CrossRef] [PubMed]
29. Heenan, D.; Lewin, L.; Mccaffery, D. Salinity tolerance in rice varieties at different growth stages. *Agric. Food Sci.* **1988**, *28*, 343–349. [CrossRef]
30. Mu, D.W.; Feng, N.J.; Zheng, D.F.; Zhou, H.; Liu, L.; Chen, G.J.; Mu, B. Physiological mechanism of exogenous brassinolide alleviating salt stress injury in rice seedlings. *Sci. Rep.* **2022**, *12*, 20439–20445. [CrossRef] [PubMed]
31. Zhou, H.; Shi, H.; Yang, Y.; Feng, X.; Chen, X.; Xiao, F.; Lin, H.; Guo, Y. Insights into plant salt stress signaling and tolerance. *J. Genet. Genom.* **2024**, *51*, 16–34. [CrossRef] [PubMed]
32. Yetgin, A. Exploring the dynamic nature of root plasticity and morphology in the face of changing environments. *Ecol. Front.* **2024**, *44*, 112–119. [CrossRef]
33. Chapman, N.; Miller, A.J.; Lindsey, K.; Whalley, W.R. Roots, water, and nutrient acquisition: Let's get physical. *Trends Plant Sci.* **2012**, *17*, 701–710. [CrossRef] [PubMed]
34. Alharbi, K.; Osman, H.S.; Rashwan, E.; Hafez, E.M.; Omara, A.E.D. Stimulating the growth, anabolism, antioxidants, and yield of rice plants grown under salt stress by combined application of bacterial inoculants and nano-dilicon. *Plants* **2022**, *11*, 3431. [CrossRef]
35. Nazari, M.; Kordrostami, M.; Ghasemi-Soloklui, A.A.; Eaton-Rye, J.J.; Pashkovskiy, P.; Kuznetsov, V.; Allakhverdiev, S.I. Enhancing photosynthesis and plant productivity through genetic modification. *Cells* **2024**, *13*, 1319. [CrossRef] [PubMed]
36. Das, K.; Roychoudhury, A. Reactive oxygen species (ROS) and response of antioxidants as ROS-scavengers during environmental stress in plants. *Front. Environ. Sci.* **2014**, *2*, 53–59. [CrossRef]
37. Wang, Y.; Tan, C.; Li, Y.; Meng, F.; Du, Y.; Zhang, S.; Jiang, W.; Feng, N.; Zhao, L.; Zheng, D. 5-ALA, DTA-6, and nitrogen mitigate NaCl stress by promoting photosynthesis and carbon metabolism in rice deedlings. *Metabolites* **2024**, *14*, 142. [CrossRef]
38. Zaki, F.S.; Elsayed, A.E.; Ahmed, A.M.; Khalid, K.A. Salinity stress and different types of nano silicon's effects on lupine morphology and biochemical accumulations. *Biocatal. Agric. Biotechnol.* **2024**, *55*, 102997–103005. [CrossRef]
39. Zhao, S.; Kamran, M.; Rizwan, M.; Ali, S.; Yan, L.; Alwahibi, M.S.; Elshikh, M.S.; Riaz, M. Regulation of proline metabolism, AsA-GSH cycle, cadmium uptake and subcellular distribution in *Brassica napus* L. under the effect of nano-silicon. *Environ. Pollut.* **2023**, *335*, 122321. [CrossRef] [PubMed]
40. Yabuta, Y.; Motoki, T.; Yoshimura, K.; Takeda, T.; Ishikawa, T.; Shigeoka, S. Thylakoid membrane-bound ascorbate peroxidase is a limiting factor of antioxidative systems under photo-oxidative stress. *Plant J.* **2002**, *32*, 915–925. [CrossRef]
41. Fan, N.; Zhao, C.; Yue, L.; Ji, H.; Wang, X.; Xiao, Z.; Rasmann, S.; Wang, Z. Nanosilicon alters oxidative stress and defence reactions in plants: A meta-analysis, mechanism and perspective. *Environ. Sci. Nano* **2022**, *9*, 3742–3755. [CrossRef]
42. Hasanuzzaman, M.; Nahar, K.; Anee, T.I.; Fujita, M. Exogenous silicon attenuates cadmium-induced oxidative stress in *Brassica napus* L. by modulating AsA-GSH pathway and glyoxalase system. *Front. Plant Sci.* **2017**, *8*, 1061–1068. [CrossRef] [PubMed]
43. Sahoo, M.R.; Devi, T.R.; Dasgupta, M.; Nongdam, P.; Prakash, N. Reactive oxygen species scavenging mechanisms associated with polyethylene glycol mediated osmotic stress tolerance in Chinese potato. *Sci. Rep.* **2020**, *10*, 5404–5412. [CrossRef] [PubMed]
44. Nazim, M.; Li, X.; Anjum, S.; Ahmad, F.; Ali, M.; Muhammad, M.; Shahzad, K.; Lin, L.; Zulfiqar, U. Silicon nanoparticles: A novel approach in plant physiology to combat drought stress in arid environment. *Biocatal. Agric. Biotechnol.* **2024**, *58*, 103190–103198. [CrossRef]
45. Kanawapee, N.; Sanitchon, J.; Lontom, W.; Threerakulpisut, P. Evaluation of salt tolerance at the seedling stage in rice genotypes by growth performance, ion accumulation, proline and chlorophyll content. *Plants* **2012**, *358*, 235–249. [CrossRef]
46. Singh, A.; Rajput, V.D.; Sharma, R.; Ghazaryan, K.; Minkina, T. Salinity stress and nanoparticles: Insights into antioxidative enzymatic resistance, signaling, and defense mechanisms. *Environ. Res.* **2023**, *235*, 1165–1185. [CrossRef]
47. Yan, G.; Fan, X.; Peng, M.; Yin, C.; Xiao, Z.; Liang, Y. Silicon improves rice salinity resistance by alleviating ionic toxicity and osmotic constraint in an organ-specific pattern. *Front. Plant Sci.* **2020**, *11*, 260–266. [CrossRef] [PubMed]
48. Duan, N.; Jia, Y.K.; Xu, J.; Chen, H.L.; Sun, P. Research progress on plant endogenous hormones. *Chin. Agric. Sci. Bull.* **2015**, *31*, 159–165.
49. Ahmad, J.; Bashir, H.; Bagheri, R.; Baig, A.; Al-Huqail, A.; Ibrahim, M.M.; Qureshi, M.I. Drought and salinity induced changes in ecophysiology and proteomic profile of parthenium hysterophorus. *PLoS ONE* **2017**, *12*, e0185118. [CrossRef]

50. Chhaya; Yadav, B.; Jogawat, A.; Gnanasekaran, P.; Kumari, P.; Lakra, N.; Lal, S.K.; Pawar, J.; Narayan, O.P. An overview of recent advancement in phytohormones-mediated stress management and drought tolerance in crop plants. *Plant Gene* **2020**, *25*, 100264–100270. [CrossRef]
51. Tripathi, D.K.; Vishwakarma, K.; Singh, V.P.; Prakash, V.; Sharma, S.; Muneer, S.; Nikolic, M.; Deshmukh, R.; Vaculík, M.; Corpas, F.J. Silicon crosstalk with reactive oxygen species, phytohormones and other signaling molecules. *J. Hazard. Mater.* **2020**, *408*, 124820–124826. [CrossRef] [PubMed]
52. Abouelsaad, I.; Renault, S. Enhanced oxidative stress in the jasmonic acid-deficient tomato mutant def-1 exposed to NaCl stress. *J. Plant Physiol.* **2018**, *226*, 136–144. [CrossRef] [PubMed]
53. Wang, J.; Cao, X.; Wang, C.; Xiao, Z.; Chen, F.; Feng, Y.; Yue, L.; Wang, Z.; Xing, B. Transcriptomics and metabolomics reveal the mechanisms of enhanced constitutive resistance in rice (*Oryza sativa* L.) by silica nanomaterials. *Environ. Sci. Nano* **2023**, *10*, 2831–2848. [CrossRef]
54. Mukarram, M.; Petrik, P.; Mushtaq, Z.; Khan, M.M.A.; Gulfishan, M.; Lux, A. Silicon nanoparticles in higher plants: Uptake, action, stress tolerance, and crosstalk with phytohormones, antioxidants, and other signalling molecules. *Environ. Pollut.* **2022**, *310*, 119855–119875. [CrossRef]
55. Zhao, S.; Zhang, Q.; Liu, M.; Zhou, H.; Ma, C.; Wang, P. Regulation of plant responses to salt stress. *Int. J. Mol. Sci.* **2021**, *22*, 4609. [CrossRef] [PubMed]
56. Dar, F.A.; Tahir, I.; Hakeem, K.R.; Rehman, R.U. Silicon application enhances the photosynthetic pigments and phenolic/flavonoid content by modulating the phenylpropanoid pathway in common buckwheat under aluminium stress. *Silicon* **2021**, *14*, 323–334. [CrossRef]
57. Naz, M.; Afzal, M.R.; Raza, M.A.; Pandey, S.; Qi, S.; Dai, Z.; Du, D. Calcium (Ca^{2+}) signaling in plants: A plant stress perspective. *S. Afr. J. Bot.* **2024**, *169*, 464–485. [CrossRef]
58. Lian, W.; Geng, A.; Wang, Y.; Liu, M.; Zhang, Y.; Wang, X.; Chen, G. The molecular mechanism of potassium absorption, transport, and utilization in rice. *Int. J. Mol. Sci.* **2023**, *24*, 16682. [CrossRef]
59. Bosnic, P.; Bosnic, D.; Jasnic, J.; Nikolic, M. Silicon mediates sodium transport and partitioning in maize under moderate salt stress. *Environ. Exp. Bot.* **2018**, *155*, 681–687. [CrossRef]
60. Gong, H.J.; Randall, D.P.; Flowers, T.J. Silicon Deposition in the Root Reduces Sodium Uptake in Rice (*Oryza sativa* L.) Seedlings by Reducing Bypass Flow. *Plant Cell Environ.* **2006**, *29*, 1970–1979. [CrossRef]
61. Zuo, G.; Huo, J.; Yang, X.; Mei, W.; Zhang, R.; Khan, A.; Feng, N.; Zehng, D. Photosynthetic mechanisms underlying NaCl-induced salinity tolerance in rice (*Oryza sativa*). *BMC Plant Biol.* **2024**, *24*, 41–50. [CrossRef]
62. Weatherley, P.E. Studies in the water relations of the cotton plant 1. The field measurement of water deficits in leaves. *New Phytol.* **2010**, *49*, 81–97. [CrossRef]
63. Ōta, Y. Diagnostic methods for the measurement of root activity in rice plant. *JARQ Jpn. Agric. Res. Q.* **1970**, *5*, 1–6.
64. Wang, X.; Huang, J. *Principles and Techniques of Plant Physiological Biochemical Experiment*; Higher Education Press: Beijing, China, 2015.
65. Sales, C.R.; Molero, G.; Evans, J.R.; Taylor, S.H.; Joynson, R.; Furbank, R.T.; Hall, A.; Carmo-Silva, E. Phenotypic variation in photosynthetic traits in wheat grown under field versus glasshouse conditions. *J. Exp. Bot.* **2022**, *73*, 10–18. [CrossRef] [PubMed]
66. Zahra, J.; Nazim, H.; Faiza, I.; Zeng, J.; Tahir, A.; Zhang, G. Physiological and antioxidant responses of cultivated and wild barley under salt stress. *Plant Soil Environ.* **2020**, *66*, 334–344. [CrossRef]
67. Elstner, E.F.; Heupel, A. Inhibition of nitrite formation from hydroxylammoniumchloride: A simple assay for superoxide dismutase. *Anal. Biochem.* **1976**, *70*, 616–620. [CrossRef] [PubMed]
68. Yoshiyuki, N.; Kozi, A. Hydrogen peroxide is scavenged by ascorbate-specific peroxidase in *Spinach chloroplasts. Plant Cell Physiol.* **1981**, *22*, 867–880.
69. Bates, L.S.; Waldren, R.P.; Teare, I.D. Rapid determination of proline for water stress studies. *Plant Soil Environ.* **1973**, *39*, 305–307. [CrossRef]
70. Bradford, M.M. A rapid and sensitive method for the quantitation of microgram quantities of protein utilizing the principle of protein-dye binding. *Anal. Biochem.* **1976**, *72*, 248–254. [CrossRef] [PubMed]
71. Ahmed, T.; Masood, H.A.; Noman, M.; Al-Huqail, A.A.; Alghanem, S.M.; Khan, M.M.; Muhammad, S.; Manzoor, N.; Rizwan, M.; Qi, X.; et al. Biogenic silicon nanoparticles mitigate cadmium (Cd) toxicity in rapeseed (*Brassica napus* L.) by modulating the cellular oxidative stress metabolism and reducing Cd translocation. *J. Hazard. Mater.* **2023**, *459*, 1320–1327. [CrossRef] [PubMed]

International Journal of
Molecular Sciences

MDPI

Article

Genome-Wide Identification, Functional Characterization, and Stress-Responsive Expression Profiling of Subtilase (*SBT*) Gene Family in Peanut (*Arachis hypogaea* L.)

Shipeng Li [1], Huiwen Fu [1,2], Yasir Sharif [1,3], Sheidu Abdullaziz [1,3,4], Lihui Wang [1,2], Yongli Zhang [1,3] and Yuhui Zhuang [1,*]

1 Centre for Legume Plant Genetics and System Biology, School of Future Technology and Haixia Institute of Science and Technology, College of Life Sciences, Fujian Agriculture and Forestry University, Fuzhou 350002, China; lishipeng0323@163.com (S.L.); fuhuiwen0702@163.com (H.F.); yasirsharif3336@gmail.com (Y.S.); sheiduabdullaziz8@gmail.com (S.A.); wanglihui3080@163.com (L.W.); zhang9745@outlook.com (Y.Z.)

2 College of Plant Protection, Fujian Agriculture and Forestry University, Fuzhou 350002, China

3 College of Agriculture, Fujian Agriculture and Forestry University, Fuzhou 350002, China

4 Department of Agronomy, Faculty of Agriculture, Nasarawa State University, Keffi P.M.B 1022, Nigeria

* Correspondence: zhuangyh@fafu.edu.cn

Abstract: Subtilases (SBTs), known as serine proteases or phytoproteases in plants, are crucial enzymes involved in plant development, growth, and signaling pathways. Despite their recognized importance in other plant species, information regarding their functional roles in cultivated peanut (*Arachis hypogea* L.) remains sparse. We identified 122 *AhSBT* genes in the STQ peanut genome, classifying them into six subgroups based on phylogenetic analysis. Detailed structural and motif analyses revealed the presence of conserved domains, highlighting the evolutionary conservation of *AhSBTs*. The collinearity results indicate that the *A. hypogea SBT* gene family has 17, 5, and 1 homologous gene pairs with *Glycine max*, *Arabidopsis thaliana*, and *Zea mays*, respectively. Furthermore, the prediction of *cis*-elements in promoters indicates that they are mainly associated with hormones and abiotic stress. GO and KEGG analyses showed that many *AhSBTs* are important in stress response. Based on transcriptome datasets, some genes, such as *AhSBT2*, *AhSBT18*, *AhSBT19*, *AhSBT60*, *AhSBT102*, *AhSBT5*, *AhSBT111*, and *AhSBT113*, showed remarkably higher expression in diverse tissues/organs, i.e., embryo, root, and leaf, potentially implicating them in seed development. Likewise, only a few genes, including *AhSBT1*, *AhSBT39*, *AhSBT53*, *AhSBT92*, and *AhSBT115*, were upregulated under abiotic stress (drought and cold) and phytohormone (ethylene, abscisic acid, paclobutrazol, brassinolide, and salicylic acid) treatments. Upon inoculation with *Ralstonia solanacearum*, the expression levels of *AhSBT39*, *AhSBT50*, *AhSBT92*, and *AhSBT115* were upregulated in disease-resistant and downregulated in disease-susceptible varieties. qRT-PCR-based expression profiling presented the parallel expression trends as generated from transcriptome datasets. The comprehensive dataset generated in the study provides valuable insights into understanding the functional roles of *AhSBTs*, paving the way for potential applications in crop improvement. These findings deepen our understanding of peanut molecular biology and offer new strategies for enhancing stress tolerance and other agronomically important traits.

Keywords: abiotic and biotic stress; gene duplication; miRNAs; peanut breeding; plant hormones; phylogenetic analysis

1. Introduction

Subtilisin is an alkaline serine protease first discovered in *Bacillus subtilis* [1]. It is found in living organisms [2]. *SBT* genes are proteases used in the feed, food, nutrition, and cosmetics sectors to hydrolyze proteins. It is commonly found in plants, bacteria, fungi, and parasites. These genes are widely distributed among the kingdom Plantae, contributing to varied functions in plant defense mechanisms and growth. However, limited information is available regarding the *A. hypogaea SBT* gene family. The MEROPS database classification standard identifies serine peptidases as a significant group of the S8 subtilase family. Hydrolysis of proteins into amino acids is catalyzed by these enzymes [3]. Subtilases are typically inactive protein precursors that include several key components: a propeptide domain (19 inhibitory regions), a signal peptide, a protease-associated (PA) domain, and a subtilisin domain. Part of the subtilases perhaps possesses a few of these structures [2]. They have a catalytic triad in their three-dimensional structure: ASP (aspartic acid), His (histidine), and Ser (serine). This triad facilitates protein hydrolysis through the nucleophilic attack of the active site Ser residues on peptide bonds [4].

Different plant species contain varying numbers of *SBT* genes, which seem to correlate with their evolutionary complexity. For instance, the moss *Physcomitrella patens* has 23 *SBT* genes [5]. Contrarily, these *SBT* genes are said to be more in land plants: *Arabidopsis thaliana* possesses 56 *SBT* genes [6]; grape (*Vitis vinifera*) has 11 [7]; rice (*Oryza sativa* L.) has 63 [8]; cotton has 122 [9]; barley (*Hordeum vulgare* L.) has 54 [10]; tomato *(Solanum lycopersicum)* has 80 [11]; and *Z. mays* has 58 [12]. Various validation methods have determined that these *SBT* genes possess a wide range of functions. For instance, in *A. thaliana*, overexpression of the subtilase gene *AtSBT3.3* triggers an innate immune response and downstream immunological signaling [13]. Upregulated expression of the Bacillus subtilis-like protease *GbSBT1* gene isolated from cotton after jasmonic acid (JA) and ethylene (Eth) treatments suggests its involvement in defense mechanisms [14]. Bacillus subtilis-like protease *HbSPA* from rubber tree leaves is possibly involved in defense mechanisms in *Phaseolus vulgaris* [15]. PvSLP2's chymotrypsin activity is amplified in dehydration, signifying its importance in senescence and drought control. [16]. Eight *SBT* genes in barley exhibited elevated expression during germination, indicating their function in barley grain germination [17].

Arachis hypogaea L. (peanut) is a significant agricultural crop that serves as a crucial source of plant oil and protein worldwide [18]. Although *SBT genes* have been identified in many cultivated species, there is still no report on their identification in peanuts. Here, we have identified 122 *SBT* genes in the peanut genome [18]. These genes are grouped based on phylogenetic relationships into six groups. By analyzing protein-conserved domains and gene structures, it was observed that different groups of peanuts had varied gene structures, although their structural domains remained relatively conserved. Additionally, we conducted chromosomal localization, homology, promoter analysis, gene expression profiling, and subcellular localization studies. The findings show that the extracellular space was the location of a sizable fraction of *AhSBTs*. Several *AhSBTs* hypothesized to reside in the nucleus, extracellular space, or plasma membrane were found to be there by experimental confirmation. Overall, this paper helps with the functional characterization and use of the peanut *SBT* gene family.

2. Results

2.1. Protein Physicochemical Property Analysis and Identification of AhSBTs in the STQ Genome

A total of 122 *AhSBTs* were identified in the peanut genome by HMM search using the Pfam: PF00082 domain, followed by de-redundant operations. These genes were named in ascending order of *AhSBT1–AhSBT122* according to chromosome locations (Supplementary Table S1). Among them, Chr04 recorded the highest gene numbers, totaling 15 members. Chr16, Chr14, and Chr19 had 14, 13, and 9 genes, respectively. Chr03/Chr05/Chr09/Chr13/Chr15 with seven genes each, Chr01/Chr12 with six genes each, Chr11 with five genes, Chr02 with four genes, Chr06/Chr08/Chr17/Chr20 with three genes each. The lowest number of *AhSBTs* (1) was found on Chr07/Chr10/Chr18 (Figure 1). Protein physicochemical characterization shows the peanut SBT proteins ranged from 250 aa (AhSBT122) to 1387 aa (AhSBT119). The expected molecular weight of 122 SBT proteins ranged from 27.77417 (AhSBT122) to 153.6648 kDa (AhSBT119), and the isoelectric point varied from 4.94 (AhSBT86) to 9.91 (AhSBT122). The shifts in MW and pI were mainly due to elevated essential amino acid content and post-translational changes. Subcellular localization prediction using BUSCA indicated diverse distributions of peanut SBT proteins across cellular compartments. Specifically, 78 SBT proteins were distributed in six different cell compartments, with 10 localized in the nucleus and 19 proteins in the plasma membrane. Three proteins were predicted in the endosperm system, including at least two found in the inner membrane of the chloroplast, and five proteins were anticipated to reside in the cytoplasm and chloroplast.

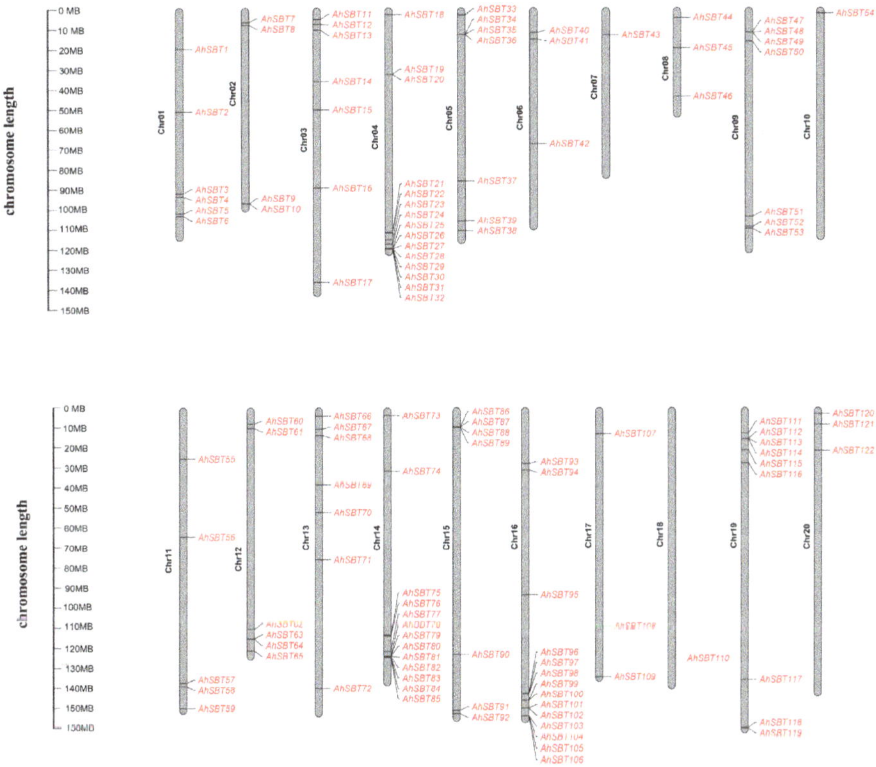

Figure 1. Chromosomal mapping of *AhSBTs* on *A. hypogea* genome. Map distribution of 20 chromosomes (grey bars). Representative chromosome numbers are shown on the left (black), and gene names are on the right side (red).

2.2. *AhSBTs Phylogenetic Tree Analysis*

To investigate the phylogenetic relationship of *AhSBT* genes, a phylogenetic tree was constructed, comprising *AhSBTs* (122) and those of *A. thaliana* (56), *G. max* (30), and *Z. mays* (58) (Figure 2). SBT proteins from these species were grouped into six clades (1–6), which contained 44, 11, 4, 26, 30, and 7 *SBT* members, respectively (Supplementary Figure S1). Except for *AhSBT3* and *AhSBT4*, the classification of *SBT* genes in *A. thaliana* aligned with previous studies suggesting the reliability of the topological structure of the phylogenetic tree [6]. These findings support the *SBT* family's strong evolutionary conservatism across plant taxa. It's interesting that Group 6 has the fewest members and Group 1 the most. As per the findings of *A. thaliana*, Group 6 is the smallest. However, compared to *A. thaliana*, *A. hypogea* has a remarkably greater number of *SBT* genes, implying that Group 1 members may have a wider variety of roles.

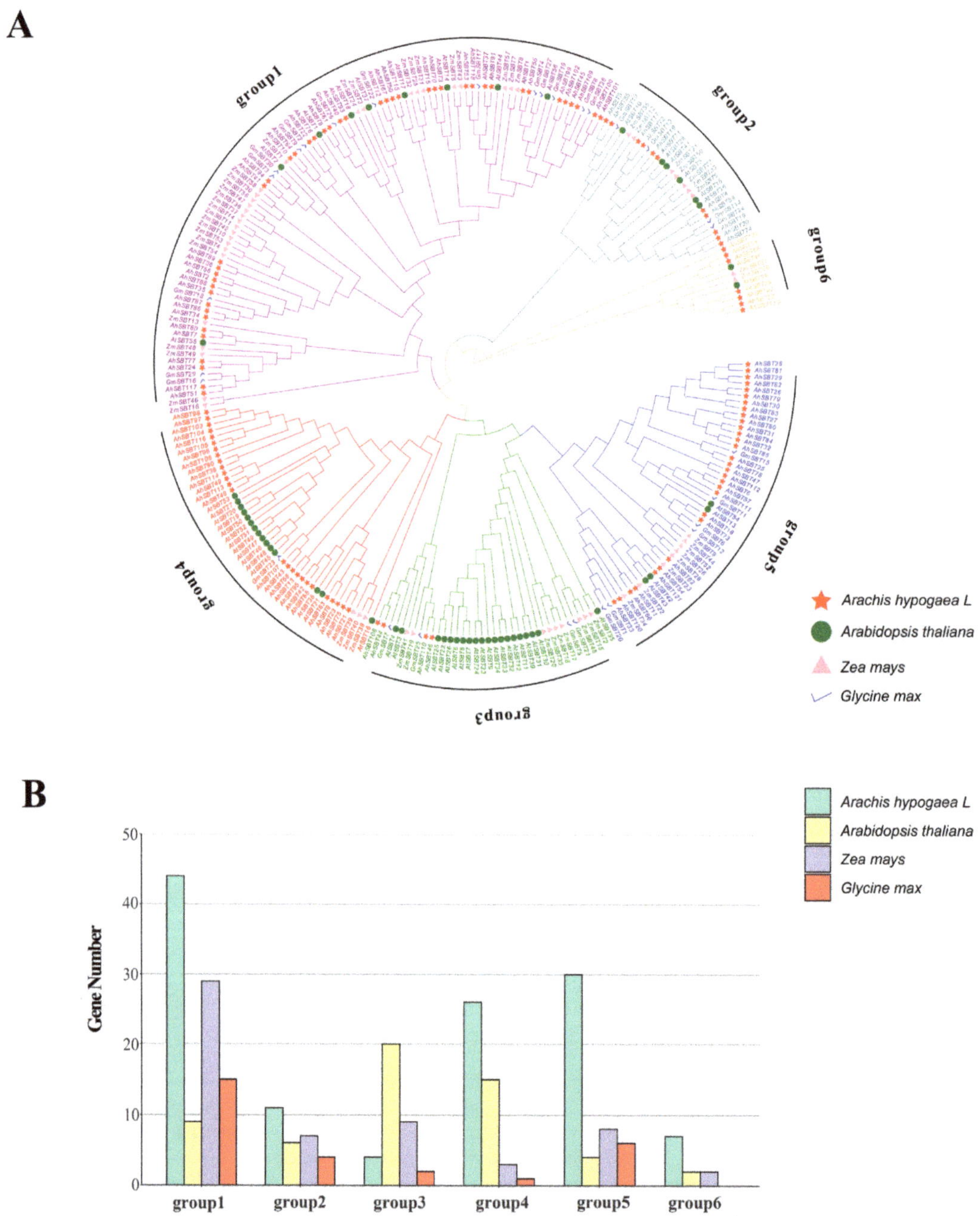

Figure 2. Phylogenetic analysis of *AhSBTs*. (**A**) The phylogenetic tree shows *AhSBTs* were classified into six groups. Shades of colors represent different branches. (**B**) Number of SBT proteins of *A. hypogea*, *Z. mays*, *G. max*, and *A. thaliana* in each group of the phylogenetic tree.

This study utilized protein structure prediction in *A. thaliana* to analyze the structure and function of various SBT subfamilies. One protein of *A. thaliana* from each of the six subgroups was selected to explore structural similarities (Supplementary Figure S2). According to prior research [19], SBT contains the peptidase S8 structural domain, which has three enzymatic active sites and is involved in the digestion of certain substrates. Comparison of structural domains among subfamilies revealed significant differences in the peptidase S8 domain of SBT6 (AtSBT40), suggesting potential functional distinctions from other subfamilies. Notably, only SBT6 members possessed a single peptidase S8 domain.

2.3. Analysis of AhSBTs Protein Conserved Domain and Structure

Using MEME's conserved motif analysis, ten motifs were found among the members of the SBT family (Supplementary Table S2). These motifs were designated motifs 1 through 10. The peptidase_S8 domain (PF00082) was present in motifs 1, 4, and 6, whereas the inhibitor_I9 domain was found in motif 7. The motif distributions among the phylogenetic subgroups exhibited similarities to gene structure (Figure 3A). However, several motifs were discovered to be exclusive to particular genes. Genes, such as *AhSBT69/14/119/52/92/39*, for example, were restricted to motifs 6, demonstrating their tight association as determined by phylogenetic tree, conserved motif analysis, and gene structure. On the other hand, *AhSBT116* had motifs 1, 4, 5, and 6, while *AhSBT122* and *AhSBT62* only included motif 1. With a few exceptions, most motifs were consistently found across nearly all genes (Figure 3A). In conclusion, phylogenetic relationships, conserved motif patterns, and gene structures confirmed the consistency of gene organizations within subgroups. This suggested a well-sustained amino acid deposition among SBT proteins and that SBT members within the same subtree might have corresponding roles.

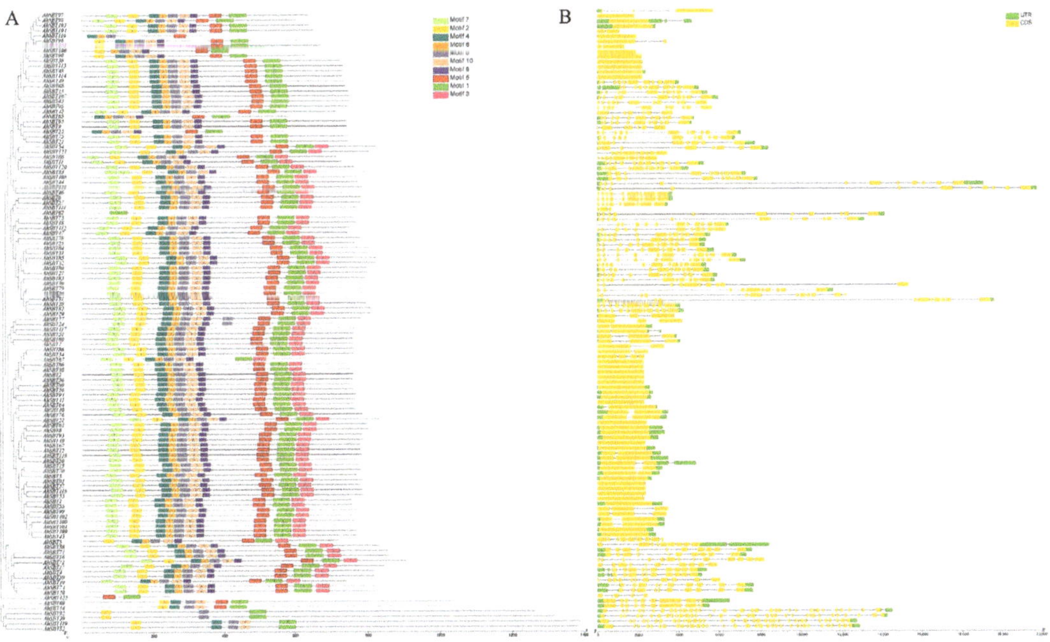

Figure 3. Gene structures and conserved domains of peanut. (**A**) *AhSBTs* evolutionary connection. The conserved domains of *AhSBTs* were shown by various colors in the right column. (**B**) *AhSBTs* structures, green, yellow, and black lines represent UTR regions, exons, and introns, respectively. The bar displays the length of *AhSBTs*.

Divergences in exon–intron architecture and amino acid substitutions can cause differences in coding regions, potentially changing gene function. We examined the exon–intron arrangement of the peanut *SBT* genes to explore the variability in gene structure. Members closely related within a phylogenetic group typically exhibit comparable exon–intron structures. The results showed that exons and introns ranged from 1–34 and 0–33, respectively. The maximum number of exons was present on *AhSBT39*, *AhSBT52*, and *AhSBT119* (Supplementary Table S1). Overall, genes within the same subgroup exhibited similar structures, with a few exceptions (Figure 3B). Among all genes analyzed, *AhSBT39* and *AhSBT92* possessed the most extended structures, while a small number of genes, such as *AhSBT52*, *AhSBT119*, and *AhSBT72*, showed complex structures (Figure 3B). There is no intron in 10 genes (*AhSBT90, AhSBT38, AhSBT113, AhSBT48, AhSBT49, AhSBT88, AhSBT35, AhSBT2, AhSBT56, AhSBT89*), suggesting that gain or loss of exon has occurred in the history of *SBT* genes. Additionally, members within specific subgroups exhibit gene structures close to their phylogenetic clusters.

2.4. Cis-Elements: Crucial Components in the Promoter Regions of AhSBTs

To gain insights into the function of *AhSBTs* in peanut development, growth, and responses to abiotic stressors and plant hormone treatment, the *cis*-regulatory components in *AhSBT* promoters were examined (Supplementary Table S3). We focused on three major categories of *cis*-regulatory elements: phytohormone responsiveness, abiotic stress responsiveness, and development and growth regulation (Figure 4). Among abiotic stress response elements, six primary types were identified, including drought, low temperature, light, defense, stress, and wounding. These elements comprise various motifs, such as ATCT, I-box, GT1, box 4, and GA, with photoresponsive motifs accounting for 90% of the total. In addition, other *cis*-elements were detected, such as MBS (3.6%), TC-rich repetitive sequences (3.2%), LTR (3%), and WUN motifs (0.2%) (Figure 5A,B, and Supplement Table S3). These elements are likely associated with specific genes, indicating their role in defense against stress conditions (Figure 5 and Supplementary Table S3).

Regarding hormone responsiveness, a total of 663 hormone-related elements were detected. Among these, two regulatory elements are linked to methyl jasmonate responsiveness—the CGTCA-motif and TGACG-motif. Two hundred and eight abscisic acid-responsive elements were present in the *SBT* promoter regions. Furthermore, it was shown that 60 gene promoters included a TCA element linked to salicylic acid responses, whereas 61 gene promoters contained a TGA element linked to auxin responses. Additionally, it was shown that the gibberellin-responsiveness-associated GARE-motif, TATC-box, and P-box were present in 36, 27, and 23 promoters (Figure 5C,D, and Supplementary Table S3), respectively. These findings suggest that genes containing specific hormone-related elements could be prioritized for further functional studies to explore their protective effects when treated with plant hormones.

Furthermore, we also identified five elements related to development and growth (regulating the expression of meristems, zein metabolism, control of the cell cycle, expression of endosperm, and circadian rhythm), and their key elements are CAT box (31%), O2-site (34%), GCN4_motif/AACA_motif (17%), MSA-like (5%), and circadian rhythm (14%) (Figure 5E,F and Supplementary Table S3), indicating their active involvement in various phases of peanut development and growth. These results reveal that certain important components are broadly and randomly distributed across multiple genes, while others are specific to particular ones. Consequently, we can infer that the expression patterns of the *AhSBTs* may vary across stages of development, besides various phytohormones and abiotic stressor conditions.

2.5. Tandem Duplication and Synteny Analyses of AhSBTs

Tandem and whole-genome duplications are important processes that increase the complexity of genomes and evolutionary innovation. Upon examining the chromosomal distribution of *AhSBTs*, we discovered they are dispersed randomly throughout 20 chromosomes, with a minimum of one *SBT* gene on each chromosome (Supplementary Table S4). Upon examining gene duplication events, we found that 96 *AhSBTs*, or 78.68% of all *AhSBTs*, were discovered to have a role in gene duplication. Tandem duplications were linked to six genes (4.9%) (Figure 6 and Supplementary Table S4). Notably, three tandemly duplicated *AhSBT* pairs were located on chromosome 16, whereas the remaining 73 *AhSBTs* showed segmental duplications, suggesting that amplification of the peanut *SBT* gene family predominantly results from segmental duplications. Group 1 comprises 24 segmentally duplicated pairs, suggesting that group 1 is expanding at the fastest rate.

Figure 4. Examination of the *AhSBTs* promoters' regions *cis*-elements. Similar colors are used to symbolize *cis*-elements that share a functional similarity.

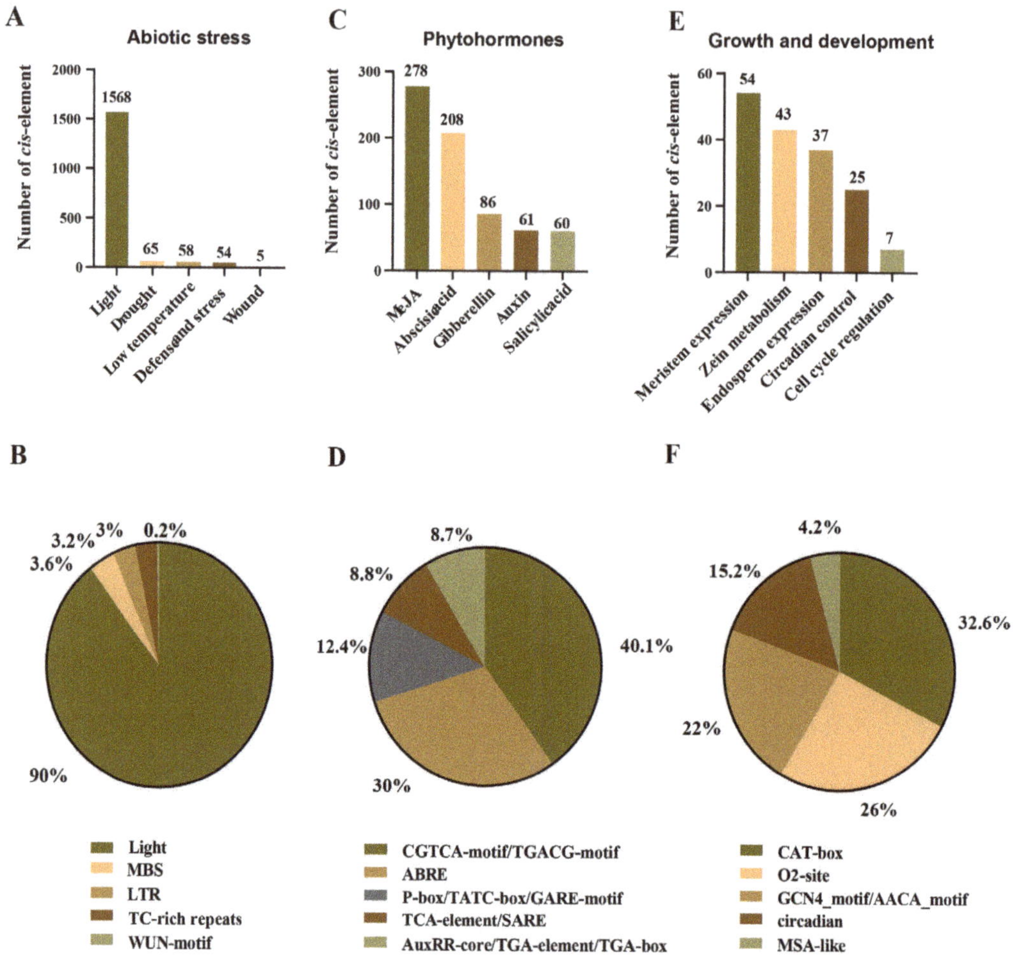

Figure 5. *Cis*-element in the promoter regions of *AhSBTs*. (**A**,**C**,**E**) The total number of elements in *AhSBT* promoters associated with abiotic stress, phytohormones, and growth and development categories, respectively. (**B**,**D**,**F**) Pie charts display the percentage (%) ratio of the various *cis*-elements from each category, such as (**B**) abiotic stress sensitive, (**D**) phytohormones responsive, and (**F**) plant growth and development responsive.

Ka represents the non-synonymous mutation rate, while *Ks* denotes the rate of synonymous mutation, which helps to determine the frequency and timing of duplication. The evolutionary rate of duplication events in the peanut *SBT* gene family was estimated by computing the *Ks* and *Ka* values for each pair of duplicated genes. Two of the three tandem duplication pairs had close *Ks* values of 0.032–0.104, indicating that duplications of these genes occurred approximately 1.97 to 6.38 million years ago (Mya). The 73 pairs of segmental duplications exhibited a wide range of *Ks* values (0.014–1.62), insinuating that duplications occurred between 0.86 to 100.02 Mya (Supplementary Table S4). To further understand the evolutionary pressures acting on *AhSBTs*, we calculated the *Ka/Ks* ratio of each pair. The *KS* values for subfamilies 1, 2, 3, 4, 5, and 6 are 0.0396–1.624, 0.027–0.634, 0.029–0.061, 0.032–1.2, 0.26–0.683, and 0.024–0.584, respectively. Based on the *KS* analysis of the subgenomes, the results show that the Group 1 subfamily has the longest and most variable divergence time, ranging from 2.437 to 100.024 Mya. The Group 3 subfamily has the most recent divergence, with a range of 1.764 to 3.731 Mya. Subfamily Groups 2, 5, and 6 mostly diverged within 1 to 42 Mya, while the Group 4 subfamily diverged between 1.980 and 73.929 Mya (Supplementary Table S4). All duplicates had a *Ka/Ks* ratio smaller than 1, meaning that the protein function of their progenitors is still present in these duplicated subtilase proteins despite being subject to intense purifying selection (Supplementary Table S4).

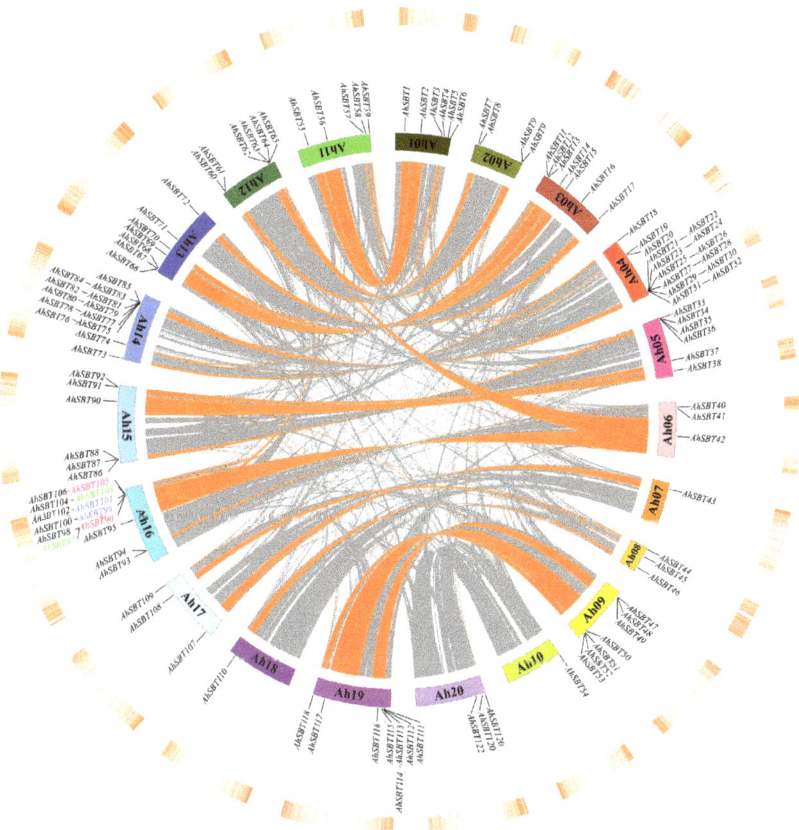

Figure 6. Distribution of *AhSBTs* on chromosomes and gene duplications. The gene density spread on the respective chromosomes is indicated by the outermost circle. Using colored lines, *AhSBTs* inside segmental duplications are connected. Tandem duplications are indicated with different colors. Orange lines show collinearity links among *AhSBTs*.

To comprehend the phylogenetic interplay of *AhSBTs* in peanut and other species, a comparative homologous map of four related genomes (*A. thaliana*, *G. max*, peanut, and *Z. mays*) was created. *AhSBTs* have 17 homologous pairs with *G. max*, five collinear gene pairs in *Arabidopsis*, and one collinear gene pair with *Z. mays* (Figure 7). The greater number of *SBT* pairs of collinear genes between peanut and other leguminous members (*G. max*) compared to distant species like *A. thaliana* or *Z. mays suggest* that the *SBT* gene family is more conserved within legume plants.

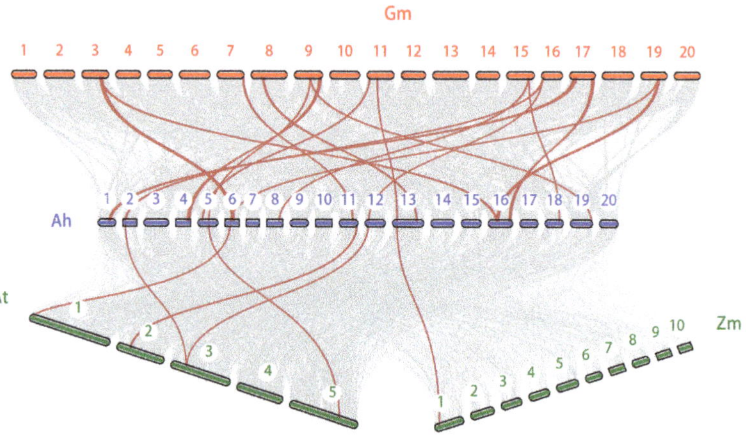

Figure 7. Multiple collinearity assessments of *SBT* genes among *A. thaliana* (At), *G. max* (Gm), *A. hypogea* (Ah), and *Z. mays* (Zm). The red lines indicate the syntenic *SBT* orthologs, whereas the grey lines in the background designate the collinear blocks within peanut and the other three genomes.

2.6. *AhSBTs Enrichment Analysis Using Kyoto Encyclopedia of Genes and Genomics and Gene Ontology*

We performed enrichment analysis using the Kyoto Encyclopaedia of Genes and Genomes (KEGG) pathways and gene ontology (GO) annotation to gain a deeper understanding of the *AhSBTs* at the molecular level (Figure 8 and Supplementary Table S5). The GO annotation revealed significant enrichment in three main categories: molecular function (MF), biological process (BP), and cellular component (CC). In the MF category, the most enriched terms included catalytic activity on proteins (GO:0140096s), hydrolase activity (GO:0016787), peptidase activity (GO:0008233), and serine-type peptidase activity (GO:0004252); serine hydrolase activity (GO:0004175); and general catalytic activity (GO:0003824). In the CC category, the most enriched terms were extracellular region (GO:0005576), exterior plant-type cell wall (GO:0009505), encapsulating structure (GO:0030312), and cell wall (GO:0005618). For the BP category, the highly enriched terms included cellular reaction to protein metabolic process (GO:0019538), proteolysis (GO:0006508), and organonitrogen compound metabolic process (GO:1901564). Several immune-related GO terms were identified, including cellular response to protein metabolic process, proteolysis, organonitrogen cell wall, extracellular region, external encapsulating structure, and compound metabolic process.

Additionally, KEGG enrichment pathway analysis identified three distinct functions of the *AhSBTs* (Figure 8B and Supplementary Table S5). Highly enriched pathways include chaperones and folding catalysts (03110); peptidases and inhibitors (01002); protein families: metabolism (B09181); and protein families: genetic information processing (B09182). GO and enrichment of KEGG validate the functional contributions of *AhSBTs* in various biochemical, molecular, and cellular processes, such as disease resistance, response to stress, and the production of several metabolites. These results illustrate the importance of *AhSBTs* in important biological processes.

2.7. *Exploration of miRNAs Targeting AhSBTs*

To understand the post-transcriptional regulation mediated by miRNAs on *AhSBTs*, we identified 24 miRNAs targeting 58 genes (Supplementary Figure S3 and Table S6). We selected *AhSBT69* and *AhSBT115* to show the representative sketch of miRNA target sites on genes (Figure 9) and provided a complete data set for all miRNA-targeting genes/sites (Supplementary Table S6). The findings indicated that ahy-miR3511-3p and ahy-miR3513-3p had the largest number of targeted genes, with ten genes each. Ahy-miR159 targets seven genes, while ahy-miR3515 and ahy-miR3517 target six genes each. In addition, ahy-miR3513-5p, ahy-miR3518, ahy-miR3519, and ahy-miR156a target five genes each. Notably, ahy-miR3513-398, ahy-miR3514-3p, and ahy-miR167-5p only target one gene each. Some genes, such as *AhSBT69*, *AhSBT115*, *AhSBT119*, *AhSBT72*, *AhSBT39*, and *AhSBT92*, were discovered to be the targets of several miRNAs. To clarify the biological roles of these anticipated miRNAs in the cultivated peanut genome, more study is necessary to validate the expression profiles between these miRNAs and the genes they target.

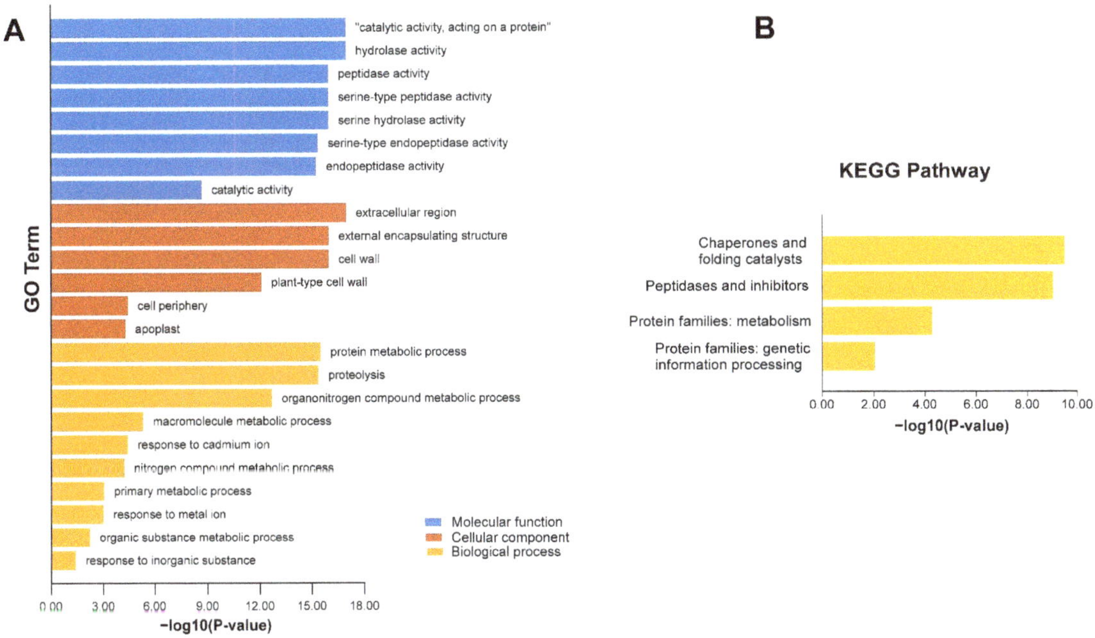

Figure 8. Gene ontology (GO) and KEGG enrichment analysis of AhSBT proteins. (**A**) The extremely rich GO terms in AhSBTs from the MF, CC, and BP categories. (**B**) Highly enriched KEGG pathways in AhSBT proteins.

2.8. Profiling the Expression of AhSBTs in Various Developmental Tissues

We examined the expression of the 122 *AhSBTs* in various peanut tissues (embryo, cotyledon, testa, pericarp, peg, root and stem, root nodule, root tip, root, step tip, stem, leaf). Overall, we found that *AhSBTs* are expressed only in certain specific tissues and not in all tissues (Supplementary Table S6 and Figure 10A). For example, 32 genes, such as *AhSBT86*, *AhSBT34*, *AhSBT90*, *AhSBT58*, *AhSBT35*, and *AhSBT21*, had more significant expression in the embryo than in other tissues. Eight genes were found to have a notable expression in root nodules. Additionally, a few genes also exhibited modest expressions in a variety of tissues. The expression dataset shows that some genes may substantially participate in peanut growth and development. Hence, the functional characterization of *AhSBTs* could potentially be conducted in future research.

2.9. Expression of AhSBTs Under Bacterial Wilt, Hormones, and Abiotic Treatments

To investigate the in-depth contribution of *AhSBTs* in peanut tolerance to abiotic, bacterial wilt, and hormonal stresses, expression levels of these genes were assessed using publicly available transcriptomic datasets (Figure 10B–D). The results show that only a few genes showed higher expressions in cold and drought stresses. For example, *AhSBT112*, *AhSBT97*, and *AhSBT98* showed increased expression levels under drought and cold conditions relative to controls. Some genes like *AhSBT67*, *AhSBT76*, *AhSBT12*, *AhSBT84*, and *AhSBT3* also showed moderate expression levels under cold and normal conditions. On the other hand, *AhSBT42*, *AhSBT115*, *AhSBT92*, *AhSBT25*, *AhSBT39*, *AhSBT51*, and *AhSBT102* displayed considerable expression under cold stress compared to control conditions (Figure 10B).

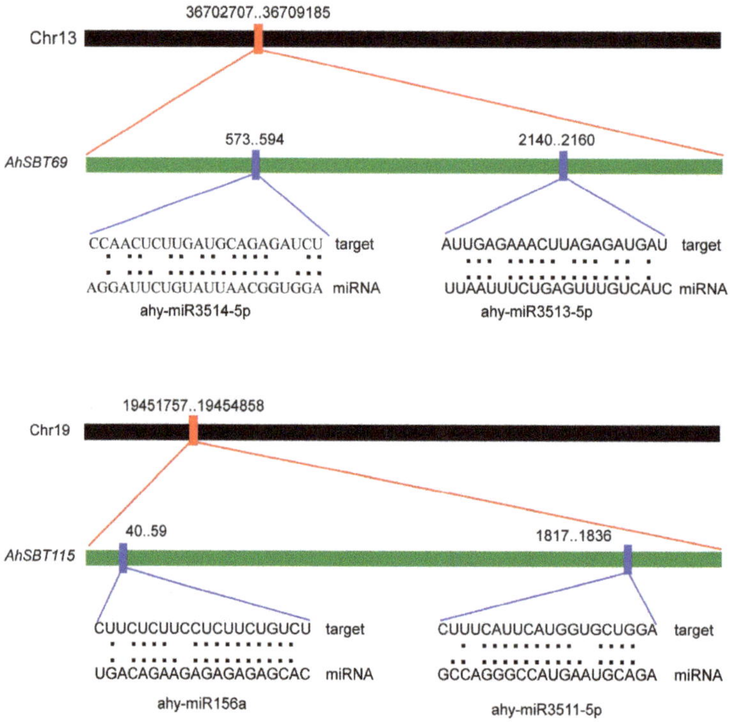

Figure 9. miRNAs targeting *AhSBTs*. The pictorial representation showing the physical target positions of miRNAs (ahy-miR3514-5p and ahy-miR3513-5p) on the *AhSBT69* and the physical target positions of miRNAs (ahy-miR156a and ahy-miR3511-5p) on *AhSBT115*. The black bar shows the chromosome; the red bar shows the gene position on the chromosome. The position of miRNAs on the gene sequence is indicated by the thick blue bar.

Under phytohormone treatments, *AhSBT78*, *AhSBT111*, *AhSBT13*, *AhSBT88*, and *AhSBT44* displayed moderate expression patterns throughout the treatments. However, certain *AhSBTs* displayed specific expression responses to hormone treatments compared to control (CK) conditions. For example, *AhSBT78*, *AhSBT88*, *AhSBT111*, and *AhSBT13* were specifically expressed under ETH treatment; *AhSBT109* and *AhSBT77* under ABA treatment; *AhSBT2* and *AhSBT14* under SA treatment; and *AhSBT23* and *AhSBT42* showed specific expression under PAC treatment (Figure 10B–D). These hormone-specific expression patterns suggest that *AhSBTs* may play diverse regulatory roles in peanut responses to hormonal cues.

Studying the expression levels of *AhSBTs* under bacterial wilt treatment revealed some important findings. Some *AhSBTs*, including *AhSBT39*, *AhSBT50*, *AhSBT92*, *AhSBT104*, and *AhSBT115*, showed a high expression level in disease-resistant and low in highly susceptible varieties. Conversely, other *AhSBTs* (*AhSBT59*, *AhSBT82*, and *AhSBT110*) showed the opposite trend. Interestingly, the expression levels of some *AhSBTs* (*AhSBT98*, *AhSBT81*, and *AhSBT97*) were upregulated in both resistant and susceptible varieties under treatment with *Ralstonia solanacearum*. These findings provide important clues and references for further research on improving peanut resistance to bacterial wilt.

2.10. *Profiling Expression of AhSBTs Under Bacterial Wilt, ABA and Cold Treatment Using qRT-PCR*

We analyzed the gene expression profiles using qRT-PCR under ABA, cold, and *Ralstonia solanacearum* treatment (Supplementary Table S7). Based on the transcriptome data, we selected six genes with specific expression patterns to verify their transcription levels under these treatments at different time points (Supplementary Table S8). Under ABA treatment, almost all genes demonstrated higher expression levels at all time points than CK, excluding a few cases (Figure 11A). Despite the elevated expression levels of *AhSBT1* and *AhSBT53*, which contradicted the transcriptome data, other *AhSBTs* are consistent with the transcriptome data. Similarly, although all the genes were upregulated in response to cold stress, some genes showed relatively low expression levels compared to CK, such as *AhSBT39*, *AhSBT92*, and *AhSBT115*. Whereas the expression levels of *AhSBT1* and *AhSBT53* are higher than CK at 9h and 12h (Figure 11B). In response to *Ralstonia solanacearum*, *AhSBT39*, *AhSBT50*, *AhSBT104*, and *AhSBT115* were downregulated in susceptible varieties and upregulated in resistant varieties. However, *AhSBT92* is upregulated in both susceptible and resistant varieties. (Figure 11C). Overall, the results of our selected genes are in accordance with the transcriptome data, which suggests the credibility of the data.

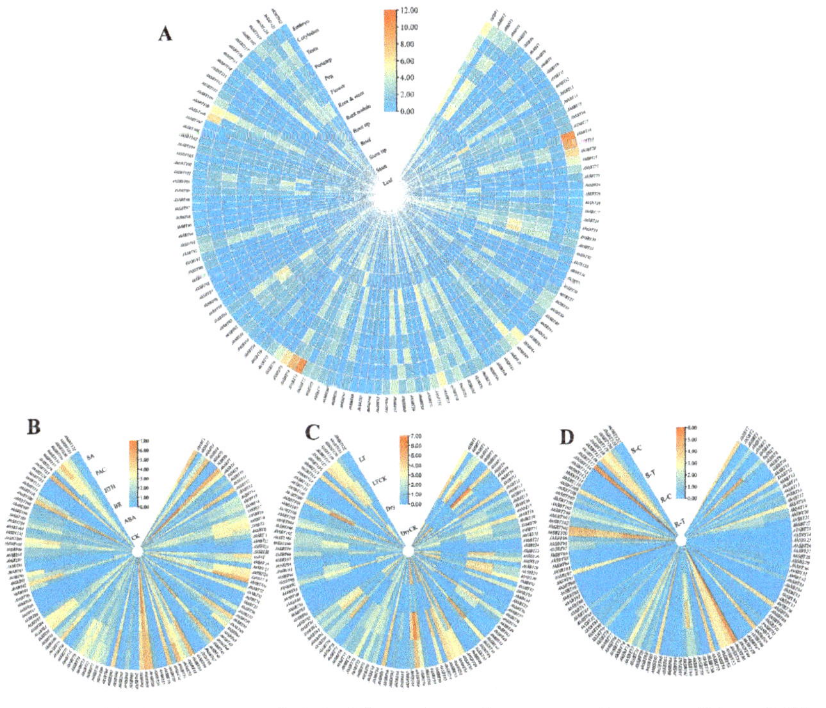

Figure 10. *AhSBTs* expression analysis in different tissues, hormones, and stress conditions. (**A**) The heatmap displays the expression levels of *AhSBTs* in 13 peanut tissues. (**B**) The expression levels of *AhSBTs* following treatment with five hormones. (**C**) The expression levels of *AhSBTs* under drought and low-temperature conditions. (**D**) Expression levels under *R. solanacearum* treatment in highly sensitive and highly resilient varieties to bacterial wilt (R/S-T: Treated resistant and susceptible varieties, R/S-C: Control for resistant/susceptible varieties).

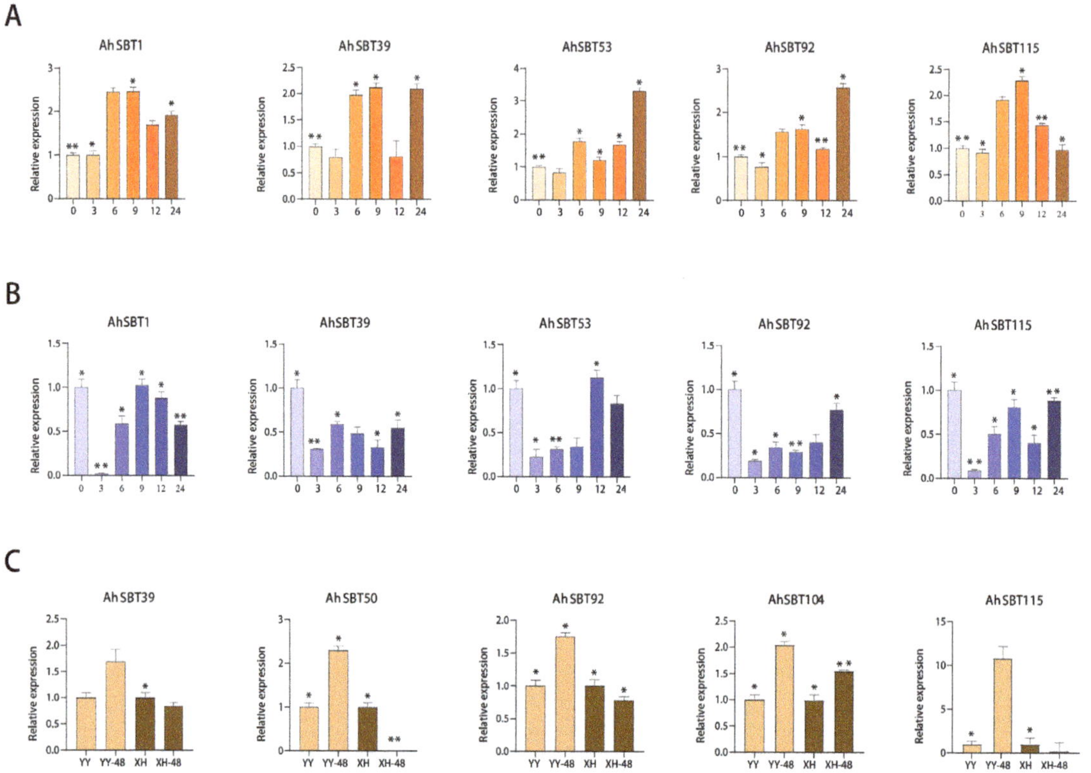

Figure 11. Quantitative expression of *AhSBTs*. The expression levels of representative *AhSBTs* under low temperature (**A**), ABA hormone treatment (**B**), and under *R. solanacearum* infection in varieties with high susceptibility and high resistance to bacterial wilt (**C**) (**: $p < 0.01$, *: $p < 0.05$).

3. Discussion

SBT genes are widespread in numerous bacteria, fungi, plants, and parasites. In plants, the multifunctional role of *SBT* genes is intimately linked to their conserved structural domains, which play a major role in signaling, protein degradation, and development regulation. So far, the *SBT* gene family has been characterized and functionally validated in several plants, including wheat [20], cotton [19] and *Z. mays* [21]. These studies have highlighted the involvement of *SBT* genes in diverse processes, including evolution, growth and development, and defensive responses to biotic and abiotic stimuli. A thorough genome-wide characterization of the *SBT* gene family in the cultivated peanut has not yet been carried out. This leaves a gap in our knowledge of the functions of these genes, particularly in the presence of biotic and abiotic stress conditions. The fully annotated reference genomes of the cultivated peanut cultivar Shitouqi (STQ) have been released [18], which now provides the opportunity to identify all the putative *SBT* genes in peanuts and analyze potential functions.

By identifying *SBT* genes in peanuts, we found that the number of *AhSBTs* (122) is much higher than *ZmSBTs* (58), *AtSBTs* (56), and *GmSBTs* (54). The high number of *AhSBTs* may be attributed to the heterozygous tetraploid nature of peanut (AABB, 2n = 40) in the cultivated species, which underwent a whole-genome duplication and a subsequent gene duplication event [22]. Genome-wide duplications and gene replication events have been established as drivers for expanding gene families in plant evolution [23]. The differences in members of *SBT* genes among different plant species may be attributed to gene duplication events, including tandem and segmental repeats, which play a role in the expansion and variation of *SBT* genses. *SBT* gene duplication has also been found in several other plant species [9,19–21]. Our findings confirmed that *AhSBTs* had undergone segmental duplications (Supplementary Table S4); this suggests that *AhSBTs* duplication events may play an essential role in gene evolution. The synonymous substitution rate (*Ks*) across gene pairs is often used to estimate the time since whole genome replication [24]. The *Ks* values of the *AhSBT* pairs ranged from 0.014 to 1.62, and all had *Ka/Ks* values less than 1.0, indicating that *AhSBTs* had undergone intense purifying selection.

The chromosomal distribution of *AhSBTs* showed non-uniform distribution across chromosomes (Figure 1). The *SBT* genes were grouped into six groups by BLAST and phylogenetic analysis (Figure 2). The recent research on wheat and cotton supports this classification, in which all *SBT* genes are divided into six subfamilies [12,19,20]. The Group 6 members in peanut had the most exons. Gene structure analysis revealed that genes within the same subtree had identical exon–intron patterns, with exons ranging from 1 to 34 and introns from 0 to 33 (Supplementary Table S1). This consistency may be due to selective pressure during peanut evolution [25]. Regarding conserved motifs and exon–intron layout, the *AhSBTs* in each group had almost similar patterns (Figure 3). Also, we predicted 3D protein structure and compared six groups of protein structures in *Arabidopsis* (Supplementary Figure S2). Notably, the sixth subfamily had a unique structure containing only a peptidasea_S8 domain, which showed significant structural differences compared with the other five groups. The peptidase_S8, along with the PA domain, is involved in substrate recognition through protein-protein interactions, facilitating homodimerization and activation of phytochelatinases [26]. In *Glycine max*, the PA domain of chymotrypsin regulates substrate length [27]. The N-terminal inhibitor_I9 domain controls chymotrypsin activity by blocking substrate access to the active site, and to increase chymotrypsin activity, this domain must be excised [27]. Additionally, the FNIII (fibronectin III-like) domain is thought to support enzymatic activity by stabilizing the ring structure near the active site, a finding confirmed in multiple studies [28].

These structural characteristics indicate that the subfamilies of *SBT* genes have undergone distinct evolutionary pathways and play unique biological roles [29]. Our analysis of the *AhSBT* gene family revealed a complex and diverse structure, which likely reflects their broad range of functions. To enhance our comprehension of how *AhSBTs* respond to various environmental factors, we predicted the *cis*-elements in promoters of *AhSBTs*. We found elements related to phytohormones, abiotic stress, and growth and development (Figure 4 and Supplementary Table S3). Recent studies showed that the *cis*-elements within *SBT* genes play a significant role in how plants respond to abiotic and biotic stress [19,29,30]. Moreover, GO enrichment and KEGG analysis further predicted the functional roles of *AhSBTs*, supporting their involvement in stress response and disease resistance (Figure 8). To validate these functions, we analyzed their expression levels under various hormone treatments and biotic and abiotic stresses. The result showed significant *AhSBTs* contributed to ABA, cold, and drought, aligning with the findings from other studies. For instance, in cotton, *GhSBT27A*-silenced plants were more susceptible to drought stress under PEG treatment [9], and *AcoSBTs* were implicated in ABA signaling [19]. Cold stress downregulated the expression of *ZmSBT17*, *ZmSBT18*, and *ZmSBT41* in *Z. mays* [21], further supporting the role of *SBT* genes in plant adaptation to environmental stresses.

Recent reports also suggest that manipulating *SBT* genes could contribute to enhancing the immune response in plants. In *A. thaliana*, overexpressing *SBT* genes can increase plant resistance to pathogens, while loss of *AtSBT3.3* activity can decrease innate immune responses [13]. In wheat, genes including *TaSBT7*, *TaSBT11*, *TaSBT213*, *TaSBT193*, *TaSBT102*, and *TaSBT26* were significantly upregulated during compatibility (CYR31) and incompatibility (CYR23) interactions, suggesting their potential roles in pathogen resistance [20]. Knocking down *TaSBT1.7* using barley stripe mosaic virus-induced gene silencing compromised the hypersensitive response and resistance against *Puccinia striiformis f.* sp. *tritici*, the causal agent of wheat stripe rust [31]. Our results also demonstrate that *SBT* genes play a significant role in the immune response of plants. For instance, *AhSBT39*, *AhSBT50*,

AhSBT92, *AhSBT104*, and *AhSBT115* showed high expression in disease-resistant varieties under *R. solanacearum* treatment. These results were verified by qRT-PCR, which agreed with the transcriptome data, strengthening the reliability of these findings. Overall, this study identified several *AhSBTs* possibly associated with *R. solanacearum*, ABA, and cold stress, laying the groundwork for future research into their specific roles and regulatory mechanisms under bacterial wilt and abiotic stresses.

We utilized publicly available transcriptome data to analyze the tissue-specific expression profiles of 122 *AhSBTs* in various tissues/organs. Overall, 122 *AhSBTs* exhibit diverse organ-specific expression patterns (Figure 10A), and most *AhSBTs* exhibit higher expression levels in underground organs like roots, embryos, and root nodules/tips. Only the leaves exhibited higher expression levels among the above-ground organs or tissues. During the reproductive development of *Arabidopsis*, *AtSBT1.4* was expressed in all above-ground organs. It was shown to down-regulate the number of branching inflorescences and the pace at which seeds were set [6]. In legumes, including *Medicago truncatula* and *Pisum sativum*, SBT1.1 proteins are highly expressed in the endosperm and localized in the endosperm, controlling seed size variation by regulating embryo cell division during reproductive development [32]. In pineapples, six genes (*AcoSBT2.4*, *AcoSBT1.24*, *AcoSBT6.2*, *AcoSBT1.6*, *AcoSBT1.13*, and *AcoSBT1.22*) are expressed in various tissues, particularly in six types of fruit developmental tissues and roots [33]. It can be concluded that *AhSBTs* may play an important role in the process of seed development, and therefore, particular attention should be paid to the *AhSBTs* (*AhSBT58*, *AhSBT72*, *AhSBT122*, and *AhSBT34*) that are specifically expressed in organs related to seed development.

miRNAs are a highly conserved class of molecules that bind precisely to messenger ribonucleic acid (mRNA) targets to suppress post-transcriptional gene expression [34]. Some progress has been made in understanding miRNA function in peanuts [35]. Several miRNAs identified are implicated in stress tolerance and plant development. For example, miRNA156 is engaged in the ABA-miRNA156 interaction, which controls the expression profile of the responsible gene for anthocyanin synthesis under drought stress in plants [36]. Another example is miR159, which regulates plant development, and loss-of-function mutations in miR159, such as those in mir159ab, a double mutant of mir159a and mir159b, enhanced seed drought tolerance and sensitivity to ABA [37]. The target genes of miR160, miR167, and miR393, namely TIR1 and ARF, play a significant role in the response to salt stress [38]. Notably, other miRNA families, including miR3513, miR3518, miR3520, miR3513, and miR3516, have yet to undergo functional characterization. As a result, future research may concentrate on these distinct miRNAs to uncover their potential role in plant development and growth. Furthermore, confirmation of the projected miRNAs' expression patterns and target genes is required to inform the biological significance of these molecules in peanut breeding initiatives.

Based on these findings, we aim to investigate the functional roles of *AhSBT30*, *AhSBT50*, *AhSBT92*, and *AhSBT115* through their heterologous overexpression in *A. thaliana*. This approach will allow us to assess their involvement in plant disease resistance, specifically against *R. solanacearum*, and determine whether overexpression leads to enhanced resistance. Additionally, we will analyze the expression dynamics of these genes and their associated pathways using qRT-PCR. To further elucidate their roles, we will perform subcellular localization studies and protein interaction screening to identify potential interacting partners involved in the plant's stress response mechanisms.

4. Materials and Methods

4.1. Identification of AhSBTs Across the Genome

We downloaded the whole genome data, including the latest protein data for identifying potential *SBT* genes in peanut species [18]. TAIR (https://arabidopsis.org/, accessed on 28 June 2024) database provided the *A. thaliana* SBT protein sequences; Soybase (https://www.soybase.org/, accessed on 28 June 2024) provided the soybean SBT protein sequences; and the ensembl (https://ensembl.gramene.org, accessed on 28 June

2024) provided the *Z. mays* SBT protein sequences. Next, the peptidase-S8 domain's HMM (hidden Markov model) file (PF00082) was obtained from the Pfam database of proteins. The SBT protein sequences in the peanut genome were identified using the HMMER 3.3 program [39]. High standard screening condition E-value < 1×10^{-5} was set to the initial results obtained using the raw peptidase-S8 HMM. Third, the obtained protein sequences were de-duplicated, and only proteins corresponding to the most extended transcript of each gene were retained. The selected proteins were further validated using the NCBI conserved domain database (CDD) (https://www.ncbi.nlm.nih.gov/Structure/cdd/wrpsb.cgi, accessed on 30 June 2024) and PFAM (phmmer search | HMMER (ebi.ac.uk, accessed on 30 June 2024). After removing sequences without typical peptidase-S8 structural domains, 122 proteins containing peptidase-S8 domains were identified as SBT proteins of peanut. The BioPerl tool was used to analyze the physicochemical characteristics, including data like molecular weight, grand average of hydropathicity (GRAVY), and isoelectric points [40]. Lastly, BUSCA software (http://busca.biocomp.unibo.it/ accessed on 30 June 2024) was used to analyze the subcellular localization data of anticipated SBT proteins.

4.2. SBTs Distribution Across Chromosomes, Evolutionary Phylogenetic Analysis, and Covariance Study

Using the MapChart 2.32 software, the chromosomal locations of *AhSBTs* were mapped based on data extracted from the peanut annotation file [41]. Afterwards, the evolutionary relationships of *SBT* genes were analyzed. The sequences of *A. hypogea* (AhSBTs), *A. thaliana* (AtSBTs), *G. max* (GmSBTs), and *Z. mays* (ZmSBTs) proteins were aligned with ClustalW in MEGA7 with default parameters, and the resulting aligned sequences were utilized to construct a neighbor-joining evolutionary tree with 1000 bootstrap replications. At the same time, the nucleotide substitution model was set to p-distance, gap/missing data treatment was set to pairwise deletion, and other settings were set to default parameters. The resulting evolutionary tree was visually depicted using the web-based tool EvolView, available at https://www.evolgenius.info/evolview (accessed on 10 July 2024). To demonstrate the three-dimensional structure of *SBT* genes, one representative protein from each of the six subgroups was selected for modeling using the InterPro database (https://www.ebi.ac.uk/interpro/, accessed on 10 July 2024)).

To determine whether the *AhSBT* genes underwent tandem or segmental duplications, we analyzed the duplication events by MCScanX [42]. To further investigate the evolutionary relationships between the peanut *SBT* gene family and those in *A. thaliana*, *G. max*, and *Z. mays*, covariate analysis was performed, and the results were visualized using Circos 0.69-8 software [43]. The KaKs calculator 3.0 was used to compute nucleotide substitution rates [44]. For that, the newick file for the NG-phylogenetic tree was exported and given as input to the KaKs calculator 3.0 software under the "Phylogenetic Tree section." The Yang–Nilson model was used to calculate the Ka and Ks values. The estimated timing of the duplication event was determined by the formula $T = Ks/2\lambda$ (λ = alike substitutions rate) for cultivated peanut was 8.12^{-9} [45].

4.3. Analysis of Gene Structure and Motifs

Information about coding sequences (CDS), exons, and UTRs of the peanut *SBT* genes was extracted from genome annotation files, and gene structures were visually represented using GSDS 2.0 [46]. MEME 5.5.7 software (https://meme-suite.org/, accessed on 12 July 2024) detected conserved protein motifs in peanut SBT proteins with default parameters for amino acid residues. Finally, these results were displayed graphically by TBtools 1.6 software [47].

4.4. Promoter Analysis of AhSBTs

The *cis*-regulatory elements of *AhSBT* promoters were identified within the 2 kb upstream region from each *AhSBT* transcription start site (ATG). The online tool PlantCARE (http://bioinformatics.psb.ugent.be/webtools/plantcare/html/, accessed on 12 July 2024) was used to identify the *cis*-acting elements, and TBtools 1.6 software was used to visualize them in picture form.

4.5. Prediction of Potential miRNAs Targeting AhSBTs and Assessment of Functional Annotation

The coding sequences (CDS) of the *AhSBTs* were utilized to predict the miRNA target sites through the psRNATarget (https://www.zhaolab.org/psRNATarget/analysis, accessed on 14 July 2024) [48]. An interactive network diagram linking the potential miRNAs and *AhSBTs* was created using Cytoscape 3.9.0 software [49]. Moreover, KEGG and GO annotations were assessed by scanning the proteins through eggNOG v.5.0 software [50]. The findings of KEGG and GO analyses were visualized using TBtools 1.6 software.

4.6. Expression Patterns of the AhSBTs

Expression of *AhSBTs* was assessed in different developmental tissues and organs, such as the testa, root, stem, cotyledon, embryo, pericarp, stem tip, peg, root nodule, root tip, leaf, and flower. The study also investigated their responses to various hormones, including ethylene, abscisic acid, salicylic acid, paclobutrazol, and brassinolide. Furthermore, the expression level of *AhSBTs* was analyzed under biotic and abiotic stressors, including cold, drought, and bacterial wilt. These evaluations were conducted using publicly available transcriptome datasets of cultivated peanuts from the PGR database. Detailed information on harvested samples and data analysis can be consulted in previous studies [18]. To address significant variations in gene expression, we normalized the data by logarithmically transforming FPKM (fragment per kilobase of transcript per million) values to base log2(FPKM + 1). Finally, heat maps were generated using R software, version 4.3.0.

4.7. Plant Material and Stress Conditions

We utilized the cultivated peanut variety "Minhua-6" from southeastern China to investigate stress responses. Viable seeds were planted in small pots containing a vermicompost mixture. Seedlings underwent cold stress at 4 °C and were treated with 10 mg/mL^{-1} of ABA at the four-leaf stage for varying durations, including 3, 6, 9, 12, and 24 h. Each treatment was replicated three times to minimize experimental errors. After treatment, all plant samples were collected, preserved in liquid nitrogen, and stored at −80 °C before RNA extraction. Four-leaf stage plants of highly resistant and susceptible varieties were inoculated with *Ralstonia solanacearum* for biotic stress, and young leaves (0 and 48 h) were sampled. Each set of experiments was conducted in triplicate.

4.8. RNA Extraction and qRT-PCR-Based Expression Analysis

The CTAB method was used to extract the total RNA. Subsequently, from 1 μg of total RNA, cDNA was synthesized using the 5x All-In-One RT MasterMix kit (ABM, Richmond, BC, Canada) as per the producers' instructions. RT-qPCR was conducted on a QuantStudio 5 system (Thermo Fisher Scientific, Waltham, MA, USA) with SYBR TB Green™ Premix (TaKaRa, Dalian, China). This study employed AhActin (housekeeping gene) to ensure gene expression stability. Data normalization was carried out using the $2^{-\Delta\Delta CT}$ method [51]. The qRT-PCR primers are listed in Supplementary Table S2, and GraphPad Prism v9.0.0 was used for graph creation.

5. Conclusions

In this study, we characterized and identified the *SBT* gene family in peanuts. A total of 122 genes were identified as *SBT* gene family members, and a comprehensive bioinformatics analysis, including characterization, evolution, structure, *cis*-elements, GO, KEGG, and miRNA, was conducted for the *AhSBTs*. Promoter *cis*-element analysis revealed key regulatory elements linked to abiotic stress responses and phytohormone signaling, while expression profiling highlighted tissue-specific and stress-induced expression patterns of key *AhSBTs*. These findings contribute significantly to the understanding of the molecular mechanisms underlying stress tolerance and developmental regulation in peanut. In particular, *AhSBT* genes showing strong expression responses to abiotic stresses (e.g., cold and drought), phytohormones (e.g., ABA and SA), and biotic challenges (e.g., *Ralstonia solanacearum* infection) represent promising candidates for further functional validation. Such genes may be exploited in peanut breeding programs to enhance stress resilience and disease resistance. In brief, this study provides a foundation for comprehending the biological roles of *SBT* gene family members under both biotic and abiotic stress. By integrating these insights into breeding pipelines, it may be possible to develop peanut cultivars with improved agronomic traits, ultimately contributing to sustainable crop production and food security.

Supplementary Materials: The following supporting information can be downloaded at https://www.mdpi.com/article/10.3390/ijms252413361/s1.

Author Contributions: Conceptualization, S.L. and Y.Z. (Yuhui Zhuang); methodology, S.L., H.F., L.W., and Y.Z. (Yongli Zhang); software, S.L., H.F., Y.S., and L.W.; formal analysis, S.L., H.F., and Y.Z. (Yongli Zhang); resources, Y.Z. (Yuhui Zhuang); writing—original draft preparation, S.L. and Y.S.; writing—review and editing, S.L., Y.S., S.A., and Y.Z. (Yongli Zhang); supervision, Y.Z. (Yuhui Zhuang); funding acquisition, Y.Z. (Yuhui Zhuang). All authors have read and agreed to the published version of the manuscript.

Funding: This work was supported by grants from the National Natural Science Foundation (NSFC) of China (32401787, U1705233, 32272155, 32072103, 31701463, and 31601337), the National Key Research and Development Program of China (2023YFD1202802 and 2023YFD1202804), the Special Fund for Scientific and Technological Innovation of Fujian Agriculture and Forestry University (KFB23002, KFb22010XA, and KFb22011XA), and the Science and Technology Foundation of Fujian Province of China (2021N5007 and 2017N0006).

Institutional Review Board Statement: Not applicable.

Informed Consent Statement: Not applicable.

Data Availability Statement: Online repositories house the datasets used in this investigation. Accession number(s) and repository names are available at https://www.ncbi.nlm.nih.gov/bioproject/PRJNA480120 (accessed on 30 June 2024).

Conflicts of Interest: There is no conflict of interest among the authors.

Abbreviations

AhSBT	Arachis hypogaea subtilases
Ah	*Arachis hypogaea*
ASP	Aspartic acid
At	*Arabidopsis thaliana*
Chr	Chromosome
FPKM	Fragments per kilobase of transcript per million mapped reads
Gm	*Glycin max*
GO	Gene ontology
HIS	Histidine
HMM	Hidden Markov model
KEGG	Kyoto Encyclopedia of Genes and Genomes
miRNAs	Micro RNAs
Mya	Million years ago
qRT-PCR	Quantitative real-time PCR
SER	Serine
STQ	Peanut cultivar Shitouqi
Zm	*Zea mays*

References

1. Schaller, A. A cut above the rest: The regulatory function of plant proteases. *Planta* **2004**, *220*, 183–197. [CrossRef] [PubMed]
2. Siezen, R.J.; Leunissen, J.A. Subtilases: The superfamily of subtilisin-like serine proteases. *Protein Sci.* **1997**, *6*, 501–523. [CrossRef] [PubMed]
3. Truskina, J.; Brück, S.; Stintzi, A.; Boeuf, S.; Doll, N.M.; Fujita, S.; Geldner, N.; Schaller, A.; Ingram, G.C. A peptide-mediated, multilateral molecular dialogue for the coordination of pollen wall formation. *Proc. Natl. Acad. Sci. USA* **2022**, *119*, e2201446119. [CrossRef] [PubMed]
4. Barrett, A.J.; Rawlings, N.D. Families and clans of serine peptidases. *Arch. Biochem. Biophys.* **1995**, *318*, 247–250. [CrossRef]
5. Schaller, A.; Stintzi, A.; Graff, L. Subtilases—Versatile tools for protein turnover, plant development, and interactions with the environment. *Physiol. Plant.* **2012**, *145*, 52–66. [CrossRef]
6. Rautengarten, C.; Steinhauser, D.; Büssis, D.; Stintzi, A.; Schaller, A.; Kopka, J.; Altmann, T. Inferring hypotheses on functional relationships of genes: Analysis of the Arabidopsis thaliana subtilase gene family. *PLoS Comput. Biol.* **2005**, *1*, e40. [CrossRef] [PubMed]
7. Roberts, I.N.; Veliz, C.G.; Criado, M.V.; Signorini, A.; Simonetti, E.; Caputo, C. Identification and expression analysis of 11 subtilase genes during natural and induced senescence of barley plants. *J. Plant Physiol.* **2017**, *211*, 70–80. [CrossRef]
8. Tripathi, L.P.; Sowdhamini, R. Cross genome comparisons of serine proteases in Arabidopsis and rice. *BMC Genom.* **2006**, *7*, 200. [CrossRef]
9. Dai, M.; Zhou, N.; Zhang, Y.; Zhang, Y.; Ni, K.; Wu, Z.; Liu, L.; Wang, X.; Chen, Q. Genome-wide analysis of the SBT gene family involved in drought tolerance in cotton. *Front. Plant Sci.* **2022**, *13*, 1097732. [CrossRef]
10. Cao, J.; Han, X.; Zhang, T.; Yang, Y.; Huang, J.; Hu, X. Genome-wide and molecular evolution analysis of the subtilase gene family in Vitis vinifera. *BMC Genom.* **2014**, *15*, 1116. [CrossRef]
11. Reichardt, S.; Repper, D.; Tuzhikov, A.I.; Galiullina, R.A.; Planas-Marquès, M.; Chichkova, N.V.; Vartapetian, A.B.; Stintzi, A.; Schaller, A. The tomato subtilase family includes several cell death-related proteinases with caspase specificity. *Sci. Rep.* **2018**, *8*, 10531. [CrossRef] [PubMed]
12. Cui, H.; Zhou, G.; Ruan, H.; Zhao, J.; Hasi, A.; Zong, N. Genome-Wide Identification and Analysis of the Maize Serine Peptidase S8 Family Genes in Response to Drought at Seedling Stage. *Plants* **2023**, *12*, 369. [CrossRef] [PubMed]
13. Ramírez, V.; López, A.; Mauch-Mani, B.; Gil, M.J.; Vera, P. An extracellular subtilase switch for immune priming in Arabidopsis. *PLoS Pathog.* **2013**, *9*, e1003445. [CrossRef] [PubMed]
14. Duan, X.; Zhang, Z.; Wang, J.; Zuo, K. Characterization of a novel cotton subtilase gene GbSBT1 in response to extracellular stimulations and its role in Verticillium resistance. *PLoS ONE* **2016**, *11*, e0153988. [CrossRef]
15. Ekchaweng, K.; Khunjan, U.; Churngchow, N. Molecular cloning and characterization of three novel subtilisin-like serine protease genes from Hevea brasiliensis. *Physiol. Mol. Plant Pathol.* **2017**, *97*, 79–95. [CrossRef]
16. Budič, M.; Sabotič, J.; Meglič, V.; Kos, J.; Kidrič, M. Characterization of two novel subtilases from common bean (*Phaseolus vulgaris* L.) and their responses to drought. *Plant Physiol. Biochem.* **2013**, *62*, 79–87. [CrossRef]
17. Galotta, M.F.; Pugliese, P.; Gutiérrez-Boem, F.H.; Veliz, C.G.; Criado, M.V.; Caputo, C.; Echeverria, M.; Roberts, I.N. Subtilase activity and gene expression during germination and seedling growth in barley. *Plant Physiol. Biochem.* **2019**, *139*, 197–206. [CrossRef]
18. Zhuang, W.; Chen, H.; Yang, M.; Wang, J.; Pandey, M.K.; Zhang, C.; Chang, W.C.; Zhang, L.; Zhang, X.; Tang, R.; et al. The genome of cultivated peanut provides insight into legume karyotypes, polyploid evolution and crop domestication. *Nat. Genet.* **2019**, *51*, 865–876. [CrossRef]
19. Xue, T.; Liu, L.; Zhang, X.; Li, Z.; Sheng, M.; Ge, X.; Xu, W.; Su, Z. Genome-Wide Investigation and Co-Expression Network Analysis of SBT Family Gene in Gossypium. *Int. J. Mol. Sci.* **2023**, *24*, 5760. [CrossRef]
20. Zhao, X.; Goher, F.; Chen, L.; Song, J.; Zhao, J. Genome-Wide Identification, Phylogeny and Expression Analysis of Subtilisin (SBT) Gene Family under Wheat Biotic and Abiotic Stress. *Plants* **2023**, *12*, 3065. [CrossRef]
21. Hou, Q.; Wang, L.; Qi, Y.; Yan, T.; Zhang, F.; Zhao, W.; Wan, X. A systematic analysis of the subtilase gene family and expression and subcellular localization investigation of anther-specific members in maize. *Plant Physiol. Biochem.* **2023**, *203*, 108041. [CrossRef] [PubMed]
22. Bertioli, D.J.; Seijo, G.; Freitas, F.O.; Valls, J.F.; Leal-Bertioli, S.C.; Moretzsohn, M.C.J.P.G.R. An overview of peanut and its wild relatives. *Plant Genet. Resour.* **2011**, *9*, 134–149. [CrossRef]
23. Conant, G.C.; Wolfe, K.H. Turning a hobby into a job: How duplicated genes find new functions. *Nat. Rev. Genet.* **2008**, *9*, 938–950. [CrossRef]
24. Ren, R.; Wang, H.; Guo, C.; Zhang, N.; Zeng, L.; Chen, Y.; Ma, H.; Qi, J. Widespread whole genome duplications contribute to genome complexity and species diversity in angiosperms. *Mol. Plant* **2018**, *11*, 414–428. [CrossRef]
25. Babenko, V.N.; Rogozin, I.B.; Mekhedov, S.L.; Koonin, E.V. Prevalence of intron gain over intron loss in the evolution of paralogous gene families. *Nucleic Acids Res.* **2004**, *32*, 3724–3733. [CrossRef]
26. Rose, R.; Schaller, A.; Ottmann, C. Structural features of plant subtilases. *Plant Signal. Behav.* **2010**, *5*, 180–183. [CrossRef] [PubMed]
27. Tan-Wilson, A.; Bandak, B.; Prabu-Jeyabalan, M. The PA domain is crucial for determining optimum substrate length for soybean protease C1: Structure and kinetics correlate with molecular function. *Plant Physiol. Biochem.* **2012**, *53*, 27–32. [CrossRef]

28. Ottmann, C.; Rose, R.; Huttenlocher, F.; Cedzich, A.; Hauske, P.; Kaiser, M.; Huber, R.; Schaller, A. Structural basis for Ca^{2+}-independence and activation by homodimerization of tomato subtilase 3. *Proc. Natl. Acad. Sci. USA* **2009**, *106*, 17223–17228. [CrossRef]
29. Xu, Y.; Wang, S.; Li, L.; Sahu, S.K.; Petersen, M.; Liu, X.; Melkonian, M.; Zhang, G.; Liu, H. Molecular evidence for origin, diversification and ancient gene duplication of plant subtilases (SBTs). *Sci. Rep.* **2019**, *9*, 12485. [CrossRef]
30. Wang, G.; Ellendorff, U.; Kemp, B.; Mansfield, J.W.; Forsyth, A.; Mitchell, K.; Bastas, K.; Liu, C.-M.; Woods-Tör, A.; Zipfel, C.; et al. A Genome-Wide Functional Investigation into the Roles of Receptor-Like Proteins in Arabidopsis. *Plant Physiol.* **2008**, *147*, 503–517. [CrossRef]
31. Yang, Y.; Zhang, F.; Zhou, T.; Fang, A.; Yu, Y.; Bi, C.; Xiao, S. In Silico Identification of the Full Complement of Subtilase-Encoding Genes and Characterization of the Role of TaSBT1.7 in Resistance Against Stripe Rust in Wheat. *Phytopathology* **2021**, *111*, 398–407. [CrossRef] [PubMed]
32. D'Erfurth, I.; Le Signor, C.; Aubert, G.; Sanchez, M.; Vernoud, V.; Darchy, B.; Lherminier, J.; Bourion, V.; Bouteiller, N.; Bendahmane, A.; et al. A role for an endosperm-localized subtilase in the control of seed size in legumes. *New Phytol.* **2012**, *196*, 738–751. [CrossRef] [PubMed]
33. Jin, X.; Liu, Y.; Hou, Z.; Zhang, Y.; Fang, Y.; Huang, Y.; Cai, H.; Qin, Y.; Cheng, Y. Genome-Wide Investigation of SBT Family Genes in Pineapple and Functional Analysis of AcoSBT1.12 in Floral Transition. *Front. Genet.* **2021**, *12*, 730821. [CrossRef]
34. Jamla, M.; Patil, S.; Joshi, S.; Khare, T.; Kumar, V. MicroRNAs and their exploration for developing heavy metal-tolerant plants. *J. Plant Growth Regul.* **2021**, *41*, 2579–2595. [CrossRef]
35. Zhao, C.; Xia, H.; Cao, T.; Yang, Y.; Zhao, S.; Hou, L.; Zhang, Y.; Li, C.; Zhang, X.; Wang, X. Small RNA and degradome deep sequencing reveals peanut microRNA roles in response to pathogen infection. *Plant Mol. Biol. Rep.* **2015**, *33*, 1013–1029. [CrossRef]
36. Yao, X.; Chen, J.; Zhou, J.; Yu, H.; Ge, C.; Zhang, M.; Gao, X.; Dai, X.; Yang, Z.-N.; Zhao, Y. An essential role for miRNA167 in maternal control of embryonic and seed development. *Plant Physiol.* **2019**, *180*, 453–464. [CrossRef] [PubMed]
37. Waadt, R.; Seller, C.A.; Hsu, P.-K.; Takahashi, Y.; Munemasa, S.; Schroeder, J.I. Plant hormone regulation of abiotic stress responses. *Nat. Rev. Mol. Cell Biol.* **2022**, *23*, 680–694. [CrossRef]
38. Li, Y.; Luo, W.; Sun, Y.; Chang, H.; Ma, K.; Zhao, Z.; Lu, L. Identification and expression analysis of miR160 and their target genes in cucumber. *Biochem. Genet.* **2022**, *60*, 127–152. [CrossRef] [PubMed]
39. Potter, S.C.; Luciani, A.; Eddy, S.R.; Park, Y.; Lopez, R.; Finn, R.D. HMMER web server: 2018 update. *Nucleic Acids Res.* **2018**, *46*, W200–W204. [CrossRef]
40. Stajich, J.E. An Introduction to BioPerl. *Methods Mol. Biol.* **2007**, *406*, 535–548. [CrossRef]
41. Voorrips, R.E. MapChart: Software for the graphical presentation of linkage maps and QTLs. *J. Hered.* **2002**, *93*, 77–78. [CrossRef] [PubMed]
42. Wang, Y.; Tang, H.; Debarry, J.D.; Tan, X.; Li, J.; Wang, X.; Lee, T.H.; Jin, H.; Marler, B.; Guo, H.; et al. MCScanX: A toolkit for detection and evolutionary analysis of gene synteny and collinearity. *Nucleic Acids Res.* **2012**, *40*, e49. [CrossRef] [PubMed]
43. Krzywinski, M.; Schein, J.; Birol, I.; Connors, J.; Gascoyne, R.; Horsman, D.; Jones, S.J.; Marra, M.A. Circos: An information aesthetic for comparative genomics. *Genome Res.* **2009**, *19*, 1639–1645. [CrossRef]
44. Zhang, Z. KaKs_Calculator 3.0: Calculating Selective Pressure on Coding and Non-coding Sequences. *Genom. Proteom. Bioinform.* **2022**, *20*, 536–540. [CrossRef]
45. Bertioli, D.J.; Cannon, S.B.; Froenicke, L.; Huang, G.; Farmer, A.D.; Cannon, E.K.; Liu, X.; Gao, D.; Clevenger, J.; Dash, S. The genome sequences of Arachis duranensis and Arachis ipaensis, the diploid ancestors of cultivated peanut. *Nat. Genet.* **2016**, *48*, 438–446. [CrossRef]
46. Hu, B.; Jin, J.; Guo, A.-Y.; Zhang, H.; Luo, J.; Gao, G. GSDS 2.0: An upgraded gene feature visualization server. *Bioinformatics* **2015**, *31*, 1296–1297. [CrossRef]
47. Chen, C.; Chen, H.; Zhang, Y.; Thomas, H.R.; Frank, M.H.; He, Y.; Xia, R. TBtools: An integrative toolkit developed for interactive analyses of big biological data. *Mol. Plant* **2020**, *13*, 1194–1202. [CrossRef]
48. Dai, X.; Zhuang, Z.; Zhao, P.X. psRNATarget: A plant small RNA target analysis server (2017 release). *Nucleic Acids Res.* **2018**, *46*, W49–W54. [CrossRef] [PubMed]
49. Shannon, P.; Markiel, A.; Ozier, O.; Baliga, N.S.; Wang, J.T.; Ramage, D.; Amin, N.; Schwikowski, B.; Ideker, T. Cytoscape: A software environment for integrated models of biomolecular interaction networks. *Genome Res.* **2003**, *13*, 2498–2504. [CrossRef]
50. Huerta-Cepas, J.; Szklarczyk, D.; Heller, D.; Hernández-Plaza, A.; Forslund, S.K.; Cook, H.; Mende, D.R.; Letunic, I.; Rattei, T.; Jensen, L.J. eggNOG 5.0: A hierarchical, functionally and phylogenetically annotated orthology resource based on 5090 organisms and 2502 viruses. *Nucleic Acids Res.* **2019**, *47*, D309–D314. [CrossRef]
51. Livak, K.J.; Schmittgen, T.D. Analysis of relative gene expression data using real-time quantitative PCR and the 2(-Delta Delta C(T)) Method. *Methods* **2001**, *25*, 402–408. [CrossRef] [PubMed]

International Journal of
Molecular Sciences

MDPI

Article

Foliar Application of Silicon Influences the Physiological and Epigenetic Responses of Wheat Grown Under Salt Stress

Renata Tobiasz-Salach [1,*], Barbara Stadnik [1,2], Marzena Mazurek [3], Jan Buczek [1] and Danuta Leszczyńska [4]

[1] Department of Crop Production, University of Rzeszow, Zelwerowicza 4 St., 35-601 Rzeszow, Poland; barbarast@dokt.ur.edu.pl (B.S.); jbuczek@ur.edu.pl (J.B.)
[2] Doctoral School of the University of Rzeszow, University of Rzeszow, Rejtana 16C St., 35-959 Rzeszow, Poland
[3] Department of Physiology and Plant Biotechnology, University of Rzeszow, Ćwiklińskiej 2 St., 35-601 Rzeszow, Poland; mmazurek@ur.edu.pl
[4] Department of Crops and Yield Quality, Institute of Soil Science and Plant Cultivation—State Research Institute, Czartoryskich 8 St., 24-100 Puławy, Poland; leszcz@iung.pulawy.pl
* Correspondence: rtobiasz@ur.edu.pl

Abstract: Soil salinity is considered a serious problem that limits agricultural productivity. Currently, solutions are being sought to mitigate the negative impact of salt on economically important crops. The aim of the study was to evaluate the effect of foliar application of silicon (Si) on the physiological and epigenetic responses of wheat grown under salt stress conditions. The experiment with wheat seedlings was established in pots with 200 mM NaCl added. After 7 days, foliar fertilizer (200 g L^{-1} SiO_2) was used at concentrations of 0.05, 0.1 and 0.2%. Physiological parameters were measured three times. The addition of salt caused a significant decrease in the values of the measured parameters in plants of all variants. In plants sprayed with Si fertilizer under salinity conditions, a significant increase in CCI and selected gas exchange parameters (P_N, C_i, E, g_s) and chlorophyll fluorescence (PI, RC/ABS, F_V/F_m, Fv/F_0) was observed. Si doses of 0.1 and 0.2% showed a better mitigating effect compared to the dose of 0.05%. The observed effect was maintained over time. The results obtained indicate a positive role for foliar silicon fertilization in mitigating salinity stress in wheat. Epigenetic mechanisms play an important role in regulating gene expression in response to stress. Changes in the status of methylation of the 5′CCGG3′ sequence of the nuclear genome of wheat plants exposed to salinity and treated with Si at different doses were determined by the MSAP approach. The obtained results showed a clear alteration of DNA methylation in plants as a response to experimental factors. The methylation changes were silicon dose-dependent. These modifications may suggest a mechanism for plant adaptation under salt stress after silicon application.

Keywords: chlorophyll fluorescence; DNA methylation; gas exchange; salinity; *Triticum aestivum* L.

1. Introduction

The changing climate negatively affects the productivity of global agriculture, which poses a global threat to food security. The negative effects of environmental factors involve disruption of growth and development of crop plants. Major environmental stresses, such as high salinity, drought, cold, and heat have a damaging impact on plant survival and biomass production. As a result, there is a decrease in the yields of basic food crops by up to 70%. The adverse effect of excess minerals, such as Na^+ and Cl^-, on plants is called salt stress [1–3].

Saline soils are difficult to remediate due to the mobility of Na and Cl ions in the soil, and their spatial variability, which is the effect of interactions between the soil, the agronomic and the climatic environment [4]. More than 6% of the world's surface is affected by salinization, and the problem of saline soil is growing constantly. The high concentration of salt in soil is estimated to cause the exclusion of approximately 1.5 million hectares of soil from production and reduces the potential for agricultural production by up to 46 million hectares per year [5]. The presence of salt in the soil solution reduces the ability of the plant to absorb water and causes osmotic, ionic and oxidative stress [1,6–11]. Excess salinity affects all cellular processes, including disruption of cellular homeostasis, impairment of photosynthesis, mRNA processing, transcription, protein synthesis, disruption of energy metabolism, amino acid biosynthesis, and lipid metabolism [12]. Under saline conditions, decreases in CO_2 assimilation in the Calvin cycle are accompanied by a decrease in the photochemical electron sink, which in the presence of light affects the functioning and efficiency of photosystems [12,13]. In plants under salt stress conditions, osmotic shock occurs. The stomata close, which limits the ability to photosynthesize by reducing the supply of CO_2 [7,12].

Currently, emphasis is placed on finding methods to reduce crop losses caused by various types of stress. One way is to use growth stimulants based on microelements in the form of spraying the above-ground organs of plants. Among the advantages of foliar fertilization, it is worth mentioning that it is a cheaper and more convenient method to use than soil fertilization. Moreover, the use of this type of application does not interfere with the soil environment and allows for quick delivery of nutrients to plants, which can be difficult, e.g., due to soil drought [14,15]. Furthermore, the use of salinity-tolerant cultivars is believed to be the most effective way to combat soil and water salinity [7]. Wheat (*Triticum aestivum* L.) is grown worldwide and currently constitutes a significant basis for human nutrition, and the demand for grain will increase in the coming years [16–18]. Therefore, it is important to look for opportunities to improve the yield of this species under stressful conditions. Silicon (Si) is a trace element that has been recognized as beneficial by the International Institute of Plant Nutrition because of its wide range of positive effects on plants. This element is not essential for the survival of most plant species, but numerous studies indicate that it increases stress tolerance and improves crop productivity [19]. Silicon is the second most abundant element on Earth after oxygen. The soil contains from 1 to 5% silicon in its active form. The form available to plants as a nutrient is mono-silicic acid (H_4SiO_4), and the amorphous form of silicon dioxide (SiO_2) can also be easily taken up by plants. It is believed that the uptake of Si by cereals is higher than in other species [11]. The positive effects of Si in plants are related to the formation of the barrier of the plant cell wall, the regulation of antioxidant enzymes, and participation in the nutrient uptake mechanisms, especially under conditions of osmotic stress. Silicon fertilization can be an agrotechnical treatment that limits the impact of environmental stress and the depletion of nutrients in the soil. Consequently, it can be an alternative to maintaining sustainable agriculture [20]. Numerous studies report a beneficial effect of foliar and soil application of silicon on the yield of crop plants. The use of silicon in different doses improves morpho-physiological parameters and causes changes in the level of plant DNA methylation [21–27]. Photosynthesis converts solar energy into chemical energy, which is the energy source of almost all living things on Earth. Photosynthetic machinery consists of various mechanisms, including gas exchange systems, photosynthetic pigments, photosystems, electron transport systems, carbon reduction pathways, and enzyme systems [24]. Leaves are the main photosynthetic organs of plants. Leaf chlorophyll is an important indicator of nutrient stress, photosynthetic capacity, and plant growth status. Chlorophyll content plays an important role in understanding how photosynthesis functions in plants. Chlorophyll is an important part of the Calvin–Benson cycle and is responsible for light harvesting during photosynthesis, resulting in the excitation of electrons that are used to drive the production of nicotinamide adenine dinucleotide phosphate and chemical energy in the form of adenosine triphosphate [28–32]. Chlorophyll meters have been successfully used to estimate the chlorophyll content in the leaves of various plant species in a nondestructive manner, and particularly to study plant stress physiology [32,33]. Chlorophyll fluorescence, in addition to biochemical methods and gas exchange, is considered a reliable noninvasive technique for evaluating electron transport and related photosynthetic processes [34].

Epigenetic regulations are essential for the growth, development, and reproduction of plants and also play a key role in the generation or acquisition of resistance to various types of environmental stress factors [35–38]. The basis of plant resistance is the ability to regulate gene expression and protein function in a different way [36]. This ability enables epigenetic modification. Epigenetic modifications are changes in gene expression that occur without a change in DNA sequence. It is one of the molecular mechanisms by which plants can silence or super-activate selected DNA templates [38,39]. Epigenetic mechanisms regulation of gene expression can be carried out by DNA methylation, histone modifications, and by non-coding small RNA and chromatin remodeling [35,40]. More important is the fact that some of these changes are inherited by the next generations as part of epigenetic memory [35,41]. DNA methylation involves the addition of a methyl group ($-CH_3$) to cytosine pyrimidine ring (5-methylcytosine, 5 mC or N6-methyldeoxyadenosine, 6 mA) [42,43], and generally occurs at CpG dinucleotides in a symmetric fashion. In this situation, cytosine (C) residue on one CpG is methylated, and the corresponding residue on the complementary strand will also be methylated. This pattern temporarily breaks down during DNA replication, when the unmethylated daughter (nascent) strand and the methylated parent strand create an asymmetrically methylated CpG dyad termed hemi-methylated DNA [43]. The methylation pattern is primarily sustained during DNA replication and cell division by DNA methyltransferases, including maintenance and de novo methylases. As a result, it is possible to faithfully transmit DNA methylation pattern to subsequent generations, which ensures transgenerational epigenetic inheritance [44–46].

Due to the need to expand knowledge on the possibility of increasing the resistance of crops to stress, studies were conducted to investigate the response of wheat seedlings to exogenous silicon application under salinity conditions. The research hypothesis was adopted, which assumes a positive effect of the application of foliar silicon on the physiological parameters of wheat seedlings.

2. Results

2.1. The Influence of Salt Stress and Foliar Si Supply on Chlorophyll Content Index

In the studies carried out, after NaCl application on each measurement date, a decrease in CCI was demonstrated compared to the control (on average 19.1% Date I, 13.2% Date II, and 31.3% Date III) (Figure 1). Exogenous Si application mitigated the negative effects of salt on all of the measured dates. On the first and second dates (Date I and Date II), each applied dose of foliar application caused a significant increase in CCI. The greatest increase was demonstrated after the application of the 0.2% Si dose (Date I—25.5% and Date II—31.3%) compared to plants of the NaCl variant. On the third measurement date (Date III), after applying the 0.2% Si dose, the increase in CCI was lower and amounted to 17.2%. A positive effect of silicon was also demonstrated with Si applications at doses of 0.1 and 0.2%. In process of time, an increase in the chlorophyll content was demonstrated in the control plants (Figure 1). On the second measurement date, plants treated with Si showed higher CCI values compared to the results obtained on the first date (Date I). On the third date (Date III), CCI decreased, except for the 0.2% silicon dose. This relationship indicates a positive effect of Si on the accumulation of CCI in wheat seedlings, especially in the first period after spraying.

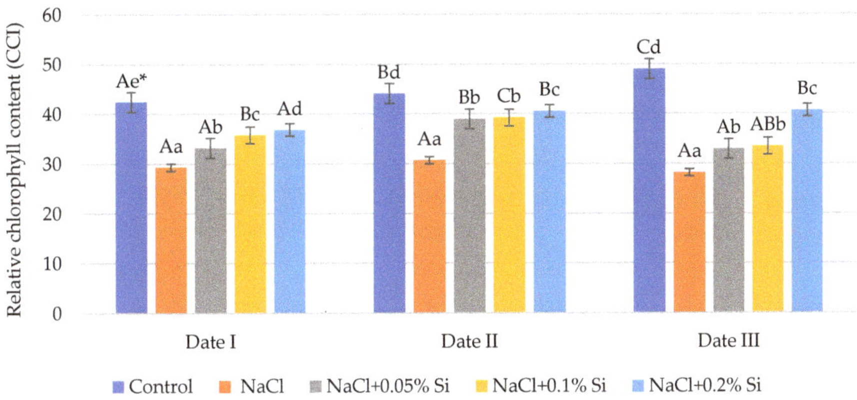

Figure 1. The impact of NaCl, concentrations of Si, and measurement date on chlorophyll content index (CCI) in wheat seedlings (Date I, Date II, Date III—7, 14, 21 days after Si application); data are as mean ± SD values. * Capital letters represent significant differences between means in subsequent measurement dates; lower case letters indicate differences between variants in a particular measurement date ($p \leq 0.05$).

2.2. *The Influence of Salt Stress and Foliar Si Supply on Gas Exchange*

In the experiment conducted, all the gas exchange parameters of the plants (P_N, E, C_i, g_s) were highest in the control. The application of NaCl without addition of Si caused a significant decrease in the values of the analyzed indices (Figure 2). The largest decrease was observed compared to the control in the values of g_s and E (Figure 2C,D). Analysis of variance indicated a beneficial effect of the foliar application of silicon on the parameters analyzed. On each measurement date, the application of silicon in the amount of 0.2% Si increased the values of the analyzed parameters compared to plants without Si application and with soil salinity. A positive effect was observed in all the analyzed indices, such as P_N, C_i, E, and g_s (except for the measurement of g_s on Date III) (Figure 2). On Date I, an increase in the P_N was shown after the application of 0.1% and 0.2% Si doses (18.2% and 26.6%, respectively) compared to the NaCl variant. On Date II, a significant and positive effect of Si was observed at the dose of 0.2% Si, an increase in P_N (by 12.5%) and C_i (by 22.7%) (Figure 2A,B). On the third measurement date (Date III), the effect of silicon was smallest. A positive effect was shown only in the case of C_i after 0.2% Si application (an increase of 10.6%, respectively, compared to plants without application and grown in saline conditions) (Figure 2B).With the passage of measurement time, the values of the P_N index decreased. On Date III, a decrease in the P_N value was observed in all plants exposed to salt and the foliar application of silicon, compared to Date I (Figure 2A). Analyzing the E and C_i indices, on Date II, a slight increase was observed, E (20.4%) and C_i (7.2%), compared to Date I (Figure 2B,C). Only for the silicon dose of 0.2% was a significant increase observed. The g_s value after the application of 0.2% Si was significantly lower at the third measurement date. The parameter decreased by 20% and 33.4% compared to Dates I and II, which was shown by the analysis of variance (Figure 2D).

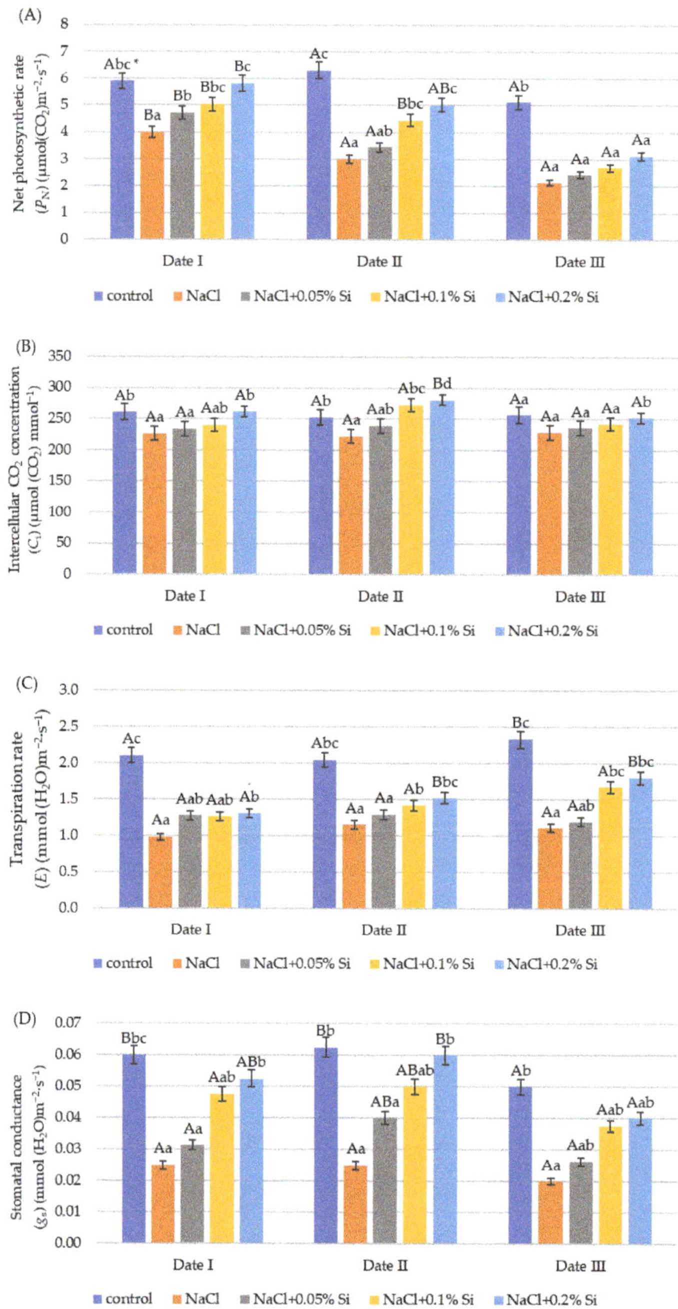

Figure 2. The impact of NaCl, concentrations of Si, and measurement date on gas exchange parameters: net photosynthetic rate (P_N) (**A**) intercellular CO_2 concentration (C_i) (**B**), transpiration rate (E) (**C**), and stomatal conductance (g_s) (**D**) in wheat seedlings (Date I, Date II, Date III—7, 14, 21 days after Si application); data are as mean ± SD values. * Capital letters represent significant differences between means in subsequent measurement dates; lower case letters indicate differences between variants in a particular measurement date ($p \leq 0.05$).

2.3. The Influence of Salt Stress and Foliar Si Supply on Chlorophyll Fluorescence

Salt stress had a deleterious effect on chlorophyll fluorescence parameters (PI, RC/ABS, F_v/F_m, F_v/F_0) (Figure 3). The negative influence of salt was observed primarily in the PI values and the RC/ABS values. In each measurement term, in plants subjected to salt stress without silicon application, the PI and RC/ABS values were significantly reduced. The greatest decrease in values, 37.3% (PI) and 7.4% (RC/ABS), compared to the control was observed on Date III (Figure 3A,B). On this date, a decrease in F_v/F_m by 4% was also noted (Figure 3C). It was shown that the foliar application of silicon had a positive effect on the chlorophyll fluorescence indices. The PI obtained higher values after applying each silicon dose compared to plants of the NaCl variant, but a statistically significant increase (26.0% and 25.6%, respectively) was confirmed at the 0.2% Si dose on the second and third measurement dates. On Date I, the foliar application of Si at levels of 0.1 and 0.2% caused a significant increase in RC/ABS by 24.9% and 31.5% compared to plants treated only with salt. On the second measurement date, the increase was 20.8% and 17.3%, respectively (Figure 3B). On the third measurement date, a positive effect of the exogenous application of Si was noted only at the highest dose. The F_v/F_m indices and F_v/F_0 on the third measurement date increased after the application of each dose of Si, which was not observed on the previous dates. In plants in the NaCl variant, a significant decrease in F_v/F_m was observed with time (Figure 3C,D). The variance analysis showed higher values of the RC/ABS parameter at Date II compared to Date I and Date III. An increase in the F_v/F_0 index was observed in the second and third dates only in control plants (Figure 3).

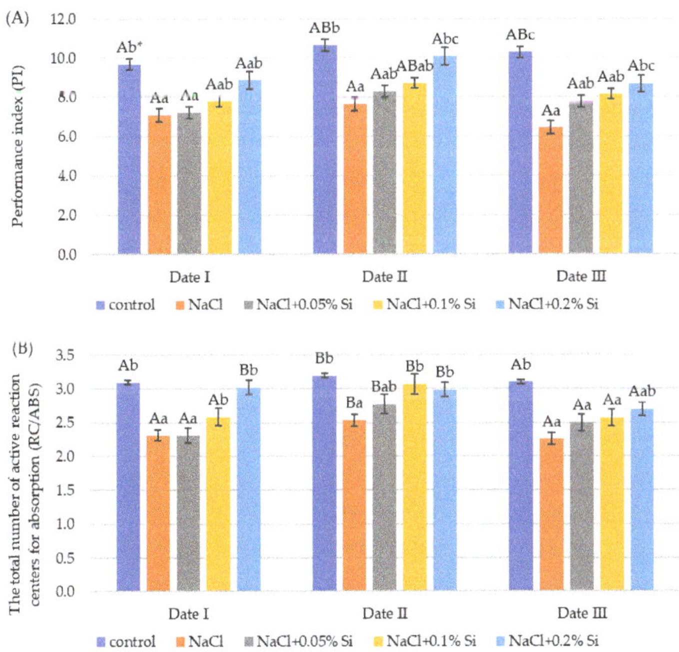

Figure 3. *Cont.*

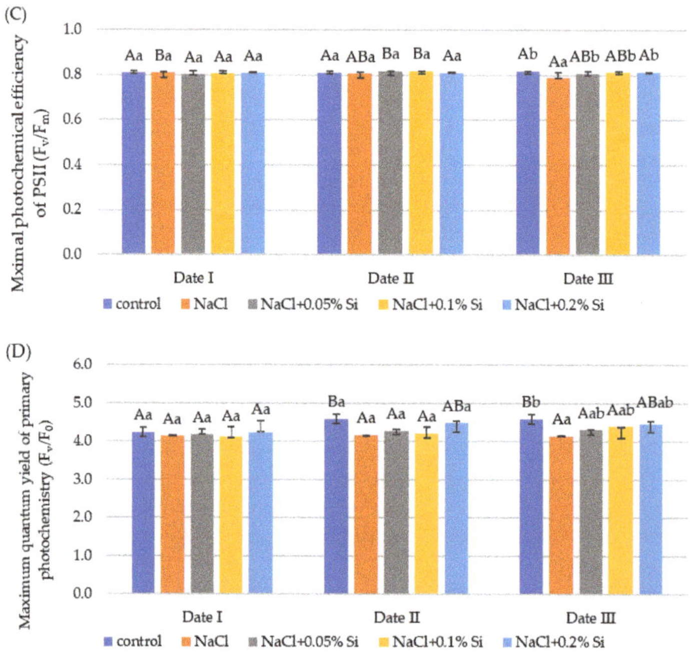

Figure 3. The impact of NaCl, concentrations of Si, and measurement date on chlorophyll fluorescence parameters: performance index (PI) (**A**), total number of active reaction centers for absorption (RC/ABS) (**B**), maximal quantum yield of PSII photochemistry (F_v/F_m) (**C**), and maximum primary photochemistry yield (F_v/F_0) (**D**) in in wheat seedlings (Date I, Date II, Date III—7, 14, 21 days after Si application); data are as mean ± SD values. * Capital letters represent significant differences between means in subsequent measurement dates; lower case letters indicate differences between variants in a particular measurement date ($p \leq 0.05$).

2.4. *The Influence of Salt Stress and Foliar Si Supply on the DNA Methylation Level*

The molecular analysis using MSAP markers indicated different levels of methylation among a group of analyzed plants. The results obtained showed a clear change in DNA methylation in wheat seedlings in response to the experimental factors used. The methylation changes were silicon dose-dependent. Using 5 combinations of primers led to obtained DNA products in a range of 217–231. The total methylation percentage was about 30% in all of the analyzed groups of plants, whereas the predominance of DNA hemi-methylation over symmetric methylation was visible in control as well as in salt-treated plants with or without silicon application (Table 1).

Table 1. Comparison of methylation level among analyzed groups of wheat.

Analysed Parameters:	Control	NaCl	NaCl + 0.05% Si	NaCl + 0.1% Si	NaCl + 0.2% Si
Number of symmetric methylation bands	29	28	25	28	28
Symmetric methylation (%)	13.1	12.9	10.8	12.5	12.5
Number of hemi-methylation bands	42	45	50	44	40
Hemi-methylation (%)	19.0	20.7	21.6	19.8	17.9
Total bands number	221	217	231	224	224
% Total methylation	32.1	33.6	32.5	32.1	30.4

The results of the level of symmetric methylation were similar for most of the analyzed wheat groups and amounted to about 12.5–13.0%. The exception was for plants subjected to NaCl stress and treated with 0.05% silicon spray. For this group, the symmetric methylation was the lowest (10.8%). In the case of hemi-methylation, greater differences among groups of analyzed plants were visible (Table 1). Plants subjected to salt stress showed a higher level of hemi-methylation compared to control. However, silicon foliar application influenced the modification of the level of hemi-methylation. The observed differences were dependent on the dose used. The highest level of hemi-methylation was characterized by wheat plants sprayed with a 0.05% Si dose. The lower level of hemi-methylation indicated occurred in plants treated 0.2% Si, whereas, plants sprayed with a 0.1% Si dose indicated a moderate value of hemi-methylation (Table 1). However, in the case of the 0.05% Si dose, hemi-methylation exceeded the values shown in both the control group and the group subjected to NaCl stress. In turn, the group treated with the highest dose of Si showed the lowest hemi-methylation values (lower than the control group and the group treated only with NaCl). In the general assessment/calculation of total methylation, the differentiation of the overall methylation level was visible also. The highest values were shown in plants subjected to NaCl stress (33.6%), while the lowest was in plants sprayed with 0.2% Si. In the remaining groups tested and the control group, the level of total methylation was similar and amounted to about 32% (Table 1).

3. Discussion

Many studies indicate that salt stress causes negative effects in plants, including physiological and biochemical changes. These are most often manifested by a decrease in plant biomass and yields. The effects of stress depend on plant species and cultivars [12,47–50]. Among various field crops, including cereals, wheat is one of the most sensitive species to salinity, which reduces crop productivity and, at extreme concentrations, even causes complete yield loss [51]. In this study with wheat, we observed changes in plant physiology under soil salinity conditions at a level of 200 mM NaCl. We noted a significant decrease in the CCI after the application of salt at each measurement date. A particularly visible difference was observed between the NaCl variant and the control at Date III. A similar reduction in chlorophyll content under salt stress has been reported in many studies, in which the authors indicate that chlorophyll concentration is used as a sensitive indicator of the metabolic state of the cell [52–55]. The relationship between CCI and chlorophyll was found to be linear in the case of wheat. Studies show that using changes in CCI values provides diagnostic results between simulated and observed crop biomass [56]. In the study by Saddiq et al. [49], the effect of salinity at 0, 100 and 200 mM on *Triticum aestivum* L. genotype seedlings was assessed and the CCI in all genotypes decreased with increasing salt stress level. Similarly, in the studies by Akhtar et al. [56] and Loudari et al. [57], a decrease was reported in CCI in wheat under high NaCl conditions on the substrate. Salinity stress alters the size and number of chloroplasts, the accumulation of starch and lipids, and the organization of the lamellar membrane and disrupts transport across the cell membrane [12,58]. The total chlorophyll content in wheat plants was also significantly reduced by imposed salinity stress in the research by Shah et al. [59]. Zhu et al. [60] investigated the ultrastructure of control wheat chloroplasts stressed with 200 mM NaCl salt using transmission electron microscopy. Under salt stress, the chloroplasts had a slender spindle shape with loosely arranged granum thylakoids. Salinity stress reduces the synthesis of chlorophylls a, b, and carotenoids and also destroys protein bonds, which can lead to a decrease in the light absorption potential of photosynthetic pigments [61]. The negative impact of salt can be alleviated by applying silicon in the foliar application [62–64]. In this pot experiment, the foliar application of silicon fertilizer increased the CCI value compared to plants of the NaCl variant. This relationship was demonstrated in the case of faba beans, cotton, and other plants. In the faba bean, the use of silicon in the form of a foliar spray significantly improved the content of photosynthetic pigments, which were grown both under normal and saline conditions. The best effect was observed after the application of the highest concentration (1000 ppm SiO_2) [23]. Thorne et al. [65] showed that the increase in chlorophyll content was a particularly noticeable result of adding Si to plants subjected to salt stress. The accumulation of chlorophyll varied in species and varieties under the influence of Si. In the study by Nisar et al. [66], foliar application of Si had a positive influence on chlorophyll content in cotton plants under saline conditions and the effect was dependent on genotype. The use of Si increases the content of photosynthetic pigments by reducing Na^+ toxicity and maintaining the proper function and structure of chloroplasts [67–69]. Tuna et al. [70] showed that the chlorophyll content was significantly reduced in wheat cultivars grown under salt stress at 100 mM NaCl, while the application of Si completely restored the chlorophyll content to the level of control plants.

Among the various physiological traits, stomatal conductance and chlorophyll content are of the greatest importance in the search for crops resistant to high salinity levels. It is generally known that salt stress causes a marked reduction in stomatal conductance [7,71]. Photosynthesis depends on the exchange of CO_2 between the atmosphere and the plant through the stomata. In addition to stomata, the photosynthetic CO_2 pathway also results from the transport of CO_2 from the mesophyll conductance to the carboxylation sites in the chloroplast stroma, where RuBisCo occurs to convert CO_2 into the end product of photosynthesis. Gaseous exchange in the plant is impaired when the water potential of the leaves decreases. Plants under salinity stress also accumulate wax in the epidermis to overcome transpiration losses, reducing the gaseous exchange attributes, and thus the photosynthetic process is limited [58].

In this study, a significant decrease in all measured gas exchange parameters was reported. Salinity stress also reduced P_N and g_s in wild-type wheat plants in the study of Zuo et al. [72]. Khoshbakht et al. [73] investigated the effects of salinity on gas exchange and physiological traits of nine citrus species. Gas exchange (P_N, g_s, C_i) was negatively affected by salinity. Photosynthesis, stomatal conductance and photosynthetic pigment concentration were substantially decreased in maize at 68 and 102 mM NaCl in the study by Hichem et al. [74], and reduction in photosynthesis was mainly caused by closure of the stomata and partly by PSII photoinhibition. The functioning of photosystem II (PS II) shows high sensitivity to salt stress [75]. Salt stress causes a decrease in the rate of CO_2 assimilation by reducing both stomatal and root conductance, leading to an excessive reduction of PSII, along with electron diversion to molecular oxygen that generates reactive oxygen species (ROS). ROS are considered to be the main source of structural damage under abiotic stress conditions. ROS are highly cytotoxic and their excess can severely disrupt metabolic pathways [12,49,58,65,76,77]. Reduced photosynthetic efficiency is considered to be the major salt-induced limitation that inhibits plant growth and ultimately crop productivity. In the experiment conducted, an increase in gas exchange parameters was observed after foliar application of Si. This relationship was demonstrated for all plant gas exchange indices, especially for the $NaCl^+$ 0.2% Si variant. An increase in P_N, g_s and E was observed after Si supplementation in sorghum plants subjected to salt stress at a level of 100 mM NaCl [78]. The mitigating role of Si involves reducing Na^+ and Cl^- uptake, increasing mineral nutrition and biosynthesis of various dissolved substances, reducing oxidative stress, modifying gas exchange attributes, and stimulating gene expression [58,62,66,79]. Si may affect stomatal function by increasing their conductance, and therefore the rate of photosynthesis, which may reduce transpiration and thus limit water loss [65]. Many authors report that silicon can improve ROS removal by influencing the antioxidant content of plants, and this effect varies depending on the plant species [79–81].

Chlorophyll fluorescence measurement is a useful indicator for quantifying salt-induced damage to the photosynthetic apparatus [82]. This test was useful in assessing wheat photosynthetic responses to abiotic stresses [83]. Leaf CO_2 assimilation, PSII efficiency, and their relationship allow the use of fluorescence to assess genotype resistance [49]. Photosystem II (PSII) has been identified as the main component of the photosynthetic system sensitive to salt stress. Salinity causes a reduction in key chlorophyll fluorescence parameters, such as F_v/F_m and PI [81]. In our study, we observed a significant reduction in

PI and RC/ABS in plants from the NaCl variant at all measurement dates and F_v/F_m and F_v/F_0 at Date III compared to the control. The reduced ratio F_v/F_m under salinity stress in wheat was also reported in the study by Yadav et al. [84]. Hichem et al. [74] reported that salinity stress caused a decrease in the parameter F_v/F_m in maize, and the observed effect was dependent on the variety. Reduced photosynthetic efficiency under salinity stress conditions can be attributed to biochemical limitations, such as impaired RuBP regeneration, chlorophyll degradation, and chloroplast ATP synthesis [58]. In the conducted research, the best effect of exogenous Si application was observed after the application of the highest dose—0.2%. The positive effect of silicon under stress conditions by improving the F_v/F_m, and F_v/F_0 values was observed in many plant species [85]. Silicon partially reduced the negative effect of NaCl stress, increasing the tolerance of tomato plants to NaCl salinity by increasing the photochemical efficiency of PSII, superoxide dismutase (SOD) and catalase (CAT) activity, chlorophyll content, and PSII photochemical efficiency [25].

One of the epigenetic mechanisms that regulate genome functioning and induce plant resistance and adaptation to abiotic stresses in plant's responses to environmental stresses is DNA methylation [86,87]. The gene body methylation may have an important role in regulating gene expression in an organ- and genotype-specific manner under stress conditions [88]. Molecular mechanisms of plant response to stress are based on the adjustment of gene expression towards, for example, the production of secondary metabolites that influence the development of signals necessary for the activation of defense mechanisms in stress conditions [89]. Modification of gene expression seems to be, therefore, the first stage of plant adaptation to stress conditions leading to the 'coping of plants in stress conditions'. Environmental signals, like salt stress, have been shown to induce epigenetic mechanisms, like DNA methylation and histone modifications, ultimately leading to plant adaptation [35,90]. Analyses performed on numerous plants, like cotton [91], *J. curcac* [92], and maize, as well as cereals like wheat [87] or barley [93], led to confirmation of the changes in DNA methylation pattern in reaction to stress conditions. The role of epigenetic modifications in plants' response to stress has also been researched by comparison of the methylation level in plants resistant and sensitive to a specific stress factor. Zhong et al. [39] showed that CCGG sequences of salt-tolerant wheat plants (Dekang-961 plants) were more methylated than those of salt-sensitive wheat plants (Lumai-15). The scientific reports above indicate that it is reasonable to analyze the level of methylation in the context of plant response and resistance to stress conditions. In the presented research, differences in methylation level were performed in wheat plants treated and untreated with salt stress induced by NaCl. Additionally, the stressed plants treated with silicon differ in percentages of methylation from each other, as well as between control or NaCl-only treated plants. The difference was silicon dose-dependent. Particularly, a visible difference was demonstrated, in the case of hemi-methylation, considered as methylation occurring during the response of plants to stress. Analysis of the white poplar *Populus alba* L. DNA methylation profiles showed that environmental conditions strongly influence internal cytosine hemi-methylation [94]. This suggests that the application of silicon during salt stress generated modification of gene expression in the direction of plants' ability to 'cope' with stress conditions. An effect of this phenomenon is the modification of the physiological and biochemical reactions of wheat plants on stress, presented in the above research.

4. Materials and Methods

4.1. Plant Growth Conditions

At the University of Rzeszów (Poland), research was conducted under controlled conditions. In pots with a diameter of 15 cm, in which 1.5 kg of soil with a grain size typical of loamy sand and a slightly acidic (pH = 5.9) were placed, wheat seeds (10 pieces per pot) of Bogatka cultivar (DANKO Hodowla Roślin Ltd., Choryń, Poland) were sown. The level of macro-elements in soil ($mg \cdot 100\ g^{-1}$) was as follows: P_2O_5 17.1, K_2O 17.0, Mg 8.87, Ca 9.46. The pot experiment was set up in a growth chamber (model GC-300/1000, JEIO Tech Co., Ltd., Seoul, Republic of Korea). The plants were grown for a 14 h day length and a day/night temperature cycle of 21 °C/18 °C. To induce salinity stress in the first leaf pair stage (BBCH 12) [95], commercial salt (NaCl) was added twice a day for two days at an amount of 50 mM to obtain a concentration of 200 mM and to avoid osmotic shock. After 7 days from the application of NaCl solution, silicon (Si) fertilizer Optysil (content 200 g·L^{-1} SiO_2) (Intermag Ltd., Olkusz, Poland) was applied foliarly. The preparation was used in three concentrations: 0.05, 0.1 and 0.2%. The control consisted of plants in pots without the addition of NaCl and Si. The experimental design was a completely randomized factorial design with three replicates. The positions of the pots in the experiment were changed randomly every week. The spraying of Optysil was performed with a handheld laboratory sprayer (Venus Super HD solvent line, Kwazar Corporation Ltd., Jaktorów, Poland). A measurement of 3 mL of solution was used for each pot. Deionized water was added to the control variant. Physiological measurements of wheat seedlings were performed on the first or second fully expanded leaf; 7 (Date I), 14 (Date II) and 21 (Date III) days after silicon fertilizer application.

4.2. Chlorophyll Content Measurement

Chlorophyll content index (CCI) was measured using a portable, nondestructive, and lightweight instrument (CCM-200; Opti-Sciences Inc., Hudson, NH, USA). CCM-200 was adopted to collect CCI values from the first fully expanded functional leaf of each plant. A total of 10 plants were randomly measured in each plot.

4.3. Measurement of Gas Exchange Parameters

The following were measured on two fully developed leaves: net photosynthetic rate (P_N), intercellular CO_2 concentration (C_i), transpiration rate (E) and stomatal conductance (g_s). The LCpro-SD photosynthesis measurement system (ADC Bioscientific Ltd., Herts, UK) was used for measurements. The LCpro-SD leaf photosynthesis chamber has a flow accuracy of ±2%. The light intensity in the measuring chamber was 350 µmol·m^{-2}·s^{-1}, and the temperature was 23 ± 2 °C. Three measurements were taken in each pot.

4.4. Measurement of Chlorophyll Fluorescence Parameters

Chlorophyll fluorescence measurements, i.e., photosynthetic efficiency index (PI), total number of active reaction centers for absorption (RC/ABS), maximum quantum yield of PSII photochemistry (F_v/F_m), and maximum quantum yield of primary photochemistry (F_v/F_0), were performed with the Handy PEA fluorimeter (Hansatech Instruments, King's Lynn, Norfolk, UK). Red actinic light (wavelength at peak 650 nm; spectral line 22 nm) with an intensity of 3500 µmol·m^{-2}·s^{-1} was used for the induction of fluorescence and 1 s of transient fluorescence was recorded. Measurement was carried out in the middle of the leaf blades away from the main leaf vein after dark adaptation (30 min) using leaf clips. Five measurements were taken in each pot.

4.5. Determination of Methylation-Sensitive Amplification Polymorphism (MSAP)

Molecular analysis was performed with Methylation Sensitive Amplification Polymorphism markers (MSAP) by Xiong et al. [96] and Xu et al. [97] with modifications. For molecular analysis, the plant material, e.g., leaves, was collected at the end of experiments and frozen in liquid nitrogen at −80 °C. For all groups of plants, the bulk of leaves were subjected to DNA isolation according to protocol followed by Doyle and Doyle [98]. To perform MSAP, only DNA of high quality (absorbance by λ260/280 in range 1.8–2.0) was used. The MSAP analysis consisted of 4 reactions: (1) restriction, (2) ligation, (3) pre-amplification, and (4) selective amplification. For the restriction enzymes like EcoRI, MspI, and HpaII (Thermo Scientific, Waltham, MA, USA), were used. MspI and HpaII are isoschizomers that recognize the same DNA sequence 5′CCGG3′. The capacity to cleave at the recognized sequence by the MspI and HpaII is dependent on the methylation state of the external or internal cytosine residues. The HpaII is inactive if one or both cytosines are fully methylated (both strands methylated; symmetric methylation) but cleaves the hemi-methylated sequence (only one DNA strand methylated), whereas MspI cleaves methylation 5′CmCGG3′ on both strands of DNA but not 5′mCCGG3′ [96]. In the restriction reaction, 500 ng of genomic DNA was digested with EcoRI/MspI (Thermo Scientific, Waltham, MA, USA) and EcoRI/HpaII (Thermo Scientific, Waltham, MA USA) restriction enzymes for 2 h at 37 °C. The products of the restriction reaction were then subjected to ligation with EcoRI and MspI or HpaII-specific adapters (Genomed, Warsaw, Poland). Subsequently, the ligated DNA was diluted to 1:10 and pre-amplified using EcoRI (Genomed, Warsaw, Poland) and MspI or HpaII (Genomed, Warsaw, Poland) primers with one selective nucleotide at the 3′ ends. The pre-amplified product was diluted 10× with TE buffer and, subsequently, selectively amplified with different combinations of MSAP primers (Genomed, Warsaw, Poland) (Table 2).

Table 2. Sequences of primers and adapters used for reaction and stages of MSAP analysis.

MSAP Stage	Primer/Adapter	Sequence
Ligation	EcoRI-Adapter	5′CTCGTAGACTGCGTACC3′ 3′CATCTGACGCATGGTTAA5′
	MspI-HpaII-Adapter	5′CGACTCAGGACTCAT3′ 3′TGAGTCCTGAGTAGCAG5′
Preamplification	Pre-EcoRI	5′GACTGCGTACCAATTC3′
	Pre-MspI-HpaII	5′GATGAGTCCTGAGTCGG3′
Selective amplification	EcoRI-ACT × MspI/HpaII-CT	5′GACTGCGTACCAATTCACT3′ 5′GATGAGTCCTGAGTCGGCT3′
	EcoRI-AG × MspI/HpaII-CTC	5′GACTGCGTACCAATTCAG3′ 5′GATGAGTCCTGAGTCGGCTC3′
	EcoRI-AC × MspI/HpaII-ATG	5′GACTGCGTACCAATTCAC3′ 5′GATGAGTCCTGAGTCGGATG3′
	EcoRI-AT × MspI/HpaII-CTC	5′GACTGCGTACCAATTCAT3′ 5′GATGAGTCCTGAGTCGGCTC3′
	EcoRI-AT × MspI/HpaII-CAT	5′GACTGCGTACCAATTCAT3′ 5′GATGAGTCCTGAGTCGGCAT3′

The PCR reactions were performed in 30 cycles with the following profile: 94 °C for 120 s pre-denaturation, 30 s denaturation at 94 °C annealing for 30 s, and extension at 72 °C for 60 s, ending to complete extension for 5 min at 72 °C. In the case of pre-amplification, the temperature of annealing was 46 °C, whereas annealing for selective amplification was initiated at 65 °C, and then reduced by 0.7 °C for the 11 cycles. The next cycles (18) were performed with annealing at 56 °C for 30 s.

Selective amplification products were separated by electrophoresis on 6.0% denaturing polyacrylamide sequencing gel. The gel was pre-run for about 30 min before loading the probes and then ran for about 70 min at 60 W. Detection of obtained products was performed with the silver nitrate method [99].

4.6. Methylation Data Analysis

MSAP techniques were performed according to the methodology described by Xu et al. [97]. The MSAP bands' gel template was transformed into a binary character matrix, using "0" and "1" to indicate the presence or absence of bands (PCR product) for particular loci. A symmetric methylation event was observed when bands present in the gel from the reaction EcoRI + MspI (M) were absent from the reaction EcoR I + HpaII (H). In this case, the internal cytosine of 5′CCGG3′ sequence was methylated (5′CmCGG 3′) ('symmetric or fully methylation' defined). Simultaneously, the simultaneous presence of a band in H and the absence in M indicated that the external cytosine of one DNA 5′CCGG3′ sequence strand was methylated (5′mCCGG3′). This is determined as the 'hemi-methylated state'.

Percentage methylation was calculated according to a formula defined by Xiangqian et al. [100]:

$$\text{Methylation (\%)} = \text{(number of methylated bands)/(total number of bands)} \times 100 \quad (1)$$

4.7. Statistical Analysis

The results obtained were statistically analyzed. The package TIBCO Statistica 13.3.0 (TIBCO Software Inc., Palo Alto, CA, USA) was used. The Shapiro–Wilk test was performed at $p = 0.05$. Then, the homogeneity of variance was assessed and a two-way ANOVA with repeated measures was performed (the date of measurement was a factor of the experiment). Tukey's post hoc test was also performed at a significance level of $p \leq 0.05$ [101].

5. Conclusions

For wheat, the most beneficial dose of silicon under salt stress conditions was 0.2% Si. After its application, the highest CCI value was shown in leaves. Wheat plants also obtained higher gas exchange rates: P_N, C_i, E, g_s and chlorophyll fluorescence parameters: PI, RC/ABS F_v/F_0, F_v/F_m, compared to lower applied doses of Si—0.05 and 0.1%. The effect of silicon was more visible after application on the first and second measurement dates compared to that on the third date.

Analyses performed on salt stress-treated wheat indicated modifications in DNA methylation in reaction to stress conditions. Modification of methylation level as an effect of silicon application confirms its impact on plant response and resistance to stress conditions. A better understanding of epigenetic regulation of plant growth and response to environmental stresses may create novel heritable variations for crop improvement.

The results of the research indicate the need to expand research on the positive role of silicon in wheat. This is particularly important in the face of climate change and the intensification of abiotic stress. Future work should focus on examining the role of foliar application of Si in the production of wheat in soils where salinity is a problem. In addition, it is important to examine the interaction of the foliar application of Si and changing weather conditions, which can modify the effects of treatments.

Author Contributions: Conceptualization, R.T.-S. and B.S.; methodology, B.S. and M.M.; formal analysis, B.S.; investigation, R.T.-S., B.S. and M.M.; data curation, R.T.-S.; writing—original draft preparation, R.T.-S., B.S., M.M. and J.B.; writing—review and editing, R.T.-S., B.S. and J.B.; visualization, R.T.-S. and D.L.; supervision, R.T.-S. and D.L. All authors have read and agreed to the published version of the manuscript.

Funding: This research was financed by the financial resources of the Ministry of Science and Higher Education for scientific activities of the Institute of Agricultural Sciences, Land Management and Environmental Protection, University of Rzeszow.

Institutional Review Board Statement: Not applicable.

Informed Consent Statement: Not applicable.

Data Availability Statement: The data presented in this study are available upon reasonable request from the corresponding author.

Conflicts of Interest: The authors declare no conflicts of interest.

References

1. Parihar, P.; Singh, S.; Singh, R.; Singh, V.P.; Prasad, S.M. Effect of salinity stress on plants and its tolerance strategies: A review. *Environ. Sci. Pollut. Res.* **2015**, *22*, 4056–4075. [CrossRef] [PubMed]
2. Munns, R. Genes and salt tolerance: Bringing them together. *New Phytol.* **2005**, *167*, 645–663. [CrossRef] [PubMed]
3. Jadidi, O.; Etminan, A.; Azizi-Nezhad, R.; Ebrahimi, A.; Pour-Aboughadareh, A. Physiological and molecular responses of barley genotypes to salinity stress. *Genes* **2022**, *13*, 2040. [CrossRef] [PubMed]
4. Arzani, A.; Ashraf, M. Smart engineering of genetic resources for enhanced salinity tolerance in crop plants. *Crit. Rev. Plant Sci.* **2016**, *35*, 146–189. [CrossRef]

5. Food and Agriculture Organization. Available online: https://www.fao.org/global-soil-partnership/areas-of-work/soil-salinity/en/ (accessed on 4 November 2024).
6. Arzani, A. Improving salinity tolerance in crop plants: A biotechnological view. *In Vitro Cell. Dev. Biol. Plant.* **2008**, *44*, 373–383. [CrossRef]
7. Munns, R.; Tester, M. Mechanisms of salinity tolerance. *Annu. Rev. Plant Biol.* **2008**, *59*, 651–681. [CrossRef]
8. Safaeighomi, J.; Batooli, H. Determination of bioactive molecules from flowers, leaves, stems and roots of *Perovskia abrotanoides* karel growing in central Iran by nano scale injection. *Dig. J. Nanomater. Biostruct.* **2010**, *5*, 551–556.
9. Negrão, S.; Schmöckel, S.; Tester, M. Evaluating physiological responses of plants to salinity stress. *Ann. Bot.* **2017**, *119*, 1–11. [CrossRef]
10. Ibrahimova, U.; Kumari, P.; Yadav, S.; Rastogi, A.; Antala, M.; Suleymanova, Z.; Ziveak, M.; Tahjib-Ul-Arif, M.; Hussain, S.; Abdelhamid, M.; et al. Progress in understanding salt stress response in plants using biotechnological tools. *J. Biotechnol.* **2021**, *329*, 180–191. [CrossRef]
11. Souri, Z.; Khanna, K.; Karimi, N.; Ahmad, P. Silicon and plants: Current knowledge and future prospects. *J. Plant Growth Regul.* **2021**, *40*, 906–925. [CrossRef]
12. Hameed, A.; Ahmed, M.Z.; Hussain, T.; Aziz, I.; Ahmad, N.; Gul, B.; Nielsen, B.L. Effects of salinity stress on chloroplast structure and function. *Cells* **2021**, *10*, 2023. [CrossRef] [PubMed]
13. Gulzar, S.; Hussain, T.; Gul, B.; Hameed, A. *Photosynthetic Adaptations and Oxidative Stress Tolerance in Halophytes from Warm Subtropical Region*; Springer: Cham, Switzerland, 2020; pp. 1–31. [CrossRef]
14. Laane, H.-M. The effects of foliar sprays with different silicon compounds. *Plants* **2018**, *7*, 45. [CrossRef]
15. Niu, J.; Liu, C.; Huang, M.; Liu, K.; Yan, D. Effects of foliar fertilization: A review of current status and future perspectives. *J. Soil Sci. Plant Nutr.* **2021**, *21*, 104–118. [CrossRef]
16. Indu, S.; Tyagi, B.; Singh, G. Enhancing wheat production-A global perspective. *Indian J. Agric. Sci.* **2015**, *85*, 3–13.
17. Igrejas, G.; Branlard, G. The importance of wheat. In *Wheat Quality for Improving Processing and Human Health*; Springer: Berlin/Heidelberg, Germany, 2020; pp. 1–7. [CrossRef]
18. Szczepanek, M.; Siwik-Ziomek, A.; Lemańczyk, G.; Lamparski, R.; Graczyk, R. Effect of reduced tillage on soil enzyme activity, pests pressure and productivity of organically grown spring wheat species. *Agronomy* **2023**, *13*, 287. [CrossRef]
19. Shivaraj, S.M.; Mandlik, R.; Bhat, J.A.; Raturi, G.; Elbaum, R.; Alexander, L.; Sonah, H. Outstanding questions on the beneficial role of silicon in crop plants. *Plant Cell Physiol.* **2022**, *63*, 4–18. [CrossRef]
20. Kovács, S.; Kutasy, E.; Csajbók, J. The multiple role of silicon nutrition in alleviating environmental stresses in sustainable crop production. *Plants* **2022**, *11*, 1223. [CrossRef]
21. Artyszak, A. Effect of silicon fertilization on crop yield quantity and quality-A literature review in Europe. *Plants* **2018**, *7*, 54. [CrossRef]
22. Ahmad, Z.; Khaliq, A.; Waraich, E.A.; Artyszak, A.; Zaman, Q.U.; Abbasi, A.; Iqbal, M.A.; Alharby, H.F.; Almaghamsi, A.A.; Qamar, M.J.; et al. Exogenously applied silicon and zinc mitigates salt stress by improving leaf pigments and antioxidant activities in canola cultivars. *Silicon* **2023**, *15*, 5435–5444. [CrossRef]
23. Hellal, F.A.; Abdelhameid, M.; Abo-Basha, D.M.; Zewainy, R.M. Alleviation of the adverse effects of soil salinity stress by foliar application of silicon on faba bean (*Vica faba* L.). *J. Appl. Sci. Res.* **2012**, *8*, 4428–4433.
24. Ashraf, M.; Harris, P.J.C. Photosynthesis under stressful environments: An overview. *Photosynthetica* **2013**, *51*, 163–190. [CrossRef]
25. Al-Aghabary, K.; Zhu, Z.; Shi, Q. Influence of silicon supply on chlorophyll content, chlorophyll fluorescence, and antioxidative enzyme activities in tomato plants under salt stress. *J. Plant Nutr.* **2005**, *27*, 2101–2115. [CrossRef]
26. Tobiasz-Salach, R.; Mazurek, M.; Jacek, B. Physiological, Biochemical, and Epigenetic Reaction of Maize (*Zea mays* L.) to Cultivation in Conditions of Varying Soil Salinity and Foliar Application of Silicon. *Int. J. Mol. Sci.* **2023**, *24*, 1141. [CrossRef] [PubMed]
27. Yeşildirek, Y.V.; Arıkan, B.; Çelik, H.; Premkumar, A.; Özden, S.; Kara, N.T. Role of Silicon in Mediating Salt Stress Responses in Arabidopsis Methylation Mutants. *J. Soil Sci. Plant Nutr.* **2024**, *24*, 4471–4482. [CrossRef]
28. Kooistra, L.; Clevers, J.G. Estimating potato leaf chlorophyll content using ratio vegetation indices. *Remote Sens. Lett.* **2016**, *7*, 611–620. [CrossRef]
29. Ali, M.M.; Al-Ani, A.; Eamus, D.; Tan, D.K.Y. Leaf nitrogen determination using non-destructive techniques—A review. *J. Plant Nutr.* **2017**, *40*, 928–953. [CrossRef]
30. Croft, H.; Chen, J.M.; Luo, X.; Bartlett, P.; Chen, B.; Staebler, R.M. Leaf chlorophyll content as a proxy for leaf photosynthetic capacity. *Glob. Chang. Boil.* **2017**, *23*, 3513–3524. [CrossRef]
31. Latifinia, E.; Eisvand, H.R. Soybean physiological properties and grain quality responses to nutrients, and predicting nutrient deficiency using chlorophyll fluorescence. *J. Soil Sci. Plant Nutr.* **2022**, *22*, 1942–1954. [CrossRef]
32. Wang, G.; Zeng, F.; Song, P.; Sun, B.; Wang, Q.; Wang, J. Effects of reduced chlorophyll content on photosystem functions and photosynthetic electron transport rate in rice leaves. *J. Plant Physiol.* **2022**, *272*, 153669. [CrossRef]
33. Gholamin, R.; Khayatnezhad, M. Assessment of the correlation between chlorophyll content and drought resistance in corn cultivars (Zea Mays). *Helix* **2020**, *10*, 93–97. [CrossRef]
34. Kalaji, H.M.; Dąbrowski, P.; Cetner, M.D.; Samborska, I.A.; Łukasik, I.; Brestic, M.; Panchal, B.M. A comparison between different chlorophyll content meters under nutrient deficiency conditions. *J. Plant Nutr.* **2017**, *40*, 1024–1034. [CrossRef]

35. Avni, D.; Ashwani, K.; Arvind, K.; Satish, S.; Mann, A.; Kulshreshtha, N. Current status and future perspectives of epigenetic gene regulation for salt tolerance in wheat. In *Salinity and Drought Tolerance in Plants*; Springer: Berlin/Heidelberg, Germany, 2023; pp. 333–345. [CrossRef]
36. Jing, J.; Chaoqun, X.; Lingyu, S.; Guo, Z.J.; Zhang, L.; Tang, H.C.; Wang, J.C.; Song, S.W.; Liu, J.W.; Zhong, Y.H.; et al. Integrative analysis of transcriptome and metabolome reveal the differential tolerance mechanisms to low and high salinity in the roots of facultative halophyte Avicennia marina. *Tree Physiol.* **2024**, *3*, 44. [CrossRef]
37. Liu, L.; Sheng, L.; Longfei, M.; Yanli, Z.; Tiantian, W.; Jicheng, W.; Xiushuo, L.; Shaowu, X. Analysis of ion transport properties of *Glycine max* HKT transporters and identifying a regulation of GmHKT1;1 by the non-functional GmHKT1, 4. *Plant Cell Physiol.* **2024**, *65*, 1399–1413. [CrossRef] [PubMed]
38. Habu, Y.; Kakutani, T.; Paszkowski, J. Epigenetic developmental mechanisms in plants: Molecules and targets of plant epigenetic regulation. *Curr. Opin. Genet. Dev.* **2001**, *11*, 215–220. [CrossRef]
39. Zhong, L.; Xu, Y.H.; Wang, J.B. DNA-methylation changes induced by salt stress in wheat *Triticum aestivum*. *Afr. J. Biotechnol.* **2009**, *8*, 6201–6207.
40. Grativol, C.; Hemerly, A.S.; Ferreira, P.C. Genetic and epigenetic regulation of stress responses in natural plant populations. *Biochim. Biophys. Acta* **2012**, *1819*, 176–185. [CrossRef]
41. Seem, K.; Kaur, S.; Kumar, S.; Mohapatra, T. Epigenome editing for targeted DNA (de)methylation: A new perspective in modulating gene expression. *Crit. Rev. Biochem. Mol. Biol.* **2024**, *9*, 69–98. [CrossRef]
42. Steward, N.; Ito, M.; Yamakuchi, Y.; Koizumi, N.; Sano, H. Periodic DNA methylation in maize nucleosomes and demethylation by environmental stress. *J. Biol. Chem.* **2002**, *277*, 37741–37746. [CrossRef]
43. Sharif, J.; Koseki, H. Hemimethylation: DNA's lasting odd couple. *Science* **2018**, *359*, 1102–1103. [CrossRef]
44. Talarico, E.; Zambelli, A.; Araniti, F.; Greco, E.; Chiappetta, A.; Bruno, L. Unravelling the epigenetic code: DNA methylation in plants and its role in stress response. *Epigenomes* **2024**, *8*, 30. [CrossRef]
45. Borges, F.; Martienssen, R.A. Establishing epigenetic variation during genome reprogramming. *RNA Biol.* **2013**, *10*, 490–494. [CrossRef] [PubMed]
46. Schmitz, R.J.; Schultz, M.D.; Lewsey, M.G.; O'Malley, R.C.; Urich, M.A.; Libiger, O.; Schork, N.J.; Ecker, J.R. Transgenerational epigenetic instability is a source of novel methylation variants. *Science* **2011**, *334*, 369–373. [CrossRef] [PubMed]
47. Zhao, X.Q.; Lu, Y.T.; Bai, M.X.; Xu, M.X.; Peng, Y.L.; Ding, Y.F.; Zhang, D.Z. Response of maize genotypes with different plant architecture to drought stress. *Acta Pratacult. Sin.* **2020**, *29*, 149–162. [CrossRef]
48. Giordano, M.; Petropoulos, S.A.; Rouphael, Y. Response and defence mechanisms of vegetable crops against drought, heat and salinity stress. *Agriculture* **2021**, *11*, 463. [CrossRef]
49. Saddiq, M.S.; Iqbal, S.; Hafeez, M.B.; Ibrahim, A.M.H.; Raza, A.; Fatima, E.M.; Baloch, H.; Jahanzaib; Woodrow, P.; Ciarmiello, L.F. Effect of salinity stress on physiological changes in winter and spring wheat. *Agronomy* **2021**, *11*, 1193. [CrossRef]
50. Farooq, M.; Rafique, S.; Zahra, N.; Rehman, A.; Siddique, K.H. Root system architecture and salt stress responses in cereal crops. *J. Agron. Crop Sci.* **2024**, *210*, e12776. [CrossRef]
51. El Sabagh, A.; Islam, M.S.; Skalicky, M.; Ali Raza, M.; Singh, K.; Anwar Hossain, M.; Arshad, A. Salinity stress in wheat (*Triticum aestivum* L.) in the changing climate: Adaptation and management strategies. *Front. Agron.* **2021**, *3*, 661932. [CrossRef]
52. Chutipaijit, S.; Cha-um, S.; Sompornpailin, K. High contents of proline and anthocyaninincrease protective response to salinity in *Oryza sativa* L. spp. indica. *Aust. J. Crop Sci.* **2011**, *5*, 1191–1198.
53. Taïbi, K.; Taïbi, F.; Abderrahim, L.A.; Ennajah, A.; Belkhodja, M.; Mulet, J.M. Effect of salt stress on growth, chlorophyll content, lipid peroxidation and antioxidant defence systems in *Phaseolus vulgaris* L. *S. Afr. J. Bot.* **2016**, *105*, 306–312. [CrossRef]
54. Ahmad, Z.; Waraich, E.A.; Akhtar, S.; Anjum, S.; Ahmad, T.; Mahboob, W.; Hafeez, O.B.; Tapera, T.; Labuschagne, M.; Rizwan, M. Physiological responses of wheat to drought stress and its mitigation approaches. *Acta Physiol. Plant.* **2018**, *40*, 1–13. [CrossRef]
55. Liu, C.; Liu, Y.; Lu, Y.; Liao, Y.; Nie, J.; Yuan, X.; Chen, F. Use of a leaf chlorophyll content index to improve the prediction of above-ground biomass and productivity. *PeerJ* **2019**, *6*, e6240. [CrossRef] [PubMed]
56. Akhtar, S.S.; Andersen, M.N.; Liu, F. Residual effects of biochar on improving growth, physiology and yield of wheat under salt stress. *Agric. Water Manag.* **2015**, *158*, 61–68. [CrossRef]
57. Loudari, A.; Mayane, A.; Zeroual, Y.; Colinet, G.; Oukarroum, A. Photosynthetic performance and nutrient uptake under salt stress: Differential responses of wheat plants to contrasting phosphorus forms and rates. *Front. Plant Sci.* **2022**, *13*, 1038672. [CrossRef] [PubMed]
58. Zahra, N.; Al Hinai, M.S.; Hafeez, M.B.; Rehman, A.; Wahid, A.; Siddique, K.H.; Farooq, M. Regulation of photosynthesis under salt stress and associated tolerance mechanisms. *Plant Physiol. Biochem.* **2022**, *178*, 55–69. [CrossRef] [PubMed]
59. Shah, S.H.; Houborg, R.; McCabe, M.F. Response of chlorophyll, carotenoid and SPAD-502 measurement to salinity and nutrient stress in wheat (*Triticum aestivum* L.). *Agronomy* **2017**, *7*, 61. [CrossRef]
60. Zhu, D.; Luo, F.; Zou, R.; Liu, J.; Yan, Y. Integrated physiological and chloroplast proteome analysis of wheat seedling leaves under salt and osmotic stresses. *J. Proteom.* **2021**, *234*, 104097. [CrossRef]
61. Hosseini, S.J.; Tahmasebi-Sarvestani, Z.; Pirdashti, H.; Modarres-Sanavy, S.A.M.; Mokhtassi-Bidgoli, A.; Hazrati, S.; Nicola, S. Investigation of yield, phytochemical composition, and photosynthetic pigments in different mint ecotypes under salinity stress. *Food Sci. Nutr.* **2021**, *9*, 2620–2643. [CrossRef]

62. Abbas, T.; Sattar, A.; Ijaz, M.; Aatif, M.; Khalid, S.; Sher, A. Exogenous silicon application alleviates salt stress in okra. *Hortic. Environ. Biotechnol.* **2017**, *58*, 342–349. [CrossRef]
63. Dhiman, P.; Rajora, N.; Bhardwaj, S.; Sudhakaran, S.S.; Kumar, A.; Raturi, G.; Chakraborty, K.; Gupta, O.P.; Devanna, B.N.; Tripathi, D.K.; et al. Fascinating role of silicon to combat salinity stress in plants: An updated overview. *Plant Physiol. Biochem.* **2021**, *162*, 110–123. [CrossRef]
64. Zhu, Y.X.; Gong, H.J.; Yin, J.L. Role of silicon in mediating salt tolerance in plants: A review. *Plants* **2019**, *8*, 147. [CrossRef]
65. Thorne, S.J.; Hartley, S.E.; Maathuis, F.J. Is silicon a panacea for alleviating drought and salt stress in crops? *Front. Plant Sci.* **2020**, *11*, 1221. [CrossRef] [PubMed]
66. Nisar, S.; Iqbal, M.; Ashraf, J.; Naeem, M.; Ahmad, Z.; Afzal, M.; Raza, A. Enhancing salt tolerance in cotton by improving its morpho-physiological and antioxidant potential through foliar applied silicon. *Silicon* **2022**, *14*, 11243–11252. [CrossRef]
67. Rizwan, M.; Ali, S.; Ibrahim, M.; Farid, M.; Adrees, M.; Bharwana, S.A.; Abbas, F. Mechanisms of silicon-mediated alleviation of drought and salt stress in plants: A review. *Environ. Sci. Pollut. Res.* **2015**, *22*, 15416–15431. [CrossRef]
68. Ibrahim, M.A.; Merwad, A.M.; Elnaka, E.A.; Burras, C.L.; Follett, L. Application of silicon ameliorated salinity stress and improved wheat yield. *J. Soil Sci. Environ. Manag.* **2016**, *7*, 81–91.
69. Singh, P.; Kumar, V.; Sharma, J.; Saini, S.; Sharma, P.; Kumar, S.; Sinhmar, Y.; Kumar, D.; Sharma, A. Silicon supplementation alleviates the salinity stress in wheat plants by enhancing the plant water status, photosynthetic pigments, proline content and antioxidant enzyme activities. *Plants* **2022**, *11*, 2525. [CrossRef] [PubMed]
70. Tuna, A.L.; Kaya, C.; Higgs, D.; Murillo-Amador, B.; Aydemir, S.; Girgin, A.R. Silicon improves salinity tolerance in wheat plants. *Environ. Exp. Bot.* **2008**, *62*, 10–16. [CrossRef]
71. Nawaz, K.; Hussain, K.; Majeed, A.; Khan, F.; Afghan, S.; Ali, K. Fatality of salt stress to plants: Morphological, physiological and biochemical aspects. *Afr. J. Biotechnol.* **2010**, *9*, 5475–5480.
72. Zuo, Z.; Ye, F.; Wang, Z.; Li, S.; Li, H.; Guo, J.; Mao, H.; Zhu, X.; Li, X. Salt acclimation induced salt tolerance in wild-type and chlrophyll b-deficient mutant wheat. *Plant Soil Environ.* **2021**, *67*, 26–32. [CrossRef]
73. Khoshbakht, D.; Ramin, A.A.; Baninasab, B. Effects of sodium chloride stress on gas exchange, chlorophyll content and nutrient concentrations of nine citrus rootstocks. *Photosynthetica* **2015**, *53*, 241–249. [CrossRef]
74. Hichem, H.; El Naceur, A.; Mounir, D. Effects of salt stress on photosynthesis, PSII photochemistry and thermal energy dissipation in leaves of two corn (*Zea mays* L.) varieties. *Photosynthetica* **2009**, *47*, 517–526. [CrossRef]
75. Allakhverdiev, S.I.; Sakamoto, A.; Nishiyama, Y.; Inaba, M.; Murata, N. Ionic and osmotic effects of NaCl-induced inactivation of photosystems I and II in Synechococcus sp. *Plant Physiol.* **2000**, *123*, 1047–1056. [CrossRef] [PubMed]
76. Stepien, P.; Klobus, G. Antioxidant defense in the leaves of C3 and C4 plants under salinity stress. *Physiol. Plant.* **2005**, *125*, 31–40. [CrossRef]
77. Bose, J.; Rodrigo-Moreno, A.; Shabala, S. ROS homeostasis in halophytes in the context of salinity stress tolerance. *J. Exp. Bot.* **2013**, *65*, 1241–1257. [CrossRef]
78. Yin, L.N.; Wang, S.W., Li, J.Y.; Tanaka, K.; Oka, M. Application of silicon improves salt tolerance through ameliorating osmotic and ionic stresses in the seedling of *Sorghum bicolour. Acta Physiol. Plant.* **2013**, *35*, 3099–3107. [CrossRef]
79. Coskun, D.; Britto, D.T.; Huynh, W.Q.; Kronzucker, H.J. The role of silicon in higher plants under salinity and drought stress. *Front. Plant Sci.* **2016**, *7*, 1072. [CrossRef]
80. Kim, Y.H.; Khan, A.L.; Waqas, M.; Lee, I.J. Silicon regulates antioxidant activities of crop plants under abiotic-induced oxidative stress: A review. *Front. Plant Sci.* **2017**, *8*, 510. [CrossRef]
81. Dabravolski, S.A.; Isayenkov, S.V. The physiological and molecular mechanisms of silicon action in salt stress amelioration. *Plants* **2024**, *13*, 525. [CrossRef]
82. Baker, N.R.; Rosenqvist, E. Applications of chlorophyll fluorescence can improve crop production strategies: An examination of future possibilities. *J. Exp. Bot.* **2004**, *55*, 1607–1621. [CrossRef]
83. Sayed, O.H. Chlorophyll fluorescence as a tool in cereal crop research. *Photosynthetica* **2003**, *41*, 321–330. [CrossRef]
84. Yadav, D.S.; Mishra, A.K.; Rai, R.; Chaudhary, N.; Mukherjee, A.; Agrawal, S.B.; Agrawal, M. Responses of an old and a modern Indian wheat cultivar to future O3 level: Physiological, yield and grain quality parameters. *Environ. Pollut.* **2020**, *259*, 113939. [CrossRef]
85. Rastogi, A.; Yadav, S.; Hussain, S.; Kataria, S.; Hajihashemi, S.; Kumari, P.; Brestic, M. Does silicon really matter for the photosynthetic machinery in plants...? *Plant Physiol. Biochem.* **2021**, *169*, 40–48. [CrossRef]
86. Sun, M.; Yang, Z.; Liu, L.; Duan, L. DNA Methylation in plant responses and adaption to abiotic stresses. *Int. J. Mol. Sci.* **2022**, *23*, 6910. [CrossRef] [PubMed]
87. Kumar, S.; Beena, A.S.; Awana, M.; Singh, A. Salt-induced tissue-specific cytosine methylation downregulates expression of HKT genes in contrasting wheat (Triticum aestivum L.) genotypes. *DNA Cell Biol.* **2017**, *36*, 283–294. [CrossRef]
88. Karan, R.; DeLeon, T.; Biradar, H.; Subudhi, K.P. Salt stress induced variation in DNA methylation pattern and its influence on gene expression in contrasting rice genotypes. *PLoS ONE* **2012**, *7*, e40203. [CrossRef] [PubMed]
89. Anjali; Kumar, S.; Korra, T.; Thakur, R.; Arutselvan, R.; Kashyap, S.A.; Nehela, N.; Chaplygin, V.; Minkina, M.; Keswani, C. Role of plant secondary metabolites in defence and transcriptional regulation in response to biotic stress. *Plant Stress* **2023**, *8*, 100154. [CrossRef]

90. Chang, Y.N.; Zhu, C.; Jiang, J.; Zhang, H.; Zhu, J.K.; Duan, C.G. Epigenetic regulation in plant abiotic stress responses. *J. Integr. Plant Biol.* **2020**, *62*, 563–580. [CrossRef] [PubMed]
91. Xue-Lin, L.; Zhong-Xu, L.; Yi-Chun, N.; Xiao-Ping, G.; Xian-Long, Z. MSAP analysis of epigentic changes in cotton (*Gossypium hirsutum* L.) under salt stress. *Acta Agron. Sin.* **2009**, *35*, 588–596. [CrossRef]
92. Mastan, S.G.; Rathore, M.S.; Bhatt, V.D.; Yadav, P.; Chikara, J. Assessment of changes in DNA methylation by methylation-sensitive amplification polymorphism in *Jatropha curcas* L. subjected to salinity stress. *Gene* **2012**, *508*, 125–129. [CrossRef]
93. Stadnik, B.; Tobiasz-Salach, R.; Mazurek, M. Physiological and epigenetic reaction of barley (*Hordeum vulgare* L.) to the foliar application of silicon under soil salinity conditions. *Int. J. Mol. Sci.* **2022**, *23*, 1149. [CrossRef]
94. Guarino, F.; Cicatelli, A.; Brundu, G.; Heinze, B.; Castiglione, S. Epigenetic diversity of clonal white poplar (*Populus alba* L.) populations: Could methylation support the success of vegetative reproduction strategy? *PLoS ONE* **2015**, *10*, e0131480. [CrossRef]
95. Lancashire, P.D.; Bleiholder, H.; Van Den Boom, T.; Langelüddeke, P.; Stauss, R.; Weber, E.; Witzenberger, A. A uniform decimal code for growth stages of crops and weeds. *Ann. Appl. Biol.* **1991**, *119*, 561–601. [CrossRef]
96. Xiong, L.Z.; Xu, C.G.; Saghai Maroof, M.A.; Zhang, Q. Patterns of cytosine methylation in an elite rice hybrid and its parental lines, detected by a methylation-sensitive amplification polymorphism technique. *Mol. Gen. Genet.* **1999**, *261*, 439–446. [CrossRef] [PubMed]
97. Xu, M.; Li, X.; Korban, S.S. AFLP based detection of DNA methylation. *Plant Mol. Biol. Rep.* **2000**, *18*, 361–368. [CrossRef]
98. Doyle, J.J.; Doyle, J.L. Isolation of plant DNA from fresh tissue. *Focus* **1990**, *12*, 13–15.
99. Bassam, B.J.; Gresshoff, P.M. Silver staining DNA in polyacrylamide gels. *Nat. Protoc.* **2007**, *2*, 2649–2654. [CrossRef]
100. Li, X.; Xu, M.; Korban, S.S. DNA methylation profiles differ between field-and in vitro-grown leaves of apple. *J. Plant Physiol.* **2002**, *159*, 1229–1234. [CrossRef]
101. Gomez, K.A.; Gomez, A.A. *Statistical Procedures for Agricultural Research*, 2nd ed.; Wiley Inter Science: New York, NY, USA, 1984; pp. 1–690.

International Journal of
Molecular Sciences

MDPI

Article

Energy Deficiency and Misdistribution Leads to Disrupted Formation in Grain Yield and Rice Quality

Yiding Wang [1,2,†], Guangyan Li [2,3,†], Jiaying Ma [2], Haoran Su [1,2], Wenfei Hu [1,2], Junjiang Lin [1,2], Weimeng Fu [2], Yvxiang Zeng [2], Longxing Tao [2], Guanfu Fu [2], Jie Xiong [1,*] and Tingting Chen [2,*]

1 College of Life Sciences and Medicine, Zhejiang Sci-Tech University, Hangzhou 310018, China; wang_1ding@163.com (Y.W.); suhaoran0412@163.com (H.S.); 15593823808@163.com (W.H.); 18022557884@163.com (J.L.)

2 State Key Laboratory of Rice Biology and Breeding, China National Rice Research Institute, 359 Tiyuchang Road, Hangzhou 310006, China; gyli@yzu.edu.cn (G.L.); majiaying1108@163.com (J.M.); fuwmeng@163.com (W.F.); zengyvxiang@caas.cn (Y.Z.); taolongxing@caas.cn (L.T.); fuguanfu@caas.cn (G.F.)

3 Agricultural College, Yangzhou University, Yangzhou 225009, China

* Correspondence: jiexiong@zju.edu.cn (J.X.); chentingting@caas.cn (T.C.)

† These authors contributed equally to this work.

Abstract: With the progress of society and the improvement of agricultural scientific technology, the single focus on high yield for rice production has gradually shifted to high quality. Coordinated development of grain yield and rice quality has become a core issue for researchers, and the underlying mechanisms remain to be solved. Two varieties, Zhongzheyou1 (ZZY1) and Zhongzheyou8 (ZZY8), were used as study materials under field conditions. The yield of ZZY1 was higher than that of ZZY8, which was mainly characterized by a higher seed-setting rate and grain weight. The rice quality of ZZY8 was better than that of ZZY1, primarily due to lower chalkiness and a higher head rice rate. The total dry matter weight of ZZY1 was lower than that of ZZY8, but the proportion of panicle dry matter weight or nonstructural carbohydrate to the total in the former was higher than that of the latter. The maximum grain-filling rate, average grain-filling rate, and key enzyme activities of ZZY1 were significantly higher than those of ZZY8, while the active grain-filling period was shorter than that of ZZY8. Furthermore, the ATP/ATPase content and energy charge values in the grains of ZZY1 were higher than those of ZZY8 at the early grain-filling stage. Transcriptome analysis showed that carbohydrate and energy metabolism were the main ways affecting the yield and quality of the two varieties. The energy production of ZZY1 was insufficient to simultaneously supply the needs thus leading to the discordant formation in its grain yield and rice quality formation.

Keywords: grain yield; rice quality; carbohydrate accumulation; energy metabolism

1. Introduction

China is the country with the largest rice demand in the world, but the impact of increasing population, decreasing land, and agricultural resource shortages on rice production is increasing. According to the "National Population Development Plan (2016–2030)", China's population is expected to reach a peak of 1.45 billion in 2030. Therefore, in order to ensure China's food security, it is necessary to significantly increase rice yield on the existing basis [1–3]. The yield of rice is composed of panicle number, grains per panicle, seed setting rate, and grain weight, which restrict each other and jointly determine the final yield level [4]. Efforts have been made to improve grain yield, and a large amount of work on the physiological foundation for high yield in rice production has also been done by previous researchers [5–8].

In the past, the goal of rice production in our country was mainly high yield in order to ensure the basic needs of people's lives, while the pursuit of high-quality rice started later. Generally, high-quality rice varieties often have the problems of low yield and poor adaptability, and the contradiction between "high yield" and "high quality" still exists widely [9,10]. As the economy develops and the social population's living standards steadily improve in recent years, people's demand for high-quality rice has increased. The high-quality rice production has developed rapidly under the condition that the yield basically meets the demand due to the continuous growth of rice yield since the 21st century [11,12]. The number of high-quality rice varieties has increased significantly, and rice production has been optimized. Among the ten rice varieties with the largest planting area in China in 2020, eight of them were high-quality varieties. Therefore, the goal of rice production at the current stage should be to improve rice yield and quality through variety selection and optimization of cultivation and management measures to achieve the coordinated development of grain yield and rice quality [13–15]. The focus should be on further improving the rice quality of high-yield rice varieties.

Some research results have been acquired on grain yield performance and rice quality changes under optimized agronomic techniques [6,13–15]. Studies have shown that the characteristics of high-yield and high-quality japonica rice varieties mainly showed a grain yield range of 8.35~9.16 t hm^{-2} and a food taste score range of 60~74 [16]. Research has also demonstrated that japonica rice varieties with excellent eating taste and high yield were characterized by large panicles with a large quantity of spikelets, low grain weight, and strong dry matter and nitrogen accumulation ability in the middle and late growth periods [17]. These varieties were also characterized by high leaf photosynthetic rate, high nitrogen absorption rate, and high activities of key enzymes involved in carbon and nitrogen metabolism, such as sucrose phosphate synthase, sucrose synthase, glutamate synthase, and glutamine synthase, all of which would promote the efficient transfer and reuse of carbon and nitrogen from the vegetative organs to the panicle [18]. However, the obstacle to synergistically improving grain yield and the underlying physiological mechanisms were unclear.

Energy is the foundation of all living activities, and the growth and development of all organisms depend on a constant mechanism of energy transfer [19]. Plants can obtain a significant amount of ATP through photosynthesis in their leaves [20]. The energy in the seeds mainly comes from the decomposition of assimilates unloaded from the source end through cellular respiration, producing ATP and other forms of energy for cellular metabolism. Previous studies have shown that the process of rice yield formation is one of dry matter accumulation and transport, and a substantial amount of energy is consumed during material transport [21–23]. Insufficient energy will inhibit pollen germination and pollen tube elongation under heat stress, hinder assimilate transport and metabolism, lead to reduced seed setting, inadequate grain filling, and ultimately constrain yield and quality formation [24]. Generally, it can be seen that the formation of crop yield and quality is a process that requires high energy consumption, which could pertinently explain the contradictoriness in simultaneously obtaining high yield and high quality. We supposed that energy deficiency was the main reason that led to the difficulty in collaborative enhancement for grain yield and rice quality. The assumption was verified in this current study by using two experimental materials, Zhongzheyou1 (ZZY1, higher grain yield and poorer rice quality) and Zhongzheyou8 (ZZY8, lower grain yield and better rice quality). Grain yield and rice quality were compared, and the related processes such as dry matter accumulation and carbohydrate metabolism were observed. The role of energy metabolism in the generation, utilization, and distribution of energy in the formation of rice yield and quality was primarily focused. Furthermore, transcriptome analysis was also performed to better understand the relevant mechanisms underlying the synergistic formation of grain yield and rice quality.

2. Results

2.1. Grain Yield, Its Components, and Rice Quality

As shown in Figure 1, ZZY1 has a theoretical yield of 8.6 t ha^{-1}, which is 17.1% higher than the theoretical yield of 7.3 t ha^{-1} of ZZY8, and the difference was significant at the

p = 0.05 level. However, the actual yield increase was reduced to only 7.9%, and the difference was not significant. In terms of yield components, ZZY1 has both a higher number of panicles per plant and a higher 1000-grain weight compared to ZZY8, although it has a lower number of grains per panicle. However, those differences were all not significant. The seed-setting rate of ZZY1 was significantly higher than that of ZZY8, and the increment was 11.5%.

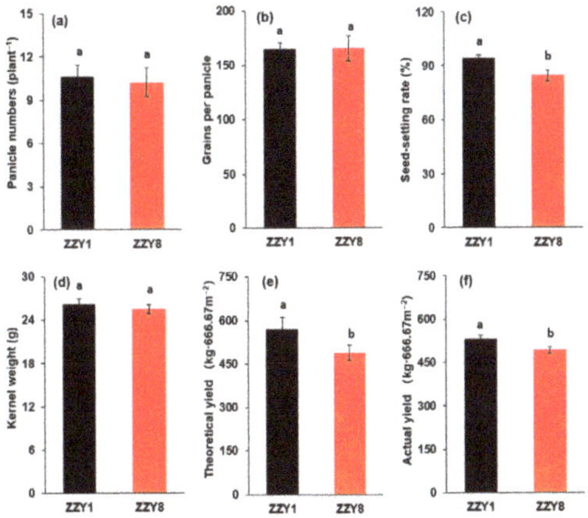

Figure 1. Grain yield and its components of two rice varieties. (**a**) Panicle numbers; (**b**) grain numbers of each panicle; (**c**) seed-setting rate; (**d**) kernel weight; (**e**) theoretical yield calculated by yield components; (**f**) actual grain yield. ZZY1 and ZZY8 indicate rice varieties, Zhongzhouyou1 and Zhongzhouyou8, respectively; error bars denote ± standard deviation (n = 3); different letters between varieties indicate significant differences at the 0.05 probability level.

Unshelled kernel and milled rice appearance are shown in Figure 2a,b. Rice quality parameters, including brown rice rate, milled rice rate, gel consistency, and protein content, showed no significant difference between varieties (Figure 2c,d,f–j). However, ZZY8 has a significantly higher head-milled rice rate than ZZY1 (Figure 2e). The chalkiness degree and chalky grain rate of ZZY1 are both higher than those of ZZY8, with an increase of 28.6% and 6.3%, respectively, although the difference was not significant (Figure 2f,g). The amylose content of ZZY1 was slightly lower than that of ZZY8, but the difference was not significant (Figure 2i).

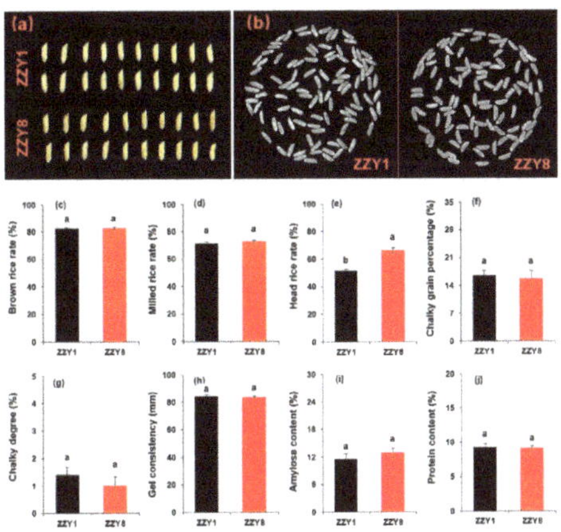

Figure 2. Grain appearance and rice quality parameters. (**a**) Unshelled kernel; (**b**) milled rice; (**c**) brown rice rate; (**d**) milled rice rate; (**e**) head rice rate; (**f**) chalky grain rate; (**g**) chalky degree; (**h**) gel

consistency; (**i**) amylose content; (**j**) protein content. ZZY1 and ZZY8 indicate rice varieties, Zhongzhouyou1 and Zhongzhouyou8, respectively; error bars denote ± standard deviation (n = 3); different letters between varieties indicate significant differences at the 0.05 probability level.

2.2. Plant Height and Dynamic Changes of Tillering

Field performance is exhibited in Figure 3a,b. During the entire determination period, the plant height of ZZY8 was consistently taller than that of ZZY1 (Figure 3a,b,d). The plant height growth rate of both varieties reached the fastest between 14–21 days after transplanting, with increases of 36.7% in ZZY1 and 34.6% in ZZY8, respectively. At 56 days after transplanting, the difference in plant height between the two varieties reached its maximum. At 63 days after transplanting, the plant heights of the two varieties were 116.5 cm and 123.7 cm, respectively, with ZZY8 being 5.82% taller than ZZY1. Both varieties reached the tiller number peak at 42 days after transplanting (Figure 3c). Throughout the whole determination period, the tiller numbers of ZZY1 were higher than ZZY8, except for 21 days after transplanting. The greatest difference occurred at 63 days after transplanting, when the tiller numbers of ZZY1 were 21.1% higher than those of ZZY8.

Figure 3. Plant morphology and the dynamic change of tiller numbers and plant height of rice plants. (**a**) Rice morphology at field condition; (**b**) plant morphology in pot experiment; (**c**) tiller numbers; (**d**) plant height. ZZY1 and ZZY8 indicate rice varieties, Zhongzhouyou1 and Zhongzhouyou8, respectively; error bars denote ± standard deviation (n = 3); different letters between varieties indicate significant differences at the 0.05 probability level.

2.3. Leaf Chlorophyll Contents and Photosynthetic Rate

The chlorophyll contents of both varieties generally followed a trend of increasing and then decreasing, as demonstrated in Figure 4a. Except for some certain dates, the chlorophyll content of ZZY1 was slightly higher than that of ZZY8, although the difference between varieties was not obvious. The net photosynthetic rate of the flag leaf showed a gradual declining trend, with the highest value at the heading stage (Figure 4b). The net photosynthetic rate of the flag leaf of ZZY8 was significantly higher than that of ZZY1 at the heading stage, while it was reversed at 10 and 25 days after heading, although the difference was not significant.

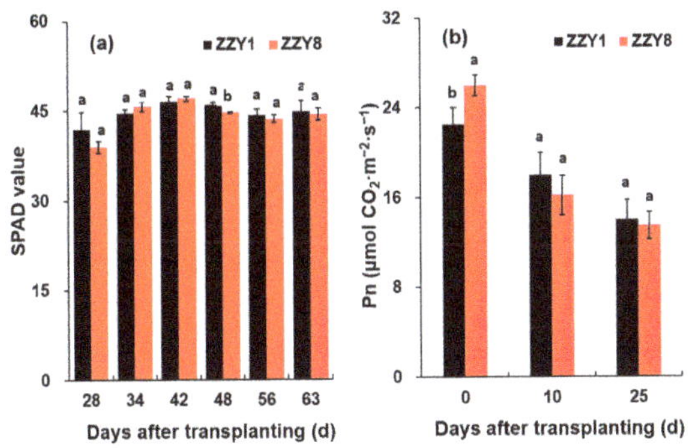

Figure 4. The chlorophyll content index (SPAD value) and net photosynthetic rate of rice leaf. (**a**) SPAD value; (**b**) net photosynthetic rate. ZZY1 and ZZY8 indicate rice varieties, Zhongzhouyou1 and Zhongzhouyou8, respectively; error bars denote ± standard deviation (n = 3). Different letters between varieties indicate significant differences at the 0.05 probability level.

2.4. Dry Matter, NSC Production, and Translocation

The total dry matter weight of ZZY1 and ZZY8 gradually increased as growth progressed (Figure 5a). The dry matter weight of ZZY8 was slightly higher than that of ZZY1 at most periods, although the difference was not significant. The proportion of panicle dry matter to total dry matter in ZZY1 was higher than ZZY8 from heading to maturity, whereas the difference at 10 DAH was significant (Figure 5b). The proportion of leaf dry matter to total dry matter showed a gradually decreasing trend (Figure 5c). There was no clear change pattern in the differences between the two varieties. The proportion of stem and sheath dry matter to total dry matter first increased and then decreased with the development growth (Figure 5d). The ratio in ZZY8 was higher than that in ZZY1 at all dates except for the panicle differentiation stage, and the difference was significant at 10 and 20 DAH.

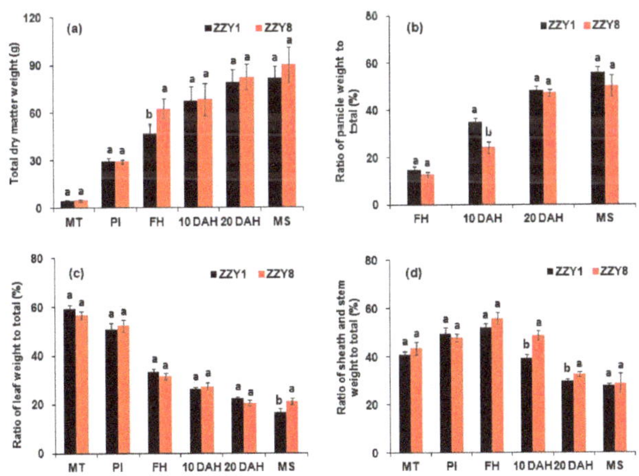

Figure 5. Total dry matter accumulation and allocation in rice plants. (**a**) Total dry weight; (**b**) ratio of panicle weight to total dry weight; (**c**) ratio of leaf weight to total dry weight; (**d**) ratio of sheath and stem weight to total dry weight. MT, PI, FH, MS, and DAH denote midtillering, panicle initiation, full-heading, maturity stage, and days after heading. ZZY1 and ZZY8 indicate rice varieties, Zhongzhouyou1 and Zhongzhouyou8, respectively; error bars denote ± standard deviation (n = 3). Different letters between varieties at the same determination time indicate significant differences at the 0.05 probability level.

The nonstructural carbohydrate (NSC) content in the stem, sheath, and leaves of ZZY1 and ZZY8 first decreased and then increased as the growth period proceeded, with the highest content in the sheath at the heading stage and in the leaves at maturity (Figure 6a,b). The NSC content in the sheath of ZZY8 was consistently higher than that of ZZY1, where the difference was significant at most dates (Figure 6a). The NSC content in the leaves of ZZY1 was slightly higher than that of ZZY8 from the heading to 20 DAH (Figure 6b). However, it was reversed at maturity, and the difference between the varieties was not significant. Except for the heading stage, the NSC content in the panicles of ZZY8 was lower than that of ZZY1 at 10 DAH, 20 DAH, and maturity (Figure 6c). The decrease ranged from 1.5% to 7.2%, and the difference was not significant. The proportion of panicle NSC to total NSC in ZZY1 was higher than in ZZY8 throughout the entire grain-filling period, and the difference reached the largest at 10 DAH (Figure 6d).

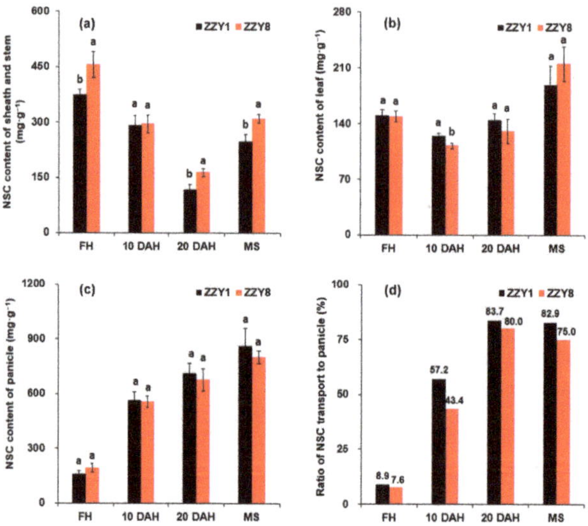

Figure 6. The nonstructural carbohydrate (NSC) content accumulation and allocation in rice plants. (**a**) NSC content in sheath and stem; (**b**) NSC content in leaf; (**c**) NSC content in panicle; (**d**) ratio of NSC content in panicle to the total. FH, MS, and DAH denote full heading, maturity stage, and days after heading; ZZY1 and ZZY8 indicate rice varieties, Zhongzhouyou1 and Zhongzhouyou8, respectively; error bars denote ± standard deviation (n = 3). Different letters between varieties at the same determination time indicate significant differences at the 0.05 probability level.

2.5. Characteristics of Grain-Filling

The grain-filling process of the two varieties estimated using Richards' growth equation is shown in Figure 7. The R2 values derived from the equation $W = A/(1+Be-kt)1/N$ were mostly above 0.95, indicating a high fitting degree, and the grain-filling process conformed well to the Richards' growth model. The final grain weights of ZZY1 were higher than ZZY8, with a difference of 0.12 g in superior grains and 0.10 g in inferior grains (Figure 7a). Figure 7b presents the grain-filling dynamics, where the grain weight of ZZY1 was higher than that of ZZY8 regardless of superior or inferior grains throughout the whole course. The superior grains of both varieties reached their maximum grain-filling rates three days after flowering and then declined rapidly (Figure 7b). The grain-filling rate of ZZY1 was obviously higher than ZZY8 during the first six days, and the former was notably higher than the latter. In comparison, the grain-filling rates of inferior grains increased slowly, reached their maximum at 15 days after anthesis, and then gradually declined in both varieties, especially for ZZY8.

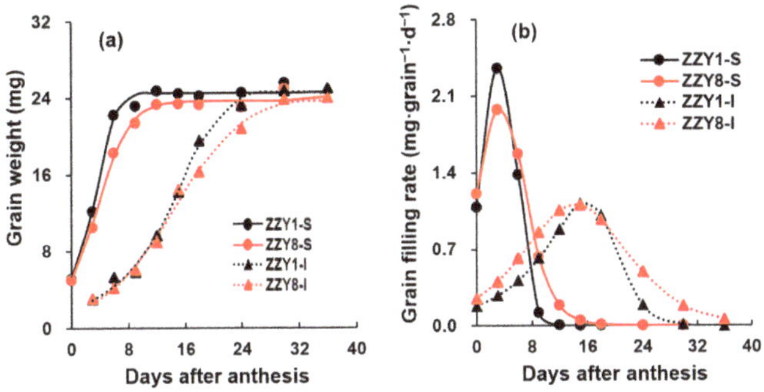

Figure 7. The grain weight and grain-filling rate dynamics of rice. (**a**) Grain weight; (**b**) grain-filling rate; ZZY1 and ZZY8 indicate rice varieties, Zhongzhouyou1 and Zhongzhouyou8, respectively; S and I indicate superior and inferior spikelets.

ZZY1 had higher mean grain-filling rates than ZZY8 for both superior and inferior grains (Table 1). In terms of the active grain-filling period, the superior grains of ZZY1 lasted for 10.6 days, while those of ZZY8 lasted for 14.5 days. For the inferior grains, ZZY1 and ZZY8 lasted 23.7 days and 30.8 days, respectively. The number of days to reach the maximum grain-filling rate of ZZY1 exceeded ZZY8 in both superior and inferior grains. A similar tendency was found in changes in the maximum grain-filling rate.

Table 1. The average grain-filling rate, the maximum grain-filling rate, and the active grain-filling period.

Variety/Grain Type	Mean Grain-Filling Rate (g·grain $^{-1}$·d $^{-1}$)	Active Grain-Filling Period (d)	Period to Reach Maximum Grain-Filling Rate (d)	Maximum Grain-Filling Rate (g·grain $^{-1}$·d $^{-1}$)
ZZY1-Superiors	2.33 a	10.6 d	3.9 c	2.54 a
ZZY1-Inferiors	1.04 c	23.7 b	15.9 a	1.14 b
ZZY8-Superiors	1.63 b	14.5 c	3.7 c	2.01 a
ZZY8-Inferiors	0.79 d	30.8 a	14.3 b	1.11 b

ZZY1 and ZZY8 indicate the rice varieties Zhongzheyou1 and Zhongzheyou8, respectively. Error bars denote ± standard deviation (n = 3). Different letters in the same column indicate significant differences at the 0.05 probability level.

2.6. Carbohydrate Content and Key Enzymes Related to Grain-Filling

As shown in Figure 8, the content of soluble sugars in the grains showed a gradually increasing trend from the heading stage to maturity. The content of soluble sugars in ZZY1 grains was lower than that in ZZY8 at most determination stages, and the largest difference was observed at maturity (Figure 8a). The starch content in the grains also showed a gradually increasing trend. The starch content in ZZY1 grains was higher than that in ZZY8 except for the heading stage, and the largest difference between varieties was observed at maturity (Figure 8b). Regarding starch transport rate, both varieties exhibited a trend of increasing first and then decreasing. The proportion of starch to nonstructural carbohydrates in ZZY1 was higher than that in ZZY8 from 10 DAH to maturity (Figure 8b).

The activities of enzymes, including invertase, SS, SBE, and AGPase, were determined at 10 DAH and 20 DAH (Figure 9). The grain acid invertase content of both varieties did not show a significant difference at 10 DAH, whereas it was significantly lower in ZZY8 than ZZY1 at 20 DAH, with a decrease of 25.0% (Figure 9a). Compared to 10 DAH, the SS activity in the grains of ZZY1 was significantly reduced, while it was slightly increased in ZZY8 at 20 DAH (Figure 9b). However, the SS activity of ZZY8 was significantly lower than that of ZZY1 at both determination dates. The starch branching enzyme activity of ZZY1 grains was significantly higher than ZZY8 at 10 DAH, while the difference between the two varieties was not significant at 20 DAH (Figure 9c). The AGPase activity in ZZY1 grains was higher than that of ZZY8 regardless of the determination date (Figure 9d). The difference between varieties was not significant at 10 DAH, while it was significantly higher in ZZY1 than in ZZY8 at 20 DAH.

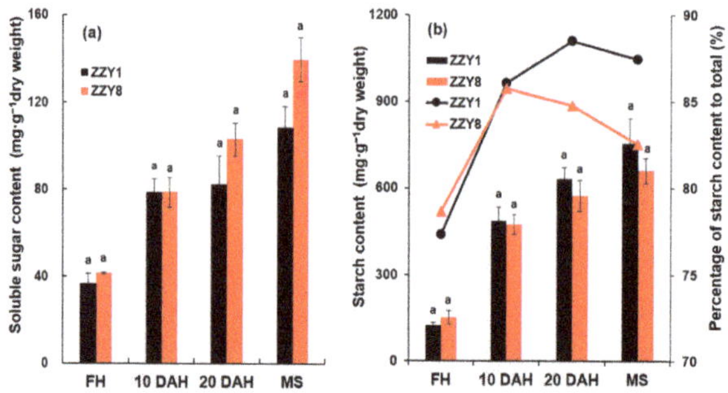

Figure 8. Soluble sugar and starch content in grains of rice. FH, MS, and DAH denote full heading, maturity stage, and days after heading. (**a**) Soluble sugar content; (**b**) starch content. FH, MS, and DAH denote full heading, maturity stage, and days after heading; ZZY1 and ZZY8 indicate rice varieties, Zhongzhouyou1 and Zhongzhouyou8, respectively; error bars denote ± standard deviation (n = 3); different letters between varieties at the same determination time indicate significant differences at the 0.05 probability level.

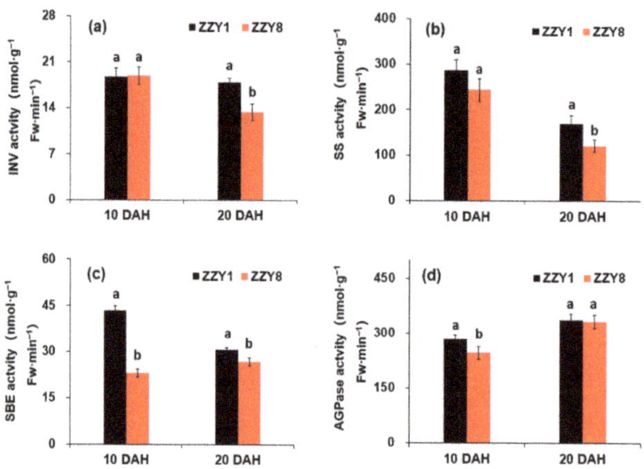

Figure 9. The activities of key enzymes related to grain-filling in grains of rice. (**a**) Invertase (INV) activity; (**b**) soluble sucrose synthase (SS); (**c**) starch branching enzyme (SBE) activity; (**d**) ADP-glucose pyrophosphorylase (AGPase) activity. DAH denotes days after heading; ZZY1 and ZZY8 indicate rice varieties, Zhongzhouyou1 and Zhongzhouyou8, respectively; error bars denote ± standard deviation (n = 3); different letters between varieties at the same determination time indicate significant differences at the 0.05 probability level.

2.7. Characteristics of Energy Metabolism

The ATP, ATPase content, and energy charge values of ZZY1 grains were significantly higher than those of ZZY8 at 10 DAH, while the difference between the varieties was not significant at 20 DAH (Figure 10a–c). The activity of the respiratory electron transport chain complex I in ZZY1 was significantly higher than in ZZY8 at 10 DAH, whereas the difference was not significant at 20 DAH (Figure 10d). The activity of the respiratory electron transport chain complex V in ZZY1 was significantly higher than in ZZY8 at both 10 and 20 DAH (Figure 10e). The PARP content in ZZY8 was only significantly higher than in ZZY1 at 20 DAH (Figure 10f).

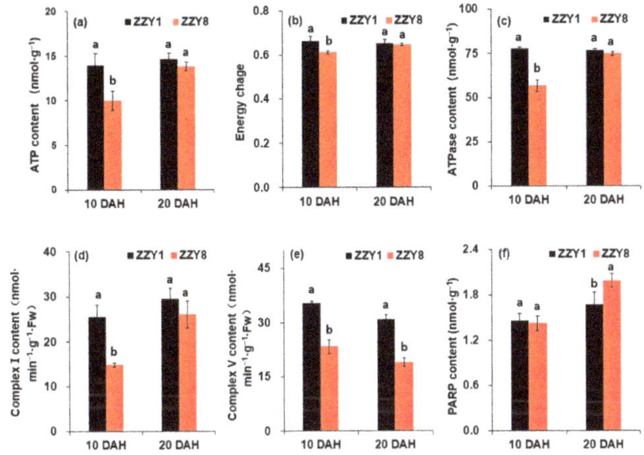

Figure 10. Energy allocation in grains of rice. (**a**) ATP content; (**b**) energy charge; (**c**) ATPase content; (**d**) Complex I content; (**e**) Complex II content; (**f**) PARP content. DAH denote days after heading; ZZY1 and ZZY8 indicate rice varieties, Zhongzhouyou1 and Zhongzhouyou8, respectively; error bars denote ± standard deviation ($n = 3$); different letters between varieties at the same determination time indicate significant differences at the 0.05 probability level.

2.8. Transcriptome Analysis of Carbohydrate and Energy Metabolism

Transcriptomic studies were used to investigate the role of energy in yield and quality formation. The number of differentially expressed genes at 20 DAH compared with 10 DAH in ZZY8 was significantly higher than that in ZZY1. The upregulated gene number was 4725 and the downregulated number was 3375 in the former, while the numbers were 3112 and 2462, respectively, in the latter (Figure 11a). This indicated that more intense metabolic changes occurred in the grains of ZZY8 at 20 DAH, whereas the changes in ZZY1 were relatively small. There were 2347 commonly upregulated genes shared by the two varieties (Figure 11b), and the functions of these genes mainly focused on stress response, abscisic acid response, dehydration response, and protein modification (Figure S1). There were 1361 commonly downregulated genes (Figure 11c), and they were mainly related to carbohydrate and energy metabolism, cellular function, and structure (Figure S2). This indicated that grain-filling at 20 DAH tended to slow down and inner metabolic function activities had reduced. Transcriptome analysis of both varieties at the same determination time indicated that the expression of differentially expressed genes was significantly higher at 10 DAH than at 20 DAH (Figure 11d). At 10 DAH, the greatest changes included carbohydrate metabolism, energy metabolism, and secondary metabolism, while the affected pathways were carbohydrate metabolism, amino acid metabolism, and secondary metabolism pathways at 20 DAH.

Gene ontology enrichment analysis of differentially expressed genes shows that these genes were primarily related to abscisic acid signaling, dehydration response, carbohydrate metabolism, and energy metabolism (Figure 12a). The Kyoto Encyclopedia of Genes and Genomes pathway annotations for the differentially expressed genes on different days after flowering for both varieties show that these genes were mainly related to carbohydrate, amino acid, and energy metabolism pathways, as well as protein processing and environmental response genes (Figure 12b). Additionally, increased responses to oxygen content and ethanol suggested an enhancement of anaerobic respiration metabolism in the grains at 20 DAH. Analysis of differentially expressed genes involved in starch and sucrose metabolism, as well as energy metabolism pathways (glycolysis, TCA, oxidative phosphorylation), revealed that genes associated with starch metabolism and energy metabolism were significantly downregulated at 20 DAH, with the number of downregulated genes in

ZZY8 being greater than in ZZY1, indicating intense inhibition of starch metabolism and energy metabolism in ZZY8 (Figure 12c).

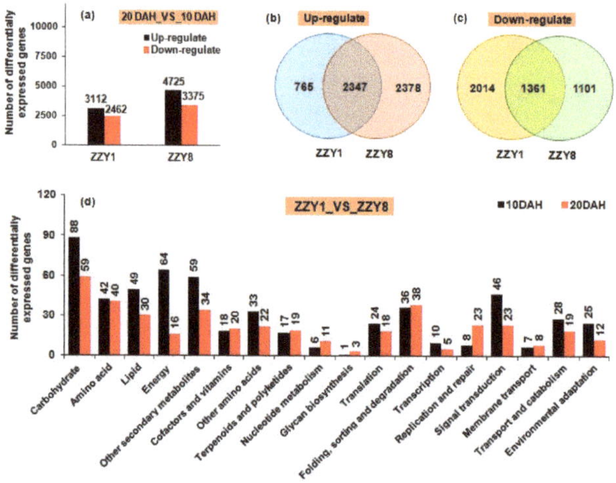

Figure 11. The transcriptome analysis of differentially expressed genes in grains of rice. (**a**) Number of differentially expressed genes at 20 days after heading (DAH) compared with 10 DAH in two varieties, with significantly different expression levels between different comparison groups; (**b**) the overlap of genes exhibiting significant upregulated expression at 20 DAH compared with 10 DAH; (**c**) the overlap of genes exhibiting significant downregulated expression at 20 DAH compared with 10 DAH; (**d**) Kyoto Encyclopedia of Genes and Genomes (KEGG) pathway annotation of differentially expressed genes in ZZY1 compared with ZZY8, respectively, at 10 DAH and 20 DAH.

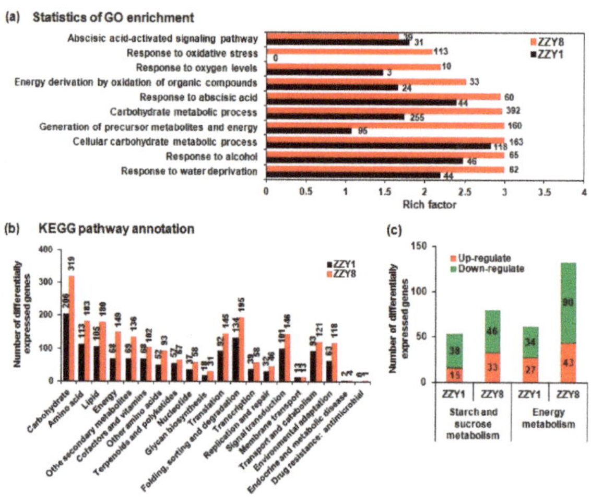

Figure 12. Distribution of genes with statistically changed expression levels in grains of two rice varieties. (**a**) Gene ontology (GO) enrichment analysis of the differentially expressed genes at 20 days after heading (DAH) compared with 10 DAH in two varieties; (**b**) Kyoto Encyclopedia of genes and genomes (KEGG) pathway annotation of the differentially expressed genes at 20 DAH compared with 10 DAH in two varieties; (**c**) differential expression gene analysis of starch and sucrose metabolism and energy metabolism pathways. ZZY1 and ZZY8 indicate rice varieties, Zhongzhouyou1 and Zhongzhouyou8, respectively.

3. Discussion

The results of this experiment indicate that both the actual and theoretical yield of ZZY1 under field conditions were higher than ZZY8, with increases of 17% and 7.84%, respectively (Figure 1). Although there have been many studies regarding the yield performance and growth characteristics in rice varieties of ZZY1 or ZZY8, there were few research results specialized in comparing the two varieties. According to the yield components, ZZY1 had higher effective panicle numbers, seed-setting rate, and grain weight when compared to ZZY8, in which the increment in seed-setting rate was 11.4% (Figure 1a,c,d). Therefore, the higher seed-setting rate was considered the main reason for the significantly higher grain yield of ZZY1 than ZZY8.

It is well known that the yield formation of rice is closely associated with the production, accumulation, and translocation of assimilates [1,25]. However, the higher yield of ZZY1 does not seem to be highly related to its photoassimilates accumulation. There were no significant differences in leaf chlorophyll content and net photosynthetic rate between the two varieties (Figure 4), and the total dry matter of ZZY1 was lower than ZZY8, especially during maturity (Figure 5a). In this experiment, the difference in spikelet numbers per panicle between the two varieties was small, but ZZY1 had a higher seed-setting rate and grain weight than ZZY8, indicating that the lower yield of ZZY8 was mainly attributable to inefficient assimilate transport. Similar research has been reported by Zhang et al. [26], in which it was indicated that the decrease in grain weight caused by high temperature during flowering was mainly due to inefficient transport or assimilate translocation rather than source and sink limitations. Furthermore, this was confirmed by the significant differences in the proportion of dry matter of the panicle to that of the total between the two varieties (Figure 5b). Additionally, the proportion of nonstructural carbohydrates in the panicles to the total in ZZY1 was higher than ZZY8 (Figure 6d). Therefore, higher assimilate transport and distribution are the main reasons for the higher yield of ZZY1.

With the improvement of living standards in recent years, high-yield and high-quality have become the primary goals of rice breeding. To improve rice quality while ensuring a certain grain yield has become a consensus in modern rice breeding [13–15]. Based on the analysis above, ZZY1 has a higher grain yield than ZZY8, while the latter has better rice quality (Figure 2). There have been many reports on the trade-off between grain yield and rice quality, but the underlying mechanisms have not been elucidated. It is believed that this may be related to energy production and utilization. Studies have shown that the formation of grain yield and rice quality is an energy-consuming process [21–23]. Insufficient energy or low utilization efficiency makes it difficult to meet the demands for both yield and quality. It was observed that the ATP content, ATPase activity, and energy charge of the grains of ZZY1 were higher than ZZY8 at 10 days after heading (DAH), while there was almost no difference between the two varieties at 20 DAH (Figure 10). This might be one of the main causes of the lower quality in ZZY1 compared to ZZY8.

Numerous studies have shown that the improper dynamics of grain filling and changes in the physiological activity of related key enzymes have a significant impact on chalkiness formation [27,28]. Enzyme activities in regions prone to form chalkiness are relatively low, resulting in an imbalance in starch synthesis and accumulation [29–31]. Varieties with a smooth grain-filling dynamic and a relatively slow grain-filling rate often have less chalkiness, while those with a fast-filling rate or large fluctuations in grain-filling dynamics tend to have more chalkiness [32]. In this experiment, ZZY1 had a significantly higher grain-filling rate than ZZY8, while it exhibited large fluctuations in grain-filling rate. The grain-filling rate of ZZY1 showed a rapid rise and then a sharp fall, and this might be the main reason for its significantly higher chalkiness compared to ZZY8 (Figure 7b). Starch synthesis in grains mainly depends on a series of enzymes such as AGPase, granule-bound starch synthase (GBSS), SS, SBE, and starch debranching enzyme (DBE), and changes in starch metabolism enzyme activity are closely related to chalkiness formation [33–35]. AGPase catalyzes the initiation of starch synthesis, in which ATP and 1-phosphate glucose are converted into ADP-glucose, transported to starch granules by the BT1 transporter protein on the starch body, and then synthesized into starch by GBSS, SS, and other enzymes [36]. It is the rate-limiting enzyme in starch synthesis, and the activity of AGPase is significantly correlated with grain-filling rate [25]. It was observed that the key enzyme activity of grain-filling in ZZY1 was higher than ZZY8 at 10 or 20 days after heading (DAH), which may be an important reason for the higher grain-filling rate of ZZY1 than ZZY8 (Figure 9). However, at the same time, the increase in grain-filling enzyme activity consumed a lot of energy, resulting in insufficient energy required for other indices of rice quality, which ultimately affected the formation of rice quality. This may also be an important reason for the increase in grain weight but the deterioration of the quality of ZZY1. Both the carbohydrate and energy metabolism differences between varieties were inferred at the gene expression level (Figures 11 and 12). The results of the current study give us the inspiration that cultivation measures such as nitrogen management or exogenous chemical spraying should be used by regulating energy metabolism [24,37]. Furthermore, the indices related to energy distribution and utilization were not determined in this experiment. Thus, the underlying mechanism of energy metabolism involved in grain yield and rice quality formation remains to be clarified. Research concerning breeding and cultivation methods to simultaneously improve grain yield and rice quality by enhancing energy supply is needed in future studies.

4. Materials and Methods

4.1. Materials and Experimental Design

The experiment was conducted in 2020 and repeated in 2021 at the Fuyang base of the China National Rice Research Institute (30.30′ N, 120.2′ E, 11 m above sea level). Two three-line *indica* hybrid rice varieties, ZZY1 (Zhongzhe A/Hanghui 570), and ZZY8 (Zhongzhe A/T-8), that are suitable for planting in the middle and lower reaches of the Yangtze River, were used as the study materials in the field experiments. The whole growth period of ZZY1 is approximately 137 days, while that of ZZY8 is slightly longer, approximately 159 days. Both of them were widely used in local production because of their high yield performances, good rice quality characteristics, and high stress resistances than other varieties. The full heading date of ZZY1 and ZZY8 was approximately 15–18 August and 4–6 September across two experimental years. After soaking the rice seeds for 48 h, they were germinated in a constant temperature box at 37 °C for 24 h, and then sown on the seedbed on 21 May. Transplanting was carried out after 25 days, on 15 June.

The test field was a loamy clay soil that contained 37.2 g kg^{-1} organic matter, 305.8 mg kg^{-1} alkali-hydrolyzable nitrogen, 20.6 mg kg^{-1} available phosphorus, and 68.9 mg k^{-1} available potassium, with pH 5.79. The area of each experimental plot was 25 m^2, with a transplanting row spacing of 20 cm × 25 cm, and each variety was replicated in three plots with a complete randomized block design. Field management followed the conventional high-yield cultivation practices for rice production. Nitrogen was applied with a total amount of 195 kg ha^{-1} in the form of urea, respectively, before transplant, at midtillering, and panicle initiation as topdressing with a proportion of 5:2:3. Phosphorus was applied in the amount of 50 kg ha^{-1} before transplanting in the form of calcium magnesium phosphate composite fertilizer. Potassium was incorporated into the experimental plots at the rate of 40 kg ha^{-1} both before transplanting and at midtillering. The paddy field was kept with a shallow water layer of approximately 2 cm until one week before harvest, except for drainage at the end of tillering. Periodic manual weeding and spraying to control pests and disease were carried to guarantee healthy growth of rice plants. The average daily air temperatures and rainfall precipitation during the rice growing season from June to October of the two study years obtained from the weather station near the experimental site are shown in Figure 13.

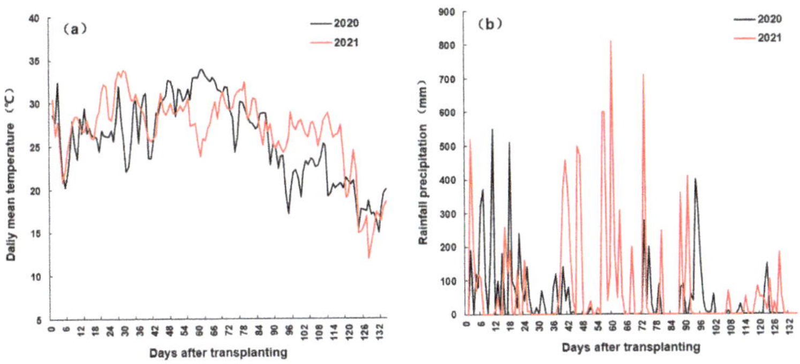

Figure 13. The daily mean air temperatures (**a**) and rainfall precipitation (**b**) during the rice growing season from June to October of 2020 and 2021 at the experimental site of Fuyang, Hangzhou.

4.2. Grain Yield, Its Components, and Rice Quality Determination

The rice plants were harvested after maturity. Twelve hills were sampled from each plot, and the yield was calculated by investigating the number of effective panicles per plant, the number of grains per panicle, seed-setting rate, and 1000-grain weight. The actual harvest from 10 m^2 area of the field trial plots was also recorded and converted into a moisture content of 13.5%. Quality testing was conducted after drying, with samples sent to the Rice Quality Testing Center of the China National Rice Research Institute, Chinese Academy of Agricultural Sciences.

4.3. Plant Height and Tiller Dynamics

Plant height and tiller numbers were surveyed every week from 7 to 63 days after transplanting. Observation points were selected in each replicate plot, and ten hills were determined at each point.

4.4. Dry Matter Production and Accumulation

Six hills of rice plants were randomly sampled from each plot at the tillering, panicle differentiation, heading, 10 days after heading (DAH), 20 DAH, and maturity stages. The plants were then divided into stem sheath, leaves, and panicles, and dried at 105 °C for 60 min and then at 80 °C for 48 h until a constant weight was reached in an oven.

4.5. Carbohydrate Content Determination

The contents of soluble sugars and starch were determined using the anthrone colorimetric method [38]. A dried sample of 0.2 g was taken, added to 10 mL of deionized water, and boiled for 30 min in a constant temperature water bath. After cooling and centrifugation, the supernatant was collected. This extraction process was repeated three times, and the extracts were combined and quantified to a certain volume for the determination of soluble sugar content. The remaining residue was dried in an oven, added to 2 mL of deionized water, boiled in a water bath for 20 min, then 2 mL of 9.2 mol L^{-1} $HClO_4$ was added. After cooling, 6 mL of deionized water was added, and the mixture was centrifuged at 2000 rpm for 20 min to collect the supernatant. The remaining residue was again treated with 2 mL of 4.6 mol L^{-1} $HClO_4$, and the process was repeated. Both extracts were then combined for starch content determination. The nonstructural carbohydrates (NSC) are the total of soluble sugar and starch contents.

4.6. Grain-Filling Dynamics

Two hundred rice panicles that flowered on the same day and grew evenly were selected and tagged to mark the flowering date. From flowering to maturity, the marked panicles were sampled every 3 days in the early grain-filling stage and every 6 days in the later grain-filling stage to investigate the weight of superior and inferior grains. The grain-filling process was fitted using the Richards' growth equation $W = A/(1 + Be^{-kt})^{1/N}$ to calculate the corresponding filling rates [39,40]. In this equation, A indicates the final growth weight (mg) of a grain calculated by the growth equation, while B, k, and N are coefficients determined by regression analysis. R^2 is the coefficient of determination.

4.7. Key Enzyme Activities During Grain-Filling

Rice grains were sampled 10 and 20 days after heading (DAH) and stored at −80 °C for the determination of various starch synthesis-related enzyme activities. Approximately thirty grains were removed from the shells, and 5 mL of 100 mmol L^{-1} Tricine-NaOH extraction solution [pH 8.0, containing 10 mmol L^{-1} $MgCl_2$, 2 mmol L^{-1} EDTA, 50 mmol L^{-1} β-mercaptoethanol, 12% (*v*/*v*) glycerol, 5% (*w*/*v*) PVP 40] was added. The mixture was ground on ice, centrifuged at 15,000× *g* for 5 min (4 °C), and the supernatant (crude enzyme solution) was collected following the method by Li et al. [41] with slight modifications. The activities of ADP-glucose pyrophosphorylase (AGPase), sucrose synthase (SS), and starch branching enzyme (SBE) are expressed in nmol g^{-1} Fw min^{-1}.

Another set of grains were finely powdered in liquid nitrogen, followed by homogenization in the 20% extraction buffer comprising 50 mmol L^{-1} HEPES (pH 7.5), 5 mmol L^{-1} $MgCl_2$, 1 mmol L^{-1} EDTA-Na_2, 0.5 mmol L^{-1} dithiothreitol, 1% Triton X-100, 2% polyvinyl pyrrolidone, and 10% glycerol. The resulting supernatant was promptly desalted using a Sephadex G-25 medium, pre-equilibrated with the extraction buffer at 4 °C. The filtered solution was promptly utilized to assess the activity of soluble acid invertase by using the test kit provided by Comin Biotechnology Co., Ltd. (Suzhou, China).

4.8. Determination of Energy Metabolism Related Indicators

ATP, ATPase, ADP, AMP, and poly-ADP-ribose polymerase (PARP) were all determined using a double-antibody sandwich ELISA method. The coated microplates with plant ATP, ATPase, ADP, AMP, and PARP capture antibodies were sequentially added with specimens, standards, and HRP-labeled detection antibodies, followed by incubation and thorough washing. Substrate TMB turned blue under the catalysis of peroxidase and finally turned yellow under acidic conditions. The intensity of the color is directly proportional to the amounts of ATP, ATPase, ADP, AMP, and PARP in the samples. Absorbance was measured at 450 nm wavelength using a Multiskan FC ELISA reader (Thermo Fisher Scientific (Shanghai) Instrument Co., Ltd., Shanghai, China). The specific operations were completed following the related reagent kit manual (Shanghai Enzyme-linked Biotechnology Co., Ltd., Shanghai, China).

Mitochondrial complex I, complex IV, and complex V were determined as referring to the instructions of the reagent kits (Suzhou Comin Biotechnology Co., Ltd., Suzhou, China).

4.9. RNA-Seq and Bioinformatics Analysis

The Plant RNA Purification Reagent was used to extract total RNA from the grains, and genomic DNA was further removed by DNase I (TaKaRa Biotechnology, Dalian, China). The TruSeq^{TM} RNA sample preparation kit (Illumina, San Diego, CA, USA) was used to construct the library. After quantification using TBS-380 (Beijing Yuanpinghao Biotechnology Co., Ltd., Beijing, China), the library was sequenced on the Illumina/NovaSeq 6000 sequencing platform with a reading length of PE150. Transcriptome sequencing and analysis services were provided by Shanghai Majorbio Bio-Pharm Technology Co., Ltd., Shanghai, China.

4.10. Statistical Analysis

Data from the experiment were statistically analyzed using Excel and SPSS software version 11.5 (IBM Corp., Armonk, NY, USA).

5. Conclusions

The grain yield of ZZY1 was higher, while the rice quality was poorer than that of ZZY8. The high yield of ZZY1 was mainly benefited from its higher seed-setting rate and grain weight. Correspondingly, the proportion of panicle dry matter weight or nonstructural carbohydrate to the total of ZZY1 was higher than that of ZZY8. The maximum grain-filling rate, average grain-filling rate, starch contents, and key enzyme activities involved in starch synthesis in the former were significantly higher than those of the latter. Furthermore, the ATP/ATPase content, ATP metabolism-related enzyme activity, and energy charge values in the grains of ZZY1 were higher than those of ZZY8 at the early grain-filling stage. The above results showed that carbohydrate and energy metabolism were the main ways affecting the yield and quality of the two varieties. It was verified by the transcriptome analysis, demonstrating that the energy production of ZZY1 was insufficient to simultaneously supply the needs thus leading to the discordant formation in its grain yield and rice quality formation.

Supplementary Materials: The supporting information can be downloaded at: https://www.mdpi.com/article/10.3390/ijms252312751/s1.

Author Contributions: Conceptualization: L.T.; methodology, Y.W. and J.M.; software, J.L. and Y.Z.; validation, H.S., W.H., G.L. and W.F.; investigation, Y.W., J.M., H.S. and W.F.; resources, J.L. and Y.Z.; data curation, G.L. and Y.Z.; writing-original draft preparation, T.C.; writing-review and editing, G.F. and J.X.; supervision, L.T.; project administration, G.F.; funding acquisition, G.F. and T.C. All authors have read and agreed to the published version of the manuscript.

Funding: This work was supported by National Key Research and Development Plan of China (2022YFD1500400), Zhejiang Provincial Natural Science Foundation of China (Z24C130019 and Z23C130001), National Natural Science Foundation of China (32301932), the "Sannongjiufang" Science and Technology Cooperation Project of Zhejiang province (2023SNJF001, 2023SNJF003), the "Pioneer" and "Leading Goose" Key R&D Program of Zhejiang (2023C02005-4), the State Key Laboratory of Rice Biology and breeding (2020ZZKT10401, 2023ZZKT20404), and the Central Public-interest Scientific Institution Basal Research Fund.

Institutional Review Board Statement: All methods were in compliance with relevant institutional, national, and international guidelines and legislation.

Informed Consent Statement: Not applicable.

Data Availability Statement: The data that support the results of this study are available from the corresponding author, upon reasonable request.

Conflicts of Interest: The authors declare no conflicts of interest.

Abbreviations

AGPase: ADP-glucose pyrophosphorylase; DAH, days after heading; DBE, starch debranching enzyme; GBSS, granule-bound starch synthase; NSC, non-structural carbohydrates; SBE, starch branching enzyme; SS, sucrose synthase; ZZY1, Zhongzheyou1; ZZY8, Zhongzheyou8.

References

1. Dong, M.H.; Jiang, Y.; Chen, P.F.; Zhao, B.H.; Gu, J.R. The relationship between non-structural carbohydrates and rice spikelet formation: A Review. *J. Agric. Sci.* **2020**, *10*, 1–6.
2. Chen, T.; Yang, S.; Wei, T.; Li, Y.; Wang, S.; Su, Y. Overexpression of OsGS1;2 for improved nitrogen use efficiency and grain yield of rice: A field test. *Field Crops Res.* **2023**, *303*, 109146. [CrossRef]
3. Zhang, X.; Guo, Z.; Xu, J.; Huang, C.; Dang, H.; Mu, W.; Zhang, L.; Hou, S.; Huang, N.; Li, C.; et al. Nutrient requirements determined by grain yield and protein content to optimize N, P, and K fertilizer management in China. *Sci. Total Environ.* **2024**, *946*, 174187. [CrossRef]
4. Wang, Y.; Wang, X.; Zhai, L.; Zafar, S.; Shen, C.; Zhu, S.; Chen, K.; Wang, Y.; Xu, J. A novel effective panicle number per plant 4 haplotype enhances grain yield by coordinating panicle number and grain number in rice. *Crop J.* **2024**, *12*, 202–212. [CrossRef]
5. Song, J.; Cui, Y.; Fan, H.; Tang, L.; Wang, J. Molecular breeding of Zheyou810, an *Indica-Japonica* hybrid rice variety with superior quality and high yield. *Agriculture* **2023**, *13*, 1807. [CrossRef]

6. Zhou, Q.; Yuan, R.; Zhang, W.; Gu, J.; Liu, L.; Zhang, H.; Wang, Z.; Yang, J. Grain yield, nitrogen use efficiency and physiological performance of *indica/japonica* hybrid rice in response to various nitrogen rates. *J. Integr. Agric.* **2023**, *22*, 63–79. [CrossRef]
7. He, H.; Hu, C.; Xu, H.; Yang, R.; You, C.; Ke, J.; Zhang, Q.; Wu, L. High yield, good eating quality, and high N use efficiency for medium hybrid indica rice: From the perspective of balanced source-sink relationships at heading. *Eur. J. Agron.* **2024**, *159*, 127281. [CrossRef]
8. Ye, S.; Liu, M.; Zhai, R.; Ye, J.; Wu, M.; Zhu, G.; Zhang, X. RL99 was essential for high yield formation by regulating leaf morphogenesis and root development in rice. *Plant Mol. Bio. Rep.* **2024**, *15*, 1–8. [CrossRef]
9. Deng, Z.Y.; Qu, Y.Q.; Wu, Y.R.; Zhang, J.S.; Wang, T. Current progress and prospect of crop quality research. *Sci. Sin. Vitae.* **2021**, *51*, 1405–1414. [CrossRef]
10. Yang, W.L. The Physiological Mechanism of High Quality and High Yield in Hybrid Rice. Master's Thesis, Huazhong Agricultural University, Wuhan, China, 2021; pp. 7–8.
11. Gao, J.; Gao, L.; Chen, W.; Huang, J.; Qing, D.; Pan, Y.; Ma, C.; Wu, H.; Zhou, W.; Li, J.; et al. Genetic effects of grain quality enhancement in indica hybrid rice: Insights for Molecular Design Breeding. *Rice* **2024**, *17*, 39. [CrossRef]
12. Zhang, X.; Wang, Y.; Liu, M.; Yan, P.; Niu, F.; Ma, F.; Hu, J.; He, S.; Cui, J.; Yuan, X.; et al. OsEXPA7 encoding an expansin affects grain size and quality traits in rice (*Oryza sativa* L.). *Rice* **2024**, *17*, 36. [CrossRef] [PubMed]
13. Hong, W.; Duan, M.; Wang, Y.; Chen, Y.; Mo, Z.; Qi, J.; Pan, S.; Tang, X. Enriching iodine and regulating grain aroma, appearance quality, and yield in aromatic rice by foliar application of sodium iodide. *Rice Sci.* **2024**, *31*, 328–342.
14. Lu, B.; Xu, C.; Li, Z.; Tang, X.; Yang, F.; Xu, D.; Zhu, G.; Zhang, J.; Jiang, Y.; Li, W.; et al. Influence of the temperature during grain filling stage and nitrogen application rate on yield and quality of indica hybrid rice. *Field Crops Res.* **2024**, *309*, 109333. [CrossRef]
15. Xiong, R.; Tan, X.; Wang, H.; Pan, X.; Zeng, Y.; Zeng, Y. Optimized nitrogen fertilizer application strategy improves grain yield and quality of high-quality late indica rice under field ambient warming. *Eur. J. Agron.* **2024**, *154*, 127091. [CrossRef]
16. Wei, P.Y.; Qiu, S.; Tang, J.; Xiao, D.D.; Zhu, Y.; Liu, G.D.; Xing, Z.P.; Hu, Y.J.; Guo, B.W.; Gao, S.Q.; et al. Screening and characterization of high-quality and high-yield japonica rice varieties in Yanhuai region of Anhui province. *Acta Agron. Sin.* **2020**, *46*, 571–585.
17. Zhang, Q.; Hu, Y.J.; Guo, B.W.; Zhang, H.C.; Xu, X.J.; Xu, Y.F.; Zhu, B.H.; Xu, J.F. Study on the Characteristics of Soft japonica Rice Varieties with Good Taste and High Yield in Taihu Lake Area. *Chin. J. Rice Sci.* **2021**, *35*, 279–290.
18. Liu, Q.Y. Good Quality, High Grain Yield and High Nitrogen Efficiency Types of Medium Japonica Rice in the Eastern Region of Jianghuai and Their Morphological and Physiological Characteristics. Ph.D. Thesis, Yangzhou University, Yangzhou, China, 2021; pp. 141–144.
19. Sousa, J.S.; Calisto, F.; Langer, J.D.; Mills, D.J.; Refojo, P.N.; Teixeira, M.; Kuhlbrandt, W.; Vonck, J.; Pereira, M.M. Structural basis for energy transduction by respiratory alternative complex III. *Nat. Commun.* **2018**, *9*, 1728. [CrossRef]
20. Cao, H.; Duncan, O.; Millar, A.H. Protein turnover in the developing *Triticum aestivum* grain. *New Phytol.* **2021**, *233*, 1188–1201. [CrossRef]
21. Chen, T.; Yang, X.; Fu, W.; Li, G.; Feng, B.; Fu, G.; Tao, L. Strengthened assimilate transport improves yield and quality of super rice. *Agronomy* **2022**, *12*, 753. [CrossRef]
22. Li, H.; Feng, B.; Li, J.; Fu, W.; Wang, W.; Chen, T.; Liu, L.; Wu, Z.; Peng, S.; Tao, L.; et al. *RGA1* alleviates low-light-repressed pollen tube elongation by improving the metabolism and allocation of sugars and energy. *Plant Cell Environ.* **2023**, *46*, 1–21. [CrossRef]
23. Feng, B.; Xu, Y.; Fu, W.; Li, H.; Li, G.; Li, J.; Wang, W.; Tao, L.; Chen, T.; Fu, G. *RGA1* negatively regulates thermo-tolerance by affecting carbohydrate metabolism and the energy supply in rice. *Rice* **2023**, *16*, 32. [CrossRef]
24. Jiang, N.; Yu, P.; Fu, W.; Li, G.; Feng, B.; Chen, T.; Li, H.; Tao, L.; Fu, G. Acid invertase confers heat tolerance in rice plants by maintaining energy homoeostasis of spikelets. *Plant Cell Environ.* **2020**, *43*, 1273–1287. [CrossRef] [PubMed]
25. Xiong, D. Perspectives of improving rice photosynthesis for higher grain yield. *Crop Environ.* **2024**, *3*, 123–137. [CrossRef]
26. Jagadish, S. Heat stress during flowering in cereals-effects and adaptation strategies. *New Phytol.* **2020**, *226*, 1567–1572. [CrossRef]
27. Xie, Q.; Xu, J.; Huang, K.; Su, Y.; Tong, J.; Huang, Z.; Huang, C.; Wei, M.; Lin, W.; Xiao, L. Dynamic formation and transcriptional regulation mediated by phytohormones during chalkiness formation in rice. *BMC Plant Biol.* **2021**, *21*, 308. [CrossRef]
28. Zhang, Y.; Liu, G.; Cheng, Y.; Xu, J.; Wang, C.; Yang, J. The effects of dry cultivation on grain-filling and chalky grains of upland rice and paddy rice. *Food Energy Secur.* **2020**, *9*, e198. [CrossRef]
29. Chen, L.; Deng, X.; Duan, H.; Tan, X.; Xie, X.; Pan, X.; Guo, L.; Gao, H.; Wei, H.; Zhang, H.; et al. Water management can alleviate the deterioration of rice quality caused by high canopy humidity. *Agric. Water Manag.* **2023**, *289*, 108567. [CrossRef]
30. Chen, Z.; Zhou, W.; Guo, X.; Lin, S.; Li, W.; Yao, J. Heat stress responsive Aux/IAA protein, OsIAA29 regulates grain filling through OsARF17 mediated Auxin signaling pathway. *Rice* **2024**, *17*, 16. [CrossRef]
31. Xiong, Q.; Lu, Y.; Gu, W.; Zhang, Y.; Li, A.; Cai, S.; Zhou, N. Biomarkers of the main nutritional components in purple rice during five successive grain filling stages. *Food Chem. X* **2024**, *23*, 101528. [CrossRef]
32. Zhang, J.; Zhang, Y.; Song, N.; Chen, Q.; Sun, H.; Peng, T.; Huang, S.; Zhao, Q. Response of grain-filling rate and grain quality of mid-season indica rice to nitrogen application. *J. Integr. Agric.* **2021**, *20*, 1465–1473. [CrossRef]
33. Crofts, N.; Abe, N.; Oitome, N.; Matsushima, R.; Hayashi, M.; Tetlow, I.; Emes, M.; Nakamura, Y.; Fujita, N. Amylopectin biosynthetic enzymes from developing rice seed form enzymatically active protein complexes. *J. Exp. Bot.* **2015**, *66*, 4469–4482. [CrossRef]

34. Seung, D.; Smith, A. Starch granule initiation and morphogenesis-progress in *Arabidopsis* and cereals. *J. Exp. Bot.* **2019**, *70*, 771–784. [CrossRef]
35. Huang, L.; Tan, H.; Zhang, C.; Li, Q.; Liu, Q. Starch biosynthesis in cereal endosperms: An updated review over the last decade. *Plant Commun.* **2021**, *2*, 100237. [PubMed]
36. Cakir, B.; Shiraishi, S.; Tuncel, A.; Matsusaka, H.; Satoh, R.; Singh, S.; Crofts, N.; Hosaka, Y.; Fujita, N.; Hwang, S.; et al. Analysis of the rice ADP-Glucose transporter (OsBT1) indicates the presence of regulatory processes in the amyloplast stroma that control ADP-Glucose flux into starch. *Plant Physiol.* **2016**, *170*, 1271–1283.
37. Ma, J.; Chen, T.; Lin, J.; Fu, W.; Feng, B.; Li, G.; Li, H.; Li, J.; Wu, Z.; Tao, L.; et al. Functions of nitrogen, phosphorus and potassium in energy status and their influences on rice growth and development. *Rice Sci.* **2022**, *29*, 166–179.
38. Shen, T.; Xiong, Q.; Zhong, L.; Shi, X.; Cao, C.; He, H.; Chen, X. Analysis of main metabolisms during nitrogen deficiency and compensation in rice. *Acat. Physiol. Plant.* **2019**, *41*, 1–14. [CrossRef]
39. Liu, Y.; Han, J.; Liu, D.; Gu, D.; Wang, Y.; Liao, Y.; Wen, X. Effect of plastic film mulching on the grain filling and hormonal changes of maize under different irrigation conditions. *PLoS ONE* **2015**, *10*, e0122791.
40. Zhang, W.; Cao, Z.; Zhou, Q.; Chen, J.; Xu, G.; Gu, J.; Liu, L.; Wang, Z.; Yang, J.; Zhang, H. Grain filling characteristics and their relations with endogenous hormones in large-and small-grain mutants of rice. *PLoS ONE* **2016**, *11*, e0165321. [CrossRef] [PubMed]
41. Li, K.; Xu, Y.; Gu, D.; Yin, X.; Jia, Y.; Wen, T.; Jiang, W.; Che, Y.; Li, Q.; Wen, Z.; et al. Response of Different Exogenous Phytohormones to Rice Yield Under Low-Temperature Stress at the Filling Stage. *Agronomy* **2024**, *14*, 2450. [CrossRef]

MDPI AG
Grosspeteranlage 5
4052 Basel
Switzerland
Tel.: +41 61 683 77 34

International Journal of Molecular Sciences Editorial Office
E-mail: ijms@mdpi.com
www.mdpi.com/journal/ijms